DNA SYNTHESIS *IN VITRO*

Proceedings of the Second Annual Harry Steenbock Symposium, held in Madison, Wisconsin, on July 10-12, 1972

DNA Synthesis
in Vitro

Edited by
R. D. Wells and R. B. Inman

UNIVERSITY PARK PRESS
Baltimore • London • Tokyo

UNIVERSITY PARK PRESS
International Publishers in Science and Medicine
Chamber of Commerce Building
Baltimore, Maryland 21202

Copyright © 1973 by University Park Press

Printed in the United States of America

All rights, including that of translation into other languages, reserved. Photomechanical reproduction (photocopy, microcopy) of this book or parts thereof without special permission of the publisher is prohibited.

Library of Congress Cataloging in Publication Data

Harry Steenbock Symposium, 2nd, University of Wisconsin, 1972.
DNA synthesis in vitro.
1. Deoxyribonucleic acid synthesis—Congresses.
I. Wells, Robert D., ed.   II. Inman, Ross B., ed.
III. Title. [DNLM: 1. DNA—Biosynthesis—Congresses.
W3 HA294 1971d.   XNLM: [QU 58 H323d 1971]]
QP551.H358 1972   574.8'732   72-12753
ISBN 0-8371-0745-5

Top row (from left to right): T. Twose, B. Weisblum, D. Perlman, M. Barnes—G. Mueller—M. Bessman, P. Englund, I. Lehman. *Middle row:* A. Worcel, R. Carroll, W. Dove, M. Ryan, L. Brewer—W. Szybalski, H. Lozeron—D. Chattoraj, M. Schnös. *Bottom row:* N. Brown, D. Ludlum—B. Olivera, T. Kornberg—M. Smith, P. Gilham.

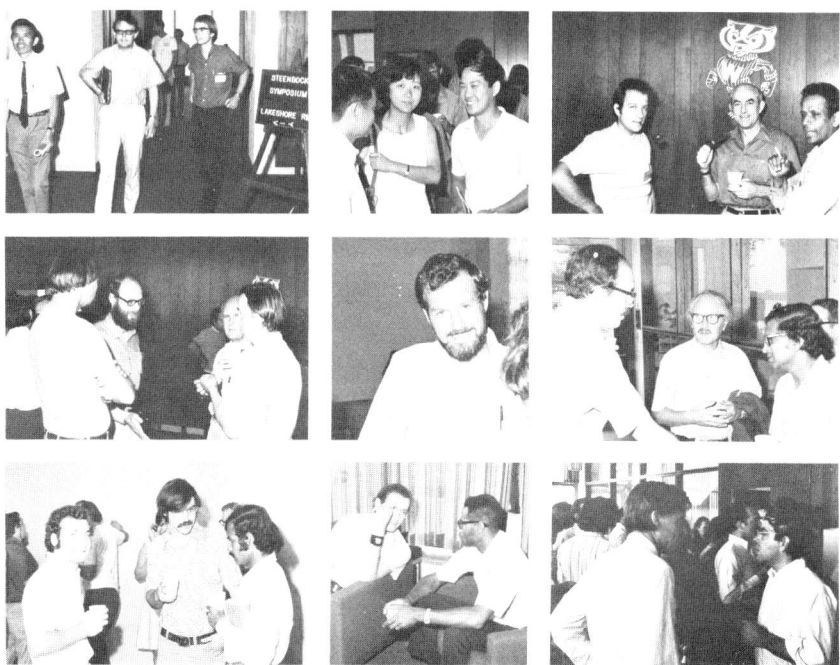

Top row (from left to right): M. Nomura, N. Kjeldgaard, F. Engbaek—R. Fujimura, L. Shih, D. Shih—R. Wells, R. Inman, A. Ganesan. *Middle row:* D. Ray, V. Nusslein, W. Szybalski, W. Dove—R. Rownd—W. Reznikoff, P. Kaesberg, S. Mitra. *Bottom row:* R. Montalaro, M. Stoltzfus, K. Medappa—J. Burd, R. Ratliff—J. Cairns, D. Dressler.

Top row (from left to right): M. Sundaralingam, D. Carlson—J. Davies, K. Ihler—B. Strauss. *Middle row:* J. Dodgson, M. Ryan, B. Beckett—K. Bartok, B. Alberts—K. Ebisuzaki. *Bottom row:* V. Paetkau—M. Goulian—L. Miller.

Photographs by Will Baumann (Molecular Biology-Biophysics Laboratory, University of Wisconsin).

Contents

Preface   ix

Participants   xi

Bacterial and Phage-Induced DNA Polymerases

DNA Polymerase I Activity in Polymerase I Mutants of *Escherichia coli*   I. R. Lehman and Janice R. Chien   **3**

Concerted Effect of Pancreatic DNase and the Large Fragment of DNA Polymerase I   Hans Klenow and Kay Overgaard-Hansen   **13**

Mechanisms of Excision-Repair   Lawrence Grossman, Barbara M. Garvick, Howard K. Ono, Andrew G. Braun, Lester D. G. Hamilton, and Inga Mahler   **27**

Nucleotide Sequencing of 3′ Termini of Duplex DNA with the T4 DNA Polymerase   Paul T. Englund, Sharon S. Price, Joan Moyer Schwing, and Paul H. Weigel   **35**

Exonuclease Activity of Wild Type and Mutant T4 DNA Polymerases: Hydrolysis during DNA Synthesis *in Vitro*   Nancy G. Nossal and Michael S. Hershfield   **47**

Bacterial DNA Synthesis

DNA Polymerases of *Escherichia coli*   Charles C. Richardson, Judith L. Campbell, John W. Chase, David C. Hinkle, Dennis M. Livingston, Henry L. Mulcahy, and Hiroaki Shizuya   **65**

Studies on DNA Polymerases II and III of *Escherichia coli*   Malcolm L. Gefter, Thomas Kornberg, Ian J. Molineux, H. G. Khorana, Lenny Mendich, and Yukinori Hirota   **71**

The Discontinuous Replication of DNA   Reiji Okazaki, Akio Sugino, Susumu Hirose, Tuneko Okazaki, Yasuo Imae, Ritsu Kainuma-Kuroda, Tohru Ogawa, Mikio Arisawa, and Yoshikazu Kurosawa   **83**

Genetics and Physiology of DNA Ligase Mutants of *Escherichia coli*   Michael M. Gottesman, Minnie L. Hicks, and Martin Gellert   **107**

Two DNase-related ATPases of *Escherichia coli*   Stuart Linn, Barnet Eskin, Alexander E. Karu, Peter J. Goldmark, and Erin Hawkins   **123**

Involvement of DNA Gene Products in the Conversion of $\phi\chi$-174 and fd DNA to Replicative Forms by Extracts of *Escherichia coli*   R. B. Wickner, M. Wright, S. Wickner, and J. Hurwitz   **137**

Factors for DNA Synthesis

Factors Stabilizing DNA Folding in Bacterial Chromosomes   David E. Pettijohn, Ralph M. Hecht, O. G. Stonington, and T. D. Stamato   145

**Protein ω: A DNA Swivelase from *Escherichia coli?*   James C. Wang   163

Initiation of DNA Synthesis   Randy Schekman, William Wickner, Ole Westergaard, Douglas Brutlag, Klaus Geider, Leroy Bertsch, and Arthur Kornberg   175

***In Vitro* Studies on *Escherichia coli* DNA Replication Factors and on the Initiation of Phage λ DNA Replication**   A. Klein, V. Nüsslein, B. Otto, and W. Powling   185

Proteins of the T4 Bacteriophage Replication Apparatus   J. Barry, H. Hama-Inaba, L. Moran, and B. Alberts   195

Discontinuous DNA Synthesis *in Vitro*: A Method for Defining the Role of Factors in Replication   Baldomero M. Olivera, K. G. Lark, Richard Herrmann, and Friedrich Bonhoeffer   215

Self-association of Gene 32 Protein   Robert B. Carroll, Kenneth E. Neet, and David A. Goldthwait   233

RNA Tumor Virus DNA Polymerase

The DNA Polymerases of RNA Tumor Viruses and Their Relationship to Cellular DNA Polymerases   Howard M. Temin and Satoshi Mizutani   239

RNA-directed and -primed DNA Polymerase Activities in Tumor Viruses and Human Lymphocytes   Robert C. Gallo, Prem S. Sarin, R. Graham Smith, Samuel N. Bobrow, Mangalasseril G. Sarngadharan, Marvin S. Reitz, Jr., and John W. Abrell   251

RNA-dependent DNA Polymerase Activity of RNA Tumor Viruses. IV. Characterization of AMV Stimulatory Protein and RNase H-associated Activity   Jonathan Leis, Ira Berkower, and Jerard Hurwitz   287

RNA-DNA Bonds Formed by DNA Polymerases from Bacteria and RNA Tumor Viruses   R. M. Flügel, J. E. Larson, P. F. Schendel, R. W. Sweet, T. R. Tamblyn, and R. D. Wells   309

Avian Myeloblastosis Virus DNA Polymerase: Initiation of DNA Synthesis and an Associated Ribonuclease   David Baltimore, Inder M. Verma, Donna F. Smoler, and Nora L. Meuth   333

Characteristics of the Transcription of RNA by the DNA Polymerase of Rous Sarcoma Virus   J. M. Bishop, A. J. Faras, A. C. Garapin, H. M. Goodman, W. E. Levinson, J. Stavnezer, J. M. Taylor, and H. E. Varmus   341

DNA Polymerase in Association with Intracisternal A-type Particles   S. H. Wilson, E. L. Kuff, E. W. Bohn, K. K. Lueders, and A. Matsukage   361

Inhibition of Leukemia Virus Replication by Vinyl Analogs of Polynucleotides   P. M. Pitha, N. M. Teich, D. R. Lowy, and J. Pitha   369

In Vivo/in Vitro DNA Synthesizing Systems

Complementation Analysis of Mutations at the *dnaB*, *dnaC*, and *dnaD* Loci   James A. Wechsler   375

In Vivo and *in Vitro* Chromosome Replication in *Bacillus subtilis*   Noboru Sueoka, Tatsuo Matsushita, Seigou Ohi, Aideen O'Sullivan, and Kalpana White   385

In Vitro DNA Synthesis and Function of DNA Polymerases in *Bacillus subtilis*   A. T. Ganesan, P. J. Laipis, and C. O. Yehle   405

D-Loops in Intracellular λ DNA   Ross B. Inman and Maria Schnös   437

Some Biochemical Elements in Bacteriophage T7 DNA Replication *in Vitro*   Rolf Knippers, Wolf Strätling, and Elke Krause   451

Autoradiographic Demonstration of Bidirectional Replication in *Escherichia coli*   P. L. Kuempel, D. M. Prescott, and P. Maglothin   463

Index   473

Preface

This book contains the papers presented at the Second Annual Harry Steenbock Symposium. The Symposium, and this book, have been entitled *DNA Synthesis in Vitro*. Knowledge gained in recent years has led to an ill-defined division between *in vitro* and *in vivo* studies as applied to DNA replication. Consequently the reader, depending on his particular point of view, will find new and useful information on many aspects of this most interesting subject.

We have divided the subject matter into five parts, and these correspond to the sessions of the Symposium. Following many of the papers we have inserted any discussion which might be of use to the reader.

The Symposium was supported by funds derived from the Harry Steenbock Symposium Trust Fund of the Department of Biochemistry, University of Wisconsin, made available by Mrs. Harry Steenbock from the Wisconsin Alumni Research Foundation. Professor Harry Steenbock was a distinguished professor of biochemistry from 1916 to 1956 and is world renowned for his pioneering work in the discovery of various vitamins. Of particular note is his discovery that irradiation of foods with ultraviolet light induces vitamin D activity, a discovery responsible for the elimination of rickets as a major medical problem.

The proceedings from the first Steenbock Symposium have been published: H. F. DeLuca and J. W. Suttie (eds.), *The Fat-Soluble Vitamins*, The University of Wisconsin Press, Madison, Wisconsin, 1969. The third Steenbock Symposium was held in Madison, June 17 to 22, 1973, in conjunction with the Second International Congress on Trace Element Metabolism in Animals. The coordinators were W. G. Hoekstra and J. W. Suttie.

We wish to express our thanks to Mrs. Harry Steenbock for the endowment that made the Symposium, and this book, possible. We also thank Doctors Lehman, Richardson, Alberts, Temin, and Sueoka, who were chairmen of the various sessions of the Symposium and who helped select the subject matter. Finally, we appreciate the secretarial assistance of Mrs. Mary Parker and Mr. Ray L. Rideout.

R. D. Wells and R. B. Inman
Department of Biochemistry
College of Agricultural and Life Sciences
University of Wisconsin
Madison, Wisconsin

Participants

Joyce Adams, Department of Biochemistry, University of Wisconsin, Madison, Wisconsin 53706

Julius Adler, Department of Biochemistry, University of Wisconsin, Madison, Wisconsin 53706

Asad Ahmed, Department of Genetics, University of Alberta, Edmonton, Alberta, Canada

***Bruce Alberts,** Department of Biological Sciences, Princeton University, Princeton, New Jersey 08540

Mary M. Allen, Department of Biological Sciences, Wellesley College, Wellesley, Massachusetts 02181

Gary Altman, Department of Molecular Biology, Vanderbilt University, Nashville, Tennessee 37235

Hiroyuki Aono, Department of Biology, Syracuse University, Syracuse, New York 13210

Basil M. Arif, Insect Pathology Research Institute, Box 490, Sault Ste. Marie, Ontario, Canada

Karen Armstrong, 1280 Raymond Avenue, Apt. 4, St. Paul, Minnesota 55108

Robert Armstrong, Department of Biological Chemistry, University of Michigan, Ann Arbor, Michigan 48104

Atilla Atalay, 905 South 47th Street, Philadelphia, Pennsylvania 19143

***David Baltimore,** 28 Donnell Street, Cambridge, Massachusetts 02138

Richard H. Baltz, Department of Microbiology, University of Illinois, Urbana, Illinois 61801

Kala Banerjee, Department of Chemistry, DePaul University, Chicago, Illinois 60614

Ralph K. Barclay, CIBA-GEIGY Corporation, Pharmaceutical Division, Summit, New Jersey 07901

Earl F. Baril, Worcester Foundation for Experimental Biology, Shrewsbury, Massachusetts 01545

Katalina Bartok, c/o Dr. M. J. Fraser, Department of Biochemistry, McGill University, Montréal, P. Q., Canada

*Invited speakers

Jacques Beaudoin, Departmente de Microbiologie et Immunologie, Université de Montréal, C. P. 6128, Montréal 101, P. Q., Canada

George S. Beaudreau, Department of Agricultural Chemistry, Oregon State University, Corvallis, Oregon 97331

John Bendler, Medical College of Wisconsin, Milwaukee, Wisconsin 53233

Claude Bernard, Cancer Research Unit, McGill University, Montréal 109, P. Q., Canada

Ora Bernard, Cancer Research Unit, McGill University, Montréal 109, P. Q., Canada

William E. Berquist, 6104 Gardendale Drive, Nashville, Tennessee 37215

Maurice J. Bessman, Mergenthaler Laboratory for Biology, Johns Hopkins University, Baltimore, Maryland 21218

Girija Bhargaua, Department of Biochemistry, Brandeis University, Waltham, Massachusetts 02154

J. K. Bhattacharjee, Department of Microbiology, Miami University, Oxford, Ohio 45056

*****J. Michael Bishop,** Department of Microbiology, University of California Medical Center, San Francisco, California 94122

Nilambar Biswal, Department of Virology, Baylor University College of Medicine, Houston, Texas 77025

Zofia Kurylo-Borowska, The Rockefeller University, New York, New York 10021

Gerard J. Bourguignon, Department of Biochemistry, State University of New York, Stony Brook, New York 11790

Karl Brackmann, Department of Molecular Virology, St. Louis University School of Medicine, St. Louis, Missouri 63110

Lorraine C. Brewer, Department of Biochemistry, University of Wisconsin, Madison, Wisconsin 53706

Neal C. Brown, Department of Cell Biology and Pharmacology, University of Maryland, Baltimore, Maryland 21201

Gilbert Brun, Faculty of Sciences, Institute of Molecular Biology, Paris, France

Clifford Brunk, Department of Zoology, University of California, Los Angeles, California 90024

Dorothy L. Buchhagen, Sloan-Kettering Institute, 410 East 68th Street, New York, New York 10021

John Burd, Department of Biochemistry, University of Wisconsin, Madison, Wisconsin 53706

Ann Burgess, Department of Oncology, University of Wisconsin, Madison, Wisconsin 53706

Richard Burgess, Department of Oncology, University of Wisconsin, Madison, Wisconsin 53706

Robert H. Burris, Department of Biochemistry, University of Wisconsin, Madison, Wisconsin 53706

Luis O. Burzio, The Population Council, The Rockefeller University, New York, New York 10021

*John Cairns, Cold Spring Harbor Laboratory, P.O. Box 100, Cold Spring Harbor, New York 11724

Judith Campbell, Department of Biological Chemistry, Harvard University School of Medicine, Boston, Massachusetts 02115

Francis Capaldo-Kimball, Department of Microbiology, Case Western Reserve University, Cleveland, Ohio 44106

Nancy Cargile, Department of Biochemistry, University of Kansas, Lawrence, Kansas 66044

Jeanine Carlson, Department of Chemistry, DePaul University, Chicago, Illinois 60614

Lucien Caro, Department de Biologie Moleculaire, 30 Quai Ecole de Medecine, Geneva, Switzerland 1211

Robert Carroll, 3186 Euclid Heights Boulevard, Cleveland, Ohio 44118

Richard V. Case, P-L Biochemicals, Inc., 1037 W. McKinley Avenue, Milwaukee, Wisconsin 53205

L. F. Cavalieri, Sloan-Kettering Institute, 410 East 68th Street, New York, New York 10021

Hardy Chan, Department of Biochemistry, University of Wisconsin, Madison, Wisconsin 53706

Che-Shen Chiu, Department of Biochemistry, University of Michigan, Ann Arbor, Michigan 48104

Don B. Clewell, University of Michigan, School of Dentistry, Ann Arbor, Michigan 48104

Joseph L. Colbourn, Miles Laboratories, 1127 Myrtle Street, Elkhart, Indiana 46514

Norman B. Coonrod, 5450 Medical Science Building I, University of Michigan, Ann Arbor, Michigan 48104

Nicholas R. Cozzarelli, Department of Biochemistry, University of Chicago, Chicago, Illinois 60637

Jane Harris Cramer, Laboratory of Molecular Biology, University of Wisconsin, Madison, Wisconsin 53706

Dennis R. Cryer, Department of Biochemistry, Albert Einstein College of Medicine, Bronx, New York 10461

J. E. Dahlberg, Department of Physiological Chemistry, University of Wisconsin, Madison, Wisconsin 53706

Ziyad Dajani, Department of Chemistry, DePaul University, Chicago, Illinois 60614

Cedric I. Davern, Division of Natural Sciences, University of California, Santa Cruz, California 95060

Daniel Devine, 1403 East Stout Street, Urbana, Illinois 60801

Jerry Dodgson, Department of Biochemistry, University of Wisconsin, Madison, Wisconsin 53706

Gary Doern, Medical College of Wisconsin, Milwaukee, Wisconsin 53233

James Dohnal, Department of Chemistry, DePaul University, Chicago, Illinois 60614

Myron Dolynink, 1205 East 60th Street, Chicago, Illinois 60637

John Dorson, Department of Biochemistry, University of Kentucky, Lexington, Kentucky 40506

William Dove, McArdle Laboratory, University of Wisconsin, Madison, Wisconsin 53706

Catherine M. Doyle, Department of Biology, University of Chicago, Chicago, Illinois 60637

David Dressler, Biological Laboratories, Harvard University, Cambridge, Massachusetts 02138

Roger F. Drong, 140 Gortner Laboratory, University of Minnesota, St. Paul, Minnesota 55101

Lawrence B. Dumas, Department of Biological Sciences, Northwestern University, Evanston, Illinois 60201

David Dunham, Department of Biological Chemistry, University of Michigan, Ann Arbor, Michigan 48104

Linda T. Dunham, Department of Biological Chemistry, University of Michigan, Ann Arbor, Michigan 48104

Kaney Ebisuzaki, Department of Cancer Research, University of Western Ontario, London, Ontario, Canada

Howard Elford, Department of Medicine, Duke University Medical Center, P.O. Box 3335, Durham, North Carolina 27710

Sarah C. R. Elgin, Division of Biology, California Institute of Technology, Pasadena, California 91109

Frode Engbaek, Department of Oncology, University of Wisconsin, Madison, Wisconsin 53706

*****Paul T. Englund,** Department of Physiological Chemistry, Johns Hopkins University School of Medicine, Baltimore, Maryland 21205

Semih Erhan, University of Pennsylvania, School of Veterinary Medicine, Philadelphia, Pennsylvania 19104

Robert Erickson, Department of Molecular Biology, Miles Laboratories, Inc., Elkhart, Indiana 46514

Helen Evans, Department of Radiology, Case Western Reserve University, Cleveland, Ohio 44106

Irene M. Evans, Department of Radiation Biology and Biophysics, University of Rochester, Rochester, New York 14620

Thomas E. Evans, Department of Radiology, Case Western Reserve University, Cleveland, Ohio 44106

Harvey Faber, 202 Genetics Building, University of Wisconsin, Madison, Wisconsin 53706

Anthony Faras, Department of Microbiology, University of California Medical Center, San Francisco, California 94122

George Fareed, National Institutes of Health, Building 5, Room 336, Bethesda, Maryland 20014

Hovsep M. Fidanian, Department of Zoology, University of California, Los Angeles, California 90024

William Firshein, Shanklin Laboratory of Biology, Wesleyan University, Middletown, Connecticut 06457

Donald A. Fischman, Department of Biology, University of Chicago, Chicago, Illinois 60637

*__Rolf M. Flügel,__ Department of Biochemistry, University of Wisconsin, Madison, Wisconsin 53706

Bertold R. Francke, c/o W. Eckhart, The Salk Institute, P.O. Box 1809, San Diego, California 92112

Elizabeth Freedlender, 406 Genetics Building, University of Wisconsin, Madison, Wisconsin 53706

Frank Freeman, Medical College of Wisconsin, Milwaukee, Wisconsin 53233

David Freifelder, Department of Biochemistry, Brandeis University, Waltham, Massachusetts 02154

Orrie M. Friedman, Collaborative Research, Inc., 1365 Main Street, Waltham, Massachusetts 02154

Robert Fujimura, Biology Division, Oak Ridge National Laboratory, Oak Ridge, Tennessee 37830

E. Gabbay, Department of Chemistry, University of Florida, Gainesville, Florida 32601

William Galley, Department of Chemistry, McGill University, Montréal, P. Q., Canada

*__Robert C. Gallo,__ National Cancer Institute, National Institutes of Health, Building 10, Room 6B17, Bethesda, Maryland 20014

*__Adayapalam Ganesan,__ Department of Genetics, Stanford University Medical School, Stanford, California 94305

Barbara Garvik, 33 Orange Street, Waltham, Massachusetts 02154

Kenneth B. Gass, 5472 South Everett Avenue, Chicago, Illinois 60615

*__Malcolm Gefter,__ Department of Biological Sciences, Columbia University, New York, New York 10027

Martin Gellert, Laboratory of Molecular Biology, National Institute of Arthritis and Metabolic Diseases, National Institutes of Health, Bethesda, Maryland 20014

Norman Gentner, Biology Branch, Atomic Energy of Canada, Chalk River, Ontario, Canada

Peter T. Gilham, Department of Biological Sciences, Purdue University, Lafayette, Indiana 47907

G. N. Godson, Department of Radiobiology, Yale University School of Medicine, New Haven, Connecticut 06510

Howard M. Goodman, Department of Biochemistry and Biophysics, University of California, San Francisco, California 94122

Myron F. Goodman, Department of Biology, Johns Hopkins University, Baltimore, Maryland 21218

Stella Gornicki, Department of Biochemistry, University of Kansas, Lawrence, Kansas 66044

*__Michael Gottesman,__ National Institute of Arthritis and Metabolic Diseases, National Institutes of Health, Building 2, Room 322, Bethesda, Maryland 20014

Mehran Goulian, Department of Hematology, University of California at San Diego, La Jolla, California 92037

Robert Grant, Division of Infectious Diseases, Department of Medicine, Stanford University, Stanford, California 94305

Ronald R. Green, Department of Microbiology, University of Illinois, Urbana, Illinois 61801

G. Robert Greenberg, Department of Biological Chemistry, University of Michigan, Ann Arbor, Michigan 48104

Josephine C. Grosch, Department of Molecular Biology, Miles Laboratories, Inc., Elkhart, Indiana 46514

Carol Gross, Institute of Molecular Biology, University of Oregon, Eugene, Oregon 97403

*****Lawrence Grossman,** Department of Biochemistry, Brandeis University, Waltham, Massachusetts 02154

Lawrence I. Grossman, Division of Biology, California Institute of Technology, Pasadena, California 91109

John A. Grunau, Department of Biological Sciences, University of Missouri, Columbia, Missouri

Richard I. Gumport, Department of Biochemistry, University of Illinois, Urbana, Illinois 61801

John Hachmann, Collaborative Research, Inc., 1365 Main Street, Waltham, Massachusetts 02154

Richard Hallick, Department of Biochemistry and Biophysics, University of California, San Francisco, California 94122

Arne Hampel, Department of Biology, Northern Illinois University, DeKalb, Illinois 60115

Michael Hanna, Department of Microbiology, University of Illinois, Urbana, Illinois 61801

Bern Hapke, Department of Biological Sciences, Northwestern University, Evanston, Illinois 60201

John E. Heinze, Department of Microbiology, University of Illinois, Urbana, Illinois 61801

Michael Hershfield, National Institutes of Health, Building 4, Room 116, Bethesda, Maryland 20014

Patrick Higgins, 5107 South Blackstone, No. 1103, Chicago, Illinois 60615

Walter Hill, Laboratory of Molecular Biology, University of Wisconsin, Madison, Wisconsin 53706

David Hinkle, Department of Biological Chemistry, Harvard University School of Medicine, Boston, Massachusetts 02115

Ronald Hitzeman, 1467 University Terrace, No. 1323, Ann Arbor, Michigan 48104

Ronald Hoess, 368 Gortner Laboratory, University of Minnesota, St. Paul, Minnesota 55101

Henry Horwitz, Department of Biology, Massachusetts Institute of Technology, Cambridge, Massachusetts 02139

Tony Hunter, The Salk Institute, P.O. Box 1809, San Diego, California 92112

Garret Ihler, Department of Biochemistry, University of Pittsburgh, Pittsburgh, Pennsylvania 15213

Karin Ihler, 706 Ivy Street, Pittsburgh, Pennsylvania 15232

Anthony Infante, Department of Biology, Wesleyan University, Middletown, Connecticut 06457

*Ross B. Inman, Biophysics Laboratory, University of Wisconsin, Madison, Wisconsin 53706

R. J. Jariwalla, Medical College of Wisconsin, Milwaukee, Wisconsin 53233

David Jensen, Institute of Molecular Biology, University of Oregon, Eugene, Oregon 97403

Paul Johnson, Division of Biology, California Institute of Technology, Pasadena, California 91109

Kazuto Kajiwara, McArdle Laboratory, University of Wisconsin, Madison, Wisconsin 53706

Pai C. Kao, Biology Division, Oak Ridge National Laboratory, Oak Ridge, Tennessee 37830

Karen Kato, Department of Microbiology, University of Chicago, Chicago, Illinois 60637

Dennis L. Kay, 208 Life Sciences Building, University of Utah, Salt Lake City, Utah 84112

William L. Kelley, Department of Biological Sciences, Mellon Institute of Science, Carnegie-Mellon University, Pittsburgh, Pennsylvania 15213

Helen Kiefer, Biochemistry Division, Department of Chemistry, Northwestern University, Evanston, Illinois 60201

Dollie Kirtikar, Department of Biochemistry, Case Western Reserve University, Cleveland, Ohio 44106

N. O. Kjeldgaard, c/o J. E. Dahlberg, Department of Physiological Chemistry, University of Wisconsin, Madison, Wisconsin 53706

*Hans Klenow, Biokemisk Institut B, Juliane Mariesvej 30, Copenhagen Ø, Denmark DK-2100

*Rolf Knippers, Fachbereich Biologie, Universität Konstanz, Postfach 733, D-7750 Konstanz, Germany

Thomas Kornberg, Department of Biology, Columbia University, New York, New York 10027

James Koziarz, Department of Chemistry, DePaul University, Chicago, Illinois 60614

Evangelia Kranias, 1570 Oak Avenue, Apt. 416, Evanston, Illinois 60201

Monty Krieger, Department of Chemistry, California Institute of Technology, Pasadena, California 91109

H. A. Kubinski, Department of Surgery, University of Wisconsin, Madison, Wisconsin 53706

Peter Kuempel, Department of Molecular, Cellular and Developmental Biology, University of Colorado, Boulder, Colorado 80302

Yankel M. Kupersztoch, Department of Biology, University of California, La Jolla, California 92037

Paul Kupferstein, Cancer Research Laboratory, University of Western Ontario, London 72, Ontario, Canada

Sidney Kushner, Department of Biochemistry, Stanford University, Stanford, California 94305

Jerold Last, National Academy of Sciences, 2101 Constitution Avenue, Washington, D.C. 20418

J. Eugene LeClerc, Biology Division, Oak Ridge National Laboratory, Oak Ridge, Tennessee 37830

Lucy Lee, Department of Biochemistry, United States Department of Agriculture, Poultry Research Laboratory, East Lansing, Michigan 48823

*****I. Robert Lehman,** Department of Biochemistry, Stanford University, Stanford, California 94305

*****Jonathan Leis,** Albert Einstein College of Medicine, Yeshiva University, Bronx, New York 10461

Robert L. Letsinger, Department of Chemistry, Northwestern University, Evanston, Illinois 60201

*****S. M. Linn,** University of California, Department of Biochemistry, Berkeley, California 94720

Elwood Linney, Department of Biology, University of California at San Diego, P.O. Box 109, La Jolla, California 92037

Rose M. Litman, Department of Molecular Biology, Vanderbilt University, Nashville, Tennessee 37203

Robert L. Low, 5614 South Blackstone, Apt. 1, Chicago, Illinois 60637

David B. Ludlum, Department of Cell Biology and Pharmacology, University of Maryland, Baltimore, Maryland 21201

Ronald Lundquist, 208 Life Science Building, University of Utah, Salt Lake City, Utah 84112

Wayne E. Magee, Department of Life Sciences, Indiana State University, Terre Haute, Indiana 47803

James C. Mao, Abbott Laboratories, North Chicago, Illinois 60064

Tatsuo Matsushita, 28 Stanworth Drive, Princeton, New Jersey, 08540

Barbara Mazur, The Rockefeller University, New York, New York 10021

K. C. Medappa, Biophysics Laboratory, University of Wisconsin, Madison, Wisconsin 53706

Leandro Medrano, 56-543 MIT, Massachusetts Institute of Technology, Cambridge, Massachusetts 02139

Virginia Merriam, Department of Biology, Loyola University of Los Angeles, Los Angeles, California 90045

Christine Milcarek, 3404 St. Paul Street, Apt. 4C, Baltimore, Maryland 21218

Christine Miller, 3-110 O. T. Hogan Building, Northwestern University, Evanston, Illinois 60201

Jackie Miller, Department of Oncology, University of Wisconsin, Madison, Wisconsin 53706

Lois K. Miller, Division of Biology, California Institute of Technology, Pasadena, California 91109

William L. Miller, Division of Biochemistry, Walter Reed Army Medical Center, Washington, D.C. 20012

Mr. and Mrs. John Milton, 3421 Drummond Street, Montréal 109, P. Q., Canada

Sankar Mitra, Biology Division, Oak Ridge National Laboratory, Oak Ridge, Tennessee 37830

A. Richard Morgan, Department of Biochemistry, University of Alberta, Edmonton, Alberta, Canada

Robb Moses, Department of Biochemistry, Baylor University College of Medicine, Houston, Texas 77025

Gisela Mosig, Department of Molecular Biology, Vanderbilt University, Nashville, Tennessee 37203

Brahm Mulstock, Cancer Research Laboratory, University of Western Ontario, London, Ontario, Canada

Subbaratnam Muthukrishnan, Department of Biophysics, University of Chicago, Chicago, Illinois 60637

D. James McCorquodale, Division of Biology, University of Texas at Dallas, P.O. Box 30365, Dallas, Texas 75230

Kevin McEntee, 5443 Woodlawn 3W, Chicago, Illinois 60615

Grant McFadden, 3655 Drummond Street, Apt. 819, Montréal, P. Q., Canada

Roger McMacken, Department of Biochemistry, University of Florida, Gainesville, Florida 32601

Ramon Naranjo, Seton Hotel, 144 East 40th Street, New York, New York 10016

R. Naylor, Research Biochemicals Division, P-L Biochemicals, Inc., 1037 W. McKinley Avenue, Milwaukee, Wisconsin 53205

David Nelson, Department of Biochemistry, University of Wisconsin, Madison, Wisconsin 53706

Paul Neuwald, Department of Biology, Marquette University, Milwaukee, Wisconsin 53233

*__Nancy Nossal,__ National Institutes of Health, Building 4, Room 106, Bethesda, Maryland 20014

N. K. Notani, Biology Division, Bhabha Atomic Research Centre, Bombay 85, India

Robert L. Novak, Department of Chemistry, DePaul University, Chicago, Illinois 60614

*__Volker Nüsslein,__ Max-Planck-Gesellschaft, Tubingen, Germany

William O'Brien, Medical College of Wisconsin, Milwaukee, Wisconsin 53233

Paul V. O'Donnell, Sloan-Kettering Institute, 410 East 68th Street, New York, New York 10021

Max Oeschger, Department of Molecular Biophysics and Biochemistry, Yale University, New Haven, Connecticut 06510

Hideyuki and Tomoko Ogawa, Room 56-445, Massachusetts Institute of Technology, Cambridge, Massachusetts 02139

*__Reiji Okazaki,__ Institute of Molecular Biology, Faculty of Science, Nagoya University, Chikusa-ku, Nagoya, Japan

Arland E. Oleson, Department of Biochemistry, North Dakota State University, Fargo, North Dakota 58102

***Baldomero Olivera,** 208 Life Science Building, University of Utah, Salt Lake City, Utah 84102

Ronald H. Olsen, Department of Microbiology, University of Michigan, Ann Arbor, Michigan 48104

Verner Paetkau, Department of Biochemistry, University of Alberta, Edmonton, Alberta, Canada

Kathleen Parson, 140 Gortner Laboratory, University of Minnesota, St. Paul, Minnesota 55101

Michael H. Patrick, University of Texas at Dallas, P.O. Box 30365, Dallas, Texas 75230

Kshitij Patwa, Medical College of Wisconsin, Milwaukee, Wisconsin 53233

Richard Pauli, Department of Microbiology, 5724 South Ellis Avenue, Chicago, Illinois 60637

Carlos Pena, University of Sonora, Sonora, Mexico

David Petering, Department of Chemistry, University of Wisconsin at Milwaukee, Milwaukee, Wisconsin 53211

Gene Petersen, Department of Chemistry, Northwestern University, Evanston, Illinois 60201

***David E. Pettijohn,** Department of Biophysics, University of Colorado Medical Center, Denver, Colorado 80222

Manfred Philipp, Department of Chemistry, Northwestern University, Evanston, Illinois 60201

Josef Pitha, Gerontology Research Center, Baltimore City Hospitals, Baltimore, Maryland 21224

M. S. Poonian, Hoffman-LaRoche, Inc., Nutley, New Jersey 07110

Charles Pratt, Institute of Molecular Biology, University of Oregon, Eugene, Oregon 97403

Alan R. Price, Department of Biological Chemistry, University of Michigan, Ann Arbor, Michigan 48104

Sharon Price, Department of Physiological Chemistry, Johns Hopkins University School of Medicine, Baltimore, Maryland 21205

Roger Radloff, Biophysics Laboratory, University of Wisconsin, Madison, Wisconsin 53706

Robert L. Ratliff, 252 LaCueva, Los Alamos, New Mexico 87544

Dan S. Ray, Department of Zoology, University of California, Los Angeles, California 90024

Harvard Reiter, Department of Microbiology, University of Illinois Medical Center, P.O. Box 6998, Chicago, Illinois 60680

Fritz Reusser, The Upjohn Company, Kalamazoo, Michigan 49001

William S. Reznikoff, Department of Biochemistry, University of Wisconsin, Madison, Wisconsin 53706

***Charles C. Richardson,** Department of Biological Chemistry, Harvard University School of Medicine, Boston, Massachusetts 02115

Arthur D. Riggs, City of Hope Medical Center, Duarte, California 91010

Monica Riley, Department of Biochemistry, State University of New York, Stony Brook, New York 11790

Barbara Rosenberg, Sloan-Kettering Institute, 410 East 68th Street, New York, New York 10021

Martin Rosenberg, Hunter Radiation 610, Yale University School of Medicine, New Haven, Connecticut 06510

Peter Rosenthal, Division of Infectious Diseases, Stanford University School of Medicine, Stanford, California 94305

Robert H. Rownd, Department of Biochemistry, University of Wisconsin, Madison, Wisconsin 53706

Roland R. Rueckert, Department of Biochemistry, University of Wisconsin, Madison, Wisconsin 53706

Gordon Ryan, Department of Chemistry, University of Pittsburgh, Pittsburgh, Pennsylvania 15213

Michael Ryan, Department of Biochemistry, University of Wisconsin, Madison, Wisconsin 53706

Oliver Ryder, Department of Biology, University of California at San Diego, La Jolla, California 92037

Barbara Sahagan, Department of Physiological Chemistry, University of Wisconsin, Madison, Wisconsin 53706

William O. Salivar, Department of Biology, Marquette University, Milwaukee, Wisconsin 53233

Yehiam Salts, Department of Genetics, University of Washington, Seattle, Washington 98105

Lois Ann Salzman, National Institute of Allergies and Infectious Diseases, National Institutes of Health, Bethesda, Maryland 20014

Prem S. Sarin, National Cancer Institute, National Institutes of Health, Building 10, Room 6B16, Bethesda, Maryland 20014

M. Sarngadharan, Litton Bionetics, Inc., 7300 Pearl Street, Bethesda, Maryland 20014

Rob Sawyer, Department of Physiological Chemistry, University of Wisconsin, Madison, Wisconsin 53706

*Randy Schekman, Department of Biochemistry, Stanford University School of Medicine, Stanford, California 94305

Paul Schendel, Department of Biochemistry, University of Wisconsin, Madison, Wisconsin 53706

Peggy Schmitt, Department of Biology, University of California at San Diego, P.O. Box 109, La Jolla, California 92037

Maria Schnös, Biophysics Laboratory, University of Wisconsin, Madison, Wisconsin 53706

Peter Schofield, Department of Biochemistry, University of Vermont School of Medicine, Burlington, Vermont 05401

Dominic A. Scudiero, 810 South Summit Avenue, Villa Park, Illinois 60181

Stanley K. Shapiro, Department of Biological Sciences, University of Illinois at Chicago Circle, Chicago, Illinois 60680

PARTICIPANTS

Paul Shapshak, Biophysics Laboratory, University of Wisconsin, Madison, Wisconsin 53706

Patricia L. Shipley, Department of Microbiology, University of Michigan, Ann Arbor, Michigan 48104

Charles Shipman, Jr., University of Michigan, School of Dentistry, Ann Arbor, Michigan 48104

H. Shizuya, Harvard University School of Medicine, Boston, Massachusetts 02115

Jack M. Siegel, Research Biochemicals Division, P-L Biochemicals, Inc., 1037 West McKinley Avenue, Milwaukee, Wisconsin 53205

Jeff Siegel, Institute of Molecular Biology, University of Oregon, Eugene, Oregon 97403

John E. Sims, Biological Laboratories, Harvard University, Cambridge, Massachusetts 02138

George Smith, 406 Genetics Building, University of Wisconsin, Madison, Wisconsin 53706

Marvin A. Smith, 427 Cabrillo Avenue, Davis, California 95616

R. Graham Smith, National Cancer Institute, National Institutes of Health, Building 10, Room 6B17, Bethesda, Maryland 20014

Joseph Speyer, University of Connecticut, Storrs, Connecticut 06268

K. S. Sriprakash, Department of Internal Medicine, Yale University School of Medicine, New Haven, Connecticut 06510

Rolf Sternglanz, Department of Biochemistry, State University of New York, Stony Brook, New York 11790

Marvin Stodolsky, Department of Microbiology, University of Chicago, Chicago, Illinois 60637

Bernard Strauss, Department of Microbiology, University of Chicago, Chicago, Illinois 60637

*****Noboru Sueoka,** B-11 Moffett Laboratories, Princeton University, Princeton, New Jersey 08540

Yoshinobu Sugino, National Institutes of Health, Building 2, Room 202, Bethesda, Maryland 20014

Paul Sullivan, 206 Genetics Building, University of Wisconsin, Madison, Wisconsin 53706

M. Sundaralingam, Department of Biochemistry, University of Wisconsin, Madison, Wisconsin 53706

Karin O. Sundquist, California Institute of Technology, Pasadena, California 91109

Ray Sweet, Institute of Cancer Research, Columbia University, New York, New York 10032

Wlodzimierz Szer, Department of Biochemistry, New York University Medical Center, New York, New York 10016

Waclaw Szybalski, McArdle Memorial Laboratory, University of Wisconsin, Madison, Wisconsin 53706

Quivo S. Tahin, Department of Physiological Chemistry, University of Wisconsin, Madison, Wisconsin 53706

Toshiya Takano, National Institutes of Health, Building 2, Room 208, Bethesda, Maryland 20014

Jacov Tal, Institute for Molecular Virology, St. Louis University School of Medicine, St. Louis, Missouri 63110

Toby Tamblyn, Department of Biochemistry, University of Wisconsin, Madison, Wisconsin 53706

Alison Taunton-Rigby, Collaborative Research, Inc., 1365 Main Street, Waltham, Massachusetts 02154

Frederick Taylor, Department of Biochemistry, University of Illinois, Urbana, Illinois 61801

John M. Taylor, Department of Microbiology, University of California at San Francisco, San Francisco, California 94122

William Taylor, 618 Life Science I Building, The Pennsylvania State University, University Park, Pennsylvania 16802

*****Howard Temin,** McArdle Laboratory, University of Wisconsin, Madison, Wisconsin 53706

Paul Tomich, Department of Biochemistry, University of Michigan, Ann Arbor Michigan 48104

Junichi Tomizawa, National Institutes of Health, Building 2, Room 304, Bethesda, Maryland 20014

David Trauber, Biochemical Genetics Laboratory, The Rockefeller University, New York, New York 10021

David Trkula, 10603 Del Monte Drive, Houston, Texas 77042

Nicole Truffaut, Institut Pasteur, Paris, France

Chris Tsiapalis, Department of Biochemistry, University of Kentucky, School of Medicine, Lexington, Kentucky 40506

Yun-Yen Tsong, The Population Council, The Rockefeller University, New York, New York 10021

Daniel J. Tutas, Department of Biochemistry, University of Minnesota School of Medicine, Minneapolis, Minnesota 55455

Trevor Twose, Laboratory of Molecular Biology, University of Wisconsin, Madison, Wisconsin 53706

Gerda Tyrsted, Department of Biochemistry B, University of Copenhagen, 2100 Copenhagen Ø, Denmark

John S. Ullman, Department of Biology, University of Utah, Salt Lake City, Utah 84112

Charles P. Van Beveren, Department of Molecular Biology and Microbiology, Tufts University, Boston, Massachusetts 02111

Mary Vander Maten, Department of Biochemistry, University of Kansas, Lawrence, Kansas 66044

Gerbrand Van Dieijen, Department of Biological Sciences, Northwestern University, Evanston, Illinois 60201

Daniel Vapnek, Yale University School of Medicine, New Haven, Connecticut 06510

Joseph Walder, Department of Chemistry, Northwestern University, Evanston, Illinois 60201

Helen Wang, Department of Biology, State University of New York, Stony Brook, New York 11790

*****James C. Wang,** Department of Chemistry, University of California, Berkeley, California 94720

Roger M. Wartell, Department of Biochemistry, University of Wisconsin, Madison, Wisconsin 53706

*****James Wechsler,** Department of Biological Sciences, Columbia University, New York, New York 10027

Paul H. Weigel, Department of Physiological Chemistry, Johns Hopkins University School of Medicine, Baltimore, Maryland 21205

Bernard Weiss, The Johns Hopkins University School of Medicine, Baltimore, Maryland 21205

Sherman M. Weissman, Department of Internal Medicine, Yale University School of Medicine, New Haven, Connecticut 06510

Robert D. Wells, Department of Biochemistry, University of Wisconsin, Madison, Wisconsin 53706

Jim West, Department of Biochemistry, University of Kansas, Lawrence, Kansas 66044

Reed B. Wickner, Department of Developmental Biology and Cancer, Albert Einstein College of Medicine, Bronx, New York 10461

John S. Wilkes, Jr., Department of Chemistry, Northwestern University, Evanston, Illinois 60201

Samuel H. Wilson, Laboratory of Biochemistry, National Cancer Institute, National Institutes of Health, Bethesda, Maryland 20014

John Wolfson, Biological Laboratories, Harvard University, Cambridge, Massachusetts 02138

Abraham Worcel, Department of Biological Sciences, Princeton University, Princeton, New Jersey 08540

Merle Wovcha, Department of Biological Chemistry, University of Michigan School of Medicine, Ann Arbor, Michigan 48104

Andrew Wright, Department of Molecular Biology, Tufts University School of Medicine, Boston, Massachusetts 02111

Michael Wright, Department of Developmental Biology and Cancer, Albert Einstein College of Medicine, Bronx, New York 10461

Mtutuzeli Xuma, National Cancer Institute, National Institutes of Health, Building 10, 6B-18, Bethesda, Maryland 20014

David Yajko, Department of Microbiology, Johns Hopkins University School of Medicine, Baltimore, Maryland 21205

Hiroshi Yamazaki, Department of Biology, Carleton University, Ottawa, Ontario KIS5B6, Canada

Wen-Kuang Yang, Oak Ridge National Laboratory, P.O. Box Y, Oak Ridge, Tennessee 37830

Koichiro Yoshihara, 415 East 64th Street, Apt. 6-D, New York, New York 10021

Pierre Yot, Department of Internal Medicine, Yale University School of Medicine, New Haven, Connecticut 06510

Judith Zyskind, 2819 Arbor Street, Ames, Iowa 60010

DNA SYNTHESIS *IN VITRO*

Bacterial and
Phage-Induced DNA Polymerases

DNA Polymerase I Activity in Polymerase I Mutants of *Escherichia coli*

I. R. Lehman and Janice R. Chien

Department of Biochemistry
Stanford University School of Medicine
Stanford, California 94305

Partially purified extracts of the DNA polymerase I-defective mutant *Escherichia coli* P3478 contain from 0.5 to 2% as much polymerase I activity as the wild type (W3110) strain. Similarly treated extracts of JG112, a derivative of P3478, contain a lower level of polymerase I, and several preparations from this strain showed no detectable polymerase I activity. Both P3478 and JG112 contain near-normal levels of 5' → 3' exonuclease activity that is responsive to polymerase I antiserum.

In the initial report by de Lucia and Cairns describing the *pol*A1 mutation in strain P3478, extracts of this mutant were shown to have from 0.5 to 1% of the polymerase activity of the wild type parent strain W3110 (2). The question of whether the residual activity was a result of some "leakiness" of the amber mutation or represented a novel polymerase was left open. Subsequent examination of *pol*A1 mutant extracts by Kornberg and Gefter (12, 13), Moses and Richardson (15), Knippers (11), and Wickner *et al.* (21) led to the discovery of polymerases II and III, but left unresolved the possibility of residual polymerase I in this mutant. We have reinvestigated the question of whether P3478 and its derivatives contain any assayable DNA polymerase I and we report here the result of this analysis.

The two reagents that have been used to identify polymerase I are antiserum prepared against the pure enzyme and N-ethylmaleimide (NEM). Each of these easily distinguishes labeled deoxynucleoside triphosphate incorporation into acid-insoluble material due to polymerase I from that due to polymerases II and III (11–13, 15, 21) even in relatively crude fractions. Thus, activity due to polymerase I is defined as that which is inhibited by polymerase I antiserum and unaffected by concentrations of NEM up to 0.01 M. On the other hand, polymerases II and III are not affected by polymerase I antiserum, but are completely inhibited by 0.01 M NEM.

Detection of Polymerase I in Partially Purified Extracts of P3478

Assay of crude French pressure cell extracts of P3478 suggested that they contained very low levels of polymerase I activity as defined by the above criteria. To investigate this point further, the extracts were partially purified and again tested for their polymerase I content. The procedure involved removal of nucleic acids as described by Hall and Lehman (6); followed by fractionation with ammonium sulfate. With such partially purified preparations, polymerase I was easily identifiable, and clearly differentiated from polymerases II and III (Table 1). The detailed response of the ammonium sulfate II fraction of polymerase I antiserum is shown in Fig. 1. The polymerase activity in P3478 is completely suppressed even by relatively low levels of polymerase I antiserum. On the other hand, T4 DNA polymerase is completely unaffected, showing that the antiserum does not contain a nonspecific inhibi-

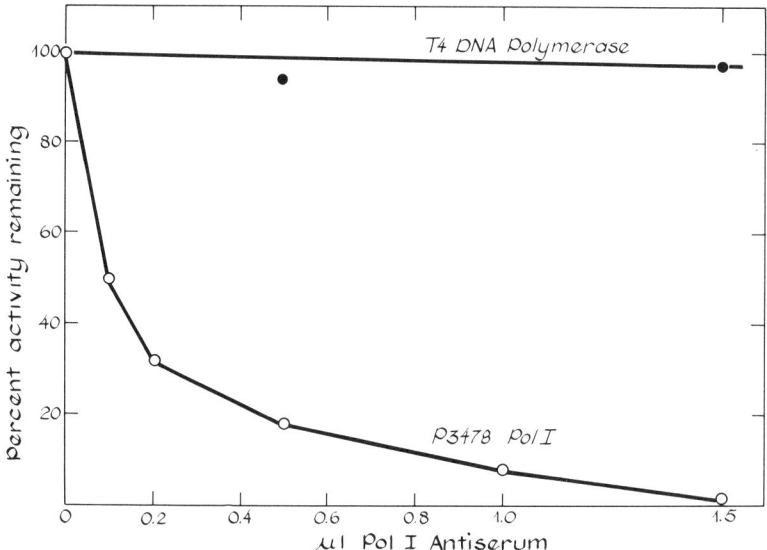

Fig. 1. Assays of P3748 polymerase I (ammonium sulfate II fraction) and T4 DNA polymerase (hydroxylapatite fraction (3)) were carried out as described in Table 1, in the presence of the indicated amounts of polymerase I antiserum.

Table 1. Separation of polymerase I from polymerase II (or III) in P3478

Conditions	^{32}P-dTTP incorporated (p moles)
Ammonium sulfate I	
$(dA)_{800} \cdot (dT)_{10}$	1.02
$(dA)_{800} \cdot (dT)_{10}$ + pol I antiserum	0.09
"Nicked" calf thymus DNA	12.9
"Nicked" calf thymus DNA + pol I antiserum	14.2
"Nicked" calf thymus DNA + 0.01 M NEM	0.8
Ammonium sulfate II	
$(dA)_{800} \cdot (dT)_{10}$	10.3
$(dA)_{800} \cdot (dT)_{10}$ + pol I antiserum	0.5
$(dA)_{800} \cdot (dT)_{10}$ + 0.01 M NEM	10.7
"Nicked" calf thymus DNA	2.2
"Nicked" calf thymus DNA + pol I antiserum	0.2

Ammonium sulfate fractions I and II were prepared in the following way: Mid-log phase cultures of P3478 (5×10^8 cells/ml) grown in H broth (19) were harvested and disrupted in a pressure cell according to Kornberg and Gefter (12). Nine milliliters of extract (approximately 20 mg of protein per ml) were treated sequentially with $MgCl_2$ and streptomycin sulfate to remove nucleic acids as described in a previous article for T4 DNA polymerase (6). To 11.2 ml of streptomycin fraction (0°C) were added 3.16 g of ammonium sulfate (40% of saturation). After 30 min at 0°C the precipitate was collected by centrifugation for 30 min at 20,000 × g. The pellet (ammonium sulfate I) was dissolved in 1 ml of 0.05 M Tris-HCl, pH 7.5; 2 mM EDTA; 1 mM dithiothreitol. Ammonium sulfate (1.69 g) was then added to the supernatant solution. After 30 min at 0°C, the suspension was centrifuged and the pellet dissolved as above (ammonium sulfate II).

Reaction mixtures (0.1 ml) contained 0.06 M Tris-HCl, pH 8.6; 8mM $MgCl_2$; 1 mM 2-mercaptoethanol; 0.06 mM each dATP, dCTP, dGTP; 0.015 mM ^{32}P-dTTP (300 to 1000 cpm/pmole); $(dA)_{800}$, 0.14 mM, and $(dT)_{10}$, 0.009 mM (in nucleotide), or "nicked" calf thymus DNA (16), 0.04 mM (in nucleotide); 1 µl of ammonium sulfate fraction (approximately 10 µg of protein). The $(dA)_{800}$ and $(dT)_{10}$ were synthesized with calf thymus terminal deoxynucleotidyl transferase as described by Kelly et al. (9). Where indicated, 0.01 M N-ethylmaleimide (NEM) or 15 µl of a 1 : 10 dilution of antiserum to homogeneous polymerase I (kindly provided by Dr. A. Kornberg) were added. (The diluted antiserum was heated for 30 min at 70°C then cooled before use.) Incubation was for 20 min at 37°C. The reaction was terminated and acid-insoluble ^{32}P was determined as described by Richardson et al. (16).

tor of polymerase activity. In a series of such preparations, the level of DNA polymerase I ranged from 0.5 to 2% of that found in the parent strain W3110 treated identically. The polymerase activity observed in P3478 was not the result of the appearance of 0.5 to 2% pol^+ revertants since each of the cultures from which the ammonium sulfate fraction was prepared plated with an efficiency of only about 10^{-6} in the presence of 0.04% methyl methane sulfonate (2).

As shown in Table 1, polymerase I can be effectively separated from polymerase II (or III) by ammonium sulfate fractionation. These enzymes can also be differentiated on the basis of their primer-template requirement. Thus, polymerase I shows a strong preference for the (dA) · (dT) homopolymer pair over "nicked" calf thymus DNA. On the other hand, polymerase II (or III) is almost inert in the presence of the (dA) · (dT), but makes effective use of "nicked" calf thymus DNA. The former, which has relatively large gaps (ratio of A : T = 20), has been shown to be a poor primer-template for purified polymerase II (20). We presume that the single strand breaks in the "nicked" calf thymus DNA have been converted to small gaps by exonuclease activity present in the ammonium sulfate fractions.

Purification of Polymerase I from P3478

The DNA polymerase I activity in P3478 has been purified approximately 1500-fold beyond the ammonium sulfate II fraction, according to the procedure described by Jovin et al. (8) (Table 2). The final preparation (Sephadex G-100 fraction) has a specific activity 0.6% that of homogeneous DNA polymerase I (kindly provided by Dr. A. Kornberg) assayed under the same conditions. Assay of the purified P3478 polymerase I for the associated 3' → 5' and 5' → 3' exonuclease activities (9) showed them to be present at 0.7% (3' → 5') and 0.4% (5' → 3') of the levels found in the homogeneous wild type enzyme.

Table 2. Purification of polymerase I from P3478

	Total activity (units)*	Units/mg protein
Ammonium sulfate II	90	0.04
DEAE-cellulose	23	0.24
Phosphocellulose	9.4	17.8
Sephadex G-100	6.3	62.8

*Ten nanomoles of ^{3}H- or ^{32}P-labeled dTTP incorporated into acid-insoluble material in 20 min at 37°C with $(dA)_{800}$ · $(dT)_{10}$ as primer-template under the assay conditons described in the footnote for Table 1.

When examined by sucrose density gradient centrifugation, the purified P3478 polymerase I showed a sedimentation coefficient of 4.9 S, a value that is significantly lower than that of the wild type enzyme (5.6 S) (8). As suggested below, the reduced sedimentation coefficient (and presumably lower molecular weight) may be the result of proteolysis during isolation of the mutant enzyme.

DNA Polymerase I Activity in Other Polymerase I Mutants of *Escherichia coli*

Fractionated extracts of JG112, a methylmethane sulfonate-sensitive strain of W3110 into which the *pol*A1 mutation had been transferred from P3478 (4) (kindly provided by T. Kornberg), contained substantially

Table 3. Polymerase activity in other polymerase I mutants

Strain	Units/mg protein
Experiment 1	
W3110 (pol I$^+$)	4.3–5.5
P3478	0.02–0.11
JG112	<0.002–0.015
polA12	0.16 (30°C)
Experiment 2	43°/30°C
polA12	0.6
W3110	3.5

Ammonium sulfate II fractions were prepared and assayed using $(dA)_{800}$ · $(dT)_{10}$ (W3110, P3478, JG112) or "nicked" calf thymus DNA (polA12) as the primer-templates. In Experiment 2, "nicked" calf thymus DNA also served as primer-template for W3110. The incubations, 20 min, in Experiment 1 were at 37°C. In Experiment 2, the incubations, 20 min, were carried out at both 30°C and 43°C; here activity is expressed as a ratio of the activity at these two temperatures.

less polymerase I activity than P3478 (Table 3). The range of values found for W3110 and P3478 is included for comparison. In fact, several independently isolated ammonium sulfate II fractions from JG112 showed no detectable polymerase I activity as judged by their polymerase I antiserum sensitivity and NEM insensitivity. On the other hand, polymerases II and III were present in amounts similar to those seen in P3478.

Earlier assays by Monk and Kinross of extracts of the temperature-sensitive polA12 mutant had indicated that there was essentially no polymerase I activity at either restrictive or permissive temperatures (14). As shown in Table 3 the ammonium sulfate II fraction prepared from this strain has approximately 3% as much polymerase I activity when assayed at 30° C as the comparable fraction derived from W3110 (assayed at 37° C). Moreover, the activity in the polA12 mutant is abnormally thermolabile, providing further evidence that the polA mutation is located in the structural gene for DNA polymerase I (10).

5′ → 3′ Exonuclease Activity in PolA1 Mutants

Although the polymerase I activity purified from P3478 through the Sephadex G-100 step showed a level of 5′ → 3′ exonuclease comparable to its polymerase I activity (0.4% of the homogeneous wild type enzyme), a nearly normal level of 5′ → 3′ exonuclease activity, suppressible by polymerase I antiserum, was detectable in the ammonium sulfate II fraction.

To minimize the possibility of proteolysis during isolation of polymerase I, ammonium sulfate II fractions were prepared from P3478 and its parent strain W3110 in the presence of the protease inhibitor phenylmethylsulfonyl chloride (PMSC), then immediately centrifuged in a 5 to 20% sucrose density gradient. As shown in Fig. 2, the sedimentation coefficient of the P3478 polymerase I is indistinguishable from that of the W3110 enzyme. This result suggests that the lower sedimentation coefficient of the purified P3478 enzyme (see above) is probably the result of proteolytic degradation in the course of the lengthy isolation procedure. Unlike the W3110 enzyme, in which the $5' \to 3'$ exonuclease and polymerase activities cosediment, the P3478 polymerase peak shows no $5' \to 3'$ exonuclease activity. On the other hand, there is a discrete peak of $5' \to 3'$ exonuclease which sediments more slowly than the polymerase activity, at about 2.8 S. In both cases, the $5' \to 3'$ exonuclease activity is $\geq 90\%$ inhibitable by antiserum to purified polymerase I but is unaffected by comparable levels of control rabbit serum. It is noteworthy that the "small fragment" derived from wild type polymerase I by proteolytic digestion, which bears the $5' \to 3'$ exonuclease but lacks the polymerase activity of the parent molecule, also has a sedimentation coefficient of approximately 2.8 S (18). Inspection of Fig. 2 further shows that although the polymerase I activity of P3478 is present at about 1% of W3110, the two strains have approximately equal levels of $5' \to 3'$ exonuclease activity. Comparable experiments carried out with JG112 indicate that it too contains normal or near-normal levels of a polymerase I antiserum-sensitive $5' \to 3'$ exonuclease activity that sediments with an S value of 2.8.

Conclusion

It is clear that *E. coli* strain P3478 contains a low but easily detectable level of DNA polymerase I activity (from 0.5 to 2% of wild type) in a form that has a sedimentation coefficient and probably a molecular weight similar to the wild type enzyme. Using the estimate of 400 polymerase molecules normally present per *E. coli* cell (16), we could say that P3478 contains an amount of polymerase I activity equivalent to 2 to 8 wild type molecules per bacterium. In view of the significant residual activity it is clear that no conclusions can be drawn as to the essentiality (or nonessentiality) of polymerase I for cell viability in this strain. In this regard, it may be significant that mutants showing a total deletion of the polymerase I gene have not been isolated (5). The analysis of polymerase I in JG112 is considerably more ambiguous than in P3478, and the activity is not reproducibly detectable. The reason for the difference between the two strains is not known.

Despite the low level of polymerase I in P3478, this strain possesses nearly normal levels of a $5' \to 3'$ exonuclease activity, separate from

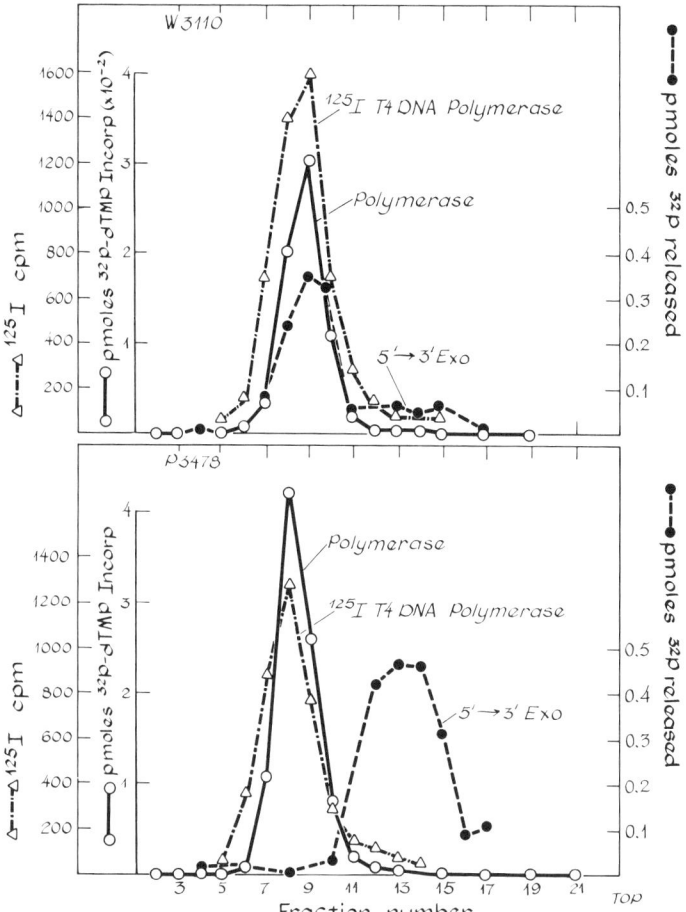

Fig. 2. Ammonium sulfate II fractions were prepared from extracts of P3478 and W3110 as described in Table 1, except that in the case of P3478 1mM phenylmethyl sulfonyl chloride (PMSC) was included in the extract buffer. After dialysis against 0.05 M Tris-HCl, pH 7.5, 5 mM EDTA, and 0.1 mM PMSC, 0.1-ml aliquots were centrifuged in a 5 to 20% sucrose density gradient containing 0.05 M Tris-HCl, pH 8.0, 1 mM EDTA, 0.05 M ammonium sulfate, and 1 mM dithiothreitol, using the SW 56 rotor in a Beckman L265B ultracentrifuge at 50,000 rpm for 16 hr at 2-3°C. ^{125}I-Labeled T4 DNA polymerase (7) and crystalline human hemoglobin (0.5 mg) were included as markers. At the end of the run, 0.19-ml fractions were collected. ^{125}I was measured in a Nuclear Chicago 1085 gamma counter; hemoglobin was determined by its absorbancy at 415 nm; polymerase I activity was assayed as described in Table 1 except that 0.01 M NEM was included. The values shown represent pmoles of dTMP incorporated in 20 min per μl of gradient fraction. 5' → 3' Exonuclease activity was assayed by the method of Setlow et al. (17) except that 0.067 M potassium phosphate, pH 7.4, was replaced with 0.067 M Tris-HCl, pH 7.5, 0.02 M potassium phosphate, pH 7.5, a 1 μmole of tRNA (nucleotide) and 0.01 M NEM were added. Activity is expressed as picomoles of ^{32}P released per 5 min per μl of gradient fraction, from ^{3}H-(dT)$_{275}$ labeled with ^{32}P (1.9 × 10^{3} g/mole) at its 5' terminus, containing a dideoxythymidylate residue at its 3' terminus and annealed to (dA)$_{4650}$.

the polymerase, that is responsive to polymerase I antiserum. A similar activity is present in JG112. Since the 5' → 3' exonuclease function of polymerase I has been shown to excise thymine dimers *in vitro*, the finding of normal levels of 5' → 3' exonuclease activity in *pol*A1 mutants is consistent with the report of Boyle *et al.* (1) that there is no gross defect *in vivo* in thymine dimer excision in cells bearing the *pol*A1 mutation.

Finally, since the *pol*A1 mutation has been shown to be an amber mutation (2), a reasonable interpretation of our findings is that the residual polymerase activity detectable in P3478 (and JG112) is due to a low level misreading of the amber mutation, and the 5' → 3' exonuclease activity with a sedimentation coefficient of 2.8 S is the result of rapid proteolytic cleavage of the amber peptide to generate the polymerase I "small fragment" (17, 18). Thus, the site at which the wild type polymerase I is particularly susceptible to proteolysis may be even more vulnerable in the case of the incomplete polypeptide. An alternative, but in our view less likely, interpretation is that the 5' → 3' exonuclease represents an intrinsic activity of the amber peptide whose sedimentation coefficient happens to be the same as that of the "small fragment." Given the finding of nearly normal levels of 5' → 3' exonuclease activity, both interpretations are, however, consistent with a positioning of this exonuclease near the N-terminal portion of the polymerase molecule.

References

1. Boyle, J. M., Paterson, M. C., and Setlow, R. B. 1970. Excision repair properties of an *Escherichia coli* mutant deficient in DNA polymerase. *Nature* 226: 708.

2. De Lucia, P., and Cairnes, J. 1969. Isolation of an *E. coli* strain with a mutation affecting DNA polymerase. *Nature* 224: 1164.

3. Goulian, M., Lucas, Z. J., and Kornberg, A. 1968. Enzymatic synthesis of deoxyribonucleic acid. XXV. Purification and properties of deoxyribonucleic acid polymerase induced by infection with phage T4. *J. Biol. Chem.* 243: 627.

4. Gross, J., and Gross, M. 1969. Genetic analysis of an *E. coli* strain with a mutation affecting DNA polymerase. *Nature* 224: 1166.

5. Gross, J. D. 1971. DNA replication in bacteria. *Current topics in microbiology and immunology* Springer Verlag, Berlin.

6. Hall, Z. W., and Lehman, I. R. 1968. An *in vitro* transversion by a mutationally altered T4 DNA polymerase. *J. Mol. Biol.* 36: 321.

7. Huang, W. M., and Lehman, I. R. 1972. On the direction of translation of the T4 DNA polymerase gene *in vivo*. *J. Biol. Chem.* 247: 7663.

8. Jovin, T. M., Englund, P. T., and Bertsch, L. 1969. Enzymatic synthesis of deoxyribonucleic acid. XXVI. Physical and chemical studies of a homogeneous deoxyribonucleic acid polymerase. *J. Biol. Chem.* 244: 2996.

9. Kelly, R. B., Cozzarelli, N. R., Deutscher, M. P., Lehman, I. R., and Kornberg, A. 1970. Enzymatic synthesis of deoxyribonucleic acid. XXXII. Replication of duplex deoxyribonucleic acid by polymerase at a single strand break. *J. Biol. Chem.* 245: 39.

10. Kelley, W. S., and Whitfield, H. J. 1971. Purification of an altered DNA polymerase from an *Escherichia coli* strain with a *pol* mutation. *Nature* 230: 33.

11. Knippers, R. 1970. DNA polymerase II. *Nature* 228: 1050.

12. Kornberg, T., and Gefter, M. 1970. DNA synthesis in cell-free extracts of a DNA polymerase defective mutant. *Biochem. Biophys. Res. Commun.* 40: 1348.

13. Kornberg, T. and Gefter, M. 1971. DNA synthesis in cell-free extracts: purification and properties of DNA polymerase II. *Proc Nat. Acad. Sci. U.S.A.* 68: 761.

14. Monk, M., and Kinross, J. 1972. Conditional lethality of recA and recB derivatives of a strain of *Escherichia coli* K-12 with a temperature-sensitive deoxyribonucleic acid polymerase I. *J. Bacteriol.* 109: 971.

15. Moses, R., and Richardson, C. C. 1970. A new DNA polymerase activity of *Escherichia coli*. I. Purification and properties of the activity present in *E. coli pol* A1. *Biochem. Biophys. Res. Commun.* 41: 1565.

16. Richardson, C. C., Schildkraut, C. L., Aposhian, H. V., and Kornberg, A. 1964. Enzymatic synthesis of deoxyribonucleic acid. XIV. Further purification and properties of deoxyribonucleic acid polymerase of *Escherichia coli*. *J. Biol. Chem.* 239: 222.

17. Setlow, P., Brutlag, D., and Kornberg, A. 1972. Deoxyribonucleic acid polymerase: Two distinct enzymes in one polypeptide. I. A proteolytic fragment containing the polymerase and $3' \rightarrow 5'$ exonuclease functions. *J. Biol. Chem.* 247: 224.

18. Setlow, P., and Kornberg, A. 1972. Deoxyribonucleic acid polymerase: Two distinct enzymes in one polypeptide. II. A proteolytic fragment containing the $5' \rightarrow 3'$ exonuclease function. Restoration of intact enzyme functions from the two proteolytic fragments. *J. Biol. Chem.* 247: 232.

19. Steinberg, C. M., and Edgar, R. S. 1962. A critical test of a current theory of genetic recombination in bacteriophage. *Genetics* 47: 187.

20. Wickner, R. B., Ginsberg, B., and Hurwitz, J. 1972. Deoxyribonucleic acid polymerase II of *Escherichia coli*. II. Studies of the template requirements and the structure of the deoxyribonucleic acid product. *J. Biol. Chem.* 247: 498.

21. Wickner, R. B., Ginsberg, B., Berkower, I., and Hurwitz, J. 1972. Deoxyribonucleic acid polymerase II of *Escherichia coli*. I. The purification and characterization of the enzyme. *J. Biol. Chem.* 247: 489.

Discussion

Stodolsky. What template was used?

Lehman. We used tritium-labeled poly (dT) about 200 residues long; it was also labeled with ^{32}P at its 5' terminus. Dideoxythymidine was present at the 3' end, so that 3' → 5' exonuclease activity would be blocked. This polymer was then annealed to poly (dA) containing about 1,000 residues.

Bessman. Did I hear you correctly when you said that in your crude fractionation the P3478 polymerase comes out in the streptomycin supernatant?

Lehman. It comes out in the supernatant, but so does the W3110. We used a procedure that was developed by Stuart Linn at the University of California for the removal of nucleic acids from cell extracts. It involves precipitation of some nucleic acids with magnesium ion followed by treatment of the magnesium supernatant with steptomycin, to yield another nucleic acid precipitate. The resulting supernatant solution has the bulk of the *Escherichia coli* polymerase I activity.

Cavalieri. Can you be more precise about the conditions which you thought lead to less proteolysis?

Lehman. Thank you for mentioning this point. We simply prepared French pressure cell extracts, went through the nucleic acid removal that I described a moment ago, then fractionated the streptomycin supernatant with ammonium sulfate, taking a 40 to 60% cut. The protease inhibitor, PMSC, was present throughout the purification. Does that answer your question?

Concerted Effect of Pancreatic DNase and the Large Fragment of DNA Polymerase I

Hans Klenow and Kay Overgaard-Hansen

Biochemical Institute B
University of Copenhagen
Copenhagen, Denmark

The large fragment of DNA polymerase I of *Escherichia coli* catalyzes the formation of poly (dT) in the presence of oligo (dT) as primer and poly (dA) or poly (rA) as template and dTTP as the only triphosphate. There is some turnover of the poly (dT) formed in this way. The rate of turnover increases with the number of nicks in the poly (dT) strand. The amount of poly (dT) formed is equivalent to the amount of poly (dA) or poly (rA) present. Nicking of a complete poly (dA) poly (dT) duplex by treatment with DNase results in formation of poly (dT) greater than the amount of poly (dA), provided that dATP is present in addition to dTTP and the large fragment of DNA polymerase I. Also, dAMP is incorporated into polymer structure, and both the poly (dA) and the poly (dT) strands grow at the nicks introduced by DNase. Nicking of the DNA strand of a complete poly (rA) poly (dT) duplex in the presence of dTTP and the large fragment of DNA polymerase I results in a rapid accumulation of poly (dT) until all dTTP has been used up. The poly (dT) which accumulates in great excess over poly (rA) is quite stable, probably due to higher affinity of DNase to poly (dT) annealed to poly (rA) than to free poly (dT). Also, the fragment may have a higher affinity for the nicks introduced in the duplex by DNase than for single stranded poly (dT). The possibility of obtaining many fragmented complementary polydeoxyribonucleotide copies of a natural single stranded RNA is discussed.

DNA polymerase I of *Escherichia coli* consists of a single polypeptide chain of molecular weight (MW) 109,000 (6). Limited proteolysis of this enzyme in the presence of DNA gives rise to the formation of two different catalytic fragments (2, 10, 11, 14). A large fragment of MW about 75,000 contains both polymerase activity and $3' \rightarrow 5'$ exonuclease activity while a smaller fragment of MW about 35,000 contains $5' \rightarrow 3'$ exonuclease activity. The cleavage of the native enzyme may under specific assay conditions be accompanied by an increase in both polymerase activity and $5' \rightarrow 3'$ exonuclease activity (about 2.2 and 6.5-fold, respectively). This increase is due to differential effects of ionic conditions on both activities depending on whether they are part of the same peptide chain or whether they exist as separate entities (12). The $5' \rightarrow 3'$ exonuclease activity of the smaller fragment is specific for double stranded DNA (13, 15). The $3' \rightarrow 5'$ exonuclease activity of the large fragment prefers single stranded DNA as substrate. Setlow and Kornberg (15) have found that $3' \rightarrow 5'$ exonuclease activity with poly d(A-T) · poly d(A-T) or poly (dA) · poly (dT) as substrate is about 25% of that with poly (dT) at 37° C. We find that there is very little if any exonuclease activity of the large fragment with double-stranded DNA as substrate at 22° C (see below). The $3' \rightarrow 5'$ exonuclease activity has been shown to be directed specifically towards a mispaired or unpaired primer terminus of the primer strand in double stranded DNA. This suggests a proofreading function of the $3' \rightarrow 5'$ exonuclease activity (3).

In addition to DNA polymerase I another DNA polymerase which lacks the $5' \rightarrow 3'$ exonuclease activity has been isolated from *E. coli* by phosphocellulose chromatography. This enzyme appears to have properties indistinguishable from those of the large fragment obtained by limited proteolysis (14, 16). Both enzymes have been used in the present investigation and no differences have been noted. These enzymes will be referred to as the large fragment.

With the large fragment at hand we were interested in devising experiments in which the enzyme might catalyze the formation of several copies of single stranded DNA or RNA. As a first approach we have worked with homopolymers as templates and with complementary oligodeoxyribonucleotides as primers. It has already been shown both for DNA polymerase I from *E. coli* and for the DNA polymerase from *Micrococcus luteus* that the amount of poly (dT) formed with poly $(dA)_n$ or poly $(rA)_n$ as template and with dTTP as the only triphosphate is equivalent to the amount of template present. This reaction requires oligo (dT) as a primer (5, 7, 16). We have confirmed these observations in similar experiments with the large fragment of DNA polymerase I and ^{14}C or ^{32}P-labeled dTTP. In the hope of obtaining further information about this reaction we have assayed not only for the formation of poly (dT), but also for dTMP and dTTP. These compounds were separated by chromatography on ion exchange paper. It has been found that there is some turnover of the poly (dT) strand as long as it contains nicks. At the nicks, strand elongation occurs and the old strands in front of synthesis are displaced and degraded. Introduction of many nicks in a poly (rA) poly (dT) duplex by treatment with DNase increases the rate of poly (dT) synthesis to such an extent that poly (dT) accumulates in excess of the amount of poly (rA) present.

It appears from Fig. 1 that with labeled dTTP as the only triphosphate the amount of labeled poly (dT) formed corresponds to the amount of poly $(dA)_n$ present as template. These experiments are performed at 22° C in phosphate buffer, pH 7.4. The labeled poly (dT) is completely stable for many hours. This is the case both when the original ratio of nucleotide residues in template and $(dT)_{10}$ primer is 9, and when it is about 1. In the former case there is only a very slow formation of labeled dTMP. This suggests that a complete duplex almost without nicks may have been formed by extension of a few oligo (dT) primer chains. This duplex is not degraded by the fragment. If, however, the initial ratio of poly $(dA)_n$ to oligo $(dT)_{10}$ is 1, there is a more rapid formation of labeled dTMP, probably due to the existence of some nicks in the newly synthesized polymer. These nicks give rise to further

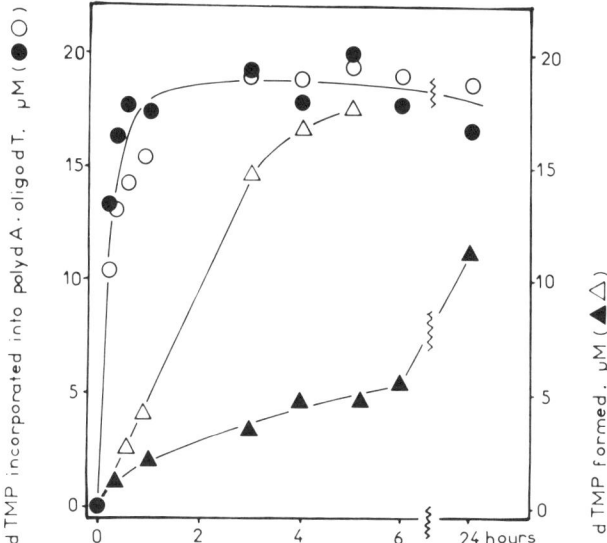

Fig. 1. Synthesis and turnover of poly (dT) with poly $(dA)_n$ as template. The reaction was carried out at 22°C and the incubation mixture was composed as follows: potassium phosphate, pH 7.4 (28 mM); MgCl$_2$ (4.0 mM); ^{32}P-dTTP (80 μM); poly (dA) (20 μM); DNA polymerase (large fragment) (33 nM). The concentration of oligo $(dT)_{10}$ was either 2.0 μM (closed symbols) or 20 μM (open symbols). The reaction mixture was preincubated at 22°C for 30 min before addition of the large fragment. Samples of the reaction mixtures were spotted on strips of DEAE-ion exchange paper together with markers of dTMP and dTTP, and the chromatograms were developed in 0.08 M ammonium formate for 16 hr. The origin and dTMP spots were cut out and counted in a liquid scintillation counter.

synthesis of polymer in excess of the amount of poly (dA) present. Any poly (dT) in excess of poly (dA) is, however, immediately degraded to dTMP by the 3' → 5' exonuclease activity. The rate of degradation is at least as high as the rate of formation of excess strands. There may, therefore, be a turnover of the newly formed poly (dT), the rate being dependent on the number of nicks. This notion is confirmed in experiments performed with a limiting amount of dTTP. If the concentration of this compound is only one-third that of the nucleotides in poly $(dA)_n$, the following results are obtained. With an excess (9-fold) of poly $(dA)_n$ over $(dT)_{10}$, all of the labeled dTTP is rapidly incorporated into polymer structure which is completely stable and there is no formation of dTMP (Fig. 2A). Upon addition of unlabeled dTTP to the reaction mixture after all the labeled dTTP has been incorporated there is only a very slow decrease in the amount of labeled poly (dT) with a corresponding very slow formation of dTMP. This suggests that there are only a few nicks in the paired poly (dT) strand. If the incorporation of limited amounts of labeled dTTP is studied with about equimolar amounts of poly $(dA)_n$ and oligo $(dT)_{10}$, the rapid incorporation of all radioactivity into polymer structure is followed by a slow depolymerization giving rise to formation of an amount of dTMP corresponding to almost one-third of the dTTP present originally (Fig. 2B). This may

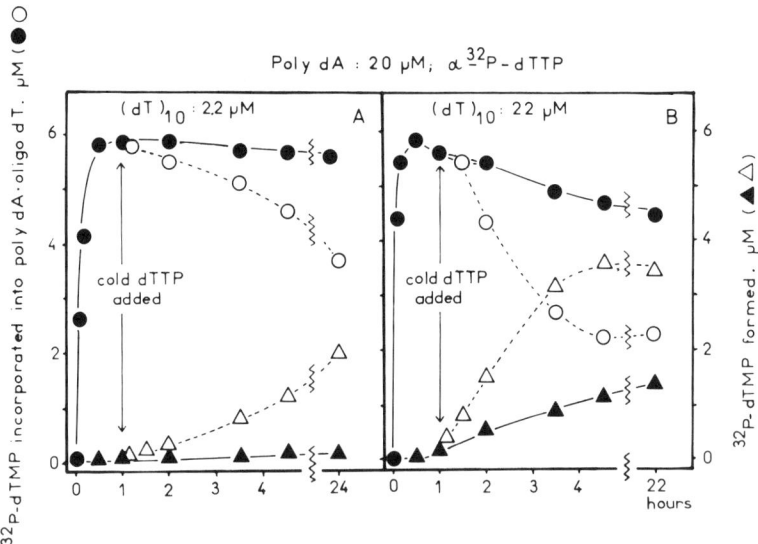

Fig. 2. Incorporation of labeling into poly (dT) and chase with cold dTTP. The reaction mixture and the assay were as described for Fig. 1 with the exception that the concentration of α-^{32}P-dTTP was only 6 μM. After 1 hr of incubation, unlabeled dTTP (28 μM) was added to an aliquot of the reaction mixture. The concentration of $(dT)_{10}$ was 2.2 μM (Fig. 2A) or 22 μM (Fig. 2B).

mean that most of the 3'-OH ends of the unlabeled oligo $(dT)_{10}$ of the original complete duplex have been extended with labeled dTMP residues. The 5' end groups of the strands in front of synthesis may simultaneously have been partially displaced by newly synthesized strands. When the labeled dTTP has been used up, a nonenzymatic slippage of the two homopolymer strands against each other may give rise to a complete duplex and some excess poly (dT) strands. The excess strands are degraded to dTMP as their 3' ends become free. The remaining duplex appears to be completely stable in the presence of the large fragment. If unlabeled dTTP is added to the reaction mixture at the point where all labeled dTTP has been incorporated into polymer structure, part of this labeling is chased out in the following hours with concomitant formation of dTMP. The more rapid chase of the label in the poly (dT) strands may in this case be due to the presence of more nicks in each poly (dT) strand than in the previous experiment.

Since there seems to be some turnover of the poly (dT) strand only when it contains nicks, we wanted to see the effect of nicking a complete poly (dA) poly (dT) duplex by treatment with pancreatic DNase. In this experiment, after a small amount of oligo (dT) strands annealed to poly (dA) has been completely repaired in the presence of labeled dTTP, pancreatic DNase is added. This treatment actually causes a slow decrease in the amount of labeled poly (dT) present (Fig. 3). If, however, both unlabeled dATP in excess over the amount of dTTP present and DNase are added, the formation of poly (dT) is resumed at a high rate and proceeds with an equivalent consumption of dTTP. The final amount of poly (dT) is much greater than the original amount of poly (dA). The dTMP formation is very slow under these conditions. Addition of unlabeled dATP alone in this system causes only a very slow increase of the amount of poly (dT). It appears, therefore, that upon nicking of both strands in a complete poly (dA) poly (dT) duplex by treatment with DNase, the generated free 3'-OH end groups of both strands may be extended provided that both deoxyribonucleoside triphosphates are present. This process probably implies a simultaneous slippage of the poly (dA) strand against the poly (dT) strand. The growth of both strands is confirmed in similar experiments performed with labeled dATP and unlabeled dTTP. Incorporation of radioactivity into the polymer structure occurs to an appreciable extent only after preincubation with dTTP and the large fragment followed by treatment with DNase.

We have performed a similar series of experiments with poly (rA) as template. It is again found that with labeled dTTP as the only triphosphate the incorporation of radioactivity into the polymer depends on the presence of oligo (dT) as primer. These experiments were performed in Tris-HCl buffer at pH 8.2. Also in this case the incorporation of dTMP into the polymer is equivalent to the amount of poly (rA) present,

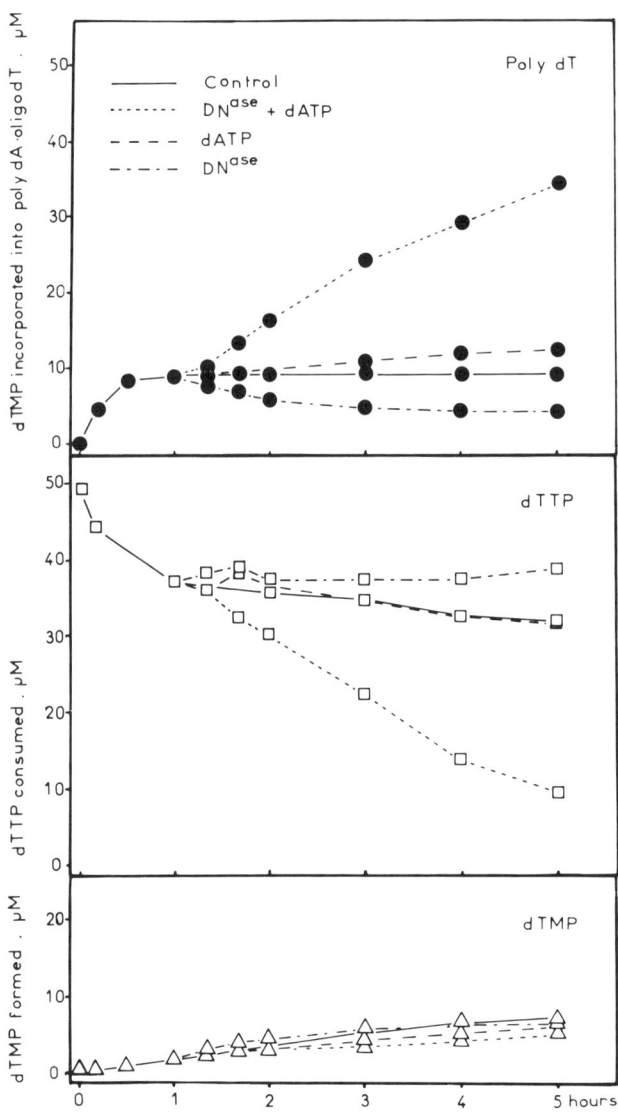

Fig. 3. The effect of DNase on the synthesis and turnover of poly (dT) with poly (dA)$_n$ as template. Reaction mixture: potassium phosphate, pH 7.4 (25 mM); MgCl$_2$ (3.6 mM); poly (dA)$_n$ (9.3 μM); oligo (dT)$_{12}$ (0.12 μM); ^{14}C-dTTP (50 μM, 22 μCi/μmole); and 0.13 μM DNA polymerase (large fragment). The reaction conditions were as described for Fig. 1. At intervals aliquots were withdrawn and analyzed for poly (dT), dTMP, and dTTP, as described in the legend for Fig. 1. After 60 min the reaction mixture was divided into four parts (a–d) and further incubated under the following conditions: (a) no additions (control); (b) dATP (0.16 mM) added; (c) dATP (0.16 mM) and DNase (0.03 μg/ml of reaction mixture) added; (d) DNase (0.03 μg/ml of reaction mixture) added.

both when the ratio of poly (rA) to oligo (dT)$_{10}$ is 6.5 : 1 and when it is 1 : 1. In the former case (Fig. 4) the initial rate of dTMP formation is lower than in the latter case (Fig. 7). This lower rate does, however, increase in the presence of KCl in the reaction mixture (Fig. 4). With poly (rA) as a template there may therefore be a turnover of the poly (dT) strand, the rate of which seems to increase with the number of nicks and in the presence of KCl.

The effect of KCl on the rates of both the 3' → 5' exonuclease reaction and the polymerase reaction was therefore investigated. The former reaction, followed by the release of acid-soluble radioactivity from ^{3}H-labeled poly (dT)$_{300}$, is completely dependent on MgCl$_2$ or MnCl$_2$ and also completely or almost completely dependent on KCl. The optimum concentration for MnCl$_2$ is about 0.2 mM and for MgCl$_2$ about 6 mM (experiment not shown). The rates are almost identical at optimal MnCl$_2$ and MgCl$_2$ concentrations. The optimal KCl concentration increases from 25 mM to about 50 mM as the MnCl$_2$ concentration

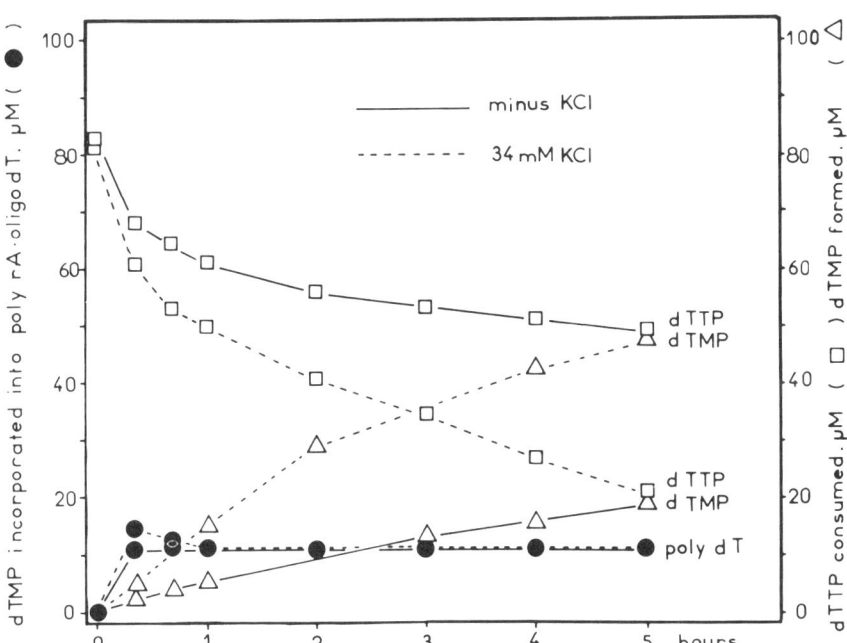

Fig. 4. The effect of KCl on the synthesis and turnover of poly (dT) with poly (rA) as template. Reaction mixture: Tris-HCl, pH 8.2 (40 mM); MnCl$_2$ (0.3 mM); poly (rA) ($S_{20,w}$ = 10) (10.4 μM); oligo (dT)$_{12}$ (1.6 μM); ^{14}C-dTTP (84 μM, 22 μCi/μmole); DNA polymerase (large fragment, 0.13 nM). The reaction was carried out at 22°C in presence or absence of KCl (34 mM). Samples were withdrawn at intervals and analyzed for poly (dT), dTMP, and dTTP, as described in the legend for Fig. 1.

increases from 0.2 to 1.7 mM. Above these concentrations of KCl the reaction rate decreases rapidly and is almost zero at 200 mM (Fig. 5). The rate of degradation is identical for chain lengths of poly (dT) from 30 to 300 in reaction mixtures with the same end group concentrations (experiment not shown).

The polymerase reaction with poly (rA) (dT)$_{10}$ is also found to be completely dependent on the presence of MgCl$_2$ or MnCl$_2$, with optimum concentrations of about 6 mM and 0.1 to 0.2 mM, respectively. KCl stimulates the polymerase reaction by a factor of about 4 at optimal MgCl$_2$ or MnCl$_2$ concentration; at 200 mM KCl the rate is about two-thirds that at optimal KCl (Fig. 6). The rate at optimal KCl and MnCl$_2$ is about 2.3-fold higher than at optimal KCl and MgCl$_2$. Similar results have recently been obtained for the polymerase activity of the native enzyme by Karkas et al. (8). The finding that dTMP formation in a poly (rA) oligo (dT) system in the presence of dTTP and the large fragment is limited but not prevented by omitting KCl from the reaction mixture (Fig. 4) suggests two mechanisms for dTMP formation. One, consisting of the breakdown of free displaced poly (dT) strands, may be KCl-dependent, while the other may depend on slippage of the two

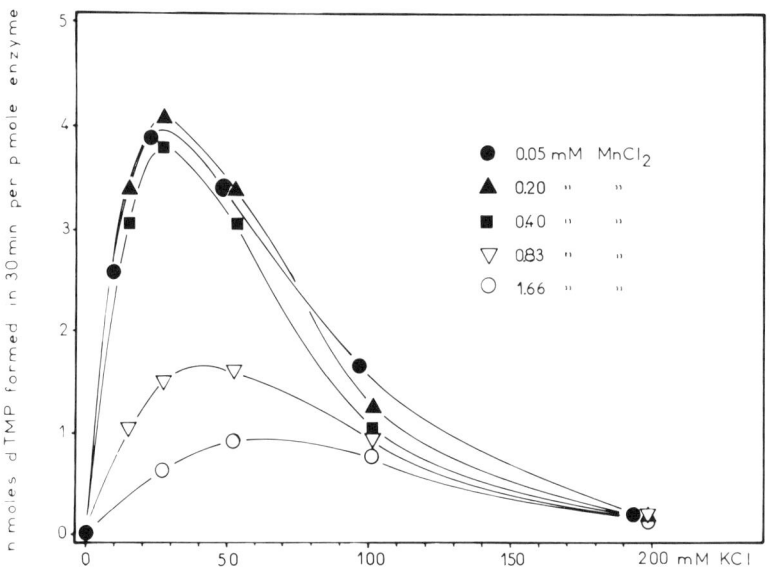

Fig. 5. Effect of KCl and MnCl$_2$ on 3' → 5' exonuclease activity of the large fragment of DNA polymerase I. The incubation mixtures contained, in a final volume of 0.310 ml: Tris-HCl, pH 8.2 (40 mM); dithiothreitol (1.3 mM); ^{3}H-labeled poly (dT)$_{300}$ (24 μM, 200 μCi/μmole) (9); DNA polymerase (large fragment) (2.1 nM); the indicated concentrations of KCl and MnCl$_2$. Samples of 70 μl were withdrawn after incubation for 10, 20, 30, and 40 min at 37°C and the rate of formation of acid-soluble radioactivity was determined as described by Klenow and Henningsen (10).

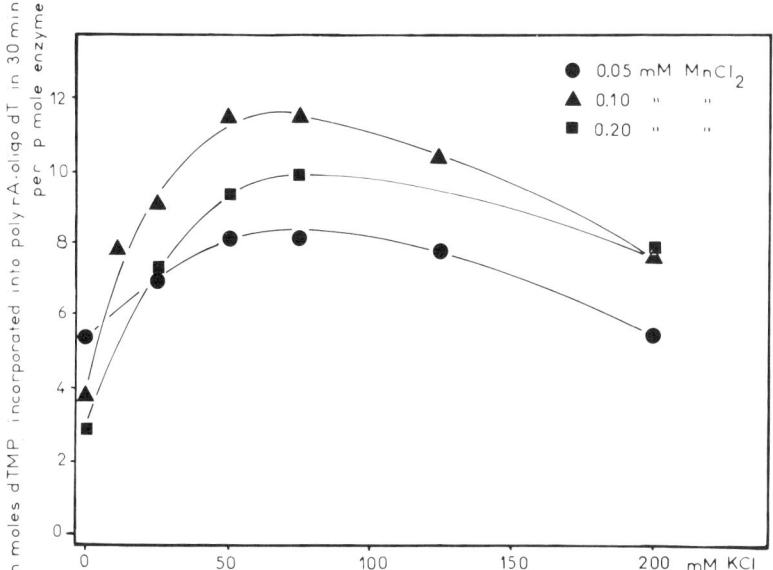

Fig. 6. Effect of KCl and $MnCl_2$ concentration on polymerase activity of large fragment. The incubation mixtures contained, in a final volume of 0.025 ml: Tris-HCl, pH 8.2 (40 mM); dithiothreitol (1.3 mM); poly (rA) (38 μM); oligo(dT)$_{12}$ (10 μM); ^{14}C-dTTP (55 μM, 22 μCi/μmole); DNA polymerase (large fragment) (2.7 nM); the indicated concentrations of KCl and $MnCl_2$. Samples of 5 μl were withdrawn after incubation for 10, 20, 30, and 40 min at 22°C and the rate of incorporation of ^{14}C-dTMP into polymer structure was assayed by the chromatographic technique described in the legend for Fig. 1.

homopolymers against each other. In the latter case enzyme molecules bound at the end of the duplex may release any mononucleotide residues protruding from the 3'-OH end. This process may occur in the absence of KCl.

Nicking of the DNA strand of poly (rA) poly (dT) with DNase, in the presence of dTTP as the only triphosphate, results in a very rapid formation of poly (dT) in a greater amount than that of the poly (rA) present. DNase is added after a duplex of poly (rA) and labeled poly (dT) has been formed. This addition gives rise to a rapid formation of excess labeled poly (dT) and to an almost equivalent consumption of labeled dTTP. This reaction progresses until all the labeled dTTP has been used up, and the poly (dT) is then almost completely stable. The rate of formation of excess poly (dT) depends on the rate of formation of nicks. These rates can be increased by a factor of nearly 2 by increasing the DNase concentration by a factor of 3 (from 0.016 to 0.048 μg/ml) (Fig. 7).

It is thus evident that the introduction of nicks into the DNA strand of the poly (rA) poly (dT) duplex by treatment with DNase results in

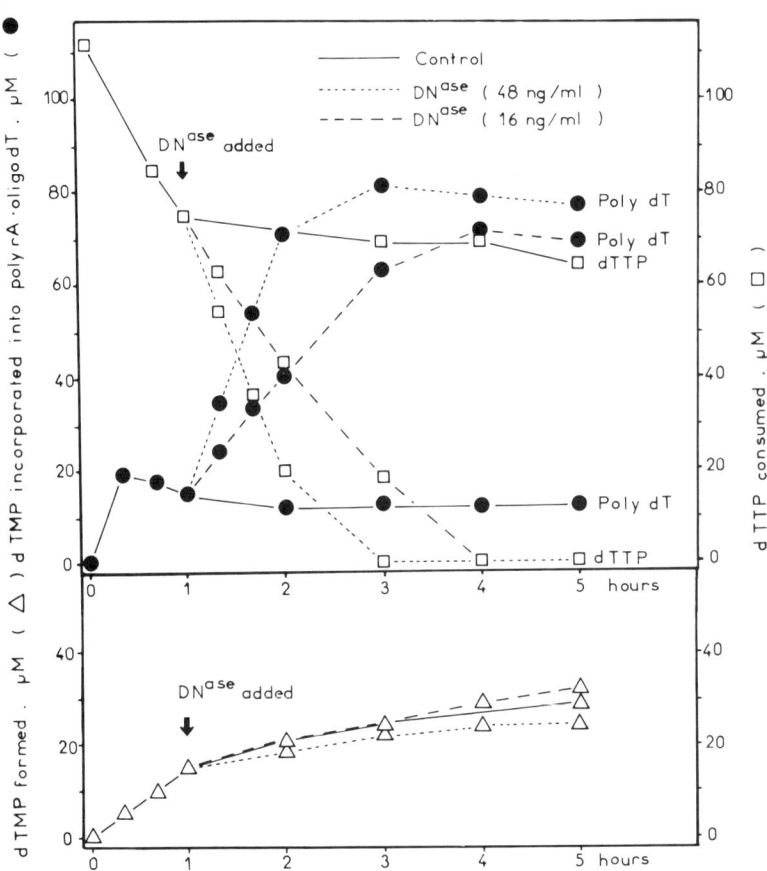

Fig. 7. The effect of DNase on the synthesis and turnover of poly (dT) with poly (rA) as template. Reaction mixture: Tris-HCl, pH 8.2 (32 mM); MnCl$_2$ (0.2 mM); poly (rA) (11 μM); oligo (dT)$_{12}$ (15 μM); ^{14}C-labeled dTTP (109 μM, 22 μCi/μmole); and DNA polymerase (large fragment) (0.13 μM). The reaction was carried out at 22°C and the reaction mixture was kept 20 min at this temperature before the addition of enzyme. At intervals samples were withdrawn and analyzed for poly (dT), dTMP, and dTTP, as described in the legend for Fig. 1. After 60 min of incubation the reaction mixture was divided into three parts which were further incubated under the following conditions: (a) no additions (control); (b) DNase (0.048 μg/ml of reaction mixture) added; (c) DNase (0.016 μg/ml of reaction mixture) added.

a rapid extension of the oligo (dT) strand at the induced 3' end groups. The strands in front of synthesis are displaced and released from the duplex. The newly synthesized poly (dT) strands are constantly being nicked, extended, and displaced by the concerted effect of DNase and the fragment. The rate of dTMP formation is very low (Fig. 7). The excess poly (dT) strands are, therefore, quite stable in spite of the presence of both 3′ → 5′ exonuclease activity and DNase. There may be several reasons for this. The fragment may have a higher affinity for the nicks introduced in the duplex by DNase than for single stranded poly (dT). The DNase may likewise have a higher affinity for poly (dT) which is annealed to poly (rA) than for poly (dT). In addition, the rate of degradation by DNase of free poly (dT) has been found to be several-fold slower than that of the double stranded poly d(A-T) · poly d(A-T).

It thus appears that using a concerted effect of pancreatic DNase and the large fragment of DNA polymerase I, it may be possible to synthesize many fragmented copies of poly (rA). A very similar effect has been obtained in analogous experiments with poly (rI) · oligo (dC). In this case also the effect of the two enzymes permits the synthesis of an amount of poly (dC) which is several-fold higher than that of the poly (rI) present, the increase depending on the amount of dCTP present. However, this reaction does not proceed in Tris buffer but only in phosphate buffer. This is in agreement with the finding of Wells *et al.* (16).

Similar experiments performed with equimolar amounts of the native DNA polymerase I instead of the large fragment have shown that in this case only a very small transient excess of the homodeoxyribonucleotide is formed. Instead a rapid formation of dTMP with a concomitant consumption of dTTP is seen. This may be ascribed to the 5′ → 3′ exonuclease activity of the native enzyme.

For experiments of this type to be performed with the large fragment and DNase and with single stranded natural RNA as template, the base sequence of the 3′ end of the RNA must be known and a corresponding complementary piece of single stranded DNA must be available as a primer for the reaction. The recent finding by two groups (1, 4) that mRNA from a number of eukaryotic cells as well as the RNA from several types of RNA virus contain poly (rA) at the 3′-OH end group may make such an approach possible. Such experiments performed with the large fragment, DNase and all four deoxyribonucleoside triphosphates may provide a means of obtaining many fragmented complementary polydeoxyribonucleotide copies of natural RNA.

Acknowledgments

This work was supported by the Danish Natural Science Research Council. The work was initiated at the Department of Chemistry and Biology at Massachusetts Institute of Technology, Cambridge, Massachusetts, while one of us (H. K.) held an EMBO-fellowship.

References

1. Armstrong, J. A., Edmonds, M., Nakazato, H., Philips, B. A., Vaughan, M. H. 1972. Polyadenylic acid sequences in the virion RNA of poliovirus and eastern equineencephalitis virus. *Science* 176: 526.
2. Brutlag, D., Atkinson, M. R., Setlow, P., and Kornberg, A. 1969. An active fragment of DNA polymerase produced by proteolytic cleavage. *Biochem. Biophys. Res. Commun.* 37: 982.
3. Brutlag, D., and Kornberg, A. 1972. Enzymatic synthesis of deoxyribonucleic acid. XXXVI. A proofreading function for the $3' \to 5'$ exonuclease activity in DNA polymerase. *J. Biol. Chem.* 247: 241.
4. Green, M., and Carkas, M. 1972. The genome of RNA tumor viruses containing polyadenylic acid sequences. *Proc. Nat. Acad. Sci. U. S. A.* 69: 791.
5. Harwood, S. J., and Wells, R. D. 1970. *Micrococcus luteus* deoxyribonucleic acid polymerase. *J. Biol. Chem.* 245: 5625.
6. Jovin, T. M., Englund, P. T., and Bertsch, L. L. 1969. Enzymatic synthesis of deoxyribonucleic acid. XXVI. Physical and chemical studies of a homogeneous deoxyribonucleic acid polymerase. *J. Biol. Chem.* 224: 2996.
7. Jovin, T. M., and Kornberg, A. 1968. Polynucleotide cellulose as solid state primers and templates for polymerases. *J. Biol. Chem.* 243: 250.
8. Karkas, J. D., Stavrianopoulos, J. G., and Chargoff, E. 1972. *Proc. Nat. Acad. Sci. U. S. A.* 1972: 398.
9. Kelly, R. B., Cozzarelli, N. R., Deutscher, M. P., Lehman, J. R., and Kornberg, A. 1970. Enzymatic synthesis of deoxyribonucleic acid. XXXII. Replication of duplex deoxyribonucleic acid by polymerase at a single strand break. *J. Biol. Chem.* 245: 39.
10. Klenow, H., and Henningsen, I. 1970. Selective elimination of the exonuclease activity of the deoxyribonucleic acid polymerase from *Escherichia coli* B by limited proteolysis. *Proc. Nat. Acad. Sci. U. S. A.* 65: 168.
11. Klenow, H., and Overgaard-Hansen, K. 1970. Proteolytic cleavage of DNA polymerase from *Escherichia coli* B into an exonuclease unit and a polymerase unit. *FEBS Lett.* 6: 25.
12. Klenow, H., Overgaard-Hansen, K., and Patkar, S. A. 1972. Proteolytic cleavage of native DNA polymerase into two different catalytic fragments. *Eur. J. Biochem.* 22: 371.
13. Overgaard-Hansen, K., and Klenow, H. 1972. Unpublished results.
14. Setlow, P., Brutlag, D., and Kornberg, A. 1972. Deoxyribonucleic acid polymerase: Two distinct enzymes. I. A proteolytic fragment containing the polymerase and $3' \to 5'$ exonuclease. *J. Biol. Chem.* 247: 224.
15. Setlow, P., and Kornberg, A. 1972. Deoxyribonucleic acid polymerase: Two distinct enzymes in one polypeptide. II. A proteolytic fragment containing the $5' \to 3'$ exonuclease function. Restoration of intact enzyme functions from the two proteolytic fragments. *J. Biol Chem.* 247: 232.

16. Wells, R. D., Flügel, R. M., Larson, J. E., Schendel, P. F. and Sweet, R. W. 1972. Comparison of some reactions catalyzed by deoxyribonucleic acid polymerase from a vian myoblastosis virus. *Biochemistry* 11: 621.

Discussion

Baltimore. In our hands at least, and I think in others, pol I, the whole pol I at least, has not copied messenger RNA at all. Do you have any indication that it will do that?

Klenow. No, we haven't done such experiments yet.

Taylor. We've done some experiments with polio and Rous sarcoma virus RNA which contained poly (A) sequences at the 3' end. With pol I, in the presence only of oligo (dT), we get synthesis, but when we analyze this synthesis, we can show conclusively that it is only the synthesis of poly (T) and then later the synthesis of poly (A). You don't get heteropolymer synthesis at all.

Mechanisms of Excision-Repair

Lawrence Grossman, Barbara M.
Garvick, Howard K. Ono, Andrew G.
Braun, Lester D. G. Hamilton, and Inga Mahler

Graduate Department of Biochemistry
Brandeis University
Waltham, Massachusetts 02154

The initial step in the excision-repair of UV-damaged DNA is perhaps the most singular and enzymatically unique step in the cycle. Once a UV-specific endonuclease incises adjacent to a thymine dimer, the resulting DNA can serve as a substrate for a variety of DNA polymerases or exonucleases. In *Escherichia coli* B, the polymerase I-associated $5' \rightarrow 3'$ exonuclease is capable of excising the dimer; whereas in *Micrococcus luteus*, a specific exonuclease can satisfy this function. The $3'$-phosphoryl group remaining on the DNA, as a result of this incision, can be removed directly either by exonuclease III in *E. coli* or by a *M. luteus* DNA polymerase-associated $3' \rightarrow 5'$ exonuclease. The formation of such a nucleophilic site provides for polymerization by DNA polymerase. The final portion of the repair cycle is potentially reactivatable by polynucleotide ligase.

The ability of cells to survive a variety of chemical or radiation-induced "insults" depends, in large part, on the cells' inherent mechanisms for repairing the resulting damage in either a direct or an indirect manner. Although many secondary and ancillary mechanisms, such as cell permeability, inactivation of the agent, or nonspecific filtration of radiation, may protect the cell, those enzymes acting directly at the damaged loci in nucleic acids probably play the primary role in such protective mechanisms.

Photoreactivation

Enzyme specificity involving direct recognition of ultraviolet photoproducts was initially discovered by Kelner (14) and Dulbecco (8), who observed that bacterial cells, initially inactivated by ultraviolet irradiation, subsequently recovered in the presence of visible light. This form of *photoreactivation* is attributable to a single enzyme (20) which in the presence of 330- to 360-nm light catalyzes the conversion of the major UV photoproduct in DNA, thymine-thymine dimers (T̂T), to thymine. It appears, therefore, that the cyclobutane ring formed between adjacent intrastrand thymidylates is broken by the photoreactivating enzyme. Although the chromophore has not been specifically identified in the photoreactivating enzyme of either yeast (17) or *Escherichia coli* (22), there is much current interest in examining such a light-dependent enzyme reaction. Enzymes with photoreactivating capabilities are not uniformly distributed in bacterial and animal species (7). This sort of enzyme reactivity may, in fact, be fortuitous, since it is present in organs unlikely to have available visible light.

Excision or Dark Repair

The most ubiquitous enzyme mechanism contributing to the life expectancy of irradiated cells literally removes photochemical damage in its DNA. Excision-repair, first described simultaneously by Setlow and Carrier (21) and Boyce and Howard-Flanders (3), results in the loss of intact thymine dimers from the DNA of cells irradiated *in vivo*. The over-all mechanism is controlled by a number of separate genetic loci located on the *E. coli* genetic map. It is this mechanism which has gained most experimental attention because of its intrinsic interest as well as its relationship to replication and recombination. Much of the enzyme characterization has been achieved with *Micrococcus luteus*, since this relatively UV-resistant organism has low endogenous levels of non-UV-specific nucleases. From studies *in vitro*, the excision-repair mechanism can be described as a cycle, pictured in Fig. 1, involving at least four separate enzymes and five defined activities. It must be assumed that this cycle, as described, is essentially heuristic because of the involvement of a series of other enzymes which can conceivably act in a concerted manner leading to the formation of DNA intermediates which would be available for other reactions.

Specific Steps in the Excision-Repair Cycle

Incision Step

The initial step in the *M. luteus* excision-repair cycle is catalyzed by an endonuclease specific for irradiated duplex DNA. The enzyme from this organism has been purifed to homogeneity, has a molecular weight of 15,000, and is stimulated by, but not dependent on, MG^{+2} (10, 11).

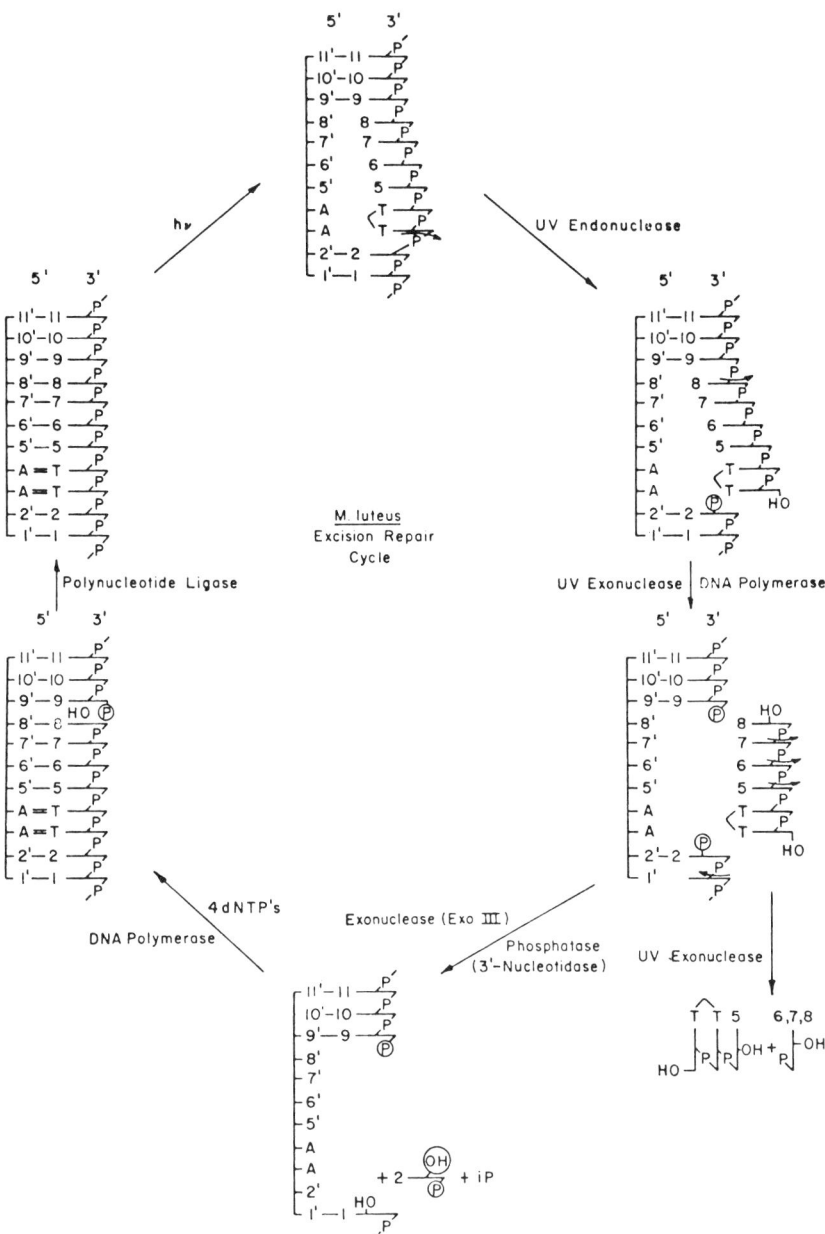

Fig. 1. Excision-repair cycle in *Micrococcus luteus*.

The enzyme is capable of binding to irradiated circular DNA duplexes ($\phi\chi_{174}$RF-I) in the presence of EDTA (5).

Approximately 1 enzyme molecule is required per thymine dimer (5) to hydrolyze the phosphodiester bond 5' to the dimer and proximal to the photoproduct, resulting in a 3'-phosphoryl group on the DNA and a 5'-hydroxyl group on the damaged single strand-like fragment (15). This step is most likely rate-limiting since the turnover number is approximately 3 min for each thymine-dimer-associated phosphodiester hydrolysis event.

Once incision has occurred, the enzyme no longer binds at this site and is free to attach to another modified but intact region of the irradiated DNA molecule. Because a 3'-phosphoryl group remains on a potential priming site for DNA polymerase-catalyzed reinsertion of nucleotides, we must posit an additional enzymatic step in the cycle. This step may involve a multifunctional DNA polymerase isolated from *M. luteus*. A further discussion of this phenomenon will be found in the section dealing with reinsertion.

Excision Step

Nucleotide and photoproduct removal *in vitro* is dependent on a second nuclease in the excision-repair cycle. This enzyme acts exonucleolytically on single stranded DNA, initiating hydrolysis from either the 5' or 3' terminus (12). Unlike most known exonucleases, the *M. luteus* exonuclease, which is capable of participating in excision *in vitro*, is largely unaffected in its rate and extent of hydrolysis by the presence of thymine dimers. Although the enzyme has a preference for initiating hydrolysis at 3' terminal, its action on incised DNA at the 5' terminus is initially *endonucleolytic*. Hence, an oligonucleotide containing the thymine dimers is initially released and subsequently degraded by the same enzyme (15). What structural constraints are necessary for this enzyme to act endonucleolytically is not known.

Although this nuclease functions predictably in catalyzing excision *in vitro*, it is not certain whether the enzyme serves such a function *in vivo*. To date, no UV-sensitive *M. luteus* mutants have been detected which lack this enzyme. Conversely, it is possible to identify the *in vivo* role for the UV endonuclease from studies of a number of *M. luteus* mutants and transformants with either completely or partially reduced endonuclease levels (16). *E. coli* polymerase I, by virtue of its 5' → 3' exonuclease activity, can assume a dual role in excision and repair. Kelly *et al.* (13) demonstrated that this homogeneous enzyme can, in fact, excise thymine dimer regions from irradiated poly (dA) · (dT) providing an initial endonuclease-catalyzed incision event has occurred. The same DNA polymerase can potentially serve in the same process by virtue of its ability to reinsert nucleotides concomitant with its error-correcting $5' \to 3'$ nuclease activity. However, pol A$^-$ mutants, although UV-sensitive, excise thymine dimers at a normal rate and to a normal extent *in vivo* (4). From such observations, it must be assumed that pol I performs the repair functions primarily at the reinsertion step of the cycle.

The DNA polymerases isolated either from T4-infected cells or from *E. coli* DNA polymerases II and III possess only $3' \to 5'$ nuclease as a potential error-correcting mechanism (2, 6). It must be assumed, as a result of such specificity, either that UV-specific endonucleases from *E. coli* or T4-infected *E. coli* incise $3'$ to photoproducts or that a specific UV exonuclease, like that in *M. luteus*, is present in *E. coli*. Recent experiments with T4-infected *E. coli* imply that the latter suggestion is the most likely. Onshima and Sekiguchi (18) identified such an activity in T4-infected *E. coli*. The V$^+$ gene product is a UV endonuclease having properties much like those described for the *M. luteus* UV endonuclease (9). It is conceivable, therefore, that T4-infected *E. coli* has an incision-excision reaction sequence catalyzed in the same fashion as described for *M. luteus*.

Prereinsertion

The action of the UV-specific endo- and exonucleases from *M. luteus* results in a small cavity in the DNA with both a $3'$- and a $5'$-phosphoryl group (15). A nucleophilic site must, therefore, be exposed for potential priming during the reinsertion step. Specific dephosphorylation at $3'$ sites can be catalyzed by polynucleotide phosphomonoesterases such as described by Becker and Hurwitz (1), by an exonuclease III-type enzyme which dephosphorylates as a prerequisite to exonucleolytic hydrolysis (19), or by a DNA polymerase with a $3' \to 5'$ exonuclease whose activity is unaffected by $3'$-phosphoryl groups. Although the $3' \to 5'$ exonuclease activity associated with DNA polymerase I of *E. coli* is inhibited by $3'$-phosphomonoesters, this does not appear to be the case with the *M. luteus* DNA polymerase-associated $3' \to 5'$ exonuclease. The latter DNA polymerase can utilize $3'$-phosphorylated double stranded primers with moderate efficiency for polymerization. It is assumed that the terminal nucleotide is removed as a $3', 5'$-deoxynucleoside diphosphate revealing the juxtaposed nucleotide with a $3'$-hydroxyl group.

This mechanism may be a dominant one in *M. luteus* since a UV-sensitive mutant, ML$_{200}$, is able to excise thymine dimers normally but appears unable to incorporate thymidine *in vivo* during the postirradiation period. It is assumed that this mutant's inability to take up a simple DNA precursor during postirradiation nongrowing conditions, as well as its inability to restore the molecular weight of its DNA *in vivo*, is attributable to a block prior to the incorporation of nucleotides. The DNA polymerase from such a mutant has been purified and is able to polymerize to the same extent as the wild type enzyme, provided that a $3'$-hydroxyl terminal priming site is available. Removal of the $3'$-phosphorylated nucleotides as part of the $3' \rightarrow 5'$ exonuclease activity appears to be an important controlling point in the reinsertion portion of the excision-repair cycle.

Reinsertion

Essentially two types of reinsertion mechanisms, depending on the nuclease-associated DNA polymerases present in the cells, can be visualized. A $5' \rightarrow 3'$ exonuclease activity might be expected to digest, in addition to the photoproducts, a large number of unaffected nucleotides while, at the same time, polymerization would belatedly follow an exorbitant loss of nucleotides. Such excessive breakdown of DNA during repair is observed in "reckless" strains of *E. coli*, while in other strains reduced levels of digestion are recorded during postirradiation periods.

For those cells which possess a $3' \rightarrow 5'$ exonuclease activity exclusively, a limited reinsertion through a finite-sized cavity is a possible mechanism. It might also be contemplated that if such a DNA polymerase could displace the damaged strand, subsequent exonucleolytic digestion would also result in uninhibited postirradiation digestion after the incision step and during the reinsertion of nascent undamaged strands in to the DNA molecule.

Acknowledgments

This work is publication No. 890 of the Graduate Department of Biochemistry, Brandeis University. It was supported by research grants from the American Cancer Society (NP-8B), National Science Foundation (GB-29172), and the National Institutes of Health (GM-15881). It was also supported by research contract AT(30-1)3449 from the Atomic Energy Commission. Lawrence Grossman is a research career development awardee (K3-GM-4845).

References

1. Becker, A., and Hurwitz, J. 1967. The enzymatic cleavage of phosphate termini from polynucleotides. *J. Biol. Chem.* 242: 936.

2. Bessman, M. J., Muzyczka, N., and Poland, R. J. 1972. Studies on the biochemical basis of spontaneous mutation rates. *Fed. Proc.* 31: 444.

3. Boyce, R. P., and Howard-Flanders, P. 1964. Release of ultraviolet light-induced thymine dimers from DNA in *E. coli* K-12. *Proc. Nat. Acad. Sci. U. S. A.* 51: 293.

4. Boyle, J. M., Paterson, M. C., and Setlow, R. B. 1970. excision-repair properties of an *Escherichia coli* mutant deficient in DNA polymerase. *Nature* 226: 708.

5. Braun, A., Ono, H., and Grossman, L. 1972. In preparation.

6. Brutlag, D., and Kornberg, A. 1972. Enzymatic synthesis of deoxyribonucleic acid. *J. Biol. Chem.* 247: 241.

7. Cook, J. S. 1972. *In* Beers, R. F., and Herriot, R. M. (eds.) *Molecular and cellular repair processes*, Johns Hopkins Press, Baltimore, Md. (In press).

8. Dulbecco, R. 1950. Experiments on photoreactivation of bacteriophages inactivated with ultraviolet radiation. *J. Bacteriol.* 59: 329.

9. Friedberg, E. C., and King, J. J. 1971. Dark repair of ultraviolet-irradiated deoxyribonucleic acid by bacteriophage T4: Purification and characterization of a dimer-specific phage-induced endonuclease. *J. Bacteriol.* 106: 500.

10. Grossman, L., Kaplan, J. C., Kushner, S. R., and Mahler, I. 1968. Enzymes involved in the early stages of repair of ultraviolet-irradiated DNA. *Cold Spring Harbor Symp. Quant. Biol.* 33: 229.

11. Kaplan, J. C., Kushner, S. R., and Grossman, L. 1969. Enzymatic repair of DNA. 1. Purification of two enzymes involved in the excision of thymine dimers frm ultraviolet-irradiated DNA. *Proc. Nat. Acad. Sci. U. S. A.* 63: 144.

12. Kaplan, J. C., Kushner, S. R., and Grossman, L. 1971. Enzymatic repair of DNA. III. Properties of the UV-endonuclease and UV-exonuclease. *Biochemistry* 10: 3315.

13. Kelly, R. B., Atkinson, M. R., Huberman, J. A., and Kornberg, A. 1969. Excision of thymine dimers and other mismatched sequences by DNA polymerase of *Escherichia coli*. *Nature* 224: 495.

14. Kelner, A. 1949. Effect of visible light on the recovery of streptomyces griseus conidia from ultra-violet irradiation injury. *Proc. Nat. Acad. Sci. U. S. A.* 35: 73.

15. Kushner, S. R., Kaplan, J. C., Ono, H., and Grossman, L. 1971. Enzymatic repair of deoxyribonucleic acid. IV. Mechanism of photoproduct excision. *Biochemistry* 10: 3325.

16. Mahler, I., Kushner, S. R., and Grossman, L. 1971. *In vivo* role of the UV-endonuclease from *Micrococcus luteus* in the repair of DNA. *Nature New Biol.* 234: 47.

17. Muhammed, A. 1966. Studies on the yeast photoreactivating enzyme. *J. Biol. Chem.* 241: 516.

18. Onshima, S., and Sekiguchi, M. 1972. Induction of a new enzyme activity to excise pyrimidine dimers in *Escherichia coli* infected with bacteriophage T4. *Biochem. Biophys. Res. Commun.* 47: 1126.

19. Richardson, C. C., and Kornberg, A. 1964. A deoxyribonucleic acid phosphatase-exonuclease from *Escherichia coli*. *J. Biol. Chem.* 239: 242.

20. Rupert, C. S. 1960. Photoreactivation of transforming DNA by an enzyme from bakers' yeast. *J. Gen. Physiol.* 43: 573.

21. Setlow, R. B., and Carrier, W. L. 1964. The disappearance of thymine

dimers from DNA: An error-correcting mechanism. *Proc. Nat. Acad. Sci. U. S. A.* 51: 226.

22. Sutherland, B. M., Court, D., and Chamberlin, M. J. 1972. Studies on the DNA photoreactivating enzyme from *Escherichia coli*. Transduction of the *phr* gene by bacteriophage lambda. *Virology* 48: 87.

Discussion

Lehman. In the experiment in which you said you were unable to find the release of phosphate, did you find it as a 3'- or 5'-nucleoside monophosphate?

Grossman. These experiments are currently in progress.

Bhattacharjee. I wondered, since the polymerase is usually contaminated with the nuclease, can this be used as an argument that the purpose of the polymerase is probably repair?

Grossman. We don't think it is a contaminant. Its temperature and pH stability seem to be about the same as the polymerase. In terms of polymerase I, I don't know if Dr. Lehman has looked at the UV sensititivity or excision properties of the polymerase I mutants.

Lehman. They've all been looked at and they are UV-sensitive.

Grossman. Yes, but the question is whether they excise thymine dimers normally.

Lehman. Well, I don't know the exact strain that Setlow looked at, but in that particular case there was no defect in thymine dimer excision.

Grossman. The argument could be that the DNA polymerase is not involved in the excision portion of the cycle, if the rate-limiting step is in fact the very first step, you may already have enough molecules of $5' \rightarrow 3'$ activity associated with pol I to carry out this process. There is this possibility, therefore, that pol I is involved only in the reinsertion step in repair.

Lehman. Based on our studies with P3478 (we haven't looked at the others yet), we find that the $5' \rightarrow 3'$ activity seems to be present at, or near, normal levels.

Stodolsky. Do you have any feeling from the dose rate data whether the dependence for AT-rich regions, which we've seen *in vitro*, also hold *in vivo*? That is, there might be another cofactor *in vivo*.

Grossman. There have been reports in which cytosine dimers are found after one or two generations in the progeny. It is possible that they aren't being excised, suggesting that this enzyme system may not excise cytosine-cytosine dimers.

Miller. Does that *Micrococcus luteus* DNA polymerase contain $5' \rightarrow 3'$ exonuclease?

Grossman. Yes. The *M. luteus* polymerase we are studying is probably identical with the one described by Bob Wells and Rose Litman. It degrades $5' \rightarrow 3'$ and $3' \rightarrow 5'$.

Nucleotide Sequencing of 3' Termini of Duplex DNA with the T4 DNA Polymerase

Paul T. Englund, Sharon S. Price,
Joan Moyer Schwing, and Paul H. Weigel

Johns Hopkins University
School of Medicine
Department of Physiological Chemistry
Baltimore, Maryland 21205

A simple procedure has been developed for determining short nucleotide sequences at 3' termini of duplex DNA molecules. The DNA is incubated at 11°C with the T4 DNA polymerase and a single α-^{32}P-labeled deoxynucleoside triphosphate. The enzyme sequentially degrades each strand from the 3' terminus until a mononucleotide is released which can be replaced by transfer from the ^{32}P-labeled triphosphate. The subsequent reaction consists of alternating removal and replacement of the ^{32}P-labeled nucleotide, and the enzyme never penetrates deeper into the strand. The result is the steady-state incorporation of about 1 mole of radioactive nucleotide per mole of strand at a unique position in the sequence at or near the original 3' terminus. In separate experiments, the DNA is labeled with each of the four ^{32}P-labeled nucleotides and the strands of the DNA are separated to permit independent analysis of the two termini. Using T7 DNA, sequence information was obtained by nearest neighbor analysis of the ^{32}P-labeled strands, by measuring the effect of unlabeled triphosphates on incorporation of radioactive nucleotides, and by determination of the structure of an oligonucleotide which arose from the terminus of a labeled strand. The r strand of T7 DNA terminates with the sequence . . . pApGpA and the l strand terminates with the sequence . . . pTpG(pT,pC)pCpCpT.

A linear duplex DNA molecule does not support net DNA synthesis catalyzed by the T4 DNA polymerase (7). Nevertheless, such a molecule is a substrate for the hydrolytic activity of this enzyme (5, 7, 8). The polymerase attacks the 3'-hydroxyl terminus of each strand and sequentially releases 5'-mononucleotides until eventually the entire molecule is degraded (4, 7, 9). However, when the DNA is treated with the polymerase in the presence of a single deoxynucleoside triphosphate, the extent of degradation of the DNA is drastically reduced. In fact, the maximum number of nucleotides released from each of the DNA strands may be as small as one or two (5).

A study of the effect of a single deoxynucleoside triphosphate on the degradation of duplex DNA has led to the mechanism illustrated in Fig. 1. The T4 DNA polymerase attacks the 3' terminus of the DNA and sequentially releases mononucleotides (Reaction I). Degradation continues until a nucleotide has been released which can be replaced by transfer from the single triphosphate. Rather than continue the degradation, the enzyme replaces the nucleotide which has just been removed (Reaction II). This added nucleotide is then subject to hydrolytic attack by the enzyme (Reaction III), but after its removal it is immediately replaced again (Reaction II). Subsequent reaction consists exclusively of alternating removal (Reaction III) and replacement (Reaction II) of the terminal nucleotide. There are two consequences of this mechanism. First, because the enzyme never penetrates any deeper into the strand after a nucleotide has been incoporated, the extent of degradation is abruptly limited to one or a few nucleotides. Second, since Reaction II is faster than Reaction III, a steady state quantity of approximately one nucleotide (which can be labeled with ^{32}P) is incorporated at a unique position either at or near the original 3' terminus.

There are several different kinds of evidence which point to the mechanism pictured in Fig. 1. First, the number of labeled nucleotides released from a uniformly labeled duplex DNA molecule in the presence of a single triphosphate is very small. Using as a substrate the fragments of T7 DNA produced by the *Hemophilus influenzae* restriction endonu-

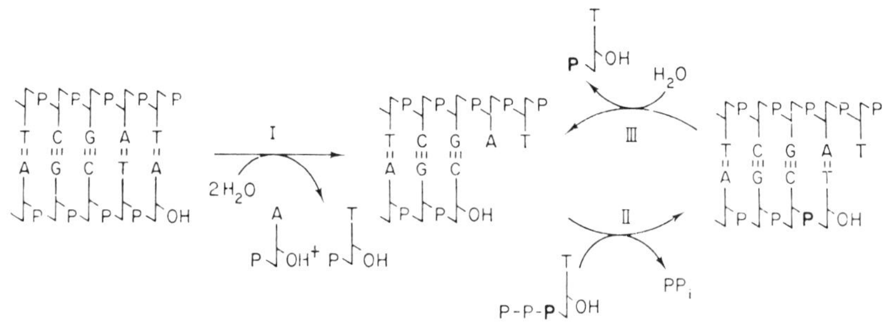

Fig. 1. Reaction catalyzed by the T4 DNA polymerase on the terminus of a linear duplex DNA molecule of hypothetical nucleotide sequence at 11°C and in the presence of a single deoxynucleoside triphosphate, α-^{32}P-dTTP. Boldface letters indicate ^{32}P-labeling.

clease (15), we find that the number (in addition to the identity) of nucleotides released in the presence of a triphosphate is in accordance with their known terminal sequence (5, 11). A second kind of evidence which supports the existence of the mechanism is the finding that, in the presence of a single α-^{32}P-deoxynucleoside triphosphate, approximately one ^{32}P-nucleotide is incorporated into each strand of the DNA (6). The fact that this nucleotide is incorporated at a unique site either at or near the original 3' terminus will be documented below from experiments with T7 DNA. A third kind of evidence is an experiment showing that the kinetics of pyrophosphate release from the single triphosphate are linear even though incorporation of a nucleotide by transfer from the triphosphate stops at the steady state level of one nucleotide per strand. This discovery was made by using a triphosphate labeled in the base with ^{3}H and in the γ-phosphate with ^{32}P (5).

Temperature has a marked effect on the reaction illustrated in Fig. 1. Only when the reaction is carried out at a rather low temperature (11°C) will the enzyme cleanly halt net degradation and incorporate a labeled nucleotide exclusively at a unique site. When the reaction is carried out at 37°C, incorporation of the labeled nucleotide seems to occur at multiple sites near the terminus (6). A possible explanation of the temperature effect is that at 37°C the ends of the DNA duplex are locally denatured. The enzyme would recognize the 3' termini as single strands, and the presence of a triphosphate would not affect their degradation (5, 8). In contrast, at 11°C the termini might be tightly base-paired and a nucleotide which was hydrolyzed could be immediately replaced by transfer from a triphosphate. Other studies of the effect of temperature on the hydrolytic activity of the T4 DNA polymerase have revealed effects which are also attributable to local denaturation of termini (2, 9).

Use of the T4 Polymerase in Sequencing

Since the reaction in Fig. 1, when carried out at 11°C, should result in a strictly limited degradation of a duplex DNA as well as the incorporation of a single nucleotide (which can be labeled with ^{32}P) at a unique site at or near the 3' terminus of each strand, it should be possible to use this reaction to identify 3'-terminal nucleotide sequences. We have used three different approaches in the application of this reaction to the problem of nucleotide sequencing. The first approach was used with fragments of T7 DNA produced by the *H. influenzae* restriction endonuclease (15). These fragments are double stranded duplex molecules, averaging about 1000 base pairs in length, which contain a unique trinucleotide terminal sequence. The identity of this sequence had been previously established by Kelly and Smith (11). Attack by the T4 DNA polymerase on uniformly labeled ^{32}P-DNA fragments in the presence of each of the four nonradioactive triphosphates resulted in the release of small amounts of ^{32}P-mononucleotides. The quantities and identities of the mononucleotides were those predicted from the previously known terminal sequence, and therefore these results independently confirmed the identity of this sequence (5). The reaction in Fig. 1 has also been applied in two other ways to investigation of the terminal nucleotide sequences of the linear duplex DNA from bacteriophage T7. Both of these approaches have involved the incorporation of ^{32}P-nucleotides at a unique site at or near the 3' terminus of each strand of unlabeled T7 DNA. As described more fully in the paragraphs that follow, sequence information was obtained from this labeled DNA in one case by nearest neighbor analysis (6) of the ^{32}P terminally labeled strands and in the other case by identifying the structure of an oligonucleotide containing the radioactive nucleotide.

Application to T7 DNA

T7 DNA was a particularly desirable molecule for terminal sequence analysis because it can be purified in large quantities almost completely free of nicks, because it contains a nonpermuted nucleotide sequence without cohesive ends (14), and because the 5'-terminal dinucleotides of its two strands were already known (21). Furthermore, the recent surge of interest in the genetics and biochemistry of T7 justifies further characterization of its DNA (18).

The incubation of nonradioactive T7 DNA with T4 DNA polymerase and a single α-^{32}P-deoxynucleoside triphosphate led to the incorporation of approximately 1 mole of nucleotide per mole of strand, as measured by the conversion of radioactivity into a form which is acid-insoluble. The kinetics of incorporation of ^{32}P-dCMP are shown in Fig. 2A. Zone sedimentation analysis of the labeled DNA in an alkaline sucrose gradient indicated that most of this labeling must have occurred at or near the termini of the molecules (Fig. 2B). Virtually all of the incorporated radioactivity sedimented in a single peak, with a sedimentation coefficient the same as that of untreated T7 DNA. If significant labeling had occurred at random internal nicks, then the alkaline gradient would have revealed radioactivity in smaller material which would have sedimented behind the major peak.

Because a radioactive nucleotide should be incorporated at both 3' termini of T7 DNA, it was desirable to study each terminus independently. This required separation of the two labeled strands, which for T7 DNA could be easily accomplished by equilibrium centrifugation in a CsCl gradient of a preparation which had been heat-denatured and cooled

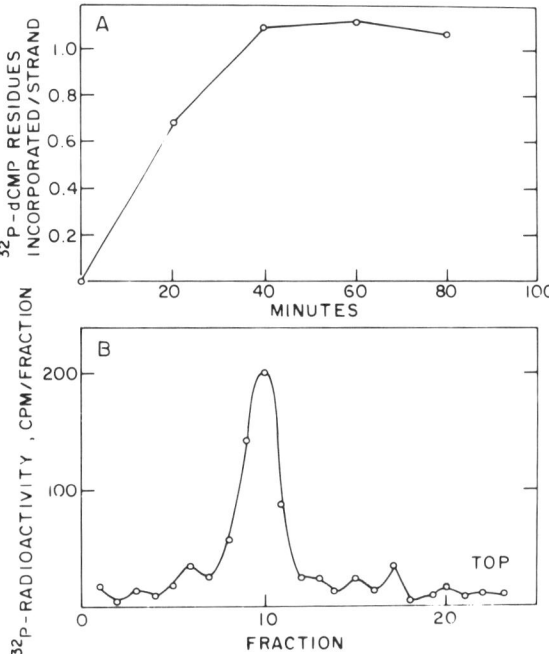

Fig. 2. A, Kinetics of incorporation of ^{32}P-dCMP into the 3' termini of T7 DNA. The reaction mixture (0.2 ml) contained 0.33 μmole of nonradioactive T7 DNA; 63 mM Tris hydrochloride, pH 8.0; 6.3 mM MgCl$_2$; 6.3 mM mercaptoethanol; 0.09 mM α-^{32}P-dCTP (4×10^9 cpm/μmole); and 150 units of T4 DNA polymerase. The solution was incubated at 11°C, and the progress of the reaction was followed by measuring the incorporation of radioactivity (in 15-μl aliquots) into a form which is insoluble in acid. The value for "^{32}P-dCMP residues incorporated per strand" was calculated by assuming that a strand of T7 DNA contains 40,000 nucleotide residues (18). B, Alkaline sucrose gradient of T7 DNA labeled at the 3' termini with α-^{32}P-dGTP. The DNA was labeled as in part A. The gradient (5.0 ml) contained 0.2 M NaOH, 0.8 M NaCl, 1.0 mM EDTA, and a linear gradient of 5 to 20% sucrose. The sample contained 19 nmoles of DNA. Centrifugation was for 90 min at 50,000 rpm at about 5°C using a Spinco SW 50.1 rotor. The sedimentation coefficient of the radioactive peak was the same as that of untreated T7 strands which were run in an identical gradient. The bottom of the gradient is at the left.

in the presence of poly (U,G) (16). A typical strand separation of DNA which had been terminally labeled with α-^{32}P-dTTP is shown in the top graph of Fig. 3. Both the *l* and the *r* strands were labeled to approximately the same extent.

Nearest Neighbor Analysis

After obtaining the isolated labeled strands from gradients similar to those shown in the top graph of Fig. 3, the simplest approach for sequence determination was nearest neighbor analysis (10). For example,

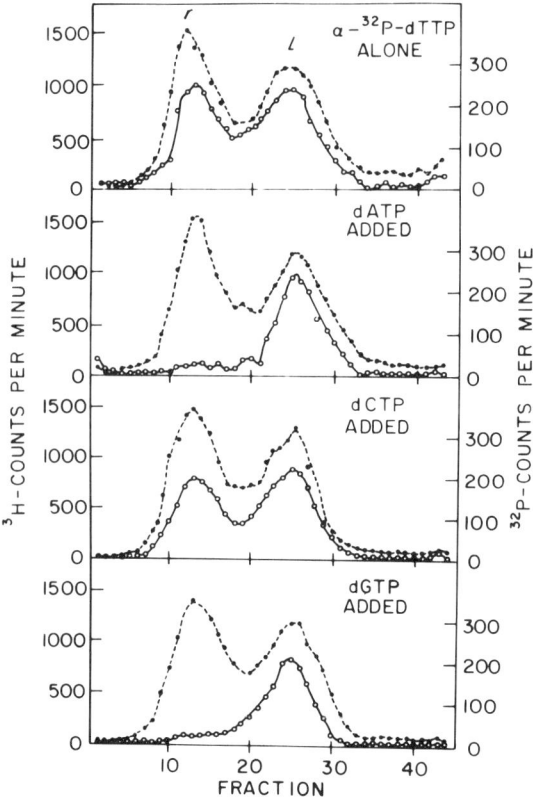

Fig. 3. Separation of the strands of T7 DNA labeled with ^{32}P-dTMP at the 3' termini. The strands were labeled as described in the legend of Fig. 2A. When unlabeled triphosphates were added, as indicated on the graphs, they were at a concentration of 0.07 mM. After labeling, the DNA was mixed T7 DNA uniformly labeled with ^{3}H, heated to 100°C for 3 min in the presence of poly (U,G), and then quickly cooled. The samples, each containing about 0.3 μmole of T7 DNA, were then centrifuged to equilibrium in a CsCl gradient (16). Fractions were collected from the bottom of the tubes and acid-insoluble radioactivity was measured in a scintillation counter. The bottoms of the gradients are at the left, and the strands designated "r" are the ones which bind poly (U,G).

if the hypothetical DNA molecule illustrated in Fig. 1 had been labeled with a ^{32}P-deoxythymidylate, a nearest neighbor analysis would indicate that this radioactive nucleotide was adjacent to a deoxycytidylate. Therefore, a CpT sequence would exist at or near the 3' terminus of that hypothetical strand. The isolated strands of T7 DNA, labeled with each of the ^{32}P-triphosphates, were then subjected to nearest neighbor analyses. After quantitative degradation by micrococcal nuclease and spleen phosphodiesterase, the resulting 3'-deoxymononucleotides were separated by paper chromatography. The chromatograms are illustrated in Fig. 4, and in each case virtually all of the radioactivity was present in

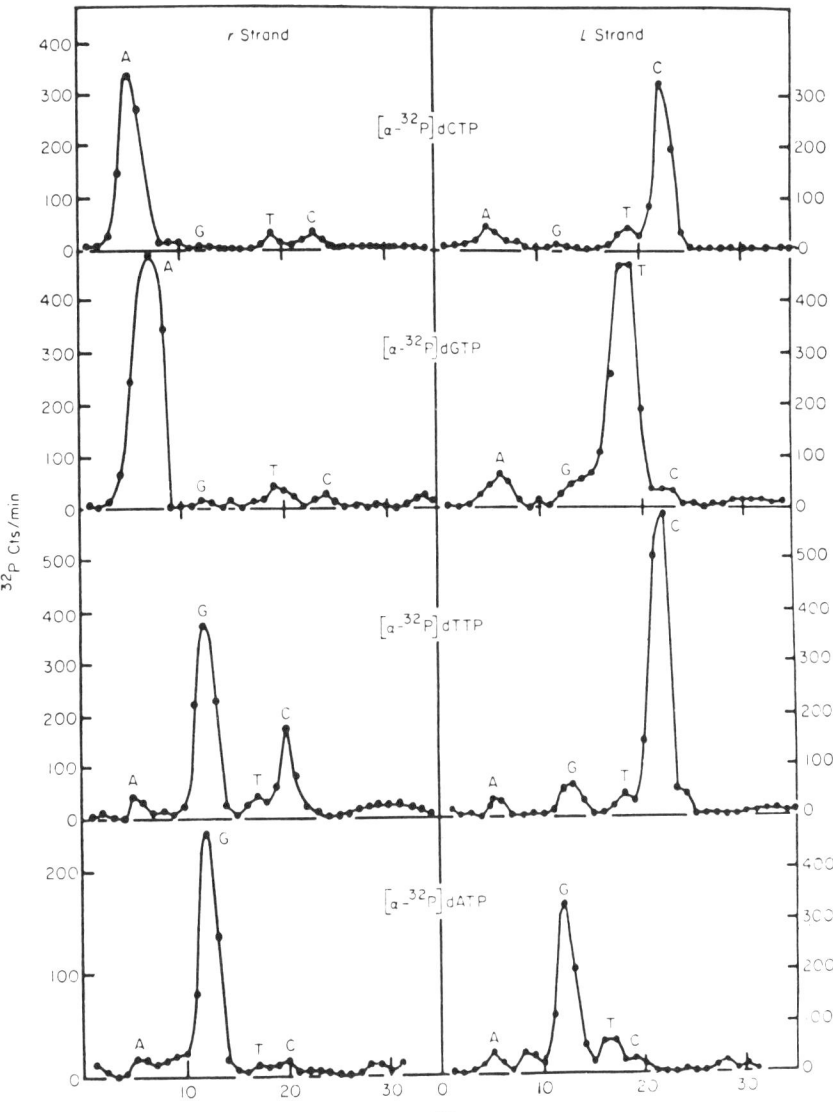

Fig. 4. Nearest neighbor analyses of strands of T7 DNA which have been labeled with a terminal ^{32}P-deoxymononucleotide. The strands were labeled as described in the legend of Fig. 2A and separated as described in the legend of Fig. 3. Enzymatic degradation of the strands and the paper chromatographic separation of the 3′-mononucleotides have been described elsewhere (6). The α-^{32}P-deoxynucleoside triphosphates used in labeling the strands are indicated on the graph. The 3′-mononucleotides are indicated by the letters A, G, T, and C.

a single ^{32}P-3'-deoxymononucleotide. These findings indicate that in each experiment the strands were labeled with a radioactive nucleotide at unique positions near the 3' termini. From the nearest neighbor analyses it follows that the sequences ApC, ApG, GpT, and GpA exist near the 3' terminus of the r strand and that the sequences CpC, TpG, CpT, and GpA exist near the 3' terminus of the l strand.

The Terminal Nucleotides

Before the sequences derived from the nearest neighbor analyses could be placed in the correct order, it was necessary to identify the terminal nucleotide of each strand. This identification was carried out by measuring the effect of unlabeled triphosphates on the incorporation of ^{32}P-nucleotides into each strand of the DNA. For example, in the hypothetical DNA molecule illustrated in Fig. 1, it would be found that unlabeled dATP, but not dGTP or dCTP, would inhibit the incorporation of ^{32}P-deoxythymidylate into the strand. The reason for the inhibition by dATP is that the enzyme would stop net degradation after the terminal deoxyadenylate had been released, and would not penetrate deeply enough into the strand to incorporate a ^{32}P-deoxythymidylate. Since neither deoxyguanylate nor deoxycytidylate occurs in the terminal sequence between the 3' terminus and the first deoxythymidylate, the presence of dGTP or dCTP would not affect the incorporation of ^{32}P-deoxythymidylate. Similarly, it would be found that neither dCTP, dGTP, nor dTTP would inhibit the incorporation of ^{32}P-dAMP into the strand, because deoxyadenylate is the terminal nucleotide. An example of an experiment of this type, applied to T7 DNA, is illustrated in Fig. 3. The graphs are strand separations of T7 DNA labeled with α-^{32}P-dTTP either alone or in the presence of one of the other three (unlabeled) triphosphates. The presence of dATP and dGTP, but not dCTP, prevented incorporation of ^{32}P-dTMP into the r strand, but no unlabeled triphosphate affected the incorporation into the l strand. This fact indicates that deoxythymidylate is the terminal nucleotide of the l strand. In a similar set of experiments it was found that neither dCTP, dGTP, nor dTTP inhibited the incorporation of ^{32}P-dAMP into the r strand. Therefore, deoxyadenylate must terminate the r strand.

Once the terminal nucleotide of each strand was known, the nearest-neighbor relationships (Fig. 5) could be placed in an order corresponding to the terminal nucleotide sequence. The terminal nucleotide of the l strand is deoxythymidylate. Its nearest neighbor is deoxycytidylate, and the nearest neighbor of deoxycytidylate is also deoxycytidylate. Therefore the terminal trinucleotide of the l strand is . . . pCpCpT. By similar reasoning, the terminal trinucleotide sequence of the r strand is . . .

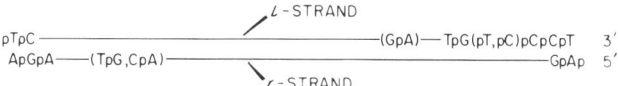

Fig. 5. Tentative structure of the 3' termini of T7 DNA. The 5'-dinucleotide sequences had been previously determined by Weiss and Richardson (21).

pApGpA. Other nearest neighbor relationships were also identified near each 3' terminus (Fig. 4) but these have not yet been linked up with the terminal sequences.

Isolation of a Terminal Oligonucleotide

Additional studies, using 3'-terminal labeling by the T4 polymerase, have made it possible to extend our knowledge of the terminal sequences of T7 DNA even further. Because the three terminal nucleotides of the *l* strand contain pyrimidines, we investigated the possibility that this sequence might be part of a longer sequence of pyrimidine nucleotides. For these studies, the terminal deoxythymidylate of the *l* strand was specifically labeled in a reaction mixture containing T7 DNA, T4 DNA polymerase, α-^{32}P-dTTP, and nonradioactive dATP. The purpose of the dATP was to prevent incorporation of radioactivity into the *r* strand. (A strand separation experiment indicated that more than 85% of the incorporated radioactivity was in the *l* strand.) After labeling, the DNA was degraded in a reaction mixture containing formic acid and diphenylamine. We know that these conditions cause depurination of the DNA and degradation into a series of pyrimidine oligonucleotides (3). However, because of the specificity of the labeling reaction, we could expect that only the oligonucleotide arising from the 3' terminus of the *l* strand would contain radioactivity. It has been possible to characterize this terminal ^{32}P-oligonucleotide. By gradient chromatography on a column of DEAE-cellulose in the presence of 7 *M* urea (20), using a pancreatic DNase digest of calf thymus DNA as a reference, the oligonucleotide was found to be a pentanucleotide. (More than 80% of the total radioactivity was found in the pentanucleotide peak.) This finding indicated that the *l* strand is terminated by a sequence of five pyrimidines. Electrophoresis of the formic acid-diphenylamine digest on DEAE-paper, using 7% formic acid as a solvent, revealed that virtually all of the radioactivity migrated as a single spot. By comparison of the mobility of this spot with the mobilities of pyrimidine oligonucleotides reported by Székely (19), it appeared likely that this oligonucleotide had the composition T_2C_3. Consequently, its sequence would be (C,T)CCT. We are presently in the process of identifying the exact structure of this oligonucleotide.

Conclusion

All the available information on the structure of the termini of T7 DNA is summarized in Fig. 5. Since we have found that nonradioactive dGTP inhibits the incorporation of ^{32}P-dAMP into the *l* strand, the purine nucleotide nearest to the terminus of the *l* strand must be deoxyguanylate. The nearest neighbor of deoxyguanylate in that strand is deoxythymidylate, and therefore the sequence pTpG must reside adjacent to the terminal sequence of five pyrimidine nucleotides. Consequently, a total of seven nucleotides have been identified at the 3' terminus of the *l* strand (except for the ambiguities in the fourth and fifth positions), and a total of three have been identified at the 3' terminus of the *r* strand. We were gratified to find that the 3'-terminal dinucleotide sequences are complementary to the 5'-terminal dinucleotide sequences previously identified by Weiss and Richardson (21). Therefore, T7 DNA may have fully base-paired termini without any single stranded projections.

Aside from its utility in terminal nucleotide sequencing, the specific labeling of the 3' termini of duplex DNA, either at ends of linear molecules or at nicks*, may prove to be as useful as has been the labeling of 5' termini by means of polynucleotide kinase (12, 13). For example, incubation of T7 DNA with T4 DNA polymerase and a mixture of α-^{32}P-dATP and α-^{32}P-dTTP should result in incorporation of a 3'-terminal label into each strand without introducing any covalent changes into the molecule. As shown in Fig. 3, if one of these triphosphates is nonradioactive, then radioactivity is incorporated into one of the strands but not into the other. This T4 polymerase reaction may also be of use in specifically removing nucleotides from the termini of duplex DNA. Incubation of T7 DNA with polymerase and a mixture of dGTP and dCTP should result in removal of a single nucleotide from each 3' terminus without causing any further covalent changes in the molecule.

---

*Evidence that the T4 polymerase will label nicks at 11°C, under conditions similar to those described in the legend of Fig. 2A, was obtained using T7 DNA which had been lightly nicked with pancreatic DNase. This nicked DNA sedimented in a neutral sucrose gradient as a single symmetrical peak with the same sedimentation coefficient as intact T7 DNA molecules. However, in an alkaline sucrose gradient, it sedimented as a broad peak with an average sedimentation coefficient about 80% that of intact T7 strands, indicating that the DNA contained an average of about one nick per strand (17). Labeling of the nicked strands with α-^{32}P-dTTP revealed that the DNA was labeled to an extent of 1.7 nucleotides per strand under conditions in which intact T7 DNA was labeled to an extent of 0.9 nucleotide per strand.

Acknowledgments

We are grateful to Nancy Catterall and Patricia Clinkenbeard for expert technical assistance and to the National Institutes of Health for financial support (Grants GM 16585-03 and CA 13602-04). S.S.P., J.M.S., and P.H.W. are supported by National Institutes of Health Training Grant 5T01GM00184-13. P.T.E. received a Faculty Research Award from the American Cancer Society.

References

1. Bancroft, F. C., and Freifelder, D. 1970. Molecular weights of coliphages and coliphage DNA. I. Measurement of the molecular weight of bacteriophage T7 by high-speed equilibrium centrifugation. *J. Mol. Biol.* 54: 537.
2. Brutlag, D., and Kornberg, A. 1972. Enzymatic synthesis of deoxyribonucleic acid. XXXVI. A proofreading function for the 3' → 5' exonuclease activity in deoxyribonucleic acid polymerases. *J. Biol. Chem.* 247: 241.
3. Burton, K. 1967. Preparation of apurinic acid and of oligodeoxyribonucleotides with formic acid and diphenylamine. *In* Grossman, L., and Moldave, K. (eds.), *Methods in enzymology*, Vol. XIIA, p. 222, Academic Press, New York.
4. Cozzarelli, N. R., Kelly, R. B., and Kornberg, A. 1969. Enzymic synthesis of DNA. XXXIII. Hydrolysis of a 5'-triphosphate-terminated polynucleotide in the active center of DNA polymerase. *J. Mol. Biol.* 45: 513.
5. Englund, P. T. 1971. Analysis of nucleotide sequences at 3' termini of duplex deoxyribonucleic acid with the use of the T4 deoxyribonucleic acid polymerase. *J. Biol. Chem.* 246: 3269.
6. Englund, P. T. 1972. The 3'-terminal nucleotide sequences of T7 DNA. *J. Mol. Biol.* 66: 209.
7. Goulian, M., Lucas, Z. J., and Kornberg, A. 1968. Enzymatic synthesis of deoxyribonucleic acid. XXV. Purification and properties of deoxyribonucleic acid polymerase induced by infection with phage T4. *J. Biol. Chem.* 243: 627.
8. Huang, W. M., and Lehman, I. R. 1972. On the exonuclease activity of phage T4 deoxyribonucleic acid polymerase. *J. Biol. Chem.* 247: 3139.
9. Hershfield, M. S. and Nossal, N. G. 1972. Hydrolysis of template and newly synthesized deoxyribonucleic acid by the 3' to 5' exonuclease activity of the T4 deoxyribonucleic acid polymerase. *J. Biol. Chem.* 247: 3393.
10. Josse, J., Kaiser, A. D., and Kornberg, A. 1961. Enzymatic synthesis of deoxyribonucleic acid. VIII. Frequencies of nearest neighbor base sequences in deoxyribonucleic acid. *J. Biol. Chem.* 236: 864.
11. Kelly, T. J., Jr., and Smith, H. O. 1970. A restriction enzyme from *Hemophilus influenzae*. II. Base sequence of the recognition site. *J. Mol. Biol.* 51: 393.
12. Novogrodsky, A., and Hurwitz, J. 1966. The enzymatic phosphorylation of ribonucleic acid and deoxyribonucleic acid. I. Phosphorylation at 5'-hydroxyl termini. *J. Biol. Chem.* 241: 2923.

13. Richardson, C. C. 1965. Phosphorylation of nucleic acid by an enzyme from T4 bacteriophage-infected *Escherichia coli*. *Proc. Nat. Acad. Sci. U. S. A.* 54: 158.
14. Ritchie, D. A., Thomas, C. A., Jr., MacHattie, L. A., and Wensink, P. C. 1967. Terminal repetition in non-permuted T3 and T7 bacteriophage DNA molecules. *J. Mol. Biol.* 23: 365.
15. Smith, H. O., and Wilcox, K. W. 1970. A restriction enzyme from *Hemophilus influenzae*. I. Purification and general properties. *J. Mol. Biol.* 51: 379.
16. Summers, W. C., and Szybalski, W. 1968. Totally asymmetric transcription of coliphage T7 *in vivo:* Correlation with poly G binding sites. *Virology* 34: 9.
17. Studier, F. W. 1965. Sedimentation studies of the size and shape of DNA. *J. Mol. Biol.* 11: 373.
18. Studier, F. W. 1972. Bacteriophage T7. *Science* 176: 367.
19. Székely, M. 1971. Fingerprinting nonradioactive nucleic acids with the aid of polynucleotide kinase. *In* Cantoni, G. L., and Davies, D. R. (eds.), *Procedures in nucleic acid research*, Vol. 2, p. 780, Harper and Row, New York.
20. Tomlinson, R. V., and Tener, G. M. 1963. The effect of urea, formamide, and glycols on the secondary binding forces in the ion-exchange chromatography of polynucleotides on DEAE-cellulose. *Biochemistry* 2: 697.
21. Weiss, B., and Richardson, C. C. 1967. The 5'-terminal dinucleotides of the separated strands of T7 bacteriophage deoxyribonucleic acid. *J. Mol. Biol.* 23: 405.

Discussion

Cozzarelli. When you have two continguous C's near the terminus of one strand, are they both labeled?

Englund. No, only the first one is labeled. The enzyme stops when it hits the first one.

Exonuclease Activity of Wild Type and Mutant T4 DNA Polymerases: Hydrolysis during DNA Synthesis *in Vitro*

Nancy G. Nossal and Michael S. Hershfield*

Laboratory of Biochemical Pharmacology
National Institute of Arthritis, Metabolism,
   and Digestive Diseases
National Institutes of Health
Bethesda, Maryland 20014

Bacteriophage T4 DNA polymerase has both polymerase and $3' \rightarrow 5'$ exonuclease activities in a single polypeptide of molecular weight (MW) 110,000. We have characterized the exonuclease activity of the wild type enzyme and compared it to that of the protein fragment made by the amber mutant, *am* B22 (gene 43), which has nuclease but no polymerase activity. The mutant nuclease differs from the parent principally in its decreased apparent affinity for DNA and oligonucleotides.

During DNA synthesis catalyzed by the wild type enzyme *in vitro*, the hydrolysis of the DNA serving as template-primer is limited to the removal of unpaired regions at the 3' chain terminus, but there is extensive hydrolysis of newly incorporated residues. With a template containing equal proportions of the four bases, each base is incorporated at the same rate and to the same extent, but dAMP is removed from newly synthesized DNA one and one-half to three times as often as dTMP and dGMP, and six to nine times as rapidly as dCMP, suggesting that the enzyme can recognize specific bases. When deoxyhomopolymers are used as the template-primer, the ratio of nucleotide incorporated to nucleotide hydrolyzed increases with increasing stability of the base pair formed. Studies with synthetic polymers of defined sequence suggest that most, but not all, of the hydrolysis of newly incorporated nucleotides on these polymers occurs at the ends of chains when polymerization cannot continue.

*Present address, School of Medicine, University of California at San Diego, La Jolla, Calif. 92037.

T4 *ts* L88 (gene 43), a temperature-sensitive polymerase mutant, has an increased rate of mutation *in vivo*. We have compared the frequency of *in vitro* copying errors of the mutant and wild type enzymes. Both enzymes misincorporate dGTP using poly (dA) · poly (dT) as the template-primer, but in each case the noncomplementary residue is subsequently completely removed by the exonuclease activity. The mutant enzyme misincorporates dGTP much more frequently than the wild type, and the rate of misincorporation by the mutant enzyme increases markedly between 25 and 30°C.

The DNA polymerase that is coded for by gene 43 of bacteriophage T4 has both polymerase and 3' → 5' exonuclease activities (9). There is no detectable 5' → 3' exonuclease or endonuclease activity (3, 19). We have measured the extent to which the polymerase-associated exonuclease is active during DNA synthesis *in vitro* with the hope of clarifying the role of this exonuclease activity.

We have compared the properties of the wild type enzyme with those of two interesting gene 43 mutants. One is an amber mutant, *am* B22 (gene 43), which in nonpermissive hosts produces a fragment of the wild type enzyme which has exonuclease but no polymerase activity (18). The second is a temperature-sensitive mutant, *ts* L88 (gene 43), which has an increased rate of mutation of alleles throughout the genome *in vivo* (21), and an increased frequency of copying errors *in vitro*.

Exonuclease Activity of the Wild Type Polymerase and the B22 Nuclease

The wild type polymerase consists of a single polypeptide chain with a molecular weight of about 110,000, as shown by sedimentation equilibrium centrifugation and disc gel electrophoresis in SDS (sodium dodecyl sulfate) and mercaptoethanol (19). The amber mutation, *am* B22, maps at a point about 80% of the way from the end of the polymerase gene (gene 43) which is closest to gene 62 (Fig. 1) (1). Assuming counterclockwise transcription of gene 43 on the *l* strand of the DNA (10, 16), an amber mutation at the B22 site would be expected to produce a protein 80% the length of that of the wild type enzyme. The mutant

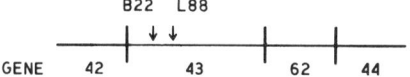

Fig. 1. Schematic drawing of the gene 43 region of the bacteriophage T4 genetic map (not to scale). Position of gene 43 mutations from reference 1.

enzyme (referred to as the B22 nuclease) was shown to consist of a single polypeptide with a molecular weight of about 80,000 (19). In addition, the B22 nuclease was shown to be an incomplete fragment of the wild type enzyme, by cochromatography on Dowex 50W of tryptic peptides of the two enzymes which had been previously labeled with radioactive iodoacetic acid (Fig. 2). Thus all of the [^{14}C] peptides from the B22 nuclease coincided with [^{3}H] peptides of the wild type protein, while there were several peaks containing predominantly peptides from the wild type enzyme. In this experiment 12.1 and 15.3 moles of iodoacetic acid were incorporated per mole of B22 nuclease and T4 polymerase, respectively, a finding which agrees well with their half-cystine contents of 11.4 (19) and 15 (9).

Under conditions where enzyme is limiting, both the wild type and the B22 nucleases remove all of the label at the 3' terminus of DNA during the initial stages of the degradation (19). This suggests that both enzymes degrade from the 3' chain end randomly rather than processively (one chain at a time) (17). The end products of the degradation in each case are 5'-mononucleotides (9) and di- and trinucleotides originating from the 5' end of the DNA chain (19).

The wild type and the mutant enzyme each degrade short oligonucleo-

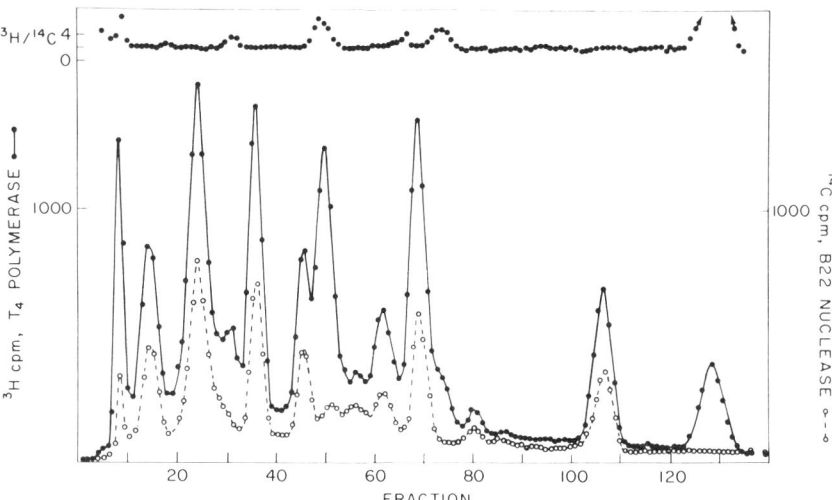

Fig. 2. Cochromatography of the carboxymethylated tryptic peptides from the B22 nuclease and the T4 DNA polymerase on Dowex 50-W. The B22 nuclease (143 μg) and the T4 DNA polymerase (95 μg) were reduced separately by incubation in 6.4 M urea, 0.4 M Tris-HCl (pH 8.3), 0.8 mM EDTA, and 5 mM dithiothreitol, and then carboxymethylated by the addition of iodo[^{14}C]- and iodo[^{3}H]acetic acid, respectively. The labeled proteins were combined, precipitated with cold acetone-1 N HCl, (39: 1, v/v), and digested exhaustively with trypsin. The total digest was chromatographed at 38°C on Dowex 50W-X2. The detailed method is described in reference 19. Reproduced from reference 19.

Table 1. Hydrolysis of DNA and DNA oligonucleotides

Substrate	B22 nuclease	T4 polymerase
Escherichia coli DNA (heat-denatured)		
V_{max} (μmoles/min/mg)	1.1	2.2
K_m (M)	3.8×10^{-4}	0.4×10^{-4}
Oligonucleotide ($n = 7$)		
V_{max} (μmoles/min/mg)	83	91
K_m (M)	7.1×10^{-4}	0.8×10^{-4}

Incubation was for 5 min at 37°C under standard conditions for the nuclease assay (19). Concentrations of DNA and oligonucleotides are nucleotide equivalents. Values shown were derived from plots of the reciprocal of the velocity against the reciprocal of the substrate concentration and are the average of triplicate determinations. Heat-denatured *E. coli* [³H]DNA was present at concentrations of 0.4, 1, 2, 4, and 6×10^{-4} M. [³H]Oligonucleotide was present at concentrations of 0.5, 1.25, 2.5, 5, and 10×10^{-4} M. Reproduced from reference 19.

tides at a much faster rate than heat-denatured DNA (Table 1), although the apparent K_m values (in nucleotide equivalents) for the two substrates are in the same range. However, for the B22 nuclease the concentration of both substrates required for half-maximal velocity is almost 10-fold higher than for the wild type enzyme. The lower affinity of the mutant enzyme for DNA was also shown by its inability to bind to DNA-cellulose under conditions which cause the parent enzyme to be tightly bound (19).

Thus the absence of 20% of the wild type protein results in a mutant enzyme which has completely lost the ability to catalyze DNA synthesis and has a lower apparent affinity for both denatured DNA and oligonucleotides. An amber mutant in the polymerase gene which maps 55% of the way from gene 62 has lost both the polymerase and exonuclease functions. Unfortunately, no amber mutants are known between this mutant and *am* B22, so we do not know how much of the wild type protein is required for its exonuclease activity. Double mutants containing the B22 mutation and a second amber mutation in gene 43 located to the right of B22 (see Fig. 1) have also lost both their exonuclease and polymerase activities (20).

Role of the Exonuclease in DNA Synthesis

Hydrolysis of Unpaired 3' Regions before Synthesis Begins

Goulian et al. (9) first reported that the nuclease activity of the T4 polymerase on single stranded DNA is inhibited in the presence of the four deoxynucleoside triphosphates required for synthesis. We and Englund have shown that this inhibition is not complete if the single stranded DNA contains unpaired regions at the 3' terminus (7, 13). A small amount

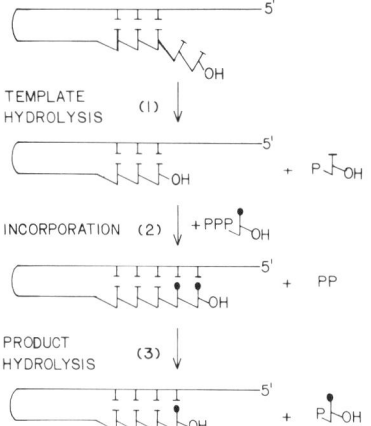

Fig. 3. Schematic drawing of the reactions catalyzed by T4 DNA polymerase.

of denatured DNA is degraded at the beginning of such a reaction apparently before the start of DNA synthesis (Reaction 1, Fig. 3, and Fig. 4). If a template-primer with a hydrogen-bonded 3' terminus is used, there is no degradation of the template prior to synthesis, as shown in Fig. 5, where native *Escherichia coli* DNA degraded to 25% acid solubility with exonuclease III was used as the template-primer. The conclusion drawn from these experiments and from those of Englund with T7 DNA is that one function of the 3' → 5' nuclease activity of the T4 polymerase is to "trim off" any unpaired 3' ends of the denatured DNA until a region of base-pairing required for copying is reached. Once synthesis begins in the 5' → 3' direction, addition of nucleotides to the 3' end of the DNA prevents further nuclease attack. Brutlag and Kornberg have demonstrated directly, using polymers of

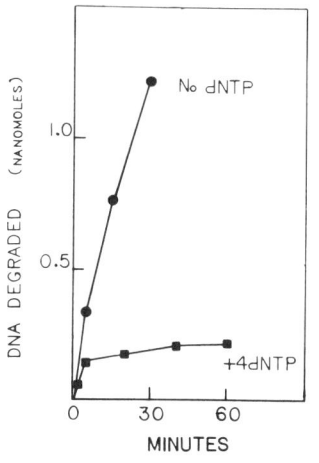

Fig. 4. Degradation of denatured *Escherichia coli* DNA during the initial stage of DNA synthesis with T4 DNA polymerase. The standard reaction mixture (13) contained [^{14}C]*E. coli* DNA (0.19 mM) (concentrations of nucleic acids are expressed as nucleotide equivalents); the four dNTPs, each at 0.125 mM; and T4 DNA polymerase (1.25 µg/ml). At the times shown, 25-µl aliquots were analyzed as described in reference 13.

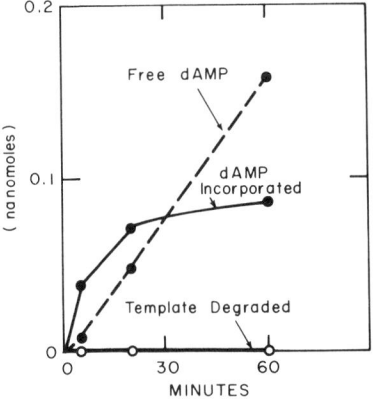

Fig. 5. Template hydrolysis and dATP utilization for *Escherichia coli* DNA treated with exonuclease III. The standard reaction mixture contained [^{14}C]*E. coli* DNA (2.4 × 10^6 cpm/μmole of nucleotide phosphate) degraded to 25% acid solubility with exonuclease III, 0.1 mM; dTTP, dCTP, dGTP, and [^{3}H]dATP (4.4 × 10^6 cpm/μmole), each at 0.1 mM; and T4 DNA polymerase, 3.15 μg/ml. Incubation was at 37°C, and 10-μl samples were removed for analysis at the times indicated. Reproduced from reference 13.

defined sequence, that only unpaired nucleotides at the 3' chain terminus are removed before the chain is extended (2).

Hydrolysis during Synthesis

In contrast to the limited hydrolysis of denatured DNA before synthesis begins, we have found that there is considerable hydrolysis of newly incorporated nucleotides. This is conveniently measured by the DNA-dependent conversion of deoxynucleoside triphosphates to free deoxynucleoside monophosphates (Reactions 1 and 2, Fig. 3, and Table 2) (13). The table shows that complementary homopolymers which can serve as template and primer are required for the conversion of each of the four dNTPs to its free dNMP. The individual homopolymers are not sufficient. In addition, a free 3'-hydroxyl group on the primer strand and a chain length of more than seven are required. The B22 protein, which cannot support incorporation, does not catalyze this conversion of dNTP to dNMP.

The large extent to which this hydrolysis of newly formed DNA occurs during synthesis *in vitro* is shown in Fig. 5. The incorporation of dAMP from dATP into DNA is approximately that expected for complete repair of the DNA removed by exonuclease III. The formation of free dAMP from dATP occurs at about 40% of the initial rate of stable incorporation, and continues at this pace until the supply of triphosphates is depleted. The rates of stable dNMP incorporation and conversion of dNTP to dNMP depend in a similar way on the concentrations of template, triphosphates, and enzyme (13). However, while stable incorporation requires all four triphosphates, considerable incorporation and subsequent hydrolysis do occur when less than a full complement is present.

Table 2. Template and primer requirements for conversion of nucleoside triphosphate to free nucleoside monophosphate

Homopolymers	^{14}C-Labeled triphosphate	Free [^{14}C]dNMP produced* (pmoles)
None	dATP, dTTP, dGTP, or dCTP	4
None	rATP	12
poly(dA) · poly(dT)	rATP	12
poly(dA) · poly(dT)	dATP	1890
poly(dA) · poly(dT)	dTTP	950
poly(dA)	dATP or dTTP	4
poly(dT)	dATP or dTTP	4
poly(dI) · poly(dC)	dGTP	40
poly(dI) · poly(dC)	dCTP	174
poly(dI)	dGTP or dCTP	4
poly(dC)	dGTP or dCTP	6

Standard reaction mixtures contained T4 DNA polymerase, 6.3 µg/ml, the ^{14}C-labeled nucleoside triphosphate indicated (6 to 12 × 10^6 cpm/µmole) at a concentration of 0.2 mM, and the homopolymers indicated at the following concentrations: poly(dA), 0.26 mM; poly(dT), 0.31 mM; poly(dI), 0.24 mM; poly(dC), 0.30 mM. Incubations were for 60 min at 37°C, after which time 10-µl samples were removed for analysis. Reproduced from reference 13.
*Blank values of 4 to 6 pmoles, obtained in the absence of enzyme, have not been subtracted from the results in this table.

Native DNA as Template

Native DNA does not support net incorporation by the T4 polymerase (9). Furthermore, in the presence of the four dNTPs, there is no net degradation of native DNA by the enzyme. Although native DNA added as template is not hydrolyzed, there is a very rapid conversion of dNTP to dNMP, and this assay is a sensitive test for both the polymerase and nuclease activities on fully native DNA, or DNA containing internal 3'-hydroxyl-terminated nicks. Englund has shown that the T4 polymerase is limited to the alternate incorporation and hydrolysis of a few nucleotides at the ends of a duplex, and has elegantly exploited this property for sequencing the 3' termini of DNA chains (6, 8). Figure 6 compares the extent of incorporation and free dAMP formation with native and denatured T7 DNA. The maximum rate of triphosphate utilization (stable incorporation plus free dNMP produced) is in fact greater with native than with denatured DNA.

If native and denatured DNA are present together at equal concentrations (Fig. 6), the kinetics of both incorporation and free dNMP formation are identical with those in the reaction with the denatured template alone. Denatured DNA at one-fortieth of the concentration of native DNA produced a 3-fold lowering of the rate of conversion of dATP

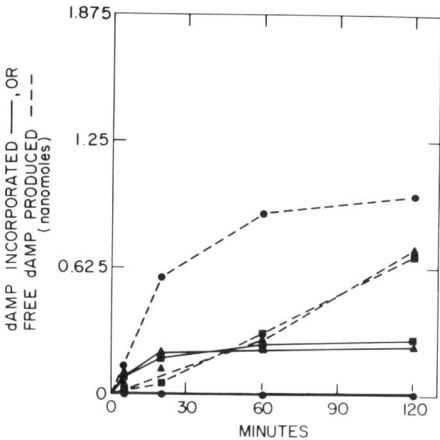

Fig. 6. Utilization of dATP in the presence of both native and denatured T7 DNA. The reaction mixtures contained denatured T7 DNA alone, 0.34 mM; native T7 DNA alone, 0.34 mM; or both native and denatured T7 DNA, each at 0.34 mM; plus dTTP, dGTP, dCTP, and [^{14}C]dATP (8.3 × 10^6 cpm/μmole), each at 0.2 mM, and T4 DNA polymerase, 15.8 μg/ml. Incubation was at 37°C and 10-μl samples were removed for analysis at the times indicated. Key: ●, native DNA alone; ■, denatured DNA alone; ▲, native plus denatured DNA; ———, dAMP incorporation; -----, free dAMP production. Reproduced from reference 13.

to dAMP. A possible explanation for these results is that the enzyme binds preferentially to denatured DNA, both nonproductively at internal nucleotides and at the growing 3' chain end.

Recognition of Specific Bases

The extent of the incorporation of an individual triphosphate depends on the distribution of bases in the DNA used as template. The rate of incorporation of each triphosphate should be equal to that of the slowest, if the bases are randomly distributed. In contrast, the rates of conversion of individual triphosphates to their monophosphates differ widely from one another, suggesting that the polymerase recognizes specific nucleotides, at least when they are located at the 3' terminus of a chain. Thus Table 3 shows that with denatured T7 DNA as the template each of the triphosphates is incorporated to the same extent and at the same rate (not shown) as the others, but that dAMP is formed two to three times as fast as dTMP and dGMP and about nine times as fast as dCMP. In other experiments the same relative rates of formation of the individual monophosphates were observed with denatured salmon sperm DNA or with native T7 DNA as the template, or with denatured T7 DNA at 20°C rather than 37°C (13). Others have proposed that DNA polymerases must play some direct role in base selection

Table 3. Comparison of incorporation and free nucleoside monophosphate production for the four deoxynucleoside triphosphates with denatured T7 DNA as template

	Extent of [^{14}C]dNMP incorporation (+3 dNTP)		Rate of free [^{14}C]dNMP production			
			+3 dNTP		−3 dNTP*	
[^{14}C]dNTP	(pmoles)	(dNMP: dAMP)	(pmoles/ 60 min)	(dNMP: dAMP)	(pmoles/ 60 min)	(dNMP: dAMP)
dATP	210	1.0	460	1.0	540	1.0
dTTP	240	1.1	150	0.33	390	0.72
dGTP	250	1.2	200	0.43	300	0.56
dCTP	180	0.86	50	0.11	90	0.17

Standard reaction mixtures contained denatured T7 DNA, 0.34 mM; T4 polymerase, 25.2 μg per ml; and the ^{14}C-labeled dNTP indicated, 0.2 mM. "+ 3 dNTP" indicates that in addition to the labeled dNTP, the three remaining unlabeled dNTPs were also present, each at 0.2 mM. "− 3 dNTP" indicates that only the labeled dNTP was present. Incubation was at 37°C for 60 min, after which time 10μl were removed for analysis. In separate experiments longer incubations were performed with each of the labeled dNTPs (plus the other three unlabeled dNTPs) to determine whether the incorporation of each in 60 min represented a yield. With each the amount incorporated in 60 min was at least 90% of that incorporated in 120 min. Conversion of each dNTP to its free dNMP under these conditions occurred almost linearly for up to 120 min, except for a slight lag during the first 5 min. Results in this table represent the average of two separate experiments. Reproduced from reference 13.

*Incorporation when a single triphosphate was present was no greater than that obtained with no enzyme present (4 pmoles). These values have been omitted.

during synthesis, since the effect of certain mutations in the T4 polymerase gene is to alter the frequency with which particular errors occur during replication (5, 21). It is not clear whether the base selection we are observing is related to this phenomenon.

Triphosphate Utilization with Homopolymers

The utilization of dATP and dTTP with poly (dA) · poly (dT) as template, and of dGTP and dCTP with poly (dI) · poly (dC), is shown in Fig. 7. The individual strands were about 800 nucleotides long and were present in approximately equal concentrations. Each labeled triphosphate was present alone (solid symbols) or with an equal concentration of its complementary triphosphate (open symbols). One can see that dAMP and dTMP incorporated into polymer are degraded (Fig. 7, A and B) in the later stages of the reaction. The degradation is somewhat inhibited by the presence of the complementary triphosphate, which would decrease the rate of hydrolysis of the strand serving as template. The rates of free dAMP and dTMP formation are more rapid than those for incorporation. (Note the difference in scales for the two reactions.)

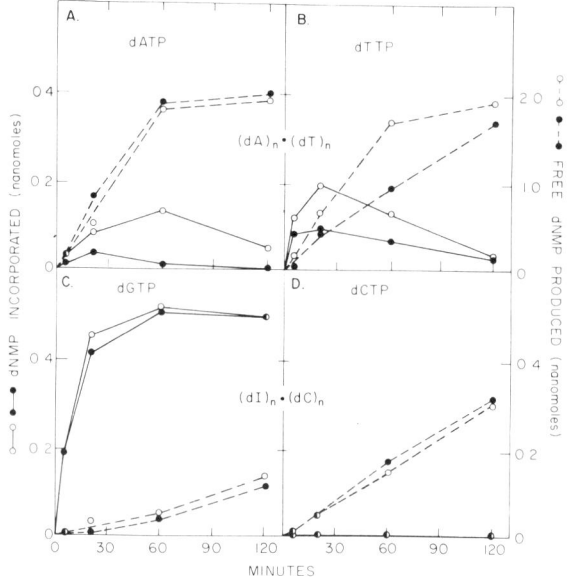

Fig. 7. Utilization of the four deoxynucleoside triphosphates with homopolymer pairs as template-primers. *A.* poly(dA) · poly(dT), [^{14}C]dATP ± dTTP. *B.* poly(dA) · poly(dT), [^{14}C]dTTP ± dATP. *C.* poly(dI) · poly(dC), [^{14}C]dGTP ± dCTP. *D.* poly(dI) · poly(dC), [^{14}C]dCTP ± dGTP. Standard reaction mixtures contained: T4 DNA polymerase, 6.3 μg/ml; the ^{14}C-labeled nucleoside triphosphate indicated in the figure (6 to 12 × 10^6 cpm/μmole), at 0.2 mM; when indicated by the key, the unlabeled complementary triphosphate, at 0.2 mM; and either poly(dA), 0.26 mM, and poly(dT), 0.31 mM; or poly(dI), 0.24 mM, and poly(dC), 0.30 mM. All additions were made at 0°C; reactions were begun by addition of enzyme, followed by incubation at 37°C. Samples of 10 μl were removed for analysis at the times indicated. *Key:* ●, only ^{14}C-labeled triphosphate present; ○, ^{14}C plus unlabeled complementary triphosphate present; ———, [^{14}C]nucleotide incorporated; -----, free [^{14}C]nucleotide produced. Reproduced from reference 13.

More than 95% of each triphosphate is eventually converted to the corresponding monophosphate (Fig. 7, A and B).

The product of dGTP incorporation with poly (dI) · poly (dC) is stable (Fig. 7C). Free dGMP is not formed until after the rate of stable incorporation has declined, and the rate of its formation is about 30-fold slower than the initial rate of incorporation. Free dCMP is formed from the beginning of the reaction (Fig. 7D), but there is no stable incorporation of dCMP at this or a wide range of other ratios of poly (dI) · poly (dC). Note that the extents of the reactions involving dGTP and dCTP are not affected by the presence of the complementary triphosphate.

These experiments with homopolymers suggest that at least one factor involved in determining the rate of removal of a newly incorporated base is the stability of the base pair formed. The base pairs formed, in order of decreasing stability, are (dG) · (dC), (dA) · (dT), and (dC) · (dI). As shown in Fig. 7, the ratio of net nucleotide incorporated to free nucleotide formed decreases in the same order as the base pair stability. When hydrogen bonding was strengthened by lowering the temperature from 37 to 20°C, dGMP incorporation decreased to 70% of that at 37°C and there was almost no dGMP formation. The dAMP incorporation was 33% and dAMP formation only 3% of that at 37°C. Similarly, decreasing the temperature to 20°C had a much larger effect on free dNMP formation than on dNMP incorporation using denatured T7 DNA as the substrate (13).

In order to assess the extent to which free dNMP formation occurs at the ends of chains when further polymerization is not possible, polymers of the type shown in Fig. 8 were synthesized, in collaboration with Dr. Fred Bollum, with the calf thymus terminal transferase (12).

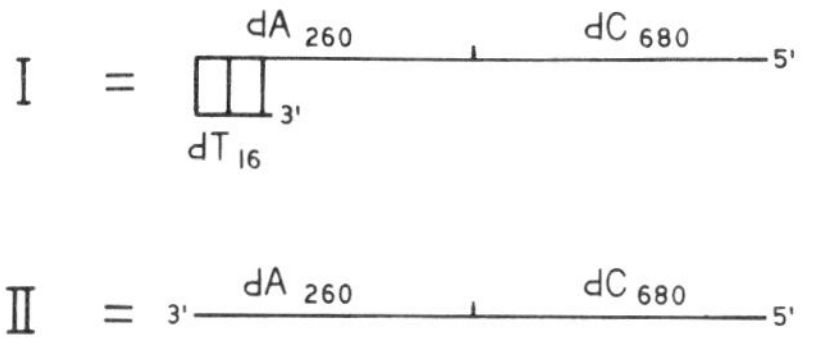

$$I = \begin{array}{c} dA_{260} \quad\quad dC_{680} \\ \rule{4cm}{0.4pt}\ \rule{4cm}{0.4pt} 5' \\ \rule{1cm}{0.4pt}_{3'} \\ dT_{16} \end{array}$$

$$II = 3'\text{---}dA_{260}\text{---}dC_{680}\text{---}5'$$

Fig. 8. Schematic drawing of deoxynucleotide polymers described in text.

When dTTP was the only triphosphate present with polymer I, there was extensive incorporation (Table 4) into polymer. Free dTMP was formed at a rate which increased with time during the first 25 min, and it continued to form until all of the dTTP had been used (data not shown). In contrast, if both dGTP and dTTP were present with polymer I, the incorporation of dTTP was similar to that observed in the absence of dGTP, but the formation of free dTMP was greatly inhibited. Thus it appears that most, but not all, of the formation of free dTMP can be eliminated if dGTP is present to allow polymerization to continue. Conversely, the incorporation of dGTP with polymer I was

Table 4. Utilization of dTTP and dGTP with polymer I (Fig. 8) as the template-primer

	dTTP + dGTP (nmoles)	dTTP alone (nmoles)	dGTP alone (nmoles)
dTMP incorporated	0.38	0.48	
dTMP free	0.08	0.58	
dGMP incorporated	1.01		0.12
dGMP free	0.10		0.01

Standard reaction mixtures containing polymer I (Fig. 8) (at approximately 0.15 mM), the indicated dNTP at 0.15 mM, and T4 polymerase (12.6 μg/ml) were incubated for 25 min at 37°C. Ten-microliter aliquots were analyzed as described in reference 13.

greatly stimulated by the addition of dTTP, and free dGMP was formed only if dTTP was present to allow synthesis to occur. The low rate of free dGMP formation was similar to that observed with poly (dC) · poly (dI) (Fig. 7C). There was no incorporation or free dNMP for either triphosphate with polymer II, since it cannot act as a primer.

DNA Polymerase from the Temperature-sensitive Mutant T4 ts L88

The T4 DNA polymerase appears to be involved in determining frequencies of mutation *in vivo* since some temperature-sensitive mutants in the polymerase gene have been found to have increased (21) or decreased (5) rates of mutation of markers throughout the genome during infection at a permissive temperature. The different mutation rates may reflect differences in the frequency of incorporation of the wrong nucleotide (misincorporation), differences in the efficiency of the removal of mismatched residues (2, 7, 13), or a combination of these factors. We have compared the DNA polymerase isolated from T4 *ts* L88 (gene 43), a mutant which promotes mutation *in vivo* (21), with that from the wild type, to determine whether differences in the frequency of misincorporation or the efficiency of the removal of the wrong base could be demonstrated *in vitro*. We hoped that total utilization of the triphosphates (incorporation plus nucleoside monophosphate production) would provide a more sensitive assay for misincorporation than stable incorporation does, since Hall and Lehman had already shown with another mutator polymerase (*ts* L56, gene 43) that stable misincorporation occurs *in vitro* at a frequency of only 10^{-5} to 10^{-6} that of the incorporation of the correct nucleotide (11).

The *ts* L88 DNA polymerase was purified to apparent homogeneity by a modification (12) of the method used for the wild type enzyme. The mutant enzyme gave a single band after acrylamide disc gel electrophoresis in the presence and absence of SDS and mercaptoethanol, had a constant ratio of polymerase to nuclease activity in individual fractions during the last two steps of the purification, and had no endonuclease activity when assayed with native or denatured λ DNA (12).

The nuclease activity of the mutant enzyme was more stable at high temperatures than was its polymerase activity. Thus when the nucleases of the mutant and wild type enzymes were compared in 15-min reactions at 30 and 39°C using denatured *E. coli* DNA as the substrate, the *ts* L88 specific activity was 85% of the wild type at 30°C, and 60% at 39°C. For incorporation with denatured T7 DNA as the template, under the same conditions used to measure the nuclease, the specific activity of the *ts* L88 enzyme was 51% of the wild type at 30°C, but only

10% at 39°C. The conversion of dATP to dAMP was also measured in this experiment. The mutant enzyme had 31% of the wild type specific activity at 30°C, and less than 1% at 39°C. In these experiments the ratio of mutant activity at 30°C to that at 39°C was 0.5 for the nuclease, 3 for polymerization, and 8 for free dAMP formation (12). The ts L88 enzyme can be seen to be relatively more active as a nuclease than as a polymerase when the levels of both activities are compared with the corresponding levels of the wild type enzyme. This observation is not consistent with the suggestion made by Bessman at this symposium that the mutator effect is due to a relative lack of nuclease activity associated with these mutant polymerases.

In view of the large differences in the rates of DNA-dependent formation of the individual monophosphates by the wild type enzyme with DNA templates (Table 3), it seemed possible that the mutant enzyme would show a different pattern of base recognition. However, when the rate of dNMP production from each of the four triphosphates by the mutant enzyme was measured with denatured T7 DNA as the template, the results were nearly identical with those shown for the wild type enzyme in Table 3 (12).

Incorporation errors can be demonstrated most clearly using homopolymers as template-primers. The wild type and mutant enzymes both misincorporate dGTP using poly (dA) · poly (dT) as the template-primer, but in each case the noncomplementary residue is subsequently completely removed by the exonuclease activity (Fig. 9). There was no stable

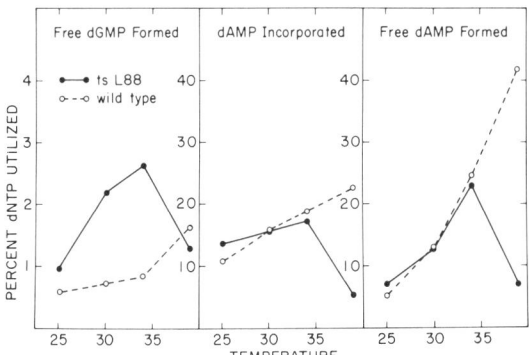

Fig. 9. Effect of temperature on the utilization of dATP and dGTP by wild type and mutator DNA polymerases, with poly(dA) · poly(dT) as the template. Standard reaction mixtures contained: poly(dA) (0.26 mM); poly(dT) (0.31 mM); wild type T4 DNA polymerase (1.25 μg/ml), or ts L88 DNA polymerase (12 μg/ml); and [^{32}P]dGTP (0.1 mM), or [^{32}P]dATP and dTTP, each at 0.1 mM. After 30 min at the indicated temperature, 5-μl aliquots were analyzed by thin layer chromatography on PEI-cellulose as in reference 13.

incorporation of dGMP under the conditions used. In the experiment shown the concentrations of enzymes were chosen so that the two enzymes incorporated equal amounts of dAMP and produced equal amounts of free dAMP at 30°C. In contrast to the results for the correct nucleotide, the wrong nucleotide was incorporated and then released much more frequently by the mutant than by the wild type enzyme at every permissive temperature tested. The increase in dGMP formation by the mutant enzyme between 25 and 30°C is especially marked. Both enzymes showed much less utilization of the wrong triphosphate than of the correct one (note the differences in the scales in Fig. 9), suggesting that discrimination against the incorrect base occurs primarily during incorporation, and only secondarily by excision or "proofreading" by the polymerase-associated exonuclease, at least under our *in vitro* conditions. Furthermore, the utilization of the wrong dNTP by both the mutant and the wild type enzyme is undetectable when the correct triphosphates are present (data not shown).

The increased rate of utilization of the wrong triphosphate by the mutant enzyme between 25 and 30°C suggested to us that a similar increase in the appearance of mutations in other genes within this temperature range might be found *in vivo*. In a preliminary experiment we grew the two phages, T4 *am* N82 (gene 44), gene 43^+ and T4 *am* N82, *ts* L88, at different temperatures in a permissive host, starting with a very small number of phages from a single plaque, and after four to five generations measured the frequency of revertants in the gene 44 amber phenotype. There were approximately nine times more gene 44 amber revertants per 10^8 phages at 25, 34, and 39°C in *ts* L88 than in wild type gene 43, as we would expect for a mutator polymerase (21). At 30°C the ratio of revertants in the mutant to that in the wild type was about five times higher than at 25°C (12).

Significance of the Hydrolysis of Newly Incorporated Residues in Vitro

The T4 DNA polymerase appears to be an inefficient enzyme for DNA synthesis *in vitro*, since a large fraction of the newly incorporated residues is hydrolyzed immediately by the polymerase-associated exonuclease. The $3' \rightarrow 5'$ exonuclease of *E. coli* DNA polymerase I apparently is much less active during synthesis, since the pyrophosphate release from nucleoside triphosphate was found to be equivalent to the nucleotide incorporation by this enzyme (4). It is possible that the extensive hydrolysis of properly base-paired residues by the T4 DNA polymerase which we have observed *in vitro* is prevented *in vivo* by factors which inhibit the nuclease activity directly, or which decrease the local unwinding at the 3' ends of strands that favors the nuclease over the polymerase activity. A good candidate for this role is the gene 32 protein, which according to recent reports stimulates DNA synthesis with the T4 polymerase (15), and inhibits the polymerase-associated nuclease (measured in the absence of nucleoside triphosphates) (14). It would be interesting to see if the gene 32 protein would decrease the nuclease activity during *in vitro* synthesis. If the exonuclease functions to help insure the fidelity of replication *in vivo*, the factors which limit the nuclease activity must not interfere with the excision of misincorporated nucleotides.

Acknowledgments

We are grateful to Doctors J. Eigner, J. W. Drake, R. S. Edgar, and J. Wiberg for furnishing the bacteria and bacteriophages used in this study.

References

1. Allen, E., Albrecht, I., and Drake, J. 1970. Properties of bacteriophage T4 mutants defective in DNA polymerase. *Genetics* 65: 187.

2. Brutlag, D., and Kornberg, A. 1972. Enzymatic synthesis of deoxyribonucleic acid. XXXVI. A proofreading function for the $3' \rightarrow 5'$ exonuclease activity in deoxyribonucleic acid polymerases. *J. Biol. Chem.* 247: 241.

3. Cozzarelli, N. R., Kelley, R. B., and Kornberg, A. 1969. Enzymatic synthesis of DNA. XXXIII. Hydrolysis of a 5'-triphosphate-terminated polynucleotide in the active center of DNA polymerase. *J. Mol. Biol.* 45: 513.

4. Deutscher, M. P., and Kornberg, A. 1969. Enzymatic synthesis of deoxyribonucleic acid. XXVIII. The pyrophosphate exchange and pyrophosphorolysis reactions of deoxyribonucleic acid polymerase. *J. Biol. Chem.* 244: 3019.

5. Drake, J. W., Allen, E. F., Forsberg, S. A., Preparata, R. M., and Greening, E. O. 1969. Spontaneous mutation: genetic control of mutation rates in bacteriophage T4. *Nature* 221: 1128.

6. Englund, P. T. 1971. Analysis of nucleotide sequences at the 3' termini of duplex deoxyribonucleic acid with the use of the T4 deoxyribonucleic acid polymerase. *J. Biol. Chem.* 246: 3269.

7. Englund, P. T. 1971. The initial step of *in vitro* synthesis of deoxyribonucleic acid by the T4 deoxyribonucleic acid polymerase. *J. Biol. Chem.* 246: 5684.

8. Englund, P. T. 1972. The 3'-terminal nucleotide sequences of T7 DNA. *J. Mol. Biol.* 66: 209.

9. Goulian, M., Lucas, Z., and Kornberg, A. 1968. Enzymatic synthesis of deoxyribonucleic acid. XXV. Purification and properties of the deoxyribonucleic acid polymerase induced by infection with phage T4$^+$. *J. Biol. Chem.* 243: 627.

10. Guha, A., Szybalski, W., Salser, W., Bolle, A., Geiduscheck, E. P., and Pulitzer, J. F. 1971. Controls and polarity of transcription during bacteriophage T4 development. *J. Mol. Biol.* 59: 329.

11. Hall, Z. W., and Lehman, I. R. 1968. An *in vitro* transversion by a mutationally altered T4-induced DNA polymerase. *J. Mol. Biol.* 36: 321.

12. Hershfield, M. S. 1973. On the role of deoxyribonucleic acid polymerase in determining mutation rates: Characterization of the defect in the T4 deoxyribonucleic acid polymerase caused by the *ts* L88 mutation. *J. Biol. Chem.* 248: 1417.

13. Hershfield, M. S., and Nossal, N. G. 1972. Hydrolysis of template and newly synthesized deoxyribonucleic acid by the $3' \rightarrow 5'$ exonuclease activity of the T4 deoxyribonucleic acid polymerase. *J. Biol. Chem.* 247: 3393.

14. Huang, W. M., and Lehman, I. R. 1972. On the exonuclease activity of phage T4 deoxyribonucleic acid polymerase. *J. Biol. Chem.* 247: 3139.

15. Huberman, J. A., Kornberg, A., and Alberts, B. M. 1971. Stimulation of T4 bacteriophage DNA polymerase by the protein product of T4 gene 32. *J. Mol. Biol.* 62: 39.

16. Jayaraman, R., and Goldberg, E. B. 1970. Transcription of the bacteriophage genome *in vivo*. *Cold Spring Harbor Symp. Quant. Biol.* 35: 197.

17. Nossal, N. G., and Singer, M. F. 1968. The processive degradation of individual polyribonucleotide chains. I. *Escherichia coli* ribonuclease II. *J. Biol. Chem.* 243: 913.

18. Nossal, N. G. 1969. A T4 bacteriophage mutant which lacks deoxyribonucleic acid polymerase but retains the polymerase-associated nuclease. *J. Biol. Chem.* 244: 218.

19. Nossal, N. G., and Hershfield, M. S. 1971. Nuclease activity in a fragment of bacteriophage T4 deoxyribonucleic acid polymerase induced by the amber mutant *am* B22. *J. Biol. Chem.* 246: 5414.

20. O'Donnell, P. V., and Karam, J. D. 1972. On the direction of reading of bacteriophage T4 gene 43 (DNA polymerase). *J. Virol.* 9: 990.

21. Speyer, J. F., Karam, J. D., and Lenny, A. B. 1966. On the role of DNA polymerase in base selection. *Cold Spring Harbor Symp. Quant. Biol.* 31: 693.

Bacterial DNA Synthesis

DNA Polymerases of *Escherichia coli*

Charles C. Richardson, Judith L. Campbell,
John W. Chase, David C. Hinkle, Dennis M.
Livingston, Henry L. Mulcahy, and Hiroaki Shizuya

Department of Biological Chemistry
Harvard Medical School
Boston, Massachusetts 02115

By assaying polymerase activity in extracts of mutagenized cells, we have isolated a mutant of *Escherichia coli* deficient in DNA polymerase II. We have been unable to detect a defect in the *polB1* mutant which can be attributed to the lack of polymerase II activity. The residual activity in a *polA$^-$ polB$^-$* mutant is polymerase III. Polymerase III has been purified from wild type cells as well as from the *polA$^-$ polB$^-$* strain. Purified polymerase III, like polymerases I and II, catalyzes a repair-type reaction. We attempted to purify polymerase III from three *dnaE* mutants, and found it to be absent in *dnaE*486, to be heat-labile in *dnaE*1026, and normal in *dnaE*293. The *dnaE* mutation has been more precisely mapped than before and found to lie to the right of *dapD*. A high frequency transducing λ phage carrying *dnaE* has been constructed. The λ *dnaE* phage can transduce *dnaE*486, 1026, and 293, and it can restore normal levels of polymerase III activity in *dnaE*486.

At present three DNA polymerases have been identified in extracts in *Escherichia coli*: polymerases I (8), II (5, 6, 9, 10, 15), and III (7). DeLucia and Cairns (2) isolated a mutant of *E. coli*, *polA1*, which had greatly reduced levels of polymerase I activity in extracts. Although the *polA1* mutant grows normally, it has a number of physiological defects including increased sensitivity to ultraviolet light and to methylmethanesulfonate. Gefter *et al.* (3) have shown that certain mutations in the *dnaE* locus result in a heat-labile polymerse III; and Nüsslein *et al.* (11) have purified the product of the *dnaE* locus by using a complementation assay *in vitro*, and have shown it to be polymerase III, proving that this enzyme is essential to DNA replication.

A Polymerase II Mutant

We have isolated a mutant of *E. coli* deficient in DNA polymerase II by mutagenizing *E. coli polA1* with N-methyl-N'-nitro-N-nitrosoguanidine and assaying the polymerase activity in extracts of the survivors (1). The *polA1* mutation was suppressed during mutagenesis by introducing the suppressor, $su7^+$, into the parental strain. Extracts of the mutant, HMS83 *polA1 polB1*, contained less than 0.5% of the normal levels of DNA polymerase II. The only polymerase activity detected in the mutant was DNA polymerase III. We found that *E. coli* HMS83 grows normally at both 25 and 42°C, and supports the growth of bacteriophages T4, T7, λ φχ-174, 186, P2, and fd. The *polB* mutation alone does not affect the sensitivity to ultraviolet irradiation or the recombination frequencies.

Although the *polB* mutant does not appear to have a physiological defect, one must be cautious in interpreting results obtained *in vivo* with mutants isolated by an *in vitro* assay. A strain whose extracts lack all or most of the assayable enzyme might retain sufficient residual activity for the cell to function normally *in vivo*. Therefore, we cannot define completely the functions of either polymerase I or II until deletion mutants are available or until mutants are isolated by selection for specific physiological defects. A defect in the *polB1* strain may become apparent when the effect of this mutation on mutations in other genes is determined.

Purification and Properties of DNA Polymerase III

The absence of polymerases I and II in extracts of the *polA1 polB1* mutant has facilitated the purification of the residual activity, polymerase III. A purification procedure has been developed which yields a highly purified polymerase III from the mutant and which is capable of separating polymerase III from the other two polymerases of *E. coli*. As a result, we have purified polymerase III from extracts of wild type cells and found it to be identical with that found in the *polA1 polB1* mutant. The purified enzyme requires DNA containing single-strand regions, the four deoxynucleoside 5'-triphosphates, and Mg^{++}. Like polymerase II, DNA polymerase III is inactive in the presence of N-ethylmaleimide and does not effectively use poly d(AT) as a template-primer.

Polymerase III catalyzes a repair-type reaction similar to that described previously for polymerases I and II (4, 12, 16). A partially single stranded T7 DNA, prepared by limited digestion of each strand with exonuclease III, is an effective primer-template for polymerase III. The extent of repair approximates the prior extent of digestion with exonuclease III, and the product is covalently attached to the primer.

Polymerase III in *dnaE* Mutants

Using our purification procedure we have purified DNA polymerase III from three strains of *E. coli* carrying the *dnaE* mutation (11, 14). In confirmation of the results of Gefter *et al.* (3) we can discover no polymerase III activity in *E. coli* E613 *dnaE*486, and we find a heat-labile polymerase III in *E. coli* BT1026 *dnaE*1026. Polymerase III, purified from *E. coli* JW108 *dnaE*293, does not appear to differ from polymerase III purified from wild type *E. coli* W3110.

A Transducing Phage-carrying *dnaE*

Inasmuch as *dnaE* is the structural gene for polymerase III, we have constructed a high frequency transducing phage carrying the *dnaE* locus by making use of the ability of phage λ to be inserted into secondary attachment sites in or near specific genes (13). We chose to insert λ into the gene for T1 and T5 sensitivity, *tonA*, since it is near *dnaE* (14) and there is a strong selection for its inactivation. On the basis of our transduction data, *dnaE* lies just to the right of *dapD*. λ was inserted into the *tonA* gene near *dnaE* by infecting *att* B · B'-deleted *E. coli* K-12 with λ_{cI857} and selecting for λ-immune, T5^r phenotype. Five T5^r λ-lysogens were isolated, and of these, two (ΔR252 and ΔR508) were found to have λ inserted into the *tonA* region. Calculating on the basis of these two lysogens, we find that the frequency of insertion of λ into the *tonA* locus is less than 10^{-7}. The prophage is inserted into the *tonA* gene in ΔR252. Low frequency transducing lysates prepared from the ΔR252 lysogen transduced the *dapC*, *dapD*, and *dnaE* genes at frequencies of 10^{-5}, 10^{-2}, and 10^{-4}, respectively.

We have prepared high frequency transducing (HFT) lysates for these three markers from the transductants. Of the three *dnaE*$^-$ mutants discussed above, all three can be transduced to *dnaE*$^+$ by the HFT λ*dnaE* phage in the presence of a helper phage at frequencies of 10^{-2} per plaque-forming unit. When *E. coli* E613 *dnaE*486, lacking polymerase III activity in extracts, is transduced by the λ*dnaE* phage, the polymerase III activity returns to normal in extracts of the lysogen. Our preliminary attempts to obtain increased levels of polymerase III by induction of lysogens carrying *dnaE* have thus far been unsuccessful.

Acknowledgments

This investigation was supported by research grants from the National Institutes of Health, United States Public Health Service Grant AI-06045, and Grant P486 from the American Cancer Society, Inc. Charles C. Richardson is the recipient of a Public Health Service Research Career Program Award, GM-13,534.

References

1. Campbell, J. L., Soll, L., and Richardson, C. C. 1972. Isolation and partial characterization of a mutant of *Escherichia coli* deficient in DNA polymerase II. *Proc. Nat. Acad. Sci. U. S. A.* 69: In press.
2. DeLucia, P., and Cairns, J. 1969. Isolation of an *E. coli* strain with a mutation affecting DNA polymerase. *Nature* 224: 1164.
3. Gefter, M. L., Hirota, Y., Kornberg, T., Wechsler, J. A., and Barnoux, C. 1971. Analysis of DNA polymerases II and III in mutants of *E. coli* thermosensitive for DNA synthesis. *Proc. Nat. Acad. Sci. U. S. A.* 68: 3150.
4. Gefter, M. L., Molineux, I. J., Kornberg, T., and Khorana, H. G. 1972. Deoxyribonucleic acid synthesis in cell-free extracts. III. Catalytic properties of deoxyribonucleic acid polymerase II. *J. Biol. Chem.* 247: 3321.
5. Knippers, R. 1970. DNA Polymerase II. *Nature* 228: 1050.
6. Kornberg, T., and Gefter, M. L. 1970. DNA synthesis in cell-free extracts of a DNA polymerase-defective mutant. *Biochem. Biophys. Res. Commun.* 40: 1348.
7. Kornberg, T., and Gefter, M. L. 1971. Purification and DNA synthesis in cell-free extracts: Properties of DNA polymerase II. *Proc. Nat. Acad. Sci. U. S. A.* 68: 761.
8. Lehman, I. R., Bessman, M. J., Simms, E. S., and Kornberg, A. 1958. Enzymatic synthesis of deoxyribonucleic acid. I. Preparation of substrates and partial purification of an enzyme from *Escherichia coli*. *J. Biol. Chem.* 233: 163.
9. Moses, R. E., and Richardson, C. C. 1970. A new DNA polymerase activity of *Escherichia coli*. I. Purification and properties of the activity present in *E. coli polA1*. *Biochem. Biophys. Res. Commun.* 41: 1557.
10. Moses, R. E., and Richardson, C. C. 1970. A new DNA polymerase activity of *Escherichia coli*. II. Properties of the enzyme purified from wild-type *E. coli* and DNA_{ts} mutants. *Biochem. Biophys. Res. Commun.* 41: 1565.
11. Nüsslein, V., Otto, B., Bonhoeffer, F., and Schaller, H. 1971. Function of DNA polymerase III in DNA replication. *Nature New Biol.* 234: 285.
12. Richardson, C. C., Inman, R. B., and Kornberg, A. 1964. Enzymatic synthesis of deoxyribonucleic acid. XVIII. The repair of partially single-stranded DNA templates by DNA polymerase. *J. Biol. Chem.* 9: 46.

13. Shimada, K., Weisberg, R. A., and Gottesman, M. E. 1972. Prophage lambda at unusual chromosomal locations. I. Location of the secondary attachment sites and the properties of the lysogens. *J. Mol. Biol.* 63: 483.
14. Wechsler, J. A., and Gross, J. D. 1971. *Escherichia coli* mutants temperature-sensitive for DNA synthesis. *Mol. Gen. Genet.* 113: 273.
15. Wickner, R. B., Ginsberg, B., Berkover, I., and Hurwitz, J. 1972. Deoxyribonucleic acid polymerase II of *Escherichia coli*. I. The purification and characterization of the enzyme. *J. Biol. Chem.* 247: 489.
16. Wickner, R. B., Ginsberg, B., and Hurwitz, J. 1972. Deoxyribonucleic acid polymerase II of *Escherichia coli*. II. Studies of the template requirements and the structure of the deoxyribonucleic acid product. *J. Biol. Chem.* 247: 498.

Discussion

Krieger. When you showed the phosphocellulose chromatogram of polymerase III purified from the temperature-sensitive pol III mutant, at the normal temperature you showed peaks for pol III and pol II and then when you increased the temperature so that it was nonpermissive, you got a little peak in an early fraction that showed up as a sort of a blip. Is this an artifact or another polymerase?

Richardson. We have not looked at that in any more detail. I think it probably is an artifact of the chromatogram, maybe it is polymerase IV, I don't know.

Szybalski. I didn't understand if your phage was plaque-forming or defective? Is it cutting into the end or not?

Richardson. We've tried to purify the transducing phage by cesium chloride density gradient and in the one experiment we carried out, we could not resolve the helper phage which was present in the lysate from the transducing phage. However, dilution experiments indicate that it is not plaque-forming and that helper phages are necessary for transduction.

Szybalski. So you don't know whether it replicates or not?

Richardson. Right.

Studies on DNA Polymerases II and III of *Escherichia coli*

Malcolm L. Gefter* and Thomas Kornberg

Department of Biology
Massachusetts Institute of Technology
Cambridge, Massachusetts 02139

Ian J. Molineux and H. G. Khorana

Departments of Chemistry and Biology
Massachusetts Institute of Technology
Cambridge, Massachusetts 02139

Lenny Mendich and Yukinori Hirota

Service de Genetique Cellulaire de l'Institut Pasteur
Paris, France

Our recent findings on the biological and biochemical properties of *Escherichia coli* DNA polymerases II and III are presented. DNA polymerase III has been purified more than 10,000-fold. The enzyme requires all four deoxytriphosphates, Mg^{++}, a reducing agent ethanol, and a DNA template-primer for maximal activity. The enzyme activity is sensitive to sulfhydryl reagents and ionic strength. Synthesis proceeds in the $5' \to 3'$ direction at rates comparable to the estimated rate of chain growth *in vivo*. Mutants temperature-sensitive for polymerase III activity are temperature-sensitive for DNA replication. In contrast, a newly isolated mutant of *E. coli*, defective in DNA polymerase II activity, is apparently without phenotype. *In vitro*, DNA polymerase II is capable of interacting with a DNA "unwinding protein" which accelerates its rate of synthesis. Synthesis catalyzed by DNA polymerase II proceeds in the $5' \to 3'$ direction by attachment to a 3'-OH end of a deoxyribonucleotide or, a synthetic or RNA polymerase-generated, ribonucleotide primer.

*Present address, Department of Biology, Massachusetts Institute of Technology, Cambridge, Mass.

The isolation of an *Escherichia coli* mutant defective in DNA polymerase I (*PolA1*⁻) (1) stimulated several investigations into the nature of the DNA synthesis capacity of such cells (6, 13, 14). Specifically, several groups attempted to isolate a new DNA polymerase, presumably the one responsible for bacterial DNA replication. The method of purification and the properties of DNA polymerase II, an enzyme distinct from DNA polymerase I, were reported by ourselves (3, 8, 9) and others (5, 11, 15). With few exceptions, this enzyme resembles DNA polymerase I in its catalytic properties. Furthermore, DNA polymerase II does not appear to be the product of a gene essential for DNA replication (2).

During the course of study on DNA polymerase II, an additional DNA polymerase, DNA polymerase III, was discovered (9). This enzyme is the product of the *dnaE* gene of *E. coli* and is essential for DNA replication (2).

This presentation will be a summary of our recent work designed to elucidate the role of DNA polymerase II in *E. coli* and to study the unique characteristics of DNA polymerase III that make it indispensable in DNA replication.

DNA Polymerase III

DNA polymerase III is present in wild type strains of *E. coli*; however, we have routinely prepared the enzyme from *PolA1*⁻ strains. We have achieved a purification of over 10,000-fold by the application of relatively standard techniques. A summary of the purification procedure is shown in Table 1, and the details are presented elsewhere (10). Because of the low concentration of DNA polymerase III in *E. coli* it is difficult

Table 1. Purification of DNA polymerase III

Fraction	Units	Protein (mg/ml)	Specific activity (units/mg protein)	Yield (%)
1. S100	5440	11.0	1.16	100*
2. DEAE-cellulose I	6048	3.4	4.2	110
3. Ammonium sulfate	3080	10.0	14.4	57
4. DEAE-cellulose II	1320	0.05	120	24
5. Phosphocellulose	220	<0.01	>12,000	4

DNA polymerase was assayed in an incubation mixture (0.3 ml) containing: 10 μM MOPS-KOH (pH 7.0); 4 μM MgCl$_2$; 15 μM 2-mercaptoethanol; 40 M (each) dCTP, dATP, dGTP; 40 nM [^{3}H]TTP (50 cpm/pmole); 32 nM calf thymus DNA treated with exonuclease III; 10% ethanol (v/v); and enzyme. Incubations were routinely for 4 min at 30°C. One unit of enzyme is defined as the amount catalyzing the incorporation of 1 nmole of TTP into an acid-insoluble product in 5 min.

*Quantitation of polymerase activity in the S100 measures DNA polymerase III, not DNA polymerase II. Mutants defective in DNA polymerase III with normal amounts of DNA polymerase II have no measurable polymerase activity in the S100 fraction.

to assess the purity of our final preparation; preliminary analysis indicates a maximum purity of 30%. Clearly, the number of enzyme molecules that can be isolated from cells is very low (this may reflect the *in vivo* concentration of enzyme or an inability to achieve efficient extraction).

The general properties of the highly purified enzyme are summarized in Table 2. For maximal activity, the enzyme requires all four deoxytriphosphates. The apparent K_m is 2×10^{-5} M and is approximately 10 times higher than that determined for DNA polymerase II. In addition, the enzyme requires Mg^{++}, a reducing agent, and a suitable DNA for template-primer. The DNA used for routine assay is calf thymus DNA treated with exonuclease III to produce internal gaps (10).

As can be seen from Table 2, the enzyme activity is stimulated by ethanol and inhibited by salt. Salt inhibition is observed for both DNA polymerase II and III activities; however, polymerase III is by far the more sensitive. Polymerase I is essentially unaffected by salt (9). Of the three *E. coli* DNA polymerases, only polymerase III shows activation in the presence of ethanol. We believe that the extreme salt sensitivity and ethanol stimulation exhibited by polymerase III may reflect a need for a low dielectric medium for maximal activity (*e.g.* the environment of cell membranes).

The use of synthetic homopolymers has facilitated determination of the direction of synthesis catalyzed by polymerase III. Oligonucleotide primers labeled at the 5' terminus with ^{32}P show that addition is exclusively to the 3'-OH terminus. The primer increases in molecular weight and the ^{32}P remains susceptible to phosphomonoesterase. The details of these experiments are presented elsewhere (3, 10).

Like DNA polymerases I and II, polymerase III contains an associated $3' \rightarrow 5'$ exonuclease activity directed against single stranded DNA. That this activity is an inherent property of the enzyme is shown by copurification of nuclease and polymerase activities as well as by the temperature sensitivity of the nuclease activity in a mutationally altered

Table 2. Properties of DNA polymerase III

Reactants	Incorporation (pmoles)
Complete system	56.0
+ Ethanol (10% v/v)	112.0
−DNA	2.0
− dATP, dGTP, dCTP	5.6
− Mg^{++}, + EDTA (3 mM)	2.0
− 2-Mercaptoethanol + N-ethylmaleimide (10 mM)	2.2
+ KCl (0.15 M)	2.0
+ DNase (15 μg/ml)	2.0

Incubations were carried out as described in Table 1.

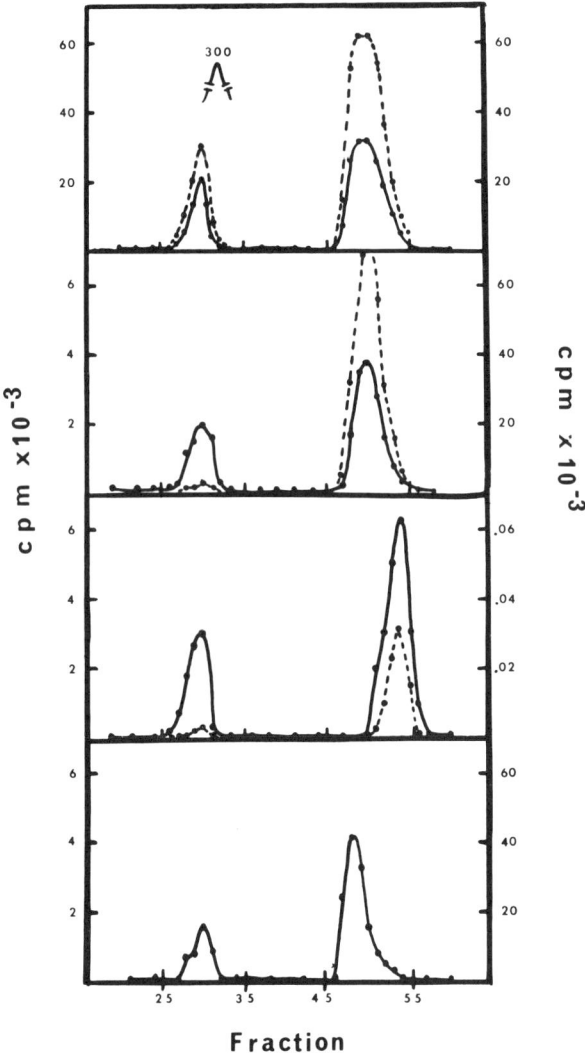

Fig. 1. Analysis of DNA polymerases II and III in *Escherichia coli* mutants. The figure represents the enzyme activities eluting from phosphocellulose treated as described in reference 2. The peak of enzyme activity eluting first (left side) represents polymerase III. The left ordinate is polymerase III activity. The peak of enzyme activity eluting last (right side) represents polymerase II. The right ordinate represents polymerase II activity. (Note the changes in scale for the different panels.) From top to bottom the panels represent enzyme activities derived from cells containing: *PolA1*⁻; *PolA1*⁻ and *dNaE*; *PolA1*⁻, *dnaE*, and*PolB*⁻; and *PolA1*⁻, *dnaE*, and *PolB*⁻ supplemented with polymerase II. The activities were measured at 30°C (solid line) and 45°C (dotted line). The inserted peak in the top panel marks the elution position of DNA polymerase I found in *PolA*⁺ strains.

enzyme having temperature-sensitive polymerase activity (T. K. and M. G., to be published). A 5' → 3' nuclease activity has not been detected in polymerase III preparations.

We have previously shown that DNA polymerase III is the product of the *dnaE* gene of *E. coli* and is essential for DNA replication (2). The *dnaE* mutants fail to replicate DNA at 42°C *in vivo* and contain temperature-sensitive DNA polymerase III *in vitro*. An example of polymerase III activity isolated from one such mutant (BT1026) is shown in Fig. 1. Polymerase III activity isolated from BT1026 (see the panel second from the top, left peak) assayed at 30°C is 10% that of wild type polymerase III (upper panel, left peak), therefore showing defectiveness even at the permissive temperature. At 45°C, the wild type enzyme activity increases by 50% while the mutant enzyme activity decreases by over 90%. Polymerase II activity (right peak) is not temperature-sensitive in *dnaE* mutants. Its activity at 45°C is twice that measured at 30°C.

DNA Polymerase II

We have previously reported on the isolation and characterization of DNA polymerase II (3, 9). The enzyme requires all four deoxytriphosphates, Mg^{++}, a reducing agent, NH_4^+, and DNA treated with exonuclease III as template-primer. The direction of polymerization catalyzed by the enzyme is 5' → 3' and the enzyme contains an associated 3' → 5' exonuclease activity.

Proof of the direction of synthesis and the fidelity of copying was achieved by the use of a defined sequence template and primer. Such a system provides unequivocal analysis of the direction of synthesis as well as sequence analysis of the product. We have conducted similar experiments which show that a ribonucleotide primer can replace a deoxyribonucleotide primer. All synthesis is primer-dependent, and proceeds by addition to the 3'-OH end. The analysis of such an experiment is shown in Table 3. Alkaline hydrolysis of the product indicated that dTTP was covalently attached to rCTP, as one would have expected from the structure of the template-primer; and the nearest neighbor analysis of the product was in excellent agreement with the known sequence of the deoxynucleotide template.

Further experiments, the details of which will be published elsewhere (I.M., M.G., and H.G.K., in preparation), show that priming of DNA synthesis directed by polymerase II may be achieved by the addition of *E. coli* RNA polymerase and all four ribotriphosphates. The single stranded circular DNA derived from bacteriophage fd is inactive in promoting deoxynucleotide incorporation under standard assay conditions.

Table 3. RNA-primed DNA synthesis

5' G-C-T-C-C-C-T-T-A-G-C-A-T-G-G-G-A-G-A-A-G-T-C-T-C-C-G-G-T-T 3'
            | | | | | | | | |
            G-A-G-G   C-A-G-A-G-G
           /   \
          C-A   G

[α-^{32}P]dXTP	Alkaline hydrolysis						3' d/r XMP			
	Origin	Ap	Cp	Gp	Up	Ap	Cp	Gp	Tp
dTTP (in cpm)	4559	168	2,527	61	23	3652	10,257	132	297
Ratio	2.8 (3.0)	– (0)	1.0 (1)	– (0)	– (0)	1.0	2.8 (3.0)	–	–
dATP (in cpm)	2360	49	57	53	88	1019	1157	580	1074
Ratio	– (4)	– (0)	– (0)	– (0)	– (0)	0.88 (1.0)	1.01 (1.0)	0.50 (1.0)	0.94 (1.0)

The conditions used for incubation with DNA polymerase II and the analysis of the produce were described in reference 3. Alkaline hydrolysis was at 37°C for 18 hr. The structure shown consists of a continuous deoxynucleotide template with annealed ribonucleotide primers. Because of the conditions of annealing and the sequence of template and primer, the structure shown is likely to be the correct one. The radioactivity remaining at the origin following alkaline hydrolysis represents deoxynucleotide.

The addition of RNA polymerase and all four ribonucleoside triphosphates leads to a stimulation of deoxynucleotide incorporation. Omission of RNA polymerase or of the ribotriphosphates or the addition of rifampicin leads to a decrease in deoxynucleotide incorporation. These results are summarized in Table 4. Alkaline hydrolysis of the product shows that as in the case of the defined sequence experiment, DNA synthesis occurs by the covalent addition of deoxynucleotide to ribonucleotide at the 3' end. Similar experiments employing DNA polymerase III indicate that it too can initiate DNA synthesis when incubated together with RNA polymerase and ribotriphosphates (T.K., unpublished).

Whether deoxy- or ribonucleotide primers are used to promote synthesis with DNA polymerases II or III, all synthesis proceeds in the $5' \rightarrow 3'$ direction. Thus, we may assume that the exonuclease III activation of "nicked" calf thymus DNA that we have observed (9) is a reflection of the production of internal gaps in DNA which are readily repaired by polymerase. During the course of studying this reaction, we observed that prolonged exonuclease III treatment rendered the DNA template-primer relatively inactive. Similarly, exonuclease III digestion of intact T7 or λ DNAs to greater than 1% of the input nucleotides rendered these molecules relatively inactive as substrates for repair. This phenomenon has also been observed with polymerase II by others (15). It therefore appears that as the template strand becomes larger than 500 nucleotides in length, the rate of repair synthesis declines. We, in collaboration with Doctors Nolan Segal and Bruce Alberts (in preparation), observed that the addition of a DNA "unwinding protein"

Table 4. RNA polymerase-dependent deoxynucleotide incorporation

Additions	dTMP incorporation (pmoles/hr)
Complete system	1420
+ fd DNA	440
− DNA, + fd DNA	0.6
Complete system*	107.6
−rCTP, −rUTP	14.8
−rCTP, −rUTP, −rGTP	6.0
−4 rNTP	2.5
−RNA polymerase	2.5
−DNA polymerase II	1.5
+rifampicin (7 μg/ml)	30.0

Incubations were standard for synthesis catalyzed by polymerase II (3). Where indicated, RNA polymerase (50 units), rXTP (10 nM), and fd DNA (10 nM) were added. Riampicin was added at zero time.

*Standard conditions for DNA polymerase II employing fd DNA in place of calf-thymus DNA.

isolated from uninfected *E. coli* stimulated the rate of repair catalyzed by DNA polymerase II. These observations are similar to those obtained with T4-induced DNA polymerase and T4-induced "32-protein" (4). DNA polymerase III could not be stimulated by the addition of the "unwinding proteins"; and the synthesis catalyzed by DNA polymerase I, which progresses quite well on such substrates, was unaffected by either the *E. coli* protein or the T4 protein. The two polymerases that could be stimulated by "unwinding proteins" showed complete specificity with regard to source of protein and polymerase (*i.e.* the T4 protein stimulates T4 polymerase only). These results are summarized in Table 5.

Given the degree of specificity linking the *E. coli* "unwinding protein" and polymerase II, it was of obvious interest to determine the physiological function of these two interacting proteins. Since polymerase II could not be shown to be a product of a gene essential for DNA replication, we initiated a genetic study designed to shed light on the biological significance of DNA polymerase II.

The rationale for the selection of a mutant defective in DNA polymerase II was based on observations of the DNA-synthetic capacity of $PolA1^-$ $dnaE^{ts}$ mutants at a nonpermissive temperature. Such cells are proficient for mating and conversion of infecting single stranded DNA to double stranded DNA. Thus mutants unable to support the growth of bacteriophage $\phi\chi$-174 were isolated and screened for their ability to mate.

We managed to isolate a derivative of strain BT1026 ($polA1^-$, $dnaE$) which, by biochemical analysis, was shown to be defective in DNA polymerase II. The analysis of polymerases II and III from this strain (10261) is displayed in Fig. 1. As shown in the third panel from the top, strain 10261 ($PolA1^-$, $dnaE$, $PolB^-$) contains a temperature-sensitive polymerase III, as expected, and 01.% of the normal level of DNA polymerase II. The isolated polymerase II is temperature-sensitive com-

Table 5. Effect of "unwinding protein" on DNA synthesis

System	Polymerase (pmoles incorporated/7 min)			
	I	II	III	T4-induced
Complete system	6.4	0.8	1.5	1.1
+ "T4-32 protein"	8.2	0.8	–	5.5
+ *Escherichia coli* "unwinding protein"	3.0	7.0	0.15	1.0

Incubations were for 7 min at 30°C. The various enzymes were added in amounts sufficient to catalyze the incorporation of 6 to 8 pmoles of TMP using calf thymus DNA treated with exonuclease III. The amount of "unwinding protein" used was titrated to give maximal stimulation in those cases where an effect was observed.

pared to the wild type enzyme. Wild type polymerase II, equivalent to the amount expected for $PolB^+$ strains, added to the crude extract of strain 10261 could be recovered following purification, at an 83% yield. This is shown in the bottom panel of Fig. 1. Thus it appears that the mutation leading to a defective polymerase II was in the structural gene (*PolB*). The biological consequences of polymerase II deficiency remain unclear. Subsequent experiments revealed that the failure of strain 10261 to support the growth of $\phi\chi$-174 was due to a second site mutation unrelated to *PolB*. Thus, we can isolate revertants capable of supporting the growth of $\phi\chi$-174 that are nevertheless polymerase II$^-$ (defective in polymerase II). In addition, polymerase II$^-$ strains are capable of supporting the replication of several R factors, and F factors; mate normally (as donors or recipients); are recombination-proficient and UV-resistant. All bacteriophages tested, including λ, T4, P1, T7, and F2, grow normally on the *PolB* mutant. Further work now in progress, including mapping of *PolB*, should provide more information regarding the biological role of DNA polymerase II. The details of the isolation and characterization of the *PolB* mutant will be presented elsewhere (L.M., Y.H., and M.G., in preparation).

A summary of relevant facts regarding the three *E. coli* DNA polymerases is shown in Table 6. These data are taken from our own results as well as those of others (7, 12, 15). The numbers concerning percentage of total *in vitro* polymerase activity, number of molecules per cell, and percentage of replication rate per enzyme molecule are approximations based on available information. With the limitations mentioned previously, it is clear that polymerase III is capable of synthesizing DNA at a much faster rate than any other known DNA polymerase and that the number of enzyme molecules per cell is very small.

Acknowledgments

This investigation was aided by Grant GMCA 18943-01 from the United States Public Health Service and a grant from the Philippe Foundation Inc. to M.G.; and by Grants 73675 and Ca05178 from the Public Health Service, and Grants 73078 and GB-21053X1 from the National Science Foundation and The Life Insurance Medical Research Fund to H.G.K.; and by a grant from the Délégation Générale à la Recherche Scientifique et Technique to Y.H.

References

1. De Lucia, P., and Cairns, J. 1969. Isolation of an *E. coli* strain with a mutation affecting DNA polymerase. *Nature* 224: 1164.

Table 6. The Escherichia coli DNA polymerases

DNA polymerase	Gene	% total activity	Molecules per cell	Mol wt	%replication rate per molecule	Direction of synthesis	Associated nuclease
I	PolA	55	400	109,000	0.25	$5' \to 3'$	$5' \to 3'$ $3' \to 5'$
II	(PolB)	2	40	90,000–120,000	0.02	$5' \to 3'$	$3' \to 5'$
III	dnaE	43	4	140,000	25.0	$5' \to 3'$	$3' \to 5'$

2. Gefter, M. L., Hirota, Y., Kornberg, T., Wechsler, J. A., and Barnoux, C. 1971. Analysis of DNA polymerases II and III in mutants of *Escherichia coli* thermosensitive for DNA synthesis. *Proc. Nat. Acad. Sci. U. S. A.* 68: 3150.

3. Gefter, M. L., Molineux, I. J., Kornberg, T., and Khorana, H. G. 1972. Deoxyribonucleic acid synthesis in cell-free extracts. III. Catalytic properties of deoxyribonucleic acid polymerase II. *J. Biol. Chem.* 247: 3321.

4. Huberman, J., Kornberg, A., and Alberts, B. 1971. Bacteriophage T4 transfer RNA. I. Isolation and characterization of two phage-coded nonsense suppressors. *J. Mol. Biol.* 62: 39.

5. Knippers, R. 1970. DNA polymerase II. *Nature* 228: 1050.

6. Knippers, R., and Stratling, W. 1970. The DNA replicating capacity of isolated *E. coli* cell wall-membrane complexes. *Nature* 226: 713.

7. Kornberg, A. 1969. Active center of DNA polymerase. *Science* 163: 1410.

8. Kornberg, T., and Gefter, M. L. 1970. DNA synthesis in cell-free extracts of a DNA polymerase-defective mutant. *Biochem. Biophys. Res. Commun.* 40: 1348.

9. Kornberg, T., and Gefter, M. L. 1971. Purification and DNA synthesis in cell-free extracts: Properties of DNA polymerase II. *Proc. Nat. Acad. Sci. U. S. A.* 68: 761.

10. Kornberg, T., and Gefter, M. L. 1972. *J. Biol. Chem.* In press.

11. Moses, R., and Richardson, C. C. 1970. A new DNA polymerase activity of *Escherichia coli*. II. Properties of the enzyme purified from wild-type *E. coli* and DNA_{ts} mutants. *Biochem. Biophys. Res. Commun.* 41: 1565.

12. Nusslein, V., Otto, B., Bonhoeffer, F., and Schaller, H. 1971. Function of DNA polymerase III in DNA replication. *Nature* 234: 285.

13. Okazaki, R., Sugimoto, K., Okazaki, T., Imae, Y., and Sugino, A. 1970. DNA chain growth: *In Vivo* and *in vitro* synthesis in a DNA polymerase-negative mutant of *E. coli*. *Nature* 228: 223.

14. Smith, D. W., Schaller, H. E., and Bonhoeffer, F. J. 1970. DNA synthesis *in vitro*. *Nature* 226: 711.

15. Wickner, R. B., Ginsberg, B., Beckower, I., and Hurwitz, J. 1972. Deoxyribonucleic acid polymerase II of *Escherichia coli*. I. The purification and characterization of the enzyme. *J. Biol. Chem.* 247: 489.

Discussion

Fidanian. If DNA synthesis catalyzed by DNA polymerase II in the fd experiment is dependent on RNA synthesis, why does rifampicin only inhibit DNA synthesis by 70%?

Gefter. We conclude that the amount of RNA synthesis needed is extremely low. Although the data were not shown, in this experiment, the addition of rifampicin reduces ribonucleotide incorporation by 99.9%. But as shown, DNA synthesis is only reduced by 70%. If we examine the nature of the product, the result of rifampicin addition is more dramatic. Approximately 50% of the DNA product is of the length of fd DNA in the absence of rifampicin. If we take only this unit-length DNA as "true" product it is almost completely (>90%) abolished by the addition of rifampicin.

Wickner. Do you know where the *PolB* (Pol II⁻) mutation maps?

Gefter. No, we have not placed it definitely. We do know, however, that it does not map near the lipopolysaccharide *(LPC)* gene which we obtained in conjunction with the *PolB* mutation.

Szybalski. You talked about the size of the gap on synthesis rate or extent. Does this also partially depend on the amount of (and the nature and condition of) unwinding protein and also on the use of alcohol?

Gefter. With DNA polymerase III, all conditions are important, all of the possible variables were tested: ethanol, salt, pH, enzyme, etc., and there was no apparent effect of the unwinding protein on that reaction

Szybalski. No, but I meant, could you unwind it with alcohol?

Gefter. I said the addition of alcohol without unwinding protein does not stimulate the rate.

The Discontinuous Replication of DNA

Reiji Okazaki, Akio Sugino, Susumu Hirose,
Tuneko Okazaki, Yasuo Imae, Ritsu Kainuma-Kuroda,
Tohru Ogawa, Mikio Arisawa, and Yoshikazu Kurosawa

Institute of Molecular Biology
Faculty of Science
Nagoya University
Chikusa-ku, Nagoya, Japan

In *Escherichia coli* mutants deficient in DNA polymerase I (*polA*$^-$), newly replicated DNA fragments are joined at about one-tenth of the rate in wild type strains. DNA polymerase I may function normally in joining the fragments, possibly by filling gaps between the fragments. Replication of a small phage P2 is inhibited in a *polA* mutant presumably because of slow joining of the nascent DNA fragments.

After brief [^{3}H]thymidine pulses of phage P2-infected cells, only half of the radioactivity incorporated into DNA is in the fragments; the rest is in long DNA chains. The pulse-labeled fragments and long DNA chains anneal predominantly with the "L" and the "H" strand of the phage DNA, respectively. Unlike *E. coli* and T4 DNA, P2 DNA might be replicated by a one-strand-discontinuous mechanism.

Contrary to the observations of Werner (42), [^{3}H]thymine and [^{3}H]thymidine pulses give essentially the same patterns of alkaline sucrose gradient sedimentation of radioactive DNA under a variety of conditions.

Synthesis and joining of DNA fragments can be demonstrated in several *in vitro* systems including the membrane fraction prepared from *polA*$^+$ strains of *E. coli* and the cellophane disc system prepared from T4-infected *polA*$^-$ cells.

Evidence is presented for the hypothesis that the short DNA fragments are formed by extension of even shorter RNA chains which are synthesized on the parental DNA strands and are removed prior to ligation of the DNA fragments. Nucleic acid that is extracted from *E. coli* labeled by a brief pulse of [^{3}H]thymidine and denatured by treatment with heat, formamide, or formaldehyde bands in a region with density higher than that of single stranded *E. coli* DNA in a Cs_2SO_4 equilibrium density gradient. Treatment with alkali or RNase shifts the density to that of single stranded DNA. The presence of RNA-linked DNA fragments is also shown by pulse labeling with [^{3}H]uridine. Analysis of nucleic acids from cells pulse-labeled for various times and of the molecules with different sizes indicates that the attachment of a short RNA chain is a unique property of the nascent DNA fragments. The size of RNA stretch attached to the DNA fragments is estimated to be 50 to 100 nucleotides.

RNA-linked DNA can also be synthesized *in vitro* with toluene-treated *E. coli* cells. Isotope transfer experiments with α-^{32}P-labeled deoxynucleoside triphosphates indicate that ribonucleotides are linked to the 5' end of DNA by a phosphodiester linkage and that the RNA-DNA junction has unique structures. The dinucleotide sequences, (rU)p(dC), (rU)p(dG), and (rC)p(dC) are found frequently but other sequences are rare. The most frequent structure in the junction is identified as . . . p(rPy)p(rA)p(rU)p(dC)p. . . .

Studies during the last 6 years provided several lines of evidence for the idea that DNA is replicated by a discontinuous mechanism involving synthesis and joining of short DNA segments. The types of evidence include the confirmation of the following three predictions of the hypothesis of discontinuous replication. (*a*) Nascent DNA is isolated as short fragments (23, 24, 28, 30, 38, 46). (*b*) These fragments accumulate upon inhibition of DNA ligase activity (10, 11, 21, 23, 27, 33, as well as personal communications from M. Gellert and I. R. Lehman). (*c*) The direction of synthesis of DNA *in vivo* is 5' → 3' (26, 36).

This paper will not review the studies on these points but will focus on more recent results and their possible implications.

Slow Joining of Nascent DNA Fragments in DNA Polymerase I-deficient Mutants of *Escherichia coli*

The nascent DNA fragments were also found in *polA* mutants (17, 22, 25). As shown in Fig. 1, in *polA* mutants the joining of the nascent fragments is very slow, so that the amount of the nascent fragments under steady-state conditions is higher than in wild-type strains (17, 22). These results suggest that DNA polymerase I plays a role in joining the fragments, possibly that of filling gaps between the fragments. As illustrated in Fig. 2, in *polA* mutants the discontinuous portion which normally corresponds to less than 1% of the entire chromosome increases to 5 to 10% of the chromosome.

Discontinuities of the daughter strands will be lethal, if they persist until that portion of the DNA becomes the template in the next cycle

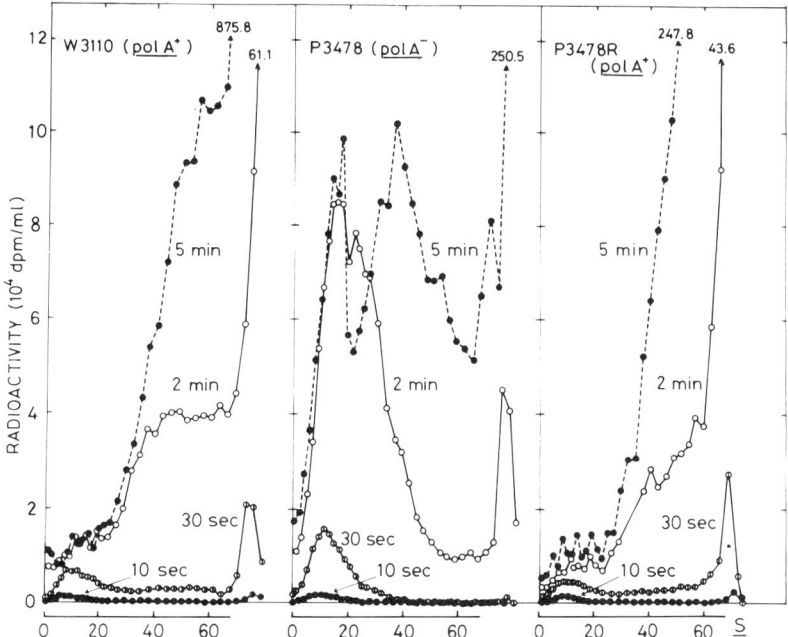

Fig. 1. Alkaline sucrose gradient sedimentation of DNA from *Escherichia coli* W3110 (thy^-, $polA^+$), P3478 (thy^-, $polA^-$), and P3478R (thy^-, $polA^+$), all pulse-labeled with [^{3}H]thymine. Cells were grown at 30°C in Medium A (24) supplemented with 0.5% Casamino Acids and 16 μM thymine up to 6×10^8 cells/ml. At this time, when the thymine concentration was 12 μM, enough [^{3}H]thymine (13.6 Ci/mmole) was added to a final concentration of 1 μM, in order to pulse label the cellular DNA. The labeling at 30°C for the times indicated was terminated by the addition of KCN-ice. DNA was extracted with 0.2 M NaOH-20 mM EDTA-0.5% sodium N-lauroyl sarcosinate and sedimented at 4°C together with ^{14}C-labeled δA DNA (19 S internal reference) through a 5 to 20% sucrose gradient in 0.1 M NaOH-0.9 M NaCl-1 mM EDTA which was prepared on an 82% sucrose cushion.

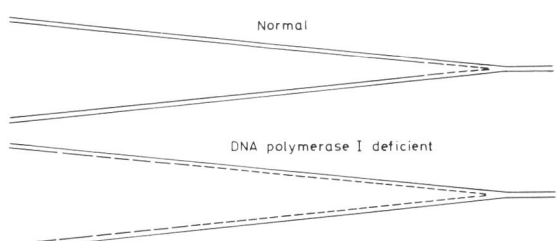

Fig. 2. Possible structures at the newly replicated regions of wild type and $polA^-$ cells.

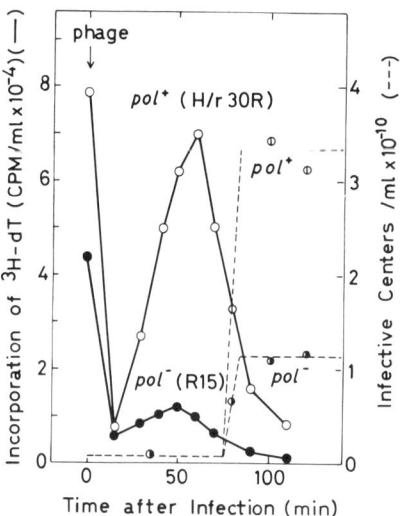

Fig. 3. DNA synthesis and phage production of P2-infected $polA^+$ and $polA^-$ cells. Cultures of Escherichia coli H/r30R ($polA^+$) or R15 ($polA^-$) grown to a titer of 5.5×10^8/ml in LC medium (1% Bacto Tryptone, 0.5% yeast extract, 0.3% NaCl, and 0.1% glucose) were infected with P2 vir_1 (adapted to the B strain of E. coli) at a multiplicity of 5. To measure the rate of DNA synthesis, portions of the culture were pulse-labeled with 0.1 μM [^{3}H]thymidine (10 Ci/m-mole) and the radioactivity incorporated into acid-insoluble material was counted.

of replication. In the polA mutants the joining of the fragments is abnormally slow yet fast enough to permit normal chromosomal replication. However, the replication of a small DNA genome could become abortive in the DNA polymerase I-deficient mutant because of incomplete joining of the daughter segments prior to the subsequent cycle of replication. It has been shown by Kingsbury and Helinski (14) that $polA^-$ strains fail to support $ColE_1$. As shown in Fig. 3, we have found that the replication of phage P2 is suppressed in a polA mutant, Escherichia coli R15. The joining of P2 nascent DNA fragments is greatly retarded in the $polA^-$ host. Analysis with Neurospora nuclease, an enzyme which specifically degrades single stranded nucleic acids, indicated that P2 DNA replicating in the $polA^-$ host contains more single stranded regions than that replicating in the $polA^+$ host (unpublished observations).

One-strand- and Two-strand-discontinuous Synthesis

Virtually all of the radioactive label incorporated into replicating E. coli and T4 DNA during a brief pulse is recovered in short DNA fragments (23, 24, 28, 46). Furthermore, almost all of the pulse label is in the accumulated fragments under the conditions of DNA ligase or DNA polymerase I deficiency (10, 17, 21-23, 33). T4 fragments found in the normal system and those accumulated under ligase-deficient conditions hybridize almost equally with the separated complementary strands of phage DNA (8, 34). These facts suggest strongly that both strands of E. coli and T4 DNA are synthesized discontinuously.

Table 1. Asymmetry of discontinuous replication of P2 DNA

Pulse labeling	Unfractionated H:L	Short chains H:L	Long chains H:L
10 sec at 8°C	1:1	17:1	1:13
5 sec at 20°C	1:1	17:1	1:11
15 sec at 20°C	1:1	23:1	1: 9

A culture of *Escherichia coli* C-1a (6 × 10^8 cells/ml) was infected with P2 vir_1 at a multiplicity of 10. After incubation for 100 min at 20°C, cells were pulse-labeled with 0.1 μM [^{3}H]thymidine as indicated. The reaction was terminated with KCN-ice, and DNA was extracted by treatment with 0.2 M NaOH-20 mM EDTA-0.5% sodium N-lauroyl sarcosinate at 37°C for 30 min and subjected to alkaline sucrose gradient sedimentation. The major components, 10 S and 32 S DNA (short and long chains) were tested for the ability to anneal with the isolated complementary strands ("L" and "H") of phage DNA by the procedure described in a previous article for T4 (34). The amounts of material which were estimated to derive from "H" and "L" strands are shown by the ratio H:L.

In cells infected with phage P2, however, only half of the radioactivity incorporated into DNA during brief pulses is in 10 S fragments, while the rest is found in long DNA chains of about one genome size. As shown in Table 1, the pulse-labeled fragments and large DNA hybridize with the "L" and the "H" strand of the phage DNA, respectively. The P2 DNA fragments that accumulate in the $polA^-$ host, like the pulse-labeled fragments, also hybridize with the "L" strand. Thus, unlike *E. coli* and T4 DNA, only the "H" strand of P2 DNA seems to be synthesized discontinuously.

P2 DNA may be unique in that its replication proceeds only in one direction (32). The one-strand-discontinuous replication of P2 DNA might be correlated with the unidirectionality of its replication. As illustrated in Fig. 4, in order for a given region of the chromosome to be replicated in either direction, both strands must have the ability to be synthesized

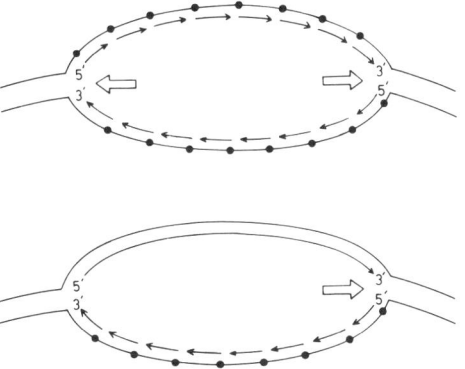

Fig. 4. Possible correlation between symmetry of discontinuous replication and direction of movement of replication fork.

discontinuously. Thus, the bidirectional movement of the replicative fork may require the mechanism for two-strand-discontinuous replication, possibly the presence of the initiation signals for synthesis of the DNA segments on both template strands. It is possible that such a requisite is not fulfilled in P2 DNA. Alternatively, the one-strand-discontinuous synthesis of P2 DNA might be correlated to possible operation of the rolling circle mechanism in its replication (32; Kainuma-Kuroda and Okazaki, manuscript in preparation).

Thymine and Thymidine Pulse

Werner (42) has recently reported on experiments in which a [^{3}H]thymine pulse, unlike a radioactive thymidine pulse, resulted in few radioactive DNA fragments and in which the fraction of the radioactivity in these fragments increased rather than decreased with pulse time. Based

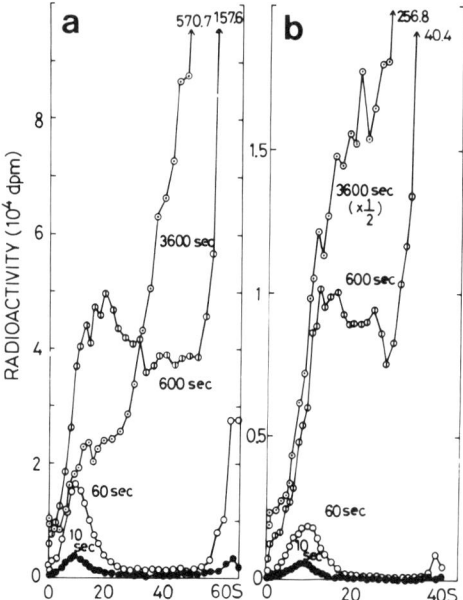

Fig. 5. Alkaline sucrose gradient sedimentation of DNA from *Escherichia coli* 15TAMT, pulse-labeled at 8°C with [^{3}H]thymidine or [^{3}H]thymine. Cells were grown at 30°C in Medium A supplemented with 0.5% Casamino Acids, 100 μg/ml of tryptophan, and 16 μM thymine up to 8×10^8 cells/ml, and transferred to 8°C. After 4 hr at 8°C, 10-ml portions of the culture (10^9 cells/ml) were pulse-labeled for the times indicated by the addition of either (a) [^{3}H]thymidine (18 Ci/mmole) to a final concentration of 0.2 μM or (b) [X^{3}H]thymine (26 Ci/mmole) to a final concentration of 1 μM. The labeling was terminated with KCN-ice. DNA extraction and sedimentation were carried out as in Fig. 1.

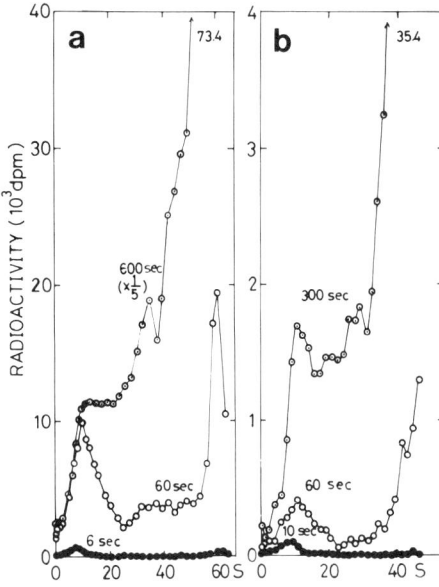

Fig. 6. Alkaline sucrose gradient sedimentation of *Escherichia coli* 15TAMT DNA, pulse-labeled at 14°C with [^{3}H]thymine. Cells were grown at 30°C in Medium A supplemented with 0.5% Casamino Acids, 100 μg/ml of tryptophan, and either (a) 16 μM (2 μg/ml) or (b) 160 μM (20 μg/ml) thymine up to 8×10^8 cells/ml, and transferred to 14°C. After 1 hr at 14°C, 10-ml portions of the culture (10^9 cells/ml) were pulse-labeled by the addition of 1 μM [^{3}H]thymine (26 Ci/mmole) and the labeling was terminated with KCN-ice. DNA extraction and sedimentation were carried out as in Fig. 1.

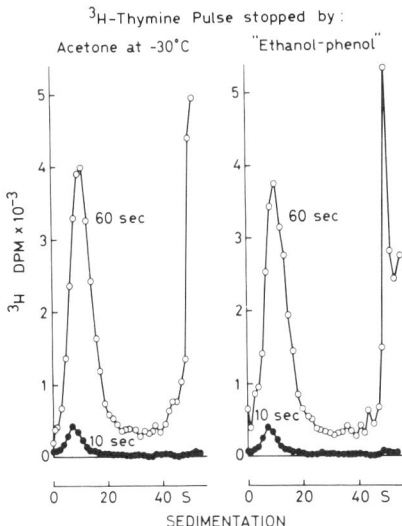

Fig. 7. Alkaline sucrose gradient sedimentation of DNA from *Escherichia coli* 15TAMT labeled at a high cell density by [^{3}H]thymine pulses terminated by "ethanol-phenol" or acetone at −30°C. Cells were grown in Medium A supplemented with 0.5% Casamino Acids, 100 μg/ml of trypotophan, and 40 μM thymine up to 8×10^8 cell/ml, transferred to 14°C, and incubated for 1 hr. The cells were collected and resuspended in the same medium at 2×10^{10} cells/ml and stirred in a beaker at 14°C for 30 min; 2.5-ml portions of the culture were then pulse-labeled by the addition of an equal volume of medium contaiining 40 μM [^{3}H]thymine (0.86 Ci/mmole). At the time indicated 2-ml samples were poured into 2 ml of an ethanol-phenol mixture (36), or into 8 ml of acetone at −30°C. DNA was extracted, as described by Werner (42), by treatment with 0.3 M NaOH-0.1 M EDTA-2% sodium N-lauroyl sarcosinate at 65°C for 5 min. The insoluble material was removed by low-speed centrifugation. Alkaline sucrose gradient sedimentation was performed as in Fig. 1.

on these and some other observations, he has suggested that the nascent DNA fragments are an artifact of thymidine pulse. We have found, contrary to Werner, that a [³H]thymine pulse gave essentially the same results as a [³H]thymidine pulse under a variety of conditions (Figs. 5 to 7). Wang and Sternglanz (40) have obtained similar results in *Bacillus subtilis*.

Discontinuous Replication *in Vitro*

Synthesis and joining of DNA fragments can be observed with *in vitro* systems such as the cellophane disc system, toluene- or ether-treated cells, and plasmolysed cells (7, 20, 29, 44). In our initial study (25) we found synthesis but not joining of the fragments with the "membrane fraction" prepared from a *polA*⁻ strain by lysozyme-EDTA and Brij treatment followed by low-speed centrifugation. But we have recently found that both synthesis and joining of the fragments occur in the

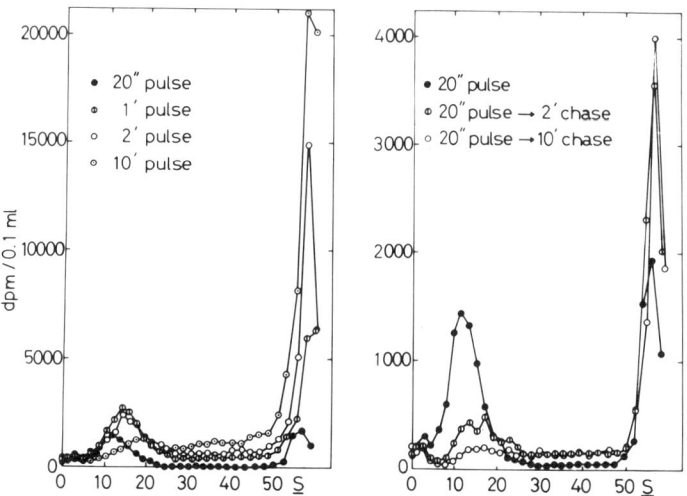

Fig. 8. Alkaline sucrose gradient sedimentation of DNA synthesized by the membrane fraction of *Escherichia coli* H/r30R. The membrane fraction was prepared as described previously (25) except that the lysozyme concentration was lowered to 100 μg/ml (final concentration) and 75 mM KCl was present during and after Brij treatment. The reaction mixture (200 μl) for DNA synthesis consisted of 25 mM Tris-HCl (pH 8.0), 12 mM MgSO$_4$, 75 mM KCl, 1 mM 2-mercaptoethanol, 200 μM ATP, 100 μM dATP, dGTP, and dCTP, 10 μM [³H]dTTP, 25 μM DPN, 130 μM *E. coli* tRNA, and 150 μl of membrane fraction. The reaction was carried out at 20°C. For "chase", cold dTTP was added up to 1.5 mM and the reaction mixture was further incubated at 20°C. The reaction was terminated by chilling and the addition of an equal volume of 0.4 M NaOH-40 mM EDTA-1% sodium N-lauroyl sarcosinate. The mixture was incubated at 37°C for 20 min and sedimented through an alkaline sucrose gradient.

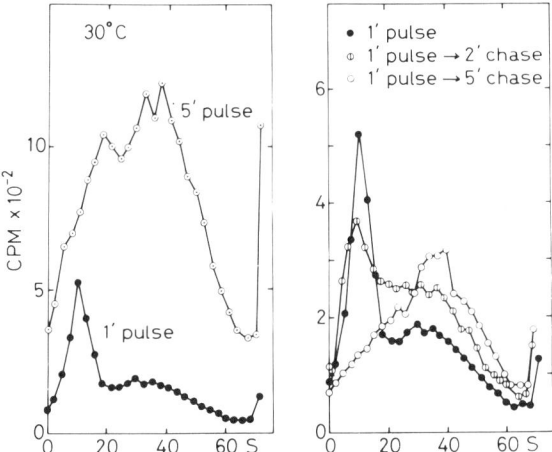

Fig. 9. Alkaline sucrose gradient sedimentation of T4 DNA synthesized in the cellophane disc system. A culture of *Escherichia coli* P3478 thy^+ (*polA1'*) (5×10^8 cells/ml) in medium A supplemented with 0.5% Casamino Acids, was infected with T4 EM7 (rII) at a multiplicity of 20, and cells were collected 20 min after infection at 30°C. A lysate on a cellophane membrane disc was prepared from 8×10^7 cells according to the method described by Schaller *et al.* (29). The reaction mixture contained 20 mM morpholinopropane sulfonic acid buffer (pH 7.2), 100 mM KCl, 5 mM MgCl$_2$, 1 mM ATP, 20 μM each of dATP, dGTP, dHMCTP, and dTTP (or [^{3}H]dTTP at 10 Ci/mmole), 200 μM thymidine, and 0.4 mM DPN. The disc was first incubated on 50 μl of the unlabeled reaction mixture for 5 min at 30°C and then transferred to 50 μl of the radioactive reaction mixture to pulse label the DNA. For the chase, the pulse-labeled disc was transferred back to an unlabeled reaction mixture and incubated. The reaction was stopped by putting the disc into 0.5 ml of 0.2 M NaOH-20 mM EDTA-0.5% sodium N-lauroyl sarcosinate. The mixture was incubated at 30°C for 10 min to extract DNA. Alkaline sucrose gradient sedimentation was carried out as in Fig. 1 without the sucrose cushion.

membrane fraction prepared from a wild type strain (Fig. 8). The inability of the membrane fraction of *polA* mutants to join the fragments may be a consequence of the DNA polymerase I deficiency. Figure 9 shows the synthesis and joining of T4 fragments observed in a cellophane disc system.

Possible Involvement of
RNA Synthesis in Discontinuous Replication

The results discussed up to this point and those reported previously leave little doubt that DNA replication occurs by a discontinuous mechanism. One of the next important questions is how the discontinuity of the chain growth is controlled, that is, what is the mechanism of initiation and termination of synthesis of the DNA fragments. We postulated at the 1968 Cold Spring Harbor Symposium that the DNA fragments might be synthesized by a mechanism in which specific structures on the template strands serve as initiation or termination signals (23). Known DNA polymerases have relatively poor template specificities, in contrast to those of RNA polymerases and other proteins involved in transcription (12, 41). Furthermore, unlike RNA polymerases, DNA polymerases do not seem to initiate new chains along templates, but catalyze the extension of a pre-existing deoxyribo- or ribopoly- or oligonucleotide chain (3, 6, 9, 13, 16, 19, 39, 41, 43). These facts, as well as several recent observations implicating transcription in DNA replication (1, 2, 4, 5, 15, 18, 31, 45), suggest that, in the discontinuous mode of replication, RNA synthesis (transcription) might be involved in the initiation of synthesis of the DNA fragments at specific sites on the template.

Figure 10 illustrates such a hypothetical mechanism. There are specif-

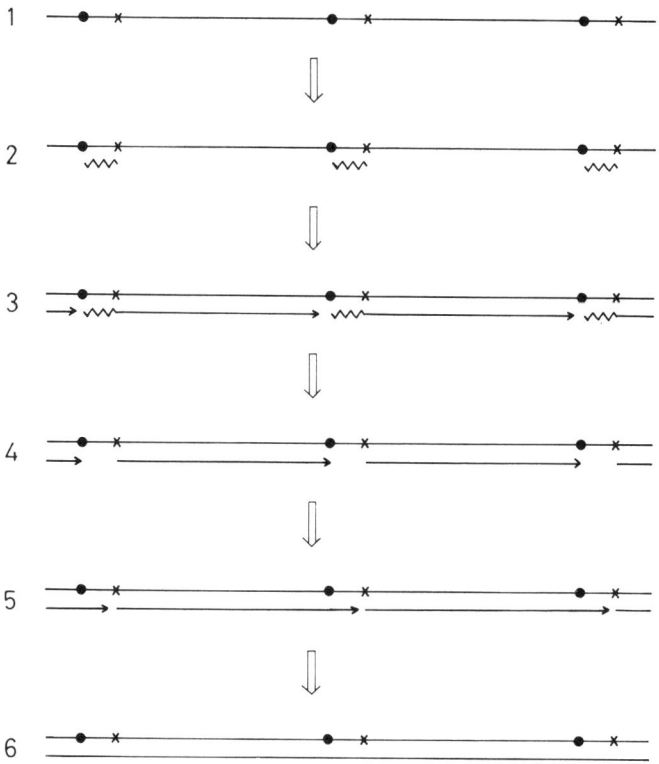

Fig. 10. A possible mechanism for discontinuous replication. (1) Unwinding of the parental strands. (2) Synthesis of short RNA chains along the DNA template by an RNA polymerase. The enzyme first binds specifically at the initiation signal (●) and stops the synthesis at the termination signal (×). (3) Elongation of the chains by a DNA polymerase using the RNA as primer. (4) Removal of the RNA segments by an RNase activity. (5) The filling of the gaps between DNA fragments by a DNA polymerase. (6) Covalent joining of the DNA fragments.

ic structures, possibly specific nucleotide sequences, on the parental DNA strands that serve as signals for chain initiation; such signals occur repeatedly along the chromosome. In the replicating region, DNA-dependent RNA polymerase molecules specifically bind to these structures and initiate synthesis of short RNA chains which terminate at certain points that may also be designated by specific structures on the template DNA. A DNA polymerase then extends the chains in the $5' \rightarrow 3'$ direction, first forming a phosphodiester bond between the ribo- and deoxyribonucleotides. The elongation of the chains by the DNA polymerase is followed by removal of the RNA portion of the chains by nuclease action. The gaps thus created between the adjacent fragments are filled by the action of a DNA polymerase. Finally, the completed DNA fragments are covalently linked to each other by DNA ligase to form long DNA. The following results (35, 37, S. Hirose, R. Okazaki, F. Tamanoi, manuscript submitted for publication) are consistent with such a hypothetical mechanism.

To test the possible attachment of short RNA strands to nascent DNA, *E. coli* Q13, a strain deficient in RNase I and polynucleotide phosphorylase, was pulse-labeled with [^{3}H]thymidine at 14°C for 15 sec, or 0.04% of the generation time. The pulse was terminated by the addition of an ethanol-phenol mixture. Nucleic acid was extracted and analyzed by Cs_2SO_4 equilibrium centrifugation after various denaturation treatments. The results presented in Fig. 11 show that after denaturation by heat, formamide, or hot formaldehyde, the pulse-labeled DNA bands heterogeneously with an average density significantly higher than that of single stranded DNA used as a reference.

As shown in Fig. 12, the pulse-labeled DNA exhibits the density of single stranded DNA after treatment with alkali or RNase.

That the high density is characteristic of "most nascent" DNA (DNA which is just synthesized) is indicated by the experiment shown in Fig. 13. The shift of labeled DNA to a high density, relative to single stranded DNA used as a reference, was evident only when the pulse time was 30 sec or less at 14°C.

Evidence for the presence of the linked RNA-DNA molecule was also obtained by pulse labeling with [^{3}H]uridine. As shown in Fig. 14, after a pulse label with [^{3}H]uridine and [^{14}C]thymidine for 15 or 30 sec at 14°C, a small portion of the [^{3}H]uridine-labeled RNA as well as a large portion of the [^{14}C]thymidine-labeled DNA banded in the region with a density a little higher than single stranded DNA used as reference. Administration of large amounts of unlabeled uridine and thymidine after the 15-sec pulse, which had an incomplete chase effect, greatly reduced the percentage of the ^{3}H label in the "light RNA."

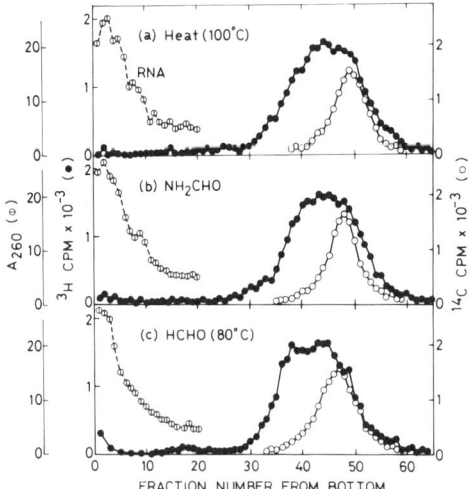

Fig. 11. Cs_2SO_4 equilibrium centrifugation of nascent *Escherichia coli* Q13 DNA after denaturation by various methods. Nucleic acid was extracted from cells pulse-labeled with [^{3}H]thymidine for 15 sec at 14°C and denatured by (a) heating in 0.1 × SSC (15mM sodium chloride-1.5mM sodium citrate) at 100°C for 5 min, (b) incubation in 90% formamide at 37°C for 30 min, or (c) heating in 12% formaldehyde at 80°C for 10 min. Each sample was subjected to equilibrium centrifugation in a Spinco 50Ti rotor together with ^{14}C-labeled denatured *E. coli* DNA.

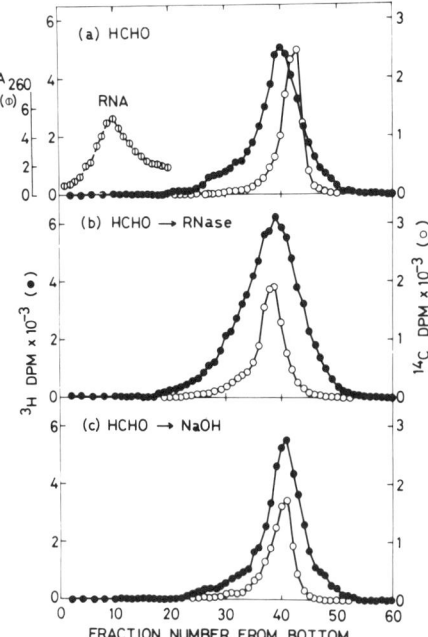

Fig. 12. Effect of RNase and alkali treatment on the density of nascent DNA. Nucleic acid, extracted from *Escherichia coli* Q13 pulse-labeled with [^{3}H]thymidine for 15 sec at 14°C and heated in 12% formaldehyde at 100°C for 10 min, was subjected to Cs_2SO_4 equilibrium centrifugation in a Beckman SW 50L rotor (a) without further treatment, (b) after incubation for 1 hr at 37°C in SSC containing 50 μg/ml of RNase IA and 10 μg/ml of RNase T1, or (c) after incubation in 0.3 M NaOH at 37°C for 15 hr. Denatured *E. coli* [^{14}C] DNA was included as a marker.

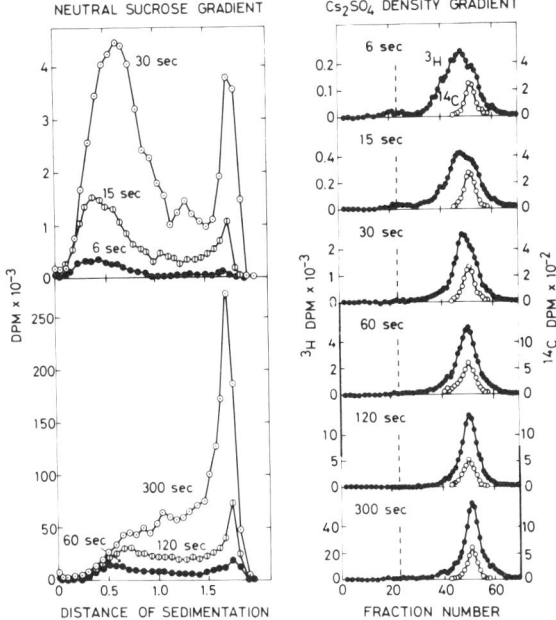

Fig. 13. Neutral sucrose gradient sedimentation and Cs_2SO_4 equilibrium centrifugation of *Escherichia coli* Q13 DNA pulse-labeled with [^{3}H] thymidine for various times at 14°C. Pulse-labeled DNA was heated in 0.1 × SSC-10 mM Tris-HCl (pH 7.6)-1 mM EDTA-0.1% sodium dodecyl sulfate (SDS) at 100°C for 5 min and chilled quickly. A 1-ml sample was layered on 31 ml of a 5 to 20% sucrose gradient containing 0.1 × SSC-10 mM Tris-HCl (pH 7.6)-1 mM EDTA-0.1% SDS; made on 82% sucrose (5 ml); and centrifuged in a Beckman SW 27 rotor for 13 hr at 24,000 rpm and 20°C. For density analysis, another 1-ml sample and denatured *E. coli* [^{14}C]DNA were centrifuged in Cs_2SO_4 solution in an SW 50L rotor. The dashed lines indicate the position of the RNA peak.

The incorporation of [^{3}H]uridine into RNA linked to DNA was much less sensitive to rifampicin than was the bulk of RNA synthesis. One possibility suggested from this observation is that an RNA polymerase different from the known RNA polymerase is involved in the discontinuous replication of the chromosomal DNA of *E. coli*.

The RNA-DNA molecules of various sizes can be isolated by Sepharose gel filtration. In the experiment shown in Fig. 15, [^{3}H]thymidine pulse-labeled nucleic acid was heat-denatured and subjected to Sepharose 2B gel filtration; fractions of a discrete size were isolated and their densities analyzed to estimate the size of the attached RNA. The molecules with a chain length of 400 to 2000 nucleotides were each found to contain an RNA chain of 60 to 100 nucleotides. No RNA seems to be attached to the molecules with a chain length of 3000 nucleotides.

The size of the RNA attached to the DNA fragments was also estimated from the difference of the elution rates of the molecule from a Se-

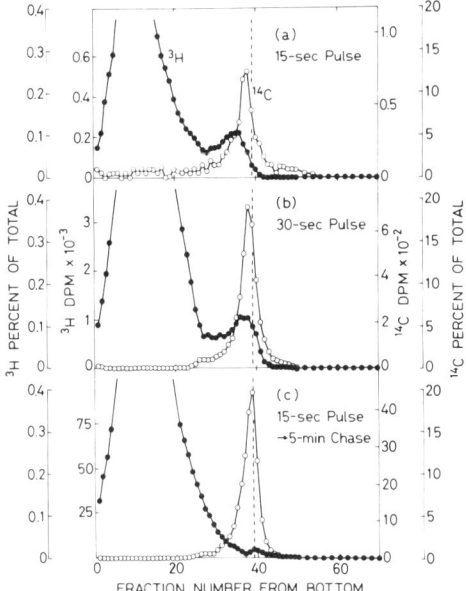

Fig. 14. Cs_2SO_4 equilibrium centrifugation of nucleic acid pulse-labeled with [^{3}H]uridine and [^{14}C]thymidine. *Escherichia coli* Q13 was pulse-labeled with [^{3}H]uridine and [^{14}C]thymidine (0.1 μM each) at 14°C for (a) 15 sec, or (b) 30 sec, or (c) pulse-labeled for 15 sec and "chased" for 5 min by the addition of 20 μM uridine and 100 μM thymidine. Nucleic acids were extracted, and rRNA was removed by precipitation with 1 M NaCl. The rest of the nucleic acids were precipitated along with carrier tRNA by ethanol, dissolved in 0.1 × SSC, and heated at 100°C for 5 min. Equilibrium centrifugation was performed as in Fig. 12, except that heat-denatured, unlabeled *E. coli* DNA was used as marker. The dashed lines indicate the position of the peak of the marker DNA.

pharose column before and after RNase or alkali treatment. The size of change produced by RNase treatment suggested that the RNA linked to the DNA fragments consists of 60 to 80 nucleotides (Fig. 16). Similar values were obtained with alkaline digestion (Fig. 17).

In the next experiment (Fig. 18), the RNA-DNA molecules doubly labeled with [^{3}H]uridine and [^{14}C]thymidine were isolated by repeated centrifugation in Cs_2SO_4 and purified by Sephadex G-200 gel filtration. After treatment with pancreatic DNase, no acid-insoluble ^{14}C label was found, and ^{3}H-labeled RNA was eluted from the Sephadex column in the region expected to show polynucleotides containing 50 to 100 nucleotides.

Thus, with three different methods the RNA chain attached to nascent DNA fragments was estimated to be in the range of 50 to 100 nucleotides. The RNA seems to be linked to an end of the DNA chain by a covalent linkage. It is not located in the middle of the molecule, nor are the ribonucleotide sequences scattered in the molecule. If either of these

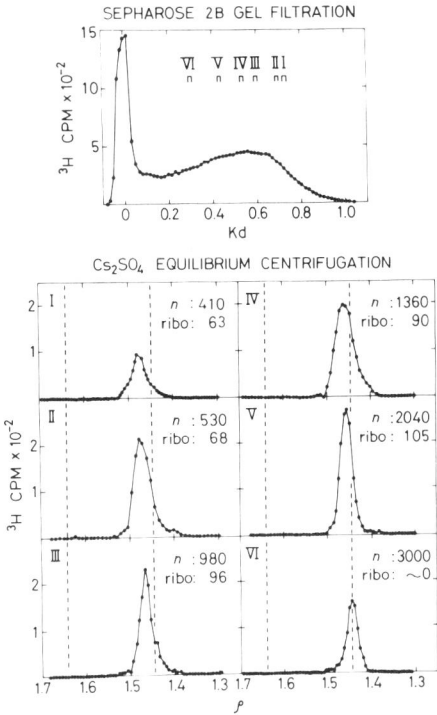

Fig. 15. Density analysis of pulse-labeled DNA fragments of various sizes. Nucleic acid was extracted from *Escherichia coli* Q13 pulse-labeled with [³H] thymidine for 30 sec at 14°C. After centrifugation at 75,000 × g for 1 hr, nucleic acid was precipitated from the supernatant with ethanol, dissolved in 5 mM Tris-HCl (pH 8.0)-50 mM NaCl-10 mM EDTA, and dialyzed against the same buffer. After heating at 100°C for 5 min and quick chilling, the nucleic acid sample was applied to a column of Sepharose 2B (1.6 × 100 cm) and eluted with upward flow of the same buffer. The indicated fractions (I–VI) were subjected, after heating at 100°C for 5 min and quick chilling, to Cs_2SO_4 equilibrium centrifugation using *E. coli* tRNA and heat-denatured *E. coli* DNA as references. The positions of these references in the density gradients are shown by dashed lines. The size of the RNA stretches (ribo) in each class of the molecules was calculated from the degree of density shift and the chain length (n), which was estimated from the K_d value using a calibration curve prepared with single stranded *E. coli* DNA of known chain length.

were the case, a drastic reduction of the size of the DNA would be produced by RNase or alkali treatment.

Positive evidence for the covalent attachment of the RNA to the 5' end of DNA was provided by the finding that one 5'-OH end of DNA is created from each RNA-DNA molecule upon alkaline hydrolysis. A preparation of purified RNA-DNA molecules was first treated with polynucleotide kinase in the presence of unlabeled ATP to mask any 5'-OH end of polynucleotide in the preparation. The sample was then subjected to alkaline hydrolysis and tested for its ability to accept ^{32}P

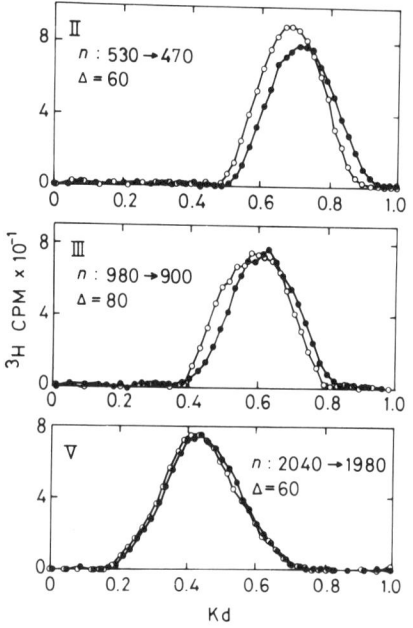

Fig. 16. Effect of RNase treatment on the size of nascent DNA fragments. Aliquots of Fractions II, III, and V from Sepharose gel filtration (Fig. 15) were each divided into two portions; one portion of each was treated with a mixture of pancreatic RNase IA (50 μg/ml) and RNase T1 (10 μg/ml) at 37°C for 15 min. After heating at 100°C for 5 min and quick chilling, the untreated (○) and RNase-treated (●) samples were subjected to gel filtration through a Sepharose 2B column (1.6 × 40 cm) in an ascending direction. The chain length was estimated from the K_d value as in Fig. 15.

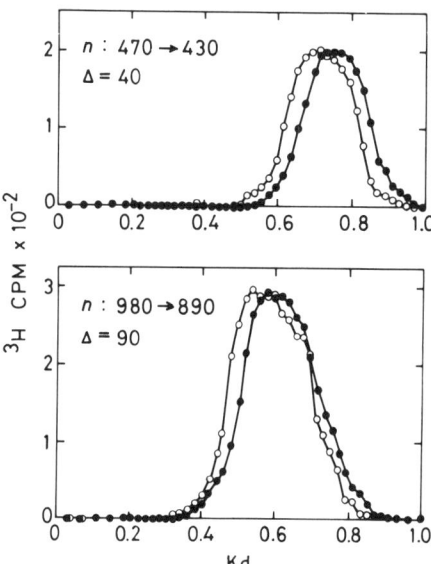

Fig. 17. Effect of alkali treatment on the size of nascent DNA fragments. Pulse-labeled DNA was fractionated by gel filtration through Sepharose 2B as in Fig. 15. Two fractions with a K_d value of 0.597 and 0.712, respectively, were each divided into two portions; one portion of each was treated with 0.3 M NaOH at 37°C for 4 hr and then neutralized. After heating at 100°C for 5 min and quick chilling, the untreated (○) and alkali-treated (●) samples were subjected to Sepharose gel filtration as in Fig. 16.

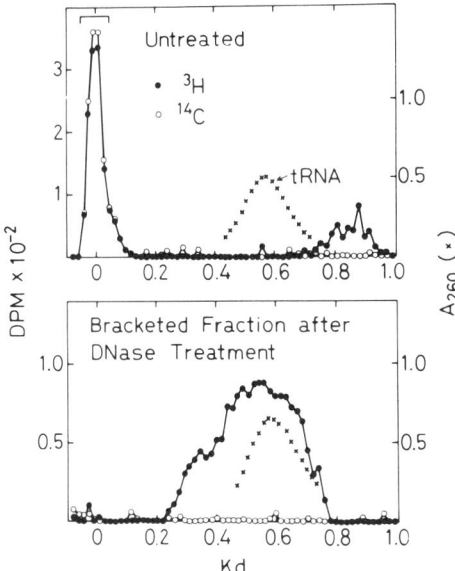

Fig. 18. Sephadex G-200 chromatography of [^{3}H]RNA-[^{14}C]DNA molecules before and after DNase treatment. Nucleic acid was extracted from *Escherichia coli* Q13 which was pulse-labeled with [^{3}H]uridine and [^{14}C]thymidine at 14°C for 15 sec. After removal of rRNA by precipitation with 1.5 M NaCl, the rest of the nucleic acid was precipitated with ethanol, dissolved in 0.1 × SSC-0.01% SDS, and dialyzed against the same buffer. After heating at 100°C and quick chilling, the sample was subjected to Cs$_2$SO$_4$ equilibrium centrifugation. The fractions containing RNA-DNA molecules were pooled and rebanded in a Cs$_2$SO$_4$ gradient. The sample of purified RNA-DNA molecules was dialyzed against 5 mM Tris-HCl (pH 7.4)-50 mM NaCl-1- mM EDTA. After heating at 100°C and quick chilling, it was applied to a column of Sephadex G-200 (1.6 × 37 cm) and eluted with upward flow of the same buffer. The material excluded from the column was incubated at 37°C for 30 min in a reaction mixture (540 µl) containing 50 mM Tris-HCl (pH 7.4), 5 mM MnCl$_2$, 54 µg of bovine serum albumin, 400 µg of *E. coli* tRNA, and 1.8 µg of pancreatic DNase, and subjected to gel filtration through the same Sephadex G-200 column.

from [γ-^{32}P]ATP in the polynucleotide kinase reaction. ^{32}P was incorporated into the DNA, indicating that new 5′-OH ends of DNA are produced by alkaline hydrolysis. The number of 5′-OH ends of DNA created by alkaline hydrolysis agreed with the number of 5′ ends of RNA originally present, as determined by the polynucleotide kinase reaction following alkaline phosphatase treatment.

The RNA-DNA molecule can also be synthesized *in vitro* by toluene-treated cells. To see if the RNA is linked to the 5′ end of the DNA chain by a phosphodiester linkage, the reaction with toluene-treated *polA1′* (*E. coli* P3478 *thy*$^+$) was performed with deoxynucleoside triphosphates, one of which was labeled with ^{32}P at the α position. The reaction mixture was incubated for 5 min at 30°C, the product was hydrolyzed with alkali, and the distribution of ^{32}P in the four ribonucleotides was

Table 2. Distribution of radioactivity in 2'(3')-ribonucleotides after alkaline hydrolysis of reaction products formed by toluene-treated cells with [α-^{32}P]deoxyribonucleoside-5'-triphosphates

α-^{32}P-dNTP substrate	% of acid-insoluble ^{32}P in			
	CMP	AMP	GMP	UMP
dCTP	0.08	0.03	0.05	0.32
dATP	0	0	0	0.01
dGTP	0	0.01	0.01	0.19
dTTP	0	0	0	0

Toluene-treated cells of *Escherichia coli* P3478 *thy*$^+$ (*polA1*') were prepared as described by Moses *et al.* (20). The reaction mixture (0.43 ml), containing 70 mM potassium phosphate buffer (pH 7.6), 13 mM MgCl$_2$, 1.3 mM ATP, 16 μM [α-^{32}P]dNTP as indicated, 33 μM each of the three other dNTPs, and 6 × 10^8 toluene-treated cells, was incubated at 30°C for 5 min. The reaction was stopped by the addition of 2 ml of an ethanol-phenol mixture (36); the cells were pelleted by low-speed centrifugation and suspended in 2 ml of 10 mM Tris-HCl (pH 8.0)-10 mM NaCl-10 mM EDTA-0.5% sodium dodecyl sulfate (SDS). The suspension was heated at 100°C for 5 min, chilled quickly, and subjected to Cs$_2$SO$_4$ equilibrium centrifugation. The material banding at the densities of RNA-DNA and DNA (total 2 to 3 × 10^5 cpm) was pooled, dialyzed, concentrated, and treated with 0.3 M KOH at 37°C for 16 hr. Material soluble in cold 3% perchloric acid was subjected to paper electrophoresis in 0.05 M sodium citrate buffer (pH 5.0). The nucleoside monophosphate region was cut out, eluted with water, and subjected to a second electrophoresis in 0.05 M sodium citrate buffer (pH 3.5), and radioactivity in the four ribonucleoside 2'(3')-monophosphates was measured. The radioactive nucleotides were identified by thin layer chromatography in isopropanol-HCl-H$_2$O (65 : 16.7 : 18.3) and saturated (NH$_4$)$_2$SO$_4$-1 M ammonium acetate-isopropanol (40 : 9 : 1).

examined. As shown in Table 2, the ^{32}P transfer from 5'-deoxynucleotide to 2'(3')-ribonucleotide did occur, indicating that ribonucleotides are covalently linked to the 5' end of DNA. A large amount of ^{32}P was transferred from [α-^{32}P]dCTP and [α-^{32}P]dGTP to 2'(3')-UMP. Some transfer from dCTP to other ribonucleotides was also observed. Thus the RNA-DNA junction has a unique structure, (rU)p(dC) and (rU)p(dG) being its most frequent sequences. There are some junctions with the sequences (rC)p(dC), (rG)p(dC), and (rA)p(dC), but other sequences are rare.

After hydrolysis of the reaction product formed with [α-^{32}P]dCTP by pancreatic RNase, 73% of the radioactivity eluted from DEAE-cellulose with 0.25 M NaCl in 7 M urea was found in a dinucleotide which was identified as ApUp. Therefore the structure that occurs most frequently at the junction in the RNA-DNA molecule synthesized *in vitro* is:

Some radioactivity (14%) was also recovered in the dinucleotide ApCp, indicating the occurrence of some junctions with the following structure:

```
           Py   A    C    C
          -OH  -OH  -OH  -OH  -H   -H
    -----P\  P\  P\  P\  P\  P\  P-----
```

When the reaction with [α-^{32}P]dCTP was followed by incubation with an excess of unlabeled dCTP, the radioactivity in these sequences disappeared, indicating an active turnover of the RNA-DNA molecules in this system.

Hopefully, further investigations into the structure and metabolism of the RNA-linked DNA fragments with *in vivo* and *in vitro* systems will contribute to our understanding of the mechanism of DNA replication.

Acknowledgments

This work was supported by grants from the Ministry of Education of Japan, the Jane Coffin Childs Memorial Fund for Medical Research, and the Toray Science Foundation. Susumu Hirose was supported by a fellowship from the Naito Foundation for the Promotion of Basic Research in Medical and Pharmaceutical Sciences.

Note added in proof: In recent experiments (Y. Kurosawa and R. Okazaki, manuscript in preparation), *E. coli* C2107, a temperature-sensitive *polA* mutant, was infected with P2 vir_1 at 20°C and shifted to 42°C during the period of active phage DNA synthesis and then pulse-labeled with [^{3}H]thymidine. Virtually all of the pulse label was found in short DNA fragments, which contained the "L" as well as the "H" strand. Therefore the asymmetry of the discontinuity of replicative P2 DNA now seems to be due to the difference of the rate of joining of the fragments of the two strands synthesized discontinuously rather than to one-strand-discontinuous replication.

References

1. Bazzicalupo, P., and Tocchini-Valentini, G. P. 1972. Curing of an *Escherichia coli* episome by rifampicin. *Proc. Nat. Acad. Sci. U. S. A.* 69: 298.

2. Brutlag, D., Schekman, R., and Kornberg, A. 1971. A possible role for RNA polymerase in the initiation of M13 DNA synthesis. *Proc. Nat. Acad. Sci. U. S. A.* 68: 2826.

3. Chang, L. M. S., and Bollum, F. J. 1972. A chemical model for transcriptional initiation of DNA replication. *Biochem. Biophys. Res. Commun.* 46: 1354.

4. Clewell, D. B., Evenchik, B., and Cranston, J. W. 1972. Direct inhibition of *ColE*$_1$ plasmid DNA replication in *Escherichia coli* by rifampicin. *Nature New Biol.* 237: 29.

5. Dove, W. F., Hargrove, E., Ohashi, M., Haugli, F., and Guha, A. 1969. Replicator activation in lambda. *Jap. J. Genet.* 44 (suppl. 1): 11.

6. Flügel, R. M., and Wells, R. D. 1972. Nucleotides at the RNA-DNA covalent bonds formed in the endogenous reaction by the avian Myeloblastosis virus DNA polymerase. *Virology* 48: 394.

7. Geider, K., and Hoffman-Berling, H. 1971. DNA synthesis in nucleotide permeable *Escherichia coli* cells: Chain elongation in specific regions of the bacterial chromosome. *Eur. J. Biochem.* 21: 374.

8. Ginsberg, B., and Hurwitz, J. 1970. Unbiased synthesis of pulse-labeled DNA fragments of bacteriophage λ and T4. *J. Mol. Biol.* 52: 265.

9. Goulian, M. 1968. Incorporation of oligodeoxynucleotides into DNA. *Proc. Nat. Acad. Sci. U. S. A.* 61: 284.

10. Hosoda, J., and Mathews, E. 1968. DNA replication *in vivo* by a temperature-sensitive polynucleotide ligase mutant of T4. *Proc. Nat. Acad. Sci. U. S. A.* 61: 997.

11. Iwatsuki, N., and Okazaki, R. 1970. Mechanism of DNA chain growth. V. Effect of chloramphenicol on the formation of T4 nascent short DNA chains. *J. Mol. Biol.* 52: 37.

12. Karkas, J. D., Starianopoulos, J. G., and Chargaff, E. 1972. Action of DNA polymerase I of *Escherichia coli* with DNA-RNA hybrids as templates. *Proc. Nat. Acad. Sci. U. S. A.* 69: 398.

13. Keller, W. 1972. RNA-primed DNA synthesis *in vitro*. *Proc. Nat. Acad. Sci. U. S. A.* 69: 1560.

14. Kingsbury, D. T., and Helinski, D. R. 1970. DNA polymerase as a requirement for the maintenance of the bacterial plasmid colicinogenic factor E$_1$. *Biochem. Biophys. Res. Commun.* 41: 1538.

15. Kline, B. C. 1972. Inhibition of plasmid DNA replication by rifampicin in *Salmonella pullorum*. *Biochem. Biophys. Res. Commun.* 46: 2019.

16. Kornberg, A. 1969. Active center of DNA polymerase. *Science* 163: 1410.

17. Kuempel, P. L., and Veomett, G. E. 1970. A possible function of DNA polymerase in chromosome replication. *Biochem. Biophys. Res. Commun.* 41: 973.

18. Lark, K. G. 1972. Evidence for the direct involvement of RNA in the initiation of DNA replication in *Escherichia coli* 15T$^-$. *J. Mol. Biol.* 64: 47.

19. Leis, B, P., and Hurwitz, J. 1972. RNA-dependent DNA polymerase activity of RNA tumor viruses II. Directing influence of RNA in the reaction. *J. Virol.* 9: 130.

20. Moses, R. E., Campbell, J. L., Fleischman, R. A., and Richardson, C. C. 1971. Enzymatic mechanisms in DNA replication. *In* Ribbons, D. W., Woessner, J. F., and Schultz, J. (eds.), *Nucleic acid-protein interactions-nucleic*

acid synthesis in viral infection, pp. 48-68, North-Holland Publishing Co., Amsterdam.

21. Newman, J., and Hanawalt, P. 1968. Role of polynucleotide ligase in T4 DNA replication. *J. Mol. Biol.* 35: 639.

22. Okazaki, R., Arisawa, M., and Sugino, A. 1971. Slow joining of newly replicated DNA chains in DNA polymerase I-deficient *Escherichia coli* mutants. *Proc. Nat. Acad. Sci. U. S. A.* 68: 2954.

23. Okazaki, R., Okazaki, T., Sakabe, K., Sugimoto, K., Kainuma, R., Sugino, A., and Iwatsuki, N. 1968. *In vivo* mechanism of DNA chain growth. *Cold Spring Harbor Symp. Quant. Biol.* 33: 129.

24. Okazaki, R., Okazaki, T., Sakabe, K., Sugimoto, K., and Sugino, A. 1968. Mechanism of DNA chain growth. I. Possible discontinuity and unusual secondary structure of newly synthesized chains. *Proc. Nat. Acad. Sci. U. S. A.* 59:598.

25. Okazaki, R., Sugimoto, K., Okazaki, T., Imae, Y., and Sugino, A. 1970. DNA chain growth: *In vivo* and *in vitro* synthesis in a DNA polymerase-negative mutant of *E. coli. Nature* 228: 223.

26. Okazaki, T., and Okazaki, R. 1969. Mechanism of DNA chain growth. IV. Direction of synthesis of T4 short DNA chains as revealed by exonucleolytic degradation. *Proc. Nat. Acad. Sci. U. S. A.* 64: 1242.

27. Pauling, C., and Hamm, L. 1969. Properties of a temperature-sensitive, radiation-sensitive mutant of *Escherichia coli.* II. DNA replication.*Proc. Nat. Acad. Sci. U. S. A.* 64:1195.

28. Sakabe, K., and Okazaki, R. 1966. A unique property of the replicating region of chromosomal DNA. *Biochim. Biophys. Acta* 129: 651.

29. Schaller, H., Otto, B., Nüsslein, V., Huf, J., Herrmann, R., and Bonhoeffer, F. 1972. Deoxyribonucleic acid replication *in vitro. J. Mol. Biol.* 63: 183.

30. Schandl, E. K., and Taylor, J. H. 1969. Early events in the replication and integration of DNA into mammalian chromosomes. *Biochem. Biophys. Res. Commun.* 34: 291.

31. Schekman, R., Wickner, W., Westergaard, O., Brutlag, D., Geider, K., Bertsch, L. L., and Kornberg, A. 1972. Initiation of DNA synthesis. III. Synthesis of φX-174 replicative form requires RNA synthesis resistant to rifampicin. *Proc. Nat. Acad. Sci. U. S. A.* 69: 2691.

32. Schnös, M., and Inman, R. B. 1971. The starting point and direction of replication in P2 DNA. *J. Mol. Biol.* 55: 31.

33. Sugimoto, K., Okazaki, T., and Okazaki, R. 1968. Mechanism of DNA chain growth, II. Accumulation of newly synthesized short chains in *E. coli* infected with ligase-defective T4 phages. *Proc. Nat. Acad. Sci. U. S. A.* 60: 1356.

34. Sugimoto, K., Okazaki, T., Imae, Y., and Okazaki, R. 1969. Mechanism of DNA chain growth. III. Equal annealing of T4 nascent short DNA chains with the separated complementary strands of the phage DNA. *Proc. Nat. Acad. Sci. U. S. A.* 63: 1343.

35. Sugino, A., Hirose, S., and Okazaki, R. 1972. RNA-linked nascent DNA

fragments in *Escherichia coli. Proc. Nat. Acad. Sci. U. S. A.* 69: 1863.
36. Sugino, A., and Okazaki, R. 1972. Mechanism of DNA chain growth. VII. Direction and rate of growth of T4 nascent short DNA chains. *J. Mol. Biol.* 64: 61.
37. Sugino, A., and Okazaki, R. 1973. RNA-linked DNA fragments *en vitro*. *Proc. Nat. Acad. Sci. U. S. A.* 70: 88.
38. Tomizawa, J., and Ogawa, T. 1968. Replication of phage lambda DNA. *Cold Spring Harbor Symp. Quant. Biol.* 33: 533.
39. Verma, I. M., Meuth, N. L., Bromfel, E., Manly, K. F., and Baltimore, D. 1971. Covalently linked RNA-DNA molecule as initial product of RNA tumor virus DNA polymerase. *Nature New Biol.* 233: 131.
40. Wang. H. F., and Sternglanz, R. 1972. DNA chain growth in *Bacillus subtilis*. *Fed. Proc.* 31: 441.
41. Wells, R. D., Flügel, R. M., Larson, J. E., Schendel, P. F., and Sweet, R. W. 1972. Comparison of some reactions catalyzed by deoxyribonucleic acid polymerase from avian Myeloblastosis virus, *Escherichia coli* and *Micrococcus luteus*. *Biochemistry* 11: 621.
42. Werner, R. 1971. Mechanism of DNA replication. *Nature* 230: 570.
43. Wickner, R. B., Ginsberg, B., and Hurwitz, J. 1972. Deoxyribonucleic acid polymerase II of *Escherichia coli*. II. Studies of the template requirements and the structure of the deoxyribonucleic acid product. *J. Biol. Chem.* 247: 498.
44. Wickner, R. B., and Hurwitz, J. 1972. DNA replication in *Escherichia coli* made permeable by treatment with high sucrose. *Biochem. Biophys. Res. Commun.* 47: 202.
45. Wickner, W., Brutlag, D., Schekman, R., and Kornberg, A. 1972. RNA synthesis initiates *in vitro* conversion of M13 DNA to its replicative form. *Proc. Nat. Acad. Sci. U. S. A.* 69: 965.
46. Yudelevich, A., Ginsberg, B., and Hurwitz, J. 1968. Discontinuous synthesis of DNA during replication. *Proc. Nat. Acad. Sci. U. S. A. 61: 1129.*

Discussion

Szybalski. I missed which mutant you used to detect the RNA at the end. If you don't use this particular DNA mutant, do you get RNA at the end of Okazaki fragments or do you get none?
Okazaki. We do get RNA.
Szybalski. Then why does Sternglanz not get anything on his short fragments? Was it because he used *subtilis?*
Okazaki. His are smaller fragments, I think, and they expose preparations to alkali. He is here and can answer your question.
Sternglanz. Helen Wang, working in my laboratory, has been studying the structure of nascent DNA in *Bacillus subtilis*. In agreement with Dr. Okazaki, and contrary to Werner's findings in *Escherichia coli*, we have found that short pulses of either [^{3}H]thymine or [^{3}H]thymidine go predominantly into short

chains (Okazaki pieces), and the short chains are precursors of larger DNA. In addition, cells grown in thymine and pulse-labeled with [^{3}H]thymine (steady state labeling) incorporate label into another component, a deoxyoligonucleotide of chain length 50 to 100. This oligonucleotide is resistant to degradation by spleen phosphodiesterase, but sensitive to a combination of spleen phosphodiesterase plus alkaline phosphatase, both before and after treatment with strong alkali. Thus, the oligonucleotide is phosphorylated and has no ribonucleotides on its 5' end, at least as we isolate it. We have not proven that it is a primer for the synthesis of Okazaki pieces. This problem is currently under investigation.

Szybalski. But your small pieces, you told us at the Gordon Conference, are the precursor to Okazaki fragments, correct?

Sternglanz. No, I didn't say that. I haven't proved that. They may be, but I haven't proved it.

Oeschger. J. Files and I have developed a system using sucrose-plasmolyzed cells in which the maturation and joining of nascent short DNA chains (Okazaki pieces) have been studied. *E. coli* cells plasmolyzed and assayed under our conditions do not initiate new nascent short chains. They do transfer existing short chains into high molecular weight DNA. The transfer is monitored using cells pulse-labeled with [^{3}H]thymidine just before harvest and plasmolysis. Using this system we have made the following observations.

1. Short chains are joined into progressively higher molecular weight multimers and finally to the chromosome as judged by alkaline sucrose velocity sedimentation analysis.
2. All four deoxynucleoside triphosphates (dNTPs) are required for the joining reaction to take place. They are incorporated into the short chains of DNA. During the incubation dNTP incorporation begins at once and precedes joining. The incorporation ceases shortly before the joining is complete.
3. When radioactive dNTPs and BdUTP (in place of TTP) are used in the reaction the newly synthesized DNA was found to be predominantly light as determined by CsCl equilibrium gradient centrifugation. Twenty per cent of the incorporated counts were observed to tail on the heavy side of the light band after shearing of the DNA to lengths of 2000 to 3000 nucleotides. Less than 2% of the incorporated radioactivity could be detected in the region of the BU-substituted reference DNA either before or after shearing. We have found the incorporation of BdUTP to be indistinguishable from that of TTP in our reactions and the incorporation of dNTPs to be completely dependent on the presence of either TTP or BdUTP.
4. At early times in the reaction all of the incorporated deoxynucleotides are associated with short chains of DNA. The incorporated nucleotides continue to be associated with the short pieces as the chains are progressively joined. We detect no direct incorporation of deoxynucleotides into bulk chromosomal DNA. The incorporation stops when the short DNA pieces have been joined.
5. The S values of the joined molecules are not affected by mild alkaline hydrolysis.

We conclude from the above data that no new short (or long) pieces of DNA are initiated under the conditions of our assay (although initiated short chains may be completed, as indicated by the dense tail on the light DNA after shearing).

If there are ribonucleotides at one end of the short chains (Okazaki, this meeting) they are removed before joining. We also conclude that there are gaps between the short DNA chains which must be filled before ligation, and that the short chains are arranged next to one another along one strand of the DNA. The short chains are joined first to one another to form an intermediate (27 to 38 S) which is then joined, as a unit, to the continuous DNA of the chromosome. These data support the model for chromosome replication proposed by Okazaki and give evidence for sizeable gaps between the short chains of DNA which must be filled before joining and attachment to the chromosome.

Speyer. I wanted to say that T4 DNA also contains RNA; we found this both in T4 DNA from mature phage, and in the T4 precursor DNA (the 400 S material which can be isolated after infection with a DNA delay mutant). The material is identified by radioactivity and degradation and is indeed RNA. It is apparently covalently linked to the DNA because, when the DNA is melted and banded, the counts representing the RNA still travel or band with the DNA. The RNA is presumably at the ends because when the DNA is treated with ribonuclease and then heated, the molecular weight of the DNA doesn't change, but this is indirect evidence. It is pretty well known that T4 DNA is not degraded in alkali, unlike, let's say T5 DNA which has gaps in it, so we have many unanswered questions about it. We presume it's at an end. We don't know which end, and we don't know the size of it either.

Genetics and Physiology of DNA Ligase Mutants of *Escherichia coli*

Michael M. Gottesman,
Minnie L. Hicks, and Martin Gellert

National Institute of Arthritis, Metabolism
  and Digestive Diseases
Laboratory of Molecular Biology
National Institutes of Health
Bethesda, Maryland 20014

This chapter describes the isolation of two classes of DNA ligase mutants of *Escherichia coli:* Class 1, overproducers of a wild type enzyme, and Class 2, synthesizers of a defective enzyme. Genetic dominance studies show that Class 1 mutations are *cis*-dominant and *trans*-recessive and therefore are probably promoter or operator mutations; and that Class 2 mutations are *trans*-recessive, as would be expected for mutations in a structural gene. The ligase overproducers have no apparent physiological defects. Bacteria carrying the ligase structural gene mutations, *lig* 4 or *lig ts*-7, have a striking defect in ligase activity *in vitro*, but only *lig ts*-7 is a conditional lethal mutation. Both mutants show delayed sealing of Okazaki fragments, this being more marked in *lig ts*-7. It is concluded that DNA ligase is an essential gene in *E. coli*, but that only very low levels of ligase activity are necessary for apparently normal DNA replication, repair, and recombination.

Most current models of DNA replication involve discontinuous synthesis of DNA on at least one strand (for a recent review see Gross (9) and other chapters in this volume). An absolute requirement of these models is that the discontinuous pieces of newly formed DNA be sealed together; the complete replication of an intact genome, therefore, should require DNA ligase. However, a recent screening of temperature-sensitive DNA mutants representing all known loci A to G failed to reveal any mutants defective in DNA ligase activity (Gellert, unpublished data), suggesting either that ligase is not involved in DNA replication, or, more likely, that ligase mutants when found might be physiologically different from other DNA temperature-sensitive mutants.

Although enzymes with DNA ligase activity have been isolated from bacteria (6, 7, 18, 24), phage-infected bacteria (1, 3, 23), and mammalian tissues (12), until recently the genetics of this enzyme had been studied only in the phage systems. In the case of T4, an ATP-dependent ligase, the product of gene 30, seems to be essential for normal DNA replication after infection. Gene 30 mutants fail to join Okazaki fragments, and stop making DNA 10 min after infection, producing only about 10% of the normal amount of phage DNA. These data strongly implicate a phage-coded ligase in T4 DNA replication (5, 16, 21).

In contrast, T7 with an amber mutation in the phage ligase gene is able to grow and replicate its DNA normally in wild type *Escherichia coli* hosts, but not in hosts defective in the bacterial ligase. Apparently, in the case of T7, the host ligase is able to compensate for lack of the phage ligase, allowing phage production (13).

Gellert and Bullock (8) have reported the isolation of a temperature-sensitive mutant of the *E. coli* DPN-dependent ligase with a relatively profound defect in activity *in vitro*, but no obvious defect in growth, sealing of Okazaki fragments, or DNA synthesis. On the other hand, Pauling and Hamm (19, 20) have described the isolation of a temperature-sensitive conditional-lethal ligase mutant of *E. coli* 15T$^-$ which has been further characterized by Modrich and Lehman (15). This mutant seals Okazaki fragments poorly, and does not form colonies above 37°C, but its genetic analysis has not been possible because the parental strain is not active in transductions or matings. Furthermore, the presence in this strain of several plasmids, including a P1-like plasmid (10), raises the possibility that its failure to grow at the nonpermissive temperature is due to induction by the ligase defect of a lysogenic bacteriophage. Before concluding that DNA ligase is essential for bacterial viability it was necessary to transfer the mutation into a well characterized genetic background.

In this chapter we shall describe and compare the genetics and physiology of two classes of DNA ligase mutants of *E. coli:* Class 1, ligase overproducers, and Class 2, synthesizers of a defective ligase. These consist of the mutants originally isolated in this laboratory as well as Pauling and Hamm's mutation *lig ts*-7 which we have been able to transduce into an *E. coli* K-12 strain. The results indicate that DNA ligase is indeed an essential enzyme, but that only very low levels of activity are necessary *in vivo* for the maintenance of growth and DNA replication.

Isolation of Ligase Mutants

Overproducers of an apparently wild type ligase were isolated in two independent ways. Initially, overproducers were selected as colonies which were "nibbled" when grown on a lawn of T4am30 phage. Since T4 defective in its own DNA ligase will not grow on wild type bacteria, "nibbling" of these colonies indicated some compensation by the host for the lack of phage ligase. Among the "nibbled" colonies were several ligase overproducers. One such ligase overproducer (*lop* 8) has been extensively characterized.

Ligase overproducers were also selected as pseudorevertants of *lig* 4 (whose isolation is described below). *Lig* 4 strains are somewhat sensitive to methylmethane sulfonate (MMS). A certain fraction of colonies growing up on MMS plates had become resistant by overproducing the defective ligase. The ligase-overproducing locus has been genetically isolated by P1 transduction from several of these pseudorevertants. The highest overproducer (*lop* 11) from this selection is described below.

Using *lop* 8, it was possible to select pseudorevertants which would no longer support the growth of T4am30 phage because they overproduced a defective ligase. The ligase mutation in one of these strains (*lop* 8 *lig* 4) has been separated from the overproducing mutation by P1 transduction and transferred to a number of other strains. The physiology of this strain (*lig* 4) will be described below.

Finally, the ligase mutation of Pauling and Hamm, *lig ts*-7, was transferred to an *E. coli* K-12 strain using P1 transduction. Even though *E. coli* 15T$^-$ strains strongly restrict incoming DNA, we were able to select for P1 lysogens by using P1clr100CM, a thermoinducible derivative of P1CM isolated and described by Rosner (22). The P1-transducing phage produced at 42°C by this thermoinducible lysogen were used to transduce an *E. coli* K strain which lacked K restriction, and from there *lig ts*-7 could be transduced into strains isogenic with the strains containing our other ligase mutations. *E. coli* K-12 strains carrying *lig ts*-7 have also been constructed by Konrad, Modrich, and Lehman (personal communication) with results similar to those we shall describe below.

Table 1 compares the specific activity and thermolability of DNA ligase in extracts of each of the mutants described above. Among 80 overproducers screened, no strains with activity greater than 6-fold wild type ligase levels were found. As shown, the activity of the representative overproducing strains, *lop* 8 and *lop* 11, has the same temperature dependence as the wild type ligase and the same ratio of ^{14}C-adenylate uptake

Table 1

Strain	^{3}H-dAT joining activity, 30°C (units/mg)	^{3}H-dAT joining activity, 42°C (units/mg)
lig$^+$	0.36	0.63
lop 8	0.70	1.53
lop 11	1.91	4.13
lig 4	0.13	≤0.01
lig ts-7	0.04	≤0.01

Assays were performed according to the method of Modrich and Lehman (14). The strains used here and throughout this paper are all pts$^+$ P1 transductants of FF7059 an *Escherichia coli* K pts$^-$ (Enzyme I$^-$) str A strain kindly provided by Dr. Wolfgang Epstein (University of Chicago). Crude extracts were prepared by sonication from cells grown to mid-log phase in tryptone broth. The absolute value of the specific activity of wild type ligase, as well as the mutant types, varies somewhat with the genetic background, hence all of these mutants are compared in the identical background.

to enzymatic activity (data not shown), strongly suggesting that they are overproducers of a wild type ligase.

Strains containing *lig* 4 have a thermolabile ligase. *Lig* 4 has about 20 to 30% of the activity of *lig*$^+$ at 30° C but only 1% at 42° C. As previously noted (8), although the activity of the enzyme isolated from this strain is highly thermolabile, its adenylate uptake from ^{14}C-DPN is normal at 42° C. *Lig ts*-7 has only 2 to 5% of wild type activity at 30° C and less than 1% of wild type activity at 42° C. Unlike *lig* 4, however, its adenylate uptake at 20° C is only 30% of wild type and at 42° C it is 10% of wild type.

The mutant enzyme from *lig* 4 has been highly purified and retains its thermolability, while the enzyme from a strain carrying *lig ts*-7 has proved to be somewhat difficult to purify, with nearly complete loss of activity beyond a few early purification steps, suggesting that the enzyme is quite subject to denaturation. This evidence strongly suggests that both *lig* 4 and *lig ts*-7 are mutations in the structural gene for ligase.

Mapping of Ligase Mutants

Using *lop* 8 and *lig* 4 we were able to map the overproducing locus and the structural gene for DNA ligase. *Lop* 8 and lop 11 are 92% contransduced with *pts* (phosphosugar transport system; 4) and *lig* 4 is 88% cotransduced with *pts* at 46 min on the *E. coli* chromosome (Gottesman *et al.*, manuscript in preparation). *Lig ts*-7 is also closely linked to *pts*. Using appropriate three-factor crosses we have been able to show that the order of genes is *dapE-ptsI-ptsH-lop-lig-aroC*, with *lig* situated counterclockwise from *pts*.

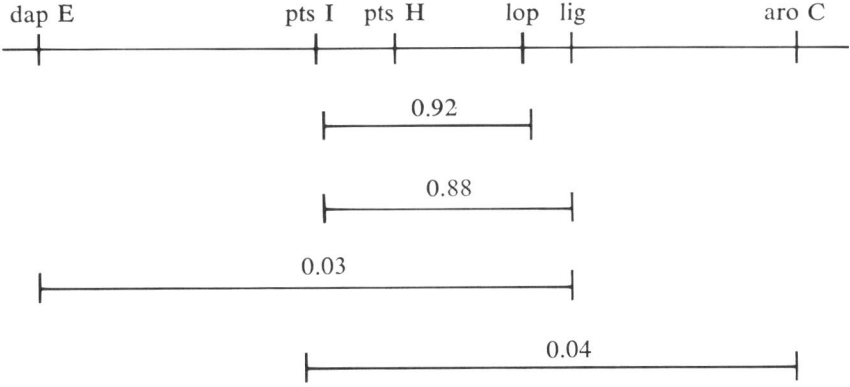

Fig. 1. Map location of *lop* and *lig* loci. Numbers refer to Pl cotransduction frequencies.

Independent confirmation of this map location has come from the isolation of a plaque-forming λ transducing phage carrying the chromosomal genes for *pts* H and *lig* but not *pts* I. This phage was isolated from a lysate produced by induction of a λ phage inserted in *pts* I (Gottesman *et al.*, manuscript in preparation). The mapping data are summarized in Fig. 1. The location of *lop* and *lig* does not correspond to any of the known DNA temperature-sensitive mutants.

Diploid Analysis of Ligase Mutants

In order to clarify the nature of these DNA ligase mutants, we undertook a series of dominance studies with the mutants *lig* 4 and *lig ts*-7. Table 2 summarizes the phage-plating phenotypes and ligase levels of a wild type strain, a strain carrying *lig* 4, and the partial diploid strains construct-

Table 2

Chromosome	Episome	T4am30 plating	T4am30rII plating	^{3}H-dAT joining activity (units/mg, 42°C)
lig$^+$		−	++	0.89
lig 4		−	−	≤0.01
lig$^+$	*lig*$^+$	+	++	0.99
lig 4	*lig*$^+$	−	++	0.96

Diploid analysis of *lig* 4 mutation. Phage-plating phenotypes and ligase activity at 42°C of haploid and partial diploid strains are shown. Strains were grown at 37°C and crude extracts prepared as described in Table 1. Minuses (−) indicate efficiency of plating $<10^{-4}$, pluses (+, ++) indicate efficiency of plating >0.1 with plating on YMC su$^+$ taken as 1.0. The F' used was F'98 derived from Hfr KL98 by Dr. Wolfgang Epstein. Diploid strains with the mutant alleles on the episome gave essentially identical results.

ed by mating in the episome F'98 which covers the region from *tyr* A to *sup* N. As shown, cells with wild type levels of DNA ligase are unable to plate T4am30, whereas the doubly mutant T4am30rII can grow on these strains, but will not grow on *lig* 4 or *lig ts-*7.

Lig 4 is clearly *trans*-recessive to *lig*$^+$, as would be expected of a mutation in a structural gene for an enzyme with one subunit. This recessivity also eliminates the possibility that strains carrying *lig* 4 actually have an inhibitor of ligase which is thermally activated. *In vitro* mixing of crude extracts from *lop* 8 *lig* 4 and *lig*$^+$ strains had already indicated that there is no inhibitor of ligase in *lig* 4 strains (8).

Studies of diploid strains containing the mutation *lig ts-*7 have so far been limited to diploids created by lysogenizing a *lig ts-*7 strain with a λ transducing phage carrying the wild type host ligase genes (λp *lig*). Such a diploid appears to have haploid wild type ligase activity *in vivo* by phage testing with T4am30. Curing of the transducing phage results in return to the *lig ts-*7 phenotype, indicating that the lysogen had both *lig*$^+$ and *lig ts-*7 genes. Thus, *lig ts-*7 is also recessive to *lig*$^+$.

Table 3 summarizes the results of dominance studies on a representative overproducer, *lop* 11. In the partial diploid *lop* 11/F'*lop*$^+$ the *lop* characteristic persists, as shown in elevated enzyme levels and efficient plating of the T4am30 phage. *Lop*, therefore, is dominant in this particular diploid strain and cannot represent loss of a repressor.

In the partial diploid strain *lop* 11 *lig* 4/F'*lop*$^+$*lig*$^+$ the *lop* characteristic is not expressed in *trans*, as shown clearly at 42° C by ligase assays and inability to plate T4am30 phage. *Cis*-dominance and *trans*-recessivity suggest strongly that *lop* is a promoter or operator mutation, although the possibility that it codes for a more efficient *cis*-acting positive control element cannot be ruled out. The identification of *lop* 11 as an apparent promoter or operator mutant which maps clockwise from the ligase structural gene indicates that the structural gene is read counterclockwise on the *E. coli* genome.

Table 3

Chromosome	Episome	30°C		42°C	
		T4am30 plating	^{3}H-dAT joining activity (units/mg)	T4am30 plating	^{3}H-dAT joining activity (units/mg)
lop$^+$		−	0.22	−	0.89
lop 11		++	1.8	++	3.6
lop$^+$	*lop*$^+$	+	0.54	+	0.99
lop 11	*lop*$^+$	++	2.2	++	3.0
lop 11 *lig* 4	.	++	0.51	−	0.06
lop 11 *lig* 4	*lop*$^+$*lig*$^+$	++	0.70	+	0.93

Diploid analysis of *lop* 11 mutation. Experiments were conducted as in Table 2.

Physiology of Ligase Mutants

From the very low but similar levels of DNA ligase activity measured by the ^{3}H-dAT joining assay in crude extracts of *lig* 4 and *lig ts*-7 it is not possible to predict which strain will be more defective *in vivo*. Growth on solid medium, such as tryptone or glucose-minimal plates, is entirely normal for *lig* 4, while above 37° C *lig ts*-7 forms colonies with an efficiency of only 1×10^{-6}. The high temperature survivors of *lig ts*-7 appear to be both true revertants and pseudorevertants which overproduce a defective ligase.

Figure 2 shows growth curves in tryptone broth at 30 and 42° C

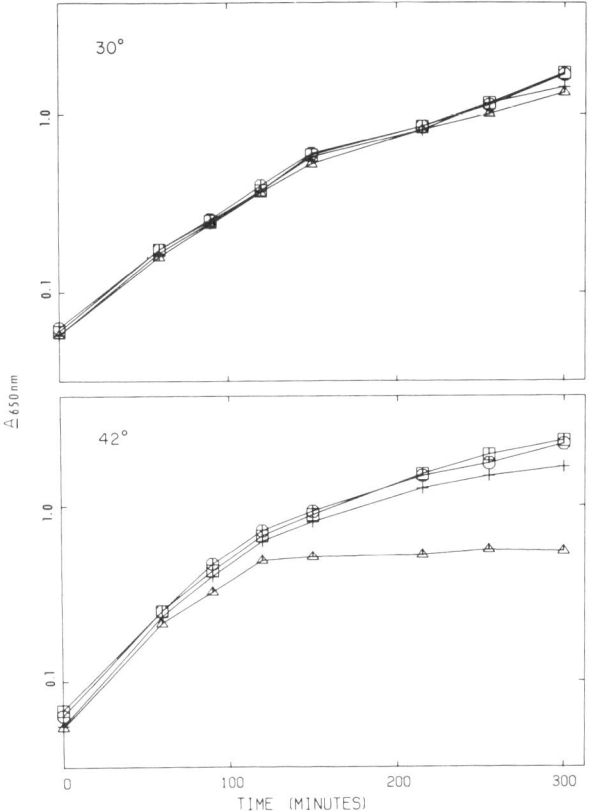

Fig. 2. Growth curves of ligase mutants and controls at 30 and 42°C. Overnight cultures grown at 30°C were transferred to fresh tryptone broth, grown to mid-log phase, and rediluted for these growth curves. Redilution of *lig ts*-7 immediately before or after cessation of increase in optical density did not reinitiate growth. (O———O) *lig*$^+$; (□———□) *lig* 4; (△———△) *lig ts*-7; (+———+) *lig ts*-7 (λ*plig*).

for *lig* 4, *lig ts*-7, a wild type parent strain, and *lig ts*-7 lysogenized with a λp *lig*. Only *lig ts*-7 has a growth defect which manifests itself about 100 min after shift to the nonpermissive temperature. Prior to this time growth is essentially logarithmic.

Figure 3 illustrates that in minimal liquid medium, with limited growth rate, *lig ts*-7 continues to divide for 2 hr as measured by increasing viable cell count. It also increases in optical density for at least 3 hr with a 10-fold increase in total DNA synthesized over this time period, as measured by ^{3}H-thymidine incorporation in the presence of deoxyadenosine. The sheer quantity of DNA synthesized at the nonpermissive temperature argues against this being merely repair synthesis by the strain carrying *lig ts*-7, but does not rule out some form of aberrant DNA synthesis.

Although there is no immediate loss of ability to make DNA in *lig ts*-7, there appears to be a very striking defect in the chasing of pulse-labeled DNA fragments into large molecular weight DNA, as might be expected if the mutant cannot seal Okazaki fragments. This defect is only manifested in our mutants with less than 5% ligase activity as

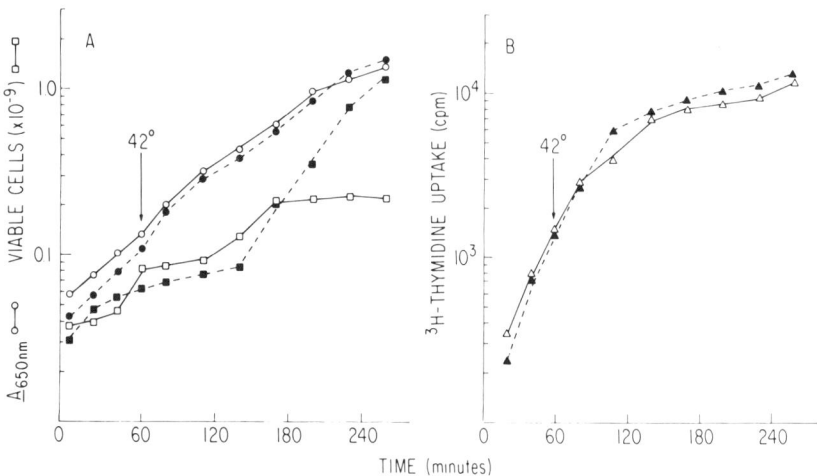

Fig. 3. Optical density and viable cell count (A), and ^{3}H-thymidine incorporation (B) in *lig ts*-7 and an isogenic *lig*⁺ control at 42°C. Overnight cultures grown at 30°C in tryptone broth were transferred to glucose minimal medium. After reinitiation of logarithmic growth, cultures were rediluted into minimal medium at 30°C containing ^{3}H-thymidine (6 μg/ml; specific activity, 15 μCi/μg) and deoxyadenosine (100 μg/ml) as described by Boyce and Setlow (2). At 60 min (shown by arrow) cultures were shifted to 42°C. At 30-min intervals aliquots were sampled for optical density, plated at appropriate dilutions for viable cell counts on tryptone plates at 30°C and collected on Millipore filters for determination of ^{3}H incorporation. (○——○) optical density; (□——□) viable cell count; (△——△) ^{3}H-thymidine incorporation, counts per min per 0.2 ml of culture. Open points connected by solid lines represent *lig ts*-7 (○——○); closed points connected by dashed lines represent *lig*⁺ (●- - -●).

measured in the ^{3}H-dAT joining reaction. Figure 4 shows a series of pulse-chase experiments on three different strains with the genotypes lig^+, lig 4, and lig ts-7.

We previously reported on the mutant lop 8 lip 4 which has 5% of wild type ligase and no defect in sealing of Okazaki fragments (8). At 42° C lig 4 shows a considerable retardation in sealing of Okazaki fragments, but lig ts-7 is even more defective. We presume that the decreased ability of lig ts-7 to seal Okazaki fragments as compared to lig 4 is responsible for the conditionally lethal effect of the former mutation.

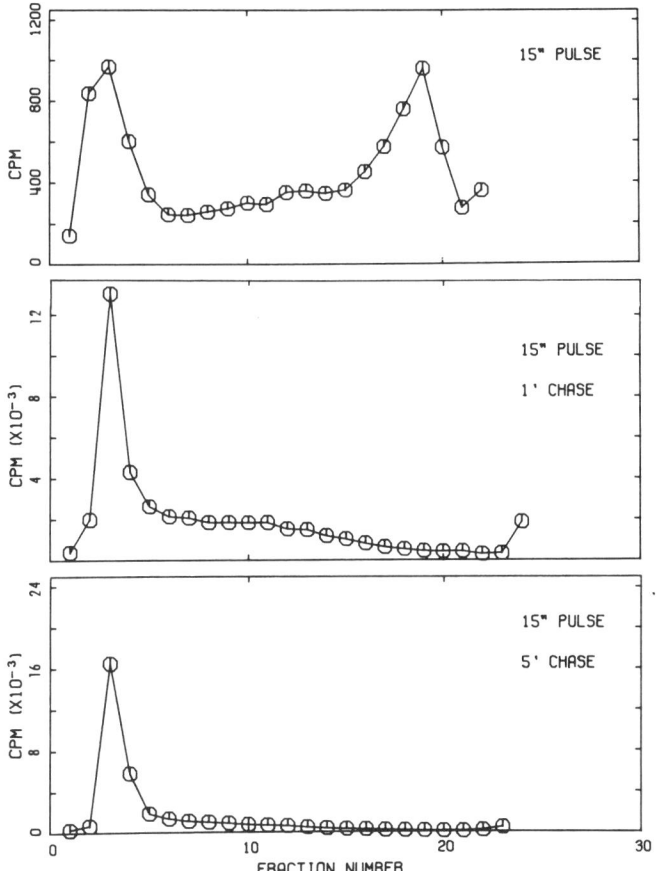

Fig. 4A

Fig. 4. Alkaline sucrose density gradients on cultures which were ^{3}H-thymidine pulse-labeled at 42°C. Experiments were performed as described by Gellert and Bullock (8). Fractions were collected from the bottoms of the gradients. (A) lig^+; (B) lig 4; (C) lig ts-7.

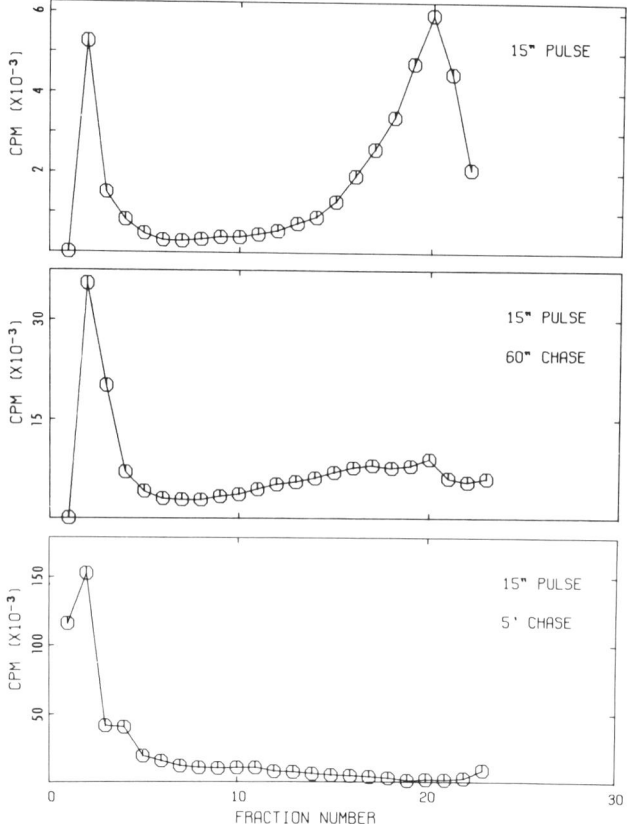

Fig. 4B

Defective sealing in *lig* 4 can also be demonstrated in experiments which measure the conversion of superinfecting λ DNA to the covalently circular form in a λ-lysogenic strain. Ten minutes after infection at 42° C, 75% of the input DNA has become covalently circular in a *lig*⁺ strain, but only 25% is covalently closed in a *lig* 4 strain.

All of the ligase overproducers studied were physiologically indistinguishable from wild type except for their ability to support growth of T4am30 phage.

Repair and Recombination

We have not been able to demonstrate major defects in recombination or repair in any of our *lop* or *lig* mutants.

Recombination was studied with Hfr crosses at 42° C into an F⁻

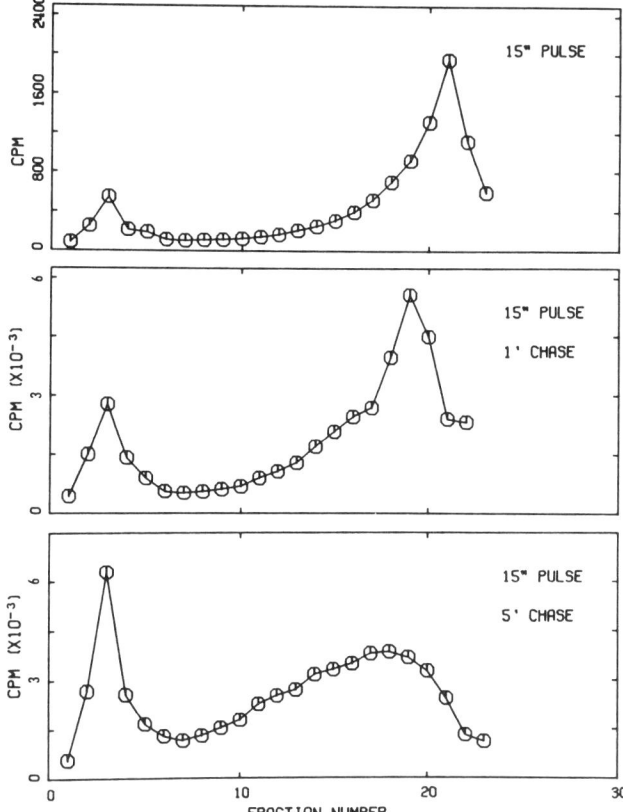

Fig. 4C

strain carrying *lig* 4. For example, in a cross of Hfr H × F⁻ *lig* 4 at 42° C, trp^+ recombinants were formed at normal frequency. Repair capability, as reflected in UV sensitivity, was studied. Strains carrying *lig* 4 were found to be only slightly UV-sensitive and strains carrying *lig ts*-7 were moderately UV-sensitive at temperatures at which the strains are viable. Both *lig* 4, to a small extent, and *lig ts*-7, to a greater extent, are MMS-sensitive.

Double Mutants

We have constructed the following double mutants by appropriate P1 transductions and matings: *rec*A56 *lig* 4; *rec*B21 *lig* 4; *rec*A56 *lop* 8; *rec*B21 *lop* 8; *pol*A1 *lig* 4. The *rec*A and *rec*B double mutants with *lop* 8 grow as well as their parent *rec*⁻ strains. The *rec*B21 *lig* 4 double mutant grows much more slowly than either parent at 42° C and the recA56 lig 4 double mutation is lethal at 42° C. The polA1 lig 4 double mutation is also lethal at 42° C. Although it is known that polA mutants, much like lig mutants, have delayed sealing of Okazaki fragment (11, 17), the double mutant seals fragments even more slowly than either parent (Gottesman et al., to be published).

Attempts to interpret the lethality of lig 4 double mutants with recA, recB, or polA in terms of known cellular processes are clouded by the fact that the somewhat more defective mutant lig ts-7 is lethal by itself.

Phage Induction by Lig ts-7

Lysogens of λcI^+ in lig ts-7 strains are induced by heating to 42° C with release of phage and complete lysis of the culture 75 min after heat induction. Lysogens of λcI ind^- are not induced at 42° C by lig ts-7, indicating that this is an effect mediated by λ repressor. Lig 4 does not induce λ. Conversely, strains with ligase levels up to 10-fold greater than wild type (diploid lop 11) do not make λ ind^+ phenotypically ind^- with respect to UV induction of lysogens; hence, the simple hypothesis that under normal conditions UV induces lysogens by titrating a limiting amount of DNA ligase is not correct.

P1 lysogens of lig ts-7 strains are slightly induced at the nonpermissive temperature, with an increase in free P1 from 10- to 20-fold over wild type controls. Further studies on the mechanism and significance of these observations are in progress. They do serve, however, to support the contention (see introduction) that a conditional lethal mutation should be isolated in a strain without potentially inducible lysogenic phage.

Conclusions

In this work we have attempted to determine whether DNA ligase is an essential enzyme in E. coli, and if so, what might be the lethal defect in ligase mutant strains. We conclude that DNA ligase is essential and that a disturbance of normal DNA replication is probably the primary reason for the lethality of lig ts-7.

The most striking fact incriminating ligase in DNA replication is the considerable accumulation of pulse-labeled small DNA fragments in ligase-defective strains. However, this defect also shows up in studies on total DNA synthesis. For lig ts-7, Modrich and Lehman (15) were able to show that after about 75 min at the nonpermissive temperature in rich liquid medium there was a decrease in the rate of ^{3}H-thymidine incorporation compared to wild type and a precipitous fall in viable cell count. However, DNA synthesis did not shut off entirely and the mutant continued to increase its total DNA by many-fold over a 3-hr period.

In minimal medium, as we have shown, the defect in DNA synthesis is less apparent. Although cells do not form colonies, they do grow logarithmically for almost 2 hr and incorporate virtually normal levels of ^{3}H-thymidine. Notwithstanding the danger of extrapolating *in vitro* assay data to *in vivo* conditions, at present the best hypothesis to explain the behavior of *lig* 4 and *lig ts*-7 is that the enzyme in *lig ts*-7 is at all temperatures more defective than *lig* 4 at 42° C and that rapid growth at temperatures 37° C and above in rich medium overburdens the limited amount of ligase in the cell, producing a major imbalance which is fatal. The conclusion that ligase is an essential enzyme is supported by the fact that despite various efforts, we have been unable to isolate deletions of the structural gene for ligase.

Our conclusions implicating ligase in DNA replication assume that Okazaki fragments are true intermediates in DNA replication. This conclusion is strongly supported by the recent finding that each pulse-labeled fragment begins with an oligoribonucleotide sequence (see R. Okazaki, this volume). We can also conclude that if Okazaki fragments are intermediates in DNA replication, this DNA is made discontinuously on *both* strands. Fig. 4 shows clearly that virtually all the pulse-labeled DNA in our ligase mutants is 10 S in size.

Furthermore, it is obvious that ligase normally exists in vast excess in the bacterial cell, since cells are insensitive to large changes in ligase levels. A several-fold increase in ligase, as in *lop* mutants, has no apparent physiological effect, and only a drop to less than 5% of wild type produces noticeable effects on cellular economy.

Finally, the fact that *lig ts*-7 continues DNA synthesis for several hours at the nonpermissive temperature explains why DNA ligase mutants were not picked up by the screening procedures used to isolate DNA temperature-sensitive mutants. These procedures were designed to detect strains in which DNA synthesis is arrested fairly soon after a shift to high temperature; thus, mutants with a conditionally lethal ligase defect would have been discarded.

Acknowledgments

We would like to thank B. Konrad, P. Modrich, and I. R. Lehman for communicating unpublished results and M. E. Gottesman, S. Gottesman, and J. L. Rosner for reviewing the manuscript. We also acknowledge the advice and encouragement of many of our other colleagues at National Institutes of Health.

References

1. Becker, A., Lyn, G., Gefter, M., and Hurwitz, J. 1967. The enzymatic repair of DNA. II. Characterization of phage-induced sealase. *Proc. Nat. Acad. Sci. U. S. A.* 58: 1996.

2. Boyce, R. P., and Setlow, R. B. 1962. A simple method of increasing the incorporation of thymidine into the deoxyribonucleic acid of *Escherichia coli*. *Biochim. Biophys. Acta* 61: 618.

3. Cozzarelli, N. R., Melechen, N. E., Jovin, T. M., and Kornberg, A. 1967. Polynucleotide cellulose as a substrate for a polynucleotide ligase induced by phage T4. *Biochem. Biophys. Res. Commun.* 28: 1967.

4. Epstein, W., Jewett, S., and Fox, C. F. 1970. Isolation and mapping of phosphotransferase mutants in *Escherichia coli*. *J. Bacteriol.* 104: 793.

5. Fareed, G. C., and Richardson, C. C. 1967. Enzymatic breakage and joining of deoxyribonucleic acid. II. The structural gene for polynucleotide ligase in bacteriophage T4. *Proc. Nat. Acad. Sci. U. S. A.* 58: 665.

6. Gefter, M. L., Becker, A., and Hurwitz, J. 1967. The enzymatic repair of DNA. I. Formation of circular λ DNA. *Proc. Nat. Acad. Sci. U. S. A.* 58: 240.

7. Gellert, M. 1967. Formation of covalent circles of lambda DNA by *E. coli* extracts. *Proc. Nat. Acad. Sci. U. S. A.* 57: 148.

8. Gellert, M., and Bullock, M. L. 1970. DNA ligase mutants of *Escherichia coli*. *Proc. Nat. Acad. Sci. U. S. A.* 67: 1580.

9. Gross, J. 1972. DNA replication in bacteria. *Curr. Top. Microbiol. Immunol.* 57: 39.

10. Ikeda, H., Inuzuka, M., and Tomizawa, J. 1970. P1-like plasmid in *Escherichia coli* 15. *J. Mol. Biol.* 50: 457.

11. Kuempel, P. L., and Veomett, G. E. 1970. A possible function of DNA polymerase in chromosome replication. *Biochem. Biophys. Res. Commun.* 41: 973.

12. Lindahl, T., and Edelman, G. M. 1968. Polynucleotide ligase from myeloid and lymphoid tissues. *Proc. Nat. Acad. Sci. U. S. A.* 61: 680.

13. Masamune, Y., Frenkel, G. D., and Richardson, C. C. 1971. A mutant of bacteriophage T7 deficient in polynucleotide ligase. *J. Biol. Chem.* 246: 6874.

14. Modrich, P., and Lehman, I. R. 1970. Enzymatic joining of polynucleotides. IX. A simple and rapid assay of polynucleotide joining (ligase) activity by measurement of circle formation from linear deoxyadenylate-deoxythymidylate copolymer. *J. Biol. Chem.* 245: 3626.

15. Modrich, P., and Lehman, I. R. 1971. Enzymatic characterization of a mutant of *Escherichia coli* with an altered DNA ligase. *Proc. Nat. Acad. Sci. U. S. A.* 68: 1002.

16. Newman, J., and Hanawalt, P. 1968. Role of polynucleotide ligase in T4 DNA replication. *J. Mol. Biol.* 35: 639.

17. Okazaki, R., Arisawa, M., and Sugino, A. 1971. Slow joining of newly replicated DNA chains in DNA polymerase I-deficient *Escherichia coli* mutants. *Proc. Nat. Acad. Sci. U. S. A.* 68: 2954.
18. Olivera, B. M., and Lehman, I. R. 1967. Linkage of polynucleotides through phosphodiester bonds by an enzyme from *Escherichia coli*. *Proc. Nat. Acad. Sci. U. S. A.* 57: 1426.
19. Pauling, C., and Hamm, L. 1968. Properties of a temperature-sensitive radiation-sensitive mutant of *Escherichia coli*. *Proc. Nat. Acad. Sci. U. S. A.* 60: 1495.
20. Pauling, C., and Hamm, L. 1969. Properties of a temperature-sensitive, radiation-sensitive mutant of *Escherichia coli*. II. DNA replication. *Proc. Nat. Acad. Sci. U. S. A.* 64: 1195.
21. Richardson, C. C., Masamune, Y., Live, T. R., Jacquemin-Sablon, A., Weiss, B., and Fareed, G. C. 1968. Studies on the joining of DNA by polynucleotide ligase of phage T4. *Cold Spring Harbor Symp. Quant. Biol.* 33: 151.
22. Rosner, J. L. 1972. Formation, induction and curing of bacteriophage P1 lysogens. *Virology* 49: 679.
23. Weiss, B., and Richardson, C. C. 1967. Enzymatic breakage and joining of deoxyribonucleic acid. I. Repair of single strand breaks in DNA by an enzyme system from *Escherichia coli* infected with T4 bacteriophage. *Proc. Nat. Acad. Sci. U. S. A.* 57: 1021.
24. Zimmerman, S. B., Little, J. W., Oshinsky, C. K., and Gellert, M. 1967. Enzymatic joining of DNA strands: A novel reaction of diphosphopyridine nucleotide. *Proc. Nat. Acad. Sci. U. S. A.* 57: 1841.

Discussion

Lehman. I might just comment on some work that Bruce Konrad has been doing in my laboratory, along lines similar to those reported here. We have independently transduced out the *ts*-7 mutation from strain 15 (using the mapping data provided by Dr. Gellert and his colleagues) and obtained results similar to those described by Dr. Gottesman. There is however one point I do wish to make regarding the difference in the defectiveness of the *lig* 4 as compared with the *ts*-7 mutation. We, too, have found that with extracts of *ts*-7, joining is almost undetectable at restrictive temperatures, so there is no apparent difference in ligase activity as measured by the poly (dA-dT) joining assay (*ts*-7 and *lig* 4 both show ≤1% of normal joining). However, if one measures enzyme-AMP formation, then a very distinct difference does appear. In the case of *lig* 4, enzyme-AMP formation proceeds almost normally at restrictive temperatures. On the other hand, there is no detectable enzyme-AMP with *ts*-7 at the higher temperature. It would therefore appear that *ts*-7 is unable to turn over even once at the restrictive temperature. Clearly then, despite the fact that there is no difference between the two ligase mutants as measured in the catalytic joining assay, there is a significant difference between the two in the stoichiometric assay measuring enzyme-AMP formation. There is therefore some basis for postulating that *ts*-7 is more defective *in vivo* than *lig* 4. The other point I would like to add is that we have tested the *ts*-7 mutation in the K background for its MMS and UV sensitivity. These experiments have been carried out at 30° C (at permissive temperature), since the organism does not grow at restrictive temperatures. We have found it to be considerably more UV- and MMS-sensitive than the wild type parent strain.

Gottesman. We have done those same experiments. We find a small amount of adenylation of *lig ts*-7 at 42°C, about 10% of normal, with approximately 30% of normal adenylation at 30°C. The adenylation phenomenon, therefore, appears to be temperature-sensitive *in vitro*. We have also found moderate UV sensitivity and considerable MMS sensitivity for strains carrying *lig ts*-7 at temperatures permissive for growth.

Two DNase-related ATPases of *Escherichia coli**

Stuart Linn, Barnet Eskin, Alexander E. Karu, Peter J. Goldmark,[†] and Erin Hawkins[‡]

Department of Biochemistry
University of California
Berkeley, California 94720

The DNase controlled by the *recB* and *recC* genes of *Escherichia coli* and the DNA restriction endonuclease of *E. coli* B are each associated with a DNA-dependent activity which hydrolyzes ATP to ADP and P_i. In both cases uncoupling of the ATP hydrolysis from the DNase has been shown, thus suggesting that ATP hydrolysis is not an integral part of the DNA hydrolysis reaction, but rather a manifestation of another enzyme function. Possibilities for such functions are discussed.

Four DNA-dependent ATPase activities have now been described in *Escherichia coli*. Two of these activities have not been found to be associated with any other enzymic activity (7), whereas the other two, the restriction enzymes of *E. coli* B or K (14, 15) and the enzyme controlled by the *recB* and *recC* genes (12, 19, 24), are associated with DNase activities. This report presents evidence that in the latter cases ATPase is not coupled directly to the DNase reaction. Hence all four DNA-dependent ATPase activities may be manifestations of the same or similar functions.

---

*The research reported here which was carried out in the authors' laboratory was supported by contract AT(04-31-34) from the United States Atomic Energy Commission and Grant GM 19020 (formerly CA11092) from the United States Public Health Service.
[†] Current address, Department of Neurobiology, Harvard University Medical School, Boston, Mass.
[‡] Current address, Department of Biochemistry, Louisiana State University, Baton Rouge, La.

There are several suggestions that one or more of these DNA-dependent ATPases could be involved in DNA synthesis. The significance of the ATP requirement during DNA synthesis in cells treated with organic solvents (17) or lysozyme (20) is unknown, although recent observations suggest that such synthesis may be implicated with repair (3). The incompatability of the *recB* or *recA* mutation with the *polA* mutation (16) might also indicate that the product of the *recB* gene takes part in DNA replication. Finally, the product of the *recB* and *recC* genes probably plays a significant role in DNA synthesis induced by γ-rays (8), and in recombinational DNA synthesis (5).

The *recBC* Enzyme

Occurrence and General Properties

Strains of *E. coli* with a mutation in the *recB* or *recB* genes show decreased DNA repair and recombinational abilities, and lack an ATP-dependent DNase (see reference 5 for a recent review). The DNase has an associated DNA-dependent activity which hydrolyzes ATP to ADP and P_i (12, 19, 24). An ATPase activity is also associated with very similar ATP-dependent DNases described in *Micrococcus lysodeikticus* (2), *Haemophilus influenzae* (21), and *Diplococcus pneumoniae* (23). This discussion will review some of our observations with the *E. coli* enzyme, but it should be noted that to date these results have been similar to findings with the enzymes from the other bacteria.

Some of the properties of the enzyme are summarized in Table 1. The enzyme degrades to small oligonucleotides linear DNA (duplex or single stranded) in an exonuclease reaction which requires ATP or another nucleoside triphosphate. Duplex circles are resistant to degrada-

Table 1. Some catalytic properties of the *recBC* enzyme (11, 12)

Linear DNA (duplex or single stranded)
   Degraded exonucleolytically
   Products are oligonucleotides (trimers to octamers)
   ATP absolutely required

Circular DNA
   Only single stranded degraded
   Random endonucleolytic breaks
   ATP stimulates degradation 7- to -fold

ATP
   Hydrolyzed to ADP and P_i
   ATPase requires DNA

tion, but single stranded circles are randomly nicked. If ATP is present, the nicked circles are subsequently reduced to oligonucleotides by the ATP-dependent exonuclease.

Factors Affecting ATP Hydrolysis

During DNA degradation, ATP is hydrolyzed to ADP and P_i, and upon cessation of DNA breakdown, ATP hydrolysis also terminates (12, 19, 24). As shown in Table 2, ATP is hydrolyzed only in the presence of degradable DNA. Hence linear *E. coli* DNA and circular, single strand fd DNA are effective, but the duplex, circular structures, fd replicative form DNA, and phage PM2 DNA are ineffective as cofactors. Likewise, $(pT)_2$ and $(pT)_3$, which are resistant to the nuclease, are not cofactors. In addition, it is observed that throughout the DNase digestion ATP is hydrolyzed at a constant rate which is directly proportional to the number of DNA phosphodiester bonds broken, but unrelated to the particular DNA substrate utilized (12, 19, 24).

Although the above observations imply that ATPase is coupled to the DNase activity, many lines of evidence show that this need not be so. First of all, the single strand-specific endonuclease activity of the *recBC* enzyme is not totally dependent on the presence of ATP (12). Secondly, the number of ATP molecules hydrolyzed per phosphodiester bond cleaved is dependent upon the reaction conditions, and may vary between 2 and 40 (12, 19, 24). This variation is most readily

Table 2. Activation of the *recBC* ATPase by various DNAs

DNA present	^{32}P released from $[\gamma-^{32}P]ATP$ (nmoles)
None	0.02
Duplex *Escherichia coli* DNA	1.57
Denatured *E. coli* DNA	0.82
fd Phage DNA	1.29
fd RFI	0.06
PM2 DNA	0.03
$(pT)_2$	0.01
$(pT)_3$	0.02
$(pT)_4$	0.17
$(pT)_5$	0.23
$(pT)_6$	0.57

Assays (0.15 ml) were for 30 min at 37°C with 4 nmoles of DNA nucleotide, 5 nmoles of $[\gamma-^{32}P]ATP$, and sufficient enzyme to degrade approximately 0.25 nmole of the *E. coli* duplex DNA to acid-soluble material. Other DNAs would be degraded to a lesser extent, in proportion to the amount of ATPase they catalyzed. "fd RFI" is the replicative form DNA of phage fd.

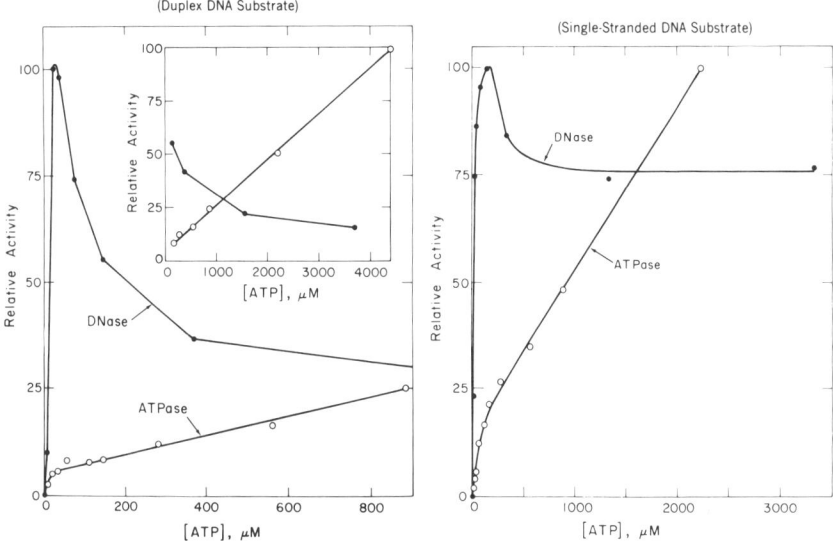

Fig. 1. Effect of ATP concentration on *recBC* DNase and ATPase activities. Assays were at pH 7.0 with duplex *Escherichia coli* DNA, or single stranded, fd DNA which had been converted to a linear form with pancreatic DNase.

observed by altering the ATP concentrations (Fig. 1). With duplex DNA (Fig. 1A), optimum DNase activity is with 30 μM ATP, and increasing the ATP concentration between 30 and 4000 μM results in successively greater inhibition of the reaction. ATPase activity, on the other hand, is stimulated over the entire range with increasing ATP concentration. With single stranded DNA (Fig. 1B), similar results are seen. DNase is optimal at 160 μM ATP, and higher levels of ATP inhibit, whereas ATPase is stimulated with increasing ATP concentrations. With 180 μM ATP, the enzyme is equally active on duplex and single stranded DNA, and approximately 25 molecules of ATP are hydrolyzed per DNA phosphodiester bond broken.

Effect of Cross-linked DNA

The most dramatic uncoupling of ATPase from DNase has been achieved with cross-linked DNA. As originally noted by Cole and Zusman (6), when DNA is exposed to 4,5′,8-trimethylpsoralen (Fig. 2) in the presence of 360-nm light, it acquires cross-links which remain intact during sedimentation in alkali. Under our conditions of exposure to psoralen (13), no breakage of the DNA is detected, and approximately one cross-link is formed per 20 psoralen molecules present. When T7 DNA, which contains approximately 10 psoralen cross-links per molecule, is treated

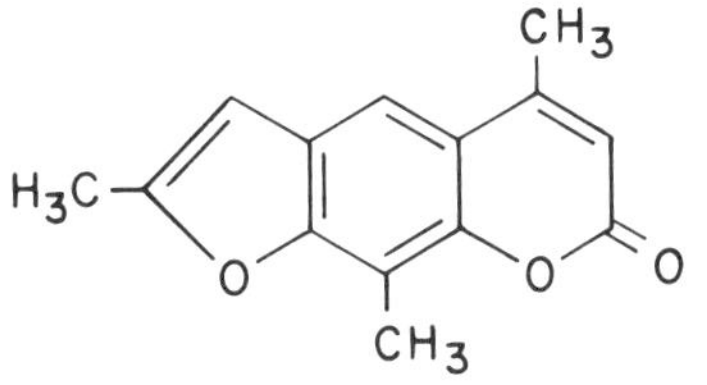

Fig. 2. The structure of 4, 5', 8-trimethyl-psoralen.

with as many as 40 molecules of *recBC* enzyme per DNA molecule, only a very slight change in sedimentation is observed, indicating that little, if any, degradation has occurred (Fig. 3). (We generally observe an amount of degradation which is consistent with the enzyme degrading the DNA from the termini up to the first cross-link, but this interpretation has not been proven experimentally.) Under these conditions untreated DNA would be degraded to completion, even if exposed to the enzyme in the presence of the psoralen-treated DNA (13).

The psoralen-treated DNA is resistant to the *recBC* DNase, in spite of the fact that the enzyme forms oligonucleotides and might have been expected to bypass cross-links. On the other hand, psoralen-treated DNA is degraded by pancreatic or micrococcal nuclease to nonsedimentable

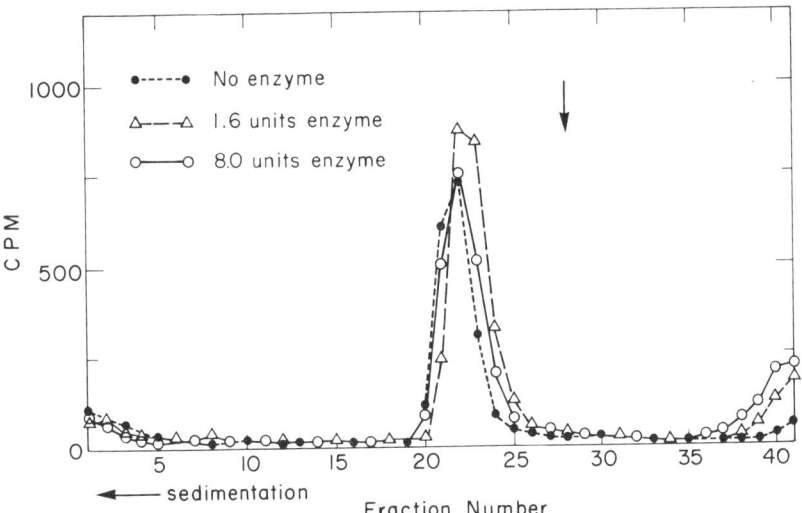

Fig. 3. Effect of the *recBC* enzyme on DNA cross-linked with psoralen. Psoralen-DNA (1.9 nmoles of DNA nucleotide, with approximately 10 cross-links per molecule) was treated with enzyme as shown then sedimented through an alkaline sucrose gradient. Two units of enzyme would degrade an equivalent amount of untreated DNA completely to nonsedimentable material. The arrow shows the position of undigested, non-cross-linked DNA.

material and is sensitive to the restriction endonuclease of *E. coli* B (13). However, it is resistant to the *Neurospora crassa* nuclease, an enzyme which is specific for single stranded DNA. By these criteria, psoralen-cross-linked DNA does not have any apparent gross distortion of its native structure. Indeed, it competes for *recBC* enzyme with single stranded or duplex DNA as though it were duplex DNA (13).

Although resistant to degradation by the *recBC* DNase activity, the psoralen-treated DNA is an excellent cofactor for the DNA-dependent ATP hydrolysis (Fig. 4). This reaction continues so long as ATP is present, whereas with noncross-linked DNA, ATP hydrolysis ceases

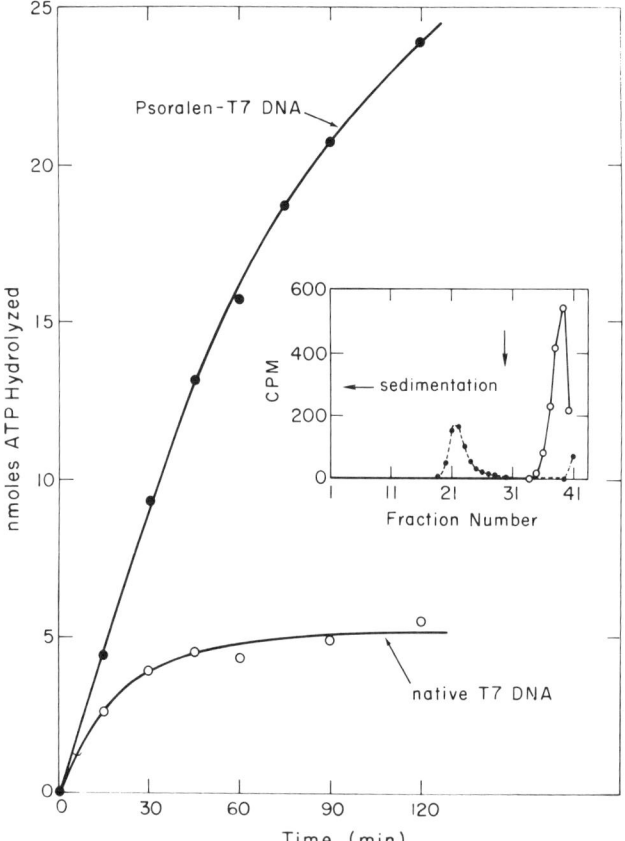

Fig. 4. Effect of cross-linked T7 DNA and untreated T7 DNA on *recBC* ATPase. Two nanomoles of DNA (^{3}H-labeled) were treated with *recBC* enzyme in the presence of [γ-^{32}P]ATP, and ATP hydolysis was monitored. After the 120-min incubation, the residual reaction mixtures were analyzed in alkaline sucrose gradients for the tritium sedimentation profile (inset). The psoralen-treated DNA (●——●) was undegraded, whereas the untreated DNA (○——○) was nonsedimentable. The arrow shows the position of sedimentation of non-cross-linked, nondegraded T7 DNA.

when the DNA is digested to a limit. The ATPase is not mediated simply by the psoralen-DNA reaction product, but it requires that the DNA be intact: a pancreatic DNase digest of psoralen-treated DNA does not mediate ATPase, and if pancreatic DNase is added after a reaction with psoralen DNA has been initiated, ATP hydrolysis is markedly reduced (13).

The unique effects of psoralen-treated DNA on the *recBC* enzyme appear to be due to cross-linking, not to intercalation by the psoralen molecule. DNA exposed to psoralen in the dark is degraded by the enzyme, as is DNA treated with the intercalating agents, ethidium bromide, 9-aminoacridine, ICR-191, or actinomycin D. On the other hand, DNA treated with the cross-linking agents, mitomycin C, or bis(2-chloroethyl)ammonium chloride (nitrogen mustard), behaves like psoralen-treated DNA (13). In addition, since these reagents cross-link guanine residues, whereas psoralen cross-links thymine residues (18), there does not appear to be a stringent requirement for a particular type of cross-linking.

Significance of the ATPase

The results presented here show that the ATPase is not an integral part of the DNase reaction, but perhaps a manifestation of another activity of the enzyme. Possible ATP-dependent activities are (*a*) putting the DNA in some peculiar, unstable secondary structure, or (*b*) tracking the enzyme along the DNA (13, 25). Both of these hypotheses would be consistent with the observations reported here, so long as they take into account the fact that duplex circular DNA does not mediate ATPase (12). In addition, these results might suggest that temporary cross-linking of DNA would prevent extensive degradation of DNA by the *recBC* enzyme *in vivo*. One would then wonder how the ATPase activity is terminated.

The *Escherichia coli* B DNA Restriction Enzyme

Occurrence and General Properties

Restriction DNases are widespread among bacteria and appear to be involved in protecting the cell against the introduction of foreign DNA (for reviews see references 1 and 4). The DNA restriction enzyme of *E. coli* B also catalyzes a DNA-dependent hydrolysis of ATP to ADP and P_i (9, 10), as do the enzymes from *E. coli* K (15), and possibly phage P1 (14, 15). However, this is not the case with the restriction enzymes associated with resistance transfer factors (4) or *H. influenzae* (22).

Table 3. Some catalytic properties of
Escherichia coli B restriction enzyme (9, 14)

Requires unmodified duplex DNA containing "B" specificity sites.
Produces double strand cleavages at these sites.
Requires S-adenosylmethionine and ATP.
ATP hydrolyzed to ADP and P_i.

Some properties of the *E. coli* B enzyme are summarized in Table 3. The enzyme introduces double strand breaks into DNA at specific sites, so long as these sites have not been methylated ("modified") by the B-modification methylase. These breaks generally reduce the DNA to fragments of approximately 10^6 daltons which are susceptible to further hydrolysis by exonuclease (9). The DNase reaction requires ATP and S-adenosylmethionine. Mutants defective in restriction are more prone to phage infection, etc., but otherwise grow normally (1, 4). Deletions of the genes responsible for restriction have not been reported, however.

Factors Affecting ATP Hydrolysis

The requirements for ATP hydrolysis reflect directly the requirements of the nuclease. The ATPase requires duplex, unmodified DNA and S-adenosylmethionine (Table 4). In addition, DNA which has been restricted, and then purified by phenol extraction and sucrose gradient sedimentation meditates little, if any, ATPase. The ATPase has the same apparent K_m for ATP as does the DNase (100 μM), and, like the DNase, it is not inhibited by higher ATP concentrations. Finally, the two reactions have the same pH optimum, divalent cation requirements, and are inhibited to the same extent by various analogs of S-adenosylmethionine (9, 10).

Although these results suggest that ATPase is coupled to the DNase reaction, this is not the case. If the time course of the reactions is examined, it is noted that the endonuclease reaction ceases after 5 to 10 min, and, in fact, the enzyme does not appear to turn over as a nuclease (9). Nevertheless, the ATPase continues linearly for at least 60 min so long as sufficient ATP remains (10). Because the nuclease and ATPase follow different time courses, the ratio of ATP molecules degraded to double strand breaks put into DNA varies with time and is approximately 200,000 after 15 min at 37°C.

State of the DNA and Enzyme during the ATPase Reaction

The above observations suggest that extensive ATP hydrolysis occurs only when preceded by the DNase reaction. These results could imply that restricted DNA is the required cofactor for ATPase. This is not

Table 4. Dependence of *Escherichia coli* B
restriction ATPase on various DNAs

DNA present*	ATP hydrolyzed per 5 min (nmoles)
None	0.04
fd RF·0	7.35
fd RF·0 (no S-adenosylmethionine)	0.06
fd RF·B	0.05
fd RF·0	6.3
fd·0 DNA	<0.5
None	<0.5
fd RF·0	11.7
Phage λ DNA·0	11.7
E. coli K-12	9.2
E. coli B	<0.5
fd RF·0	22.2
Restricted fd RF	0.7
fd RF·0 + restricted fd RF	19.1

Assays (70 μl) were at 37°C with 1.4 nmoles of DNA nucleotide and 100 nmoles of [γ-^{32}P]ATP. The restricted RF had been purified by sucrose gradient sedimentation from residual unrestricted DNA.

*Abbreviations: fd RF: the double stranded, circular replicative form DNA of phage fd; fd RF·0 and fd RF·B: fd replicative form DNA originating on a strain of *E. coli* which lacks the B-modification activity, and on *E. coli* B, respectively; λ DNA·0: DNA of phage λ which was grown on a strain of *E. coli* which lacks B-modification.

the case, however, since restricted DNA does not effectively mediate ATPase in the presence of fresh enzyme (Table 4). (In addition, it does not inhibit ATPase.) The results of Table 4 could then indicate that unmodified DNA is required only to *initiate* ATPase; they do not demonstrate whether DNA is required for *maintaining* ATPase. To answer this question an ATPase reaction can be initiated, then pancreatic DNase added. Upon addition of pancreatic DNase, ATP hydrolysis immediately ceases (10). Hence DNA must also be present to maintain ATPase activity. To test whether it is the restricted DNA (as opposed to residual unrestricted DNA) which is mediating the ATPase, an ATPase reaction was initiated using the circular, duplex fd replicative form DNA, then *recBC* enzyme was added. The *recBC* nuclease degrades the linear, restricted DNA, but leaves intact the residual circular, unrestricted DNA. Upon addition of *recBC* enzyme, the restriction ATPase again ceased (10). Hence the restricted DNA is required for maintaining ATP hydrolysis, even though it is not sufficient to initiate such hydrolysis.* Since the "initiation" of ATPase requires the presence of unmodified, unre-

---

*It is still possible that both unrestricted and restricted DNA must be present to maintain ATPase. This possibility is unlikely, however, since when high levels of enzyme are used, no inhibition of ATPase is observed, even though the DNA is at least 90% restricted.

stricted DNA, the enzyme apparently must first act as a nuclease in order to be able to catalyze the ATPase reaction which is dependent upon restricted DNA.

Significance of the ATPase

The results presented here indicate that the ATPase of the *E. coli* B restriction enzyme is clearly distinct from the DNase reaction. The obvious hypothesis that ATP is hydrolyzed in order to provide energy for finding the particular DNA sequence susceptible to the enzyme is ruled out: there is no ATP hydrolysis with modified DNA, and restricted DNA is required for the ATPase reaction. The ATPase activity does, however, reflect an altered state of the enzyme. The fact that the enzyme does not turn over as a nuclease, but does turn over as an ATPase, implies that after making an ATP-dependent scission in the DNA, the enzyme assumes a modified form which can no longer serve as a nuclease, but which is an active ATPase. Initially the enzyme cannot use restricted DNA as an ATPase cofactor, but after it has carried out restriction it is dependent upon restricted DNA to mediate ATP hydrolysis. We have not yet succeeded in determining whether the restricted DNA which is utilized is the exact molecule which the enzyme has cleaved, and, if so, whether the enzyme is irreversibly attached to this DNA. Although there is no obvious *in vivo* function for the ATP hydrolysis, it is presumably terminated upon further degradation of the DNA, perhaps by the *recBC* enzyme.

Conclusions

Although the *recBC* and the restriction enzymes have different functions and appear to catalyze quite different nuclease reactions, the results presented here show remarkable similarities between the ATPase activities associated with the two enzymes. In both cases ATP hydrolysis can proceed long after nuclease action has ceased. In addition the restriction enzyme does not turn over as a nuclease (9), and the *recBC* enzyme, although capable of hydrolyzing more than one DNA phosphodiester bond, is very reluctant to initiate degradation of a DNA molecule (12), perhaps indicating that it may not normally act upon more than one DNA molecule. Both ATPase activities require that the DNA be intact to maintain the ATPase reaction, even when it is not being simultaneously degraded. Finally, in both cases the DNA-dependent ATPase reactions are not accompanied by adenylylation or phosphorylation of the DNA cofactor (9, 12).

Both enzymes are multimeric proteins containing relatively large subunits. The *recBC* enzyme contains one each of two subunits of molecular weights approximately 140,000 and 128,000 (12). The restriction enzyme exists in at least two active forms, each of which possesses at least one each of three subunits of molecular weights 135,000, 60,000, and 55,000 (9). In both cases we are attempting to isolate the species of enzyme catalyzing the ATPase reactions after DNase action has ceased in order to (a) determine whether the DNA product is bound to the enzyme, and (b) observe whether all enzyme subunits are taking part in the ATPase reaction.

Neither enzyme appears to use ATP hydrolysis for locating a specific DNA site. The restriction enzyme finds the required DNA sequence and recognizes it (or rejects it, if modified) before massive ATP hydrolysis begins. Likewise, the recBC nuclease appears only to require an end to start DNA hydrolysis. Therefore a "tracking" function for the ATPase, either for finding a specific site on the DNA or for propelling the enzymes *during* DNA hydrolysis, appears unlikely. On the other hand, the requirement of the two ATPase activities for intact DNA could be taken to support a tracking mechanism with some unknown function.

An alternative hypothesis for the ATPase-mediated function would be putting the DNA into some special, unstable configuration for nuclease reaction. In the case of the restriction enzyme, the enzyme might be fooled: even though it has already broken the DNA, it might continue to support this modified configuration. In the case of the *recBC* enzyme in the presence of cross-linked DNA, the enzyme might be stalled at a cross-link, attempting to form the hypothetical structure. This latter function would not be simply the separation of the strands of a DNA duplex, since during the degradation of single stranded DNA the *recBC* enzyme requires ATP and degrades it to the same extent as with duplex DNA. Should the ATPase activities be a consequence of a modification of DNA secondary structure, it would be peculiar that both of the enzymes could be so easily "fooled" into the uncoupled ATPase reaction. (They have not been made to act on RNA, for example.) Hence it is possible that the ATPase activities are not manifestations of such simple functions as hypothesized here, but perhaps they reflect a new type of function which is unrelated to the nuclease action. One approach to understanding this function might be to search for factors in the cell which would control the ATPase reaction once the DNase has reached a limit.

References

1. Arber, W., and Linn, S. 1969. DNA modification and restriction. *Annu. Rev. Biochem.* 38: 467.

2. Anai, M., Hirahashi, T., Yamanaka, M., and Takagi, Y. 1970. A deoxyribonuclease which requires nucleoside triphosphate from *Micrococcus lysodeikticus*. *J. Biol. Chem.* 245: 775.

3. Bazil, G. W., Hall, R., and Gross, J. D. 1971. DNA synthesis in lysates of $RecB^-$ and Rec^+ *E. coli* cells. *Nature New Biol.* 233: 281.

4. Boyer, H. W. 1971. DNA restriction and modification mechanisms in bacteria. *Annu. Rev. Microbiol.* 25: 153.

5. Clark, A. J. 1971. Toward a metabolic interpretation of genetic recombination of *E. coli* and its phages. *Annu. Rev. Microbiol.* 25: 437.

6. Cole, R. S., and Zusman, D. 1970. Sedimentation properties of phage DNA molecules containing light-induced psoralen cross-links. *Biochim. Biophys. Acta* 224: 660.

7. Ebisuzaki, K., Behme, M. T., Senior, C., Shannon, D., and Dunn, D. 1972. An alternative approach to the study of new enzymatic reactions involving DNA. *Proc. Nat. Acad. Sci. U. S. A.* 69: 515.

8. Emerson, P. T., and Kohiyama, M. 1971. Effect of the mutations *recB* amd *polA* on γ-ray-induced DNA synthesis in a temperature-sensitive *dnaA* mutant of *Escherichia coli* K-12. *Nature New Biol.* 233: 214.

9. Eskin, B., and Linn, S. 1972. The DNA modification and restriction enzymes of *Escherichia coli* B. II. Purification, subunit structure, and catalytic properties of the restriction endonuclease. *J. Biol. Chem.* 247: 6183.

10. Eskin, B., and Linn, S. 1972. The DNA modification enzymes of *Escherichia coli* B. III. Studies of the restriction ATPase. *J. Biol. Chem.* 247: 6192.

11. Goldmark, P. J., and Linn, S. 1970. An endonuclease activity from *Escherichia coli* absent from certain rec^- strains. *Proc. Nat. Acad. Sci. U. S. A.* 67: 434.

12. Goldmark, P. J., and Linn, S. 1972. Purification and properties of the *recBC* DNase of *Escherichia coli* K-12. *J. Biol. Chem.* 247: 1849.

13. Karu, A. E., and Linn, S. 1972. Uncoupling of the *recBC* ATPase from DNase by DNA cross linked with psoralen. *Proc. Nat. Acad. Sci. U. S. A.* 69: 2855.

14. Linn, S., and Arber, W. 1968. Host specificity of DNA produced by *Escherichia coli*. X. *In vitro* restriction of phage fd replicative form. *Proc. Nat. Acad. Sci. U. S. A.* 59: 1300.

15. Meselson, M., and Yuan, R. 1968. DNA restriction enzyme from *E. coli*. *Nature* 217: 1110.

16. Monk, M., and Kinross, J. 1972. Conditional lethality of *recA* and *recB* derivatives of a strain of *Escherichia coli* K-12 with a temperature-sensitive deoxyribonucleic acid polymerase I. *J. Bacteriol.* 109: 971.

17. Moses, R. E., and Richardson, C. C. 1970. Replication and repair of DNA in cells of *Escherichia coli* treated with toluene. *Proc. Nat. Acad. Sci. U. S. A.* 67: 674.

18. Musago, L. 1967. Photochemical interaction between skin-photosensitizing furocoumarins and DNA. *In* Silini, G. (ed.) *Radiation Research (Proceedings of the Third International Congress of Radiation Research)*, pp. 803-812, John

Wiley and Sons, New York.
19. Nobrega, F. G., Rola, F. H., Pasetto-Nobrega, M., and Oishi, M. 1972. Adenosine triphosphatase associated with adenosine triphosphate-dependent deoxyribonuclease. *Proc. Nat. Acad. Sci. U. S. A.* 69: 15.
20. Smith, D. W., Schaller, H. E., and Bonhoeffer, F. J. 1970. DNA synthesis in vitro. *Nature* 226: 711.
21. Smith, H. O., and Friedman, E. A. 1972. An adenosine triphosphate-dependent deoxyribonuclease from *Hemophilus influenzae* rd. *J. Biol. Chem.* 247: 2854.
22. Smith, H. O., and Wilcox, K. W. 1970. A restriction enzyme from *Hemophilus influenzae*. I. Purification and general properties. *J. Mol. Biol.* 51: 379.
23. Vovis, G. F., and Buttin, G. 1970. An ATP-dependent deoxyribonuclease from *Diplococcus pneumoniae*. I. Partial purification and some biochemical properties. *Biochim. Biophys. Acta* 224: 29.
24. Wright, M., and Buttin, G. 1971. The isolation and characterization from *Escherichia coli* of an adenosine triphosphate-dependent deoxyribonuclease directed by *recB,C* genes. *J. Biol. Chem.* 246: 6543.
25. Winder, F. G. 1972. Role of ATP in ATP-dependent deoxyribonuclease activity. *Nature New Biol.* 236: 75.

Discussion

Fareed. We have examined products after treatment of replicating molecules of SV-40 DNA by *Escherichia coli* restriction endonuclease and have observed that cleavage occurs in the unreplicated region of the molecules, and that under nondenaturing conditions, an unusual dissociation reaction takes place in which the newly synthesized strands separate from the template strands. Would you have any comments about a possible mechanism for this latter phenomenon?

Linn. Well, that's kind of interesting, because I just received a preprint from Hamilton Smith's laboratory concerning an enzyme that seems to be an analogous enzyme to the *recBC* nuclease, and he proposes an unwinding function for the ATPase of this enzyme. We haven't found such a reaction with the *E. coli* enzyme, but, I guess, this would imply that the restriction enzyme might also have some sort of unwinding function or strand separation function.

Stodolsky. Have you looked to see whether the residue left, after the psoralen-treated DNA is digested as far as it will go with the *recBC* nuclease, has a single stranded end?

Linn. No, we're just going to start looking at susceptibility of the psoralen-treated-*recBC* nuclease-treated DNA with various nucleases after labeling the ends with kinase or with DNA polymerase.

Stodolsky. Are the covalently closed templates that have been used in these reactions supercoiled or relaxed?

Linn. We have tried both. We tried supercoiled, nicked, and gapped circles; all give basically the same result.

Involvement of dna Gene Products in the Conversion of ϕχ-174 and fd DNA to Replicative Forms by Extracts of *Escherichia coli*[*]

Reed B. Wickner, Michel Wright,
Sue Wickner, and Jerard Hurwitz

Department of Developmental Biology and Cancer
Albert Einstein College of Medicine
Bronx, New York 10461

ϕχ-174 and M13 (fd) single stranded circular DNAs are converted to their replicative forms by extracts of *Escherichia coli* pol A1 cells. We find that ϕχ DNA-dependent reaction requires Mg^{++}, ATP, and all four deoxynucleoside triphosphates, but not CTP, UTP, or GTP. This reaction also involves the products of the dna C, dna D, dna E (DNA polymerase III), and dna G genes, but not that of dna F. The *in vitro* conversion of fd single stranded DNA to the replicative form requires all four ribonucleoside triphosphates, Mg^{++}, and all four deoxynucleoside triphosphates. The reaction involves the product of gene dnaE, but not those of genes dna C, dna D, dna F, or dna G. The reaction with fd DNA is inhibited by rifampicin or antibody to RNA polymerase, while the reaction with ϕχ DNA is unaffected by either. Using the ϕχ DNA-dependent reaction, activities have been detected which specifically complement extracts of dna A, dna B, dna C, dna D, or dna G.

The conversion of M13 (fd) and ϕχ-174 DNA to their replicative forms by crude extracts of *Escherichia coli pol A1* strains was first described by W. T. Wickner et al. (2, 4, 5). They have shown that the products of

---

[*]This research was supported by grants from the National Institutes of Health (GM-1334-07) and the American Cancer Society (P-561B). R. B. W. is the recipient of a National Institutes of Health special fellowship, M. W. is a postdoctoral fellow of the Jane Coffin Childs Foundation, and S. W. is a trainee of the National Institutes of Health.

the φχ or fd DNA-dependent reactions are RFII and that both systems require DNA, Mg^{++}, and all four deoxynucleoside triphosphates (2, 4, 5). While the fd DNA-dependent system requires all four common ribonucleoside triphosphates (reference 5 and Table 1), we find that, using the same 0 to 40% ammonium sulfate fraction of the crude extract, the φχ DNA-dependent reaction requires only ATP of the ribonucleoside triphosphates. AMP, ADP, dATP, the methylene analogues of ATP, and elevated levels of other ribonucleoside triphosphates all failed to substitute for ATP.

As previously shown (5), the fd reaction was sensitive to rifampicin, while the φχ DNA-dependent activity was not. In addition, the fd DNA-dependent reaction was sensitive to antibody to RNA polymerase, but not to control γ-globulin, while the φχ DNA-dependent reaction was insensitive to both (Table 2). Thus, it seems unlikely that RNA polymerase is involved in the φχ reaction, although it is conceivable that it could be in a form inaccessible to rifampicin and antibody.

Schekman et al. (3) have shown that the dna A and dna B gene products are involved in the φχ DNA-dependent reaction, but not in the fd DNA-dependent reaction. We have extended these studies to other dna ts mutants.

The φχ DNA-dependent activity of an extract of dna G ts cells was thermolabile, on incubation at 30°C, in comparison with a temperature-insensitive revertant (Fig. 1). An equal mixture of mutant and rever-

Table 1. Requirements of φχ and fd DNA-dependent activities

Additions or omissions	φχ DNA-dependent activity (%)	fd DNA-dependent Activity (%)
Complete	100	100
− DNA	5	2
− Mg^{++}	4	4
− dATP, dCTP, dGTP	0	0
− dATP	1	
− dCTP	4	
− dGTP	1	
− ATP	0	10
− UTP, CTP, GTP	100	12

In reactions containing fd DNA the additions were as follows. Each reaction mixture (0.05 ml), containing 20 mM Tris chloride, pH 7.6, 10 mM $MgCl_2$, 4 mM dithiothreitol, 500 pmoles of fd single stranded circular DNA, 2 mM ATP, 0.08 mM each of UTP, GTP, and CTP, 0.04 mM each of dATP, dCTP, dGTP, and (α-^{32}P)dTTP(300 cpm/pmole), 0.5 units of RNA polymerase, 0.08 μg of Escherichia coli unwinding protein (B.M. Alberts, personal communication), and 50 μg of protein of a 0 to 40% ammonium sulfate fraction, was incubated for 20 min at 25°C. Acid-precipitable radioactivity was measured. E. coli unwinding protein was added prior to the addition of the ammonium sulfate fraction while RNA polymerase was added last. In experiments with φχ DNA, all additions were the same except that RNA polymerase and unwinding protein were omitted and fd DNA was replaced by an equal amount of φχ DNA. One hundred per cent activity in the case of φχ DNA-dependent activity was equivalent to the incorporation of 24 pmoles of dTMP; in the case of fd DNA, to 45 pmoles of dTMP.

Table 2. Effect of inhibition of RNA polymerase on $\phi\chi$ and fd *in vitro* DNA synthesis

Inhibitor	Activity (%)	
	$\phi\chi$	fd
None	100	100
Rifampicin (10 μg/ml)	138	8
Control γ-globulin (30 μg)	100	91
Anti-RNA polymerase γ-globulin (40 μg)	90	13

Crude extract of *Escherichia coli* (400 μg of protein) was incubated 15 min at 25°C with rifampicin, control γ-globulin, or anti-RNA polymerase γ-globulin. The reaction mixture as described in Table 1 was added and the incubation continued for 15 min at 25°C. One hundred per cent activity with $\phi\chi$ DNA was equivalent to the incorporation of 68 pmoles of dTMP, and with fd DNA, to 44 pmoles.

ROLE OF dnaG IN ϕx AND fd DNA-PRIMED REACTIONS

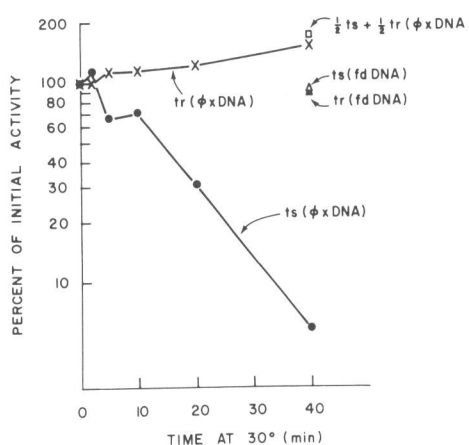

CONCLUSION:
dnaG PRODUCT INVOLVED IN ϕx-PRIMED REACTION
BUT NOT IN fd-PRIMED REACTION

Fig. 1. Heat inactivation of extracts of NY73 (pol Al, dna G ts) and a temperature-insensitive revertant of NY73. Extracts were prepared as described by W. T. Wickner *et al.* (5) except that the cultures were not chilled while harvesting the cells, and the frozen and thawed cells were lysed with 0.5% Brij-58 in place of the 37°C warming step. Mutant and revertant extracts were incubated at 30°C and at the times indicated, 0.03-ml aliquots were withdrawn and placed in assay tubes at 0°C. The other components of the incubation mixture were then added and all samples were assayed for activity at 20°C as described in Table 1 except that the ATP concentration was 0.6 m*M*. The mixing experiment was done with 0.015 ml of each extract. One hundred per cent activity for the extract of *Escherichia coli* strain NY73 was 22 pmoles of dTMP with $\phi\chi$ DNA and 48 pmoles with fd DNA; for the temperature-insensitive revertant of NY73, 100% activity was 48 pmoles with $\phi\chi$ DNA and 43 pmoles with fd DNA. All values represent rates of synthesis since incorporation under the conditions was essentially linear for 60 min. ●——●, $\phi\chi$ DNA-dependent activity in extracts of dna G ts cells; ×——×, $\phi\chi$ DNA-dependent activity in extracts of revertant cells; □, $\phi\chi$ DNA-dependent activity in an equal mixture of extracts of dna G ts cells and revertant cells; △, fd DNA-dependent activity in extract of dna G ts cells; ▲, fd DNA-dependent activity in extract of revertant cells.

tant extracts showed about twice the expected activity, suggesting that the dna G product may be present in excess in the revertant extract. The fd DNA-dependent activity, however, showed no differential thermolability between mutant and revertant in the same experiment.

Similar experiments comparing extracts of dna F ts cells with extracts of either the wild type or a revertant failed to show any differential thermolability of either the φχ or fd DNA-dependent reactions in the mutant.

Using extracts of a dna E ts strain (Fig. 2), differential thermolability of both the φχ and fd DNA-dependent activities was observed in comparison with either the wild type or a temperature-insensitive revertant. Thus, DNA polymerase III, which has been shown by Gefter *et al.* (1) to be the product of the dna E gene, appears to be involved in both the φχ and fd DNA-dependent reactions. The time delay in appear-

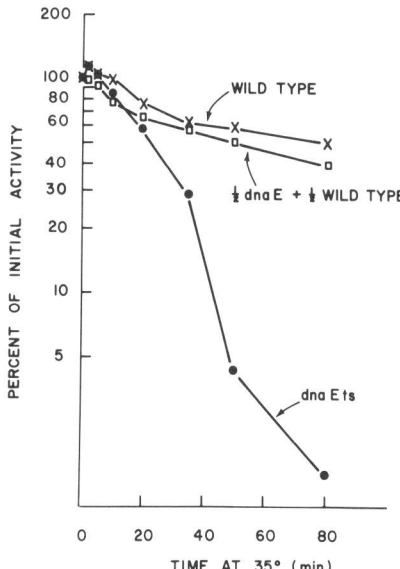

Fig. 2. Heat inactivation at 37°C of extracts of *Escherichia coli* strain BT1026 (pol Al, end dna E ts) and strain H560 (pol Al, end). Experimental details are as in Fig. 1. One hundred per cent activity for the BT1026 extract was 14 pmoles of dTMP incorporated per 20 min with φχ DNA and 16 pmoles with fd DNA; for H560, 100% activity was 15 pmoles with φχ DNA and 17 pmoles with fd DNA.

ance of this temperature sensitivity (Fig. 2) and the more than additivity of the mixing experiment suggest the initial presence in both extracts of an excess of DNA polymerase III with respect to its requirement in these reactions.

Extracts of dna C or dna D ts cells showed normal levels of the fd DNA-dependent activity, but showed no φχ DNA-dependent activity (Fig. 3). Extracts of temperature-insensitive revertants of either dna C or dna D showed normal amounts of both activities. Pending further studies with other alleles of these genes, we tentatively conclude that the dna C and dna D gene products are involved in the φχ DNA-dependent reaction, but not in the fd DNA-dependent reaction.

We have obtained complementation in the φχ DNA-dependent system by mixing extracts of dna G ts cells with extracts of wild type cells and by mixing extracts of dna B ts cells with extracts of either dna C or D ts cells. Under the conditions used, these extracts were inactive when assayed alone. Using complementation assays, we have begun to purify some elements of the φχ DNA-dependent system.

In summary (Table 3), our results, and those of Wickner *et al.* (5) and Schekman *et al.* (3), indicate that while the fd DNA-dependent reaction requires the activity of RNA polymerase and DNA polymerase

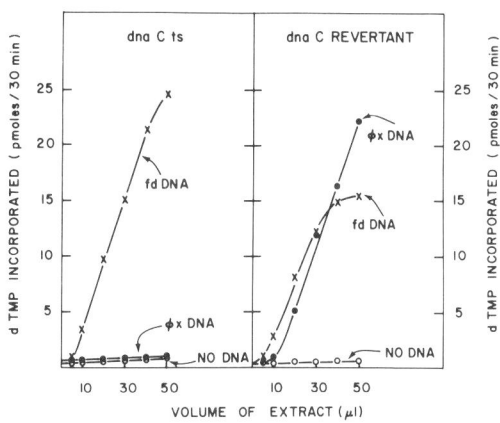

Fig. 3. φχ and fd DNA-dependent activity of extracts of *Escherichia coli* PC22 (pol Al, dna C ts) and a temperature-sensitive revertant. Extracts of each strain were prepared as described in Fig. 1 and tested for activity with φχ DNA (●——●), fd DNA (×——×), or without DNA (○——○).

Table 3. Summary

Component	φχ DNA reaction	fd DNA reaction
RNA polymerase	−	+
dna A	+	−
dna B	+	−
dna C	+	−
dna D	+	−
dna E	+	+
dna F	−	−
dna G	+	−

+, required; −, not required.

III, the φχ DNA-dependent activity does not involve RNA polymerase and does require the activity of the products of dna A, B, C, D, E, and G. The role of these gene products in the synthesis of φχ DNA will be ascertained by further purification of this system.

Acknowledgments

We wish to thank Doctors J. Wechsler, M. Gefter, Y. Hirota, P. Carl, and F. Bonhoeffer for making available the bacterial strains used in this study. We are grateful to Doctors W. T. Wickner, R. Schekman, D. Brutlag, and A. Kornberg for communicating their results prior to publication.

References

1. Gefter, M. L., Hirota, Y., Kornberg, T., Wechsler, J. A., and Barnoux, C. 1971. Analysis of DNA polymerases II and III in mutants of *Escherichia coli* thermosensitive for DNA synthesis. *Proc. Nat. Acad. Sci. U.S.A.* 68: 3150.
2. Schekman, R. 1972. Replication of M13 DNA in extracts of infected *E. coli*. *Fed. Proc.* 31: 442.
3. Schekman, R., Wickner, W. T., Westergaard, O., Brutlag, D., Geider, K., Bertsch, L. L., and Kornberg, A. 1972. Initiation of DNA synthesis. III. synthesis of φχ-174 replicative form requires RNA synthesis resistant to rifampicin. *Proc. Nat. Acad. Sci. U.S.A.* 69: 2691.
4. Wickner, W. T., Brutlag, D., and Kornberg, A. 1972. The role of RNA priming in M13 DNA synthesis. *Fed. Proc.* 31: 441.
5. Wickner, W. T., Brutlag, D., Schekman, R., and Kornberg, A. 1972. RNA synthesis initiates *in vitro* conversion of M13 DNA to its replicative form. *Proc. Nat. Acad. Sci. U.S.A.* 69: 965.

Factors for DNA Synthesis

Factors Stabilizing
DNA Folding in Bacterial Chromosomes

David E. Pettijohn, Ralph M. Hecht,
O. G. Stonington, and T. D. Stamato

Department of Biophysics and Genetics
University of Colorado Medical Center
Denver, Colorado 80220

A procedure for isolating the nucleoid of *Escherichia coli* is described. The DNA in this structure has a large molecular weight (probably in excess of 10^9 daltons) but it is folded into a compact particle sedimenting at 1600 S. No membranes can be detected in association with the purified body and the protein bound to it is predominantly the RNA polymerase subunits α, β, and β'. The nascent RNA chains of the bacteria, coupled to the bound RNA polymerase, are isolated with the nucleoid. A modification of the isolation procedure, involving more gentle treatment with detergents, isolates a membrane-associated nucleoid sedimenting at about 3200 S. Analysis of the membrane proteins from the isolated nucleoid indicates that the bound membrane is not a random sample of total cellular membrane but rather a specific fraction.

Treatment of either of the isolated nucleoids with RNase or sodium dodecyl sulfate (SDS) causes the DNA to unfold. The conformation change is manifested by a large increase in relative viscosity, a decrease in sedimentation rate of the DNA, and the appearance of the structure under the electron microscope. The nucleoids can be fixed with formaldehyde, after which the DNA cannot be unfolded with SDS but is still unfolded by RNase. These and other observations on the unfolding process have led to several tentative conclusions concerning the RNA and probably protein components which stabilize DNA folding.

Folded DNA as isolated in the nucleoids is an excellent template for added RNA polymerase, whereas unfolded DNA isolated by conventional procedures or unfolded from the nucleoid is less efficient as a template. The total template capacity of folded DNA (measured at near-saturating RNA polymerase to DNA ratios) is at least as great as unfolded E. coli DNA or purified T_4 DNA, suggesting that nearly all potential template in the tightly folded DNA is accessible to interaction with large macromolecules.

An RNA fraction which is tightly bound to the DNA of the nucleoid body has been isolated. Agents which completely disrupt the binding to DNA of nascent RNA chains (in ternary complexes) do not disrupt the DNA binding of this RNA. The RNA is dissociated from DNA by heating. There is a possibility that the RNA fraction is involved in stabilizing DNA folding, although there is no conclusive evidence to support this idea.

The conformation of double helical DNA as it exists in the cell is distinct from that of DNA as it is usually isolated for *in vitro* studies. Purified DNA in solution is extended in highly asymmetrical fibers, while DNA in the cell is folded into compact structures the organization of which is poorly understood. In the bacterium *Escherichia coli*, for example, the DNA unit of 2.5×10^9 daltons (about 1100 μ in length) is condensed into a nucleoid body having a large dimension of about 0.5 μ (5, 8). The relative dimensions suggest that the DNA molecule (or molecules) composing the nucleoid must be folded back on itself a minimum of 2000 times or about 1-fold per 1×10^6 daltons of DNA. In this compact state DNA acts as a template for RNA synthesis and it is from this state that the DNA must be dissociated for DNA replication and the accompanying segregation and recondensation of the daughter nucleoids (5, 7). It seems likely that such structural reorganizations of folded DNA occurring as DNA is synthesized are intimately coupled to the machinery of DNA replication.

Investigation of the molecular interactions which organize and stabilize DNA packaging is only just beginning. Such studies have been hindered, until recently, by the lack of a simple experimental system. We describe here a procedure for isolating bacterial DNA in a folded, compact structure resembling the nucleoid. Also described are the molecular composition of the isolated body and investigations of the component (or components) which stabilize its folded conformation.

Isolation of Nucleoids and Membrane-associated Nucleoids

At physiological pH, DNA is a negatively charged polymer, so that condensed, folded conformations are destabilized by charge repulsion, *i.e.* charge repulsion is minimal when DNA assumes the conformation of a straight, unfolded rod. In the cell such repulsions are reduced by counter ions associated with the condensed DNA; di- and monovalent cations, polyamines, and certain basic proteins probably provide this ionic shield *in vivo*. The usual methods for isolating DNA lead to dilution and extraction of the natural counterions of the cell and would, therefore, be expected to promote electrostatic destabilization of compact DNA. The procedure described here (Fig. 1) avoids this condition by providing high concentrations of monovalent salts during the cell lysis and during the subsequent purification steps. The lysis conditions described in Fig. 1 lead to efficient cell disruption of many different bacterial species, but the lysates do not become viscous, indicating that the DNA has

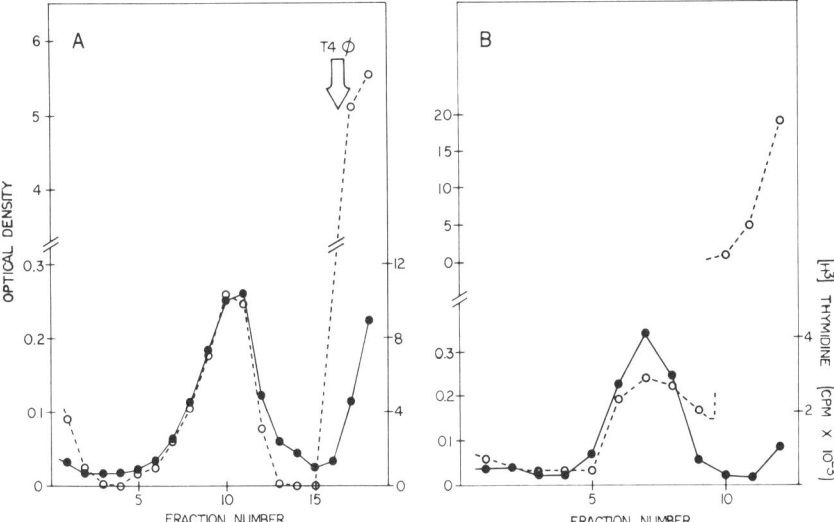

Fig. 1. Isolation of the nucleoid by centrifugation on a sucrose gradient. *A*, A 20-ml culture of *Escherichia coli* (strain D-10, about 2 × 10⁸ cells/ml) which had been labeled with [³H]thymidine for about one generation time was quick chilled to 0°C and the cells were rapidly harvested by a brief centrifugation at 4°C. The bacteria were resuspended in 0.2 ml of a cold solution containing 0.01 M Tris (pH = 8.1 at 4°C), 10 mM sodium azide, 20% w/v sucrose, and 0.10 M NaCl; then 0.05 ml of a solution containing 50 mM EDTA, 4 mg of lysozyme per ml, 0.12 M Tris was added. After 30 sec at 0°C, 0.25 ml of 1% Brij-58-0.4% deoxycholate-2.0 M NaCl-10 mM EDTA was added. The mixture was held at 0°C for several minutes until it had cleared considerably, but had not become more viscous. Debris was removed from the lysate by centrifugation for 2 to 5 min at 4°C and 4,000 × g. The supernatant was layered on a 10 to 30% w/v sucrose gradient containing 0.01 M Tris (pH = 8.1), 1.0 M NaCl, 1 mM EDTA, and 1 mM β-mercaptoethanol and centrifuged for 9 min at 17,000 rpm and 4°C in an SW 50 rotor of a Beckman ultracentrifuge. In a separate identical gradient, marker T₄ phage (s = 1025 S) was run at the same time; the position of the phage peak is indicated by the arrow. Direction of sedimentation was right to left. *B*, In a separate experiment the procedure was as described in A except that lysis in the Brij-dexoycholate-NaCl mixture was for 4 to 5 min at 20°C, debris was not removed, and centrifugation was at 17,000 rpm for 26 min. ●——●, ³H radioactivity; ○- - -○, optical density at 260 nm (note the discontinuities in the optical density scales). The small amount of radioactivity remaining at the top of the gradients is due predominantly to unincorporated thymidine, *i.e.* it is mostly acid-soluble.

not been unfolded into highly asymmetrical conformations. The condensed DNA has a large and nearly homogeneous sedimentation rate and it can be easily separated by centrifugation from other components of the crude lysate.

The two variations of the isolation procedure described in Fig. 1 differ in the temperatures at which the cells are incubated with the nonionic detergents during lysis. When the temperature is kept at 0-4°C (Fig. 1A) the resulting DNA complex sediments at about 3200 S relative to marker T_4 phage and includes a larger fraction of the total UV-absorbing components of the lysate. When the detergents act on the complex at 20°C, much UV-absorbing material is stripped away from the folded DNA and its sedimentation rate is reduced to 1600 S (Fig. 1B). The removed material has the properties of bacterial membrane, as will be described below. For this reason we refer to the two structures as the membrane-associated nucleoid and the nucleoid, respectively. Further details of the isolation procedure and precautions to be observed have been previously published (13, 21).

Molecular Components of the Isolated Nucleoid

In previous experiments (21) the membrane-associated nucleoid was isolated under conditions designed to minimize even further the possibility of deoxyribonuclease and ribonuclease activities during isolation. The DNA in these isolated structures was then unfolded (by methods described below) and sedimented at concentration <0.3 µg/ml on sucrose gradients under conditions described by Burgi and Hershey (3). The sedimentation rate of the DNA was approximately 2.2 times the rate of an internal marker intact T_4 DNA. Assuming that the emperical Burgi-Hershey relationship (3) can be extrapolated to higher molecular weights, this suggested a molecular weight in excess of 10^9 daltons, a figure approaching the size of the intact *E. coli* genome.

The two forms of the nucleoid were isolated from bacteria which had grown for many generations in media containing radioactive labeled amino acids, ^{14}C-oleic acid or [5-^{3}H]uridine. Total RNA and protein were also isolated from the bacteria to determine specific radioactivities. From these figures and the amounts of radioactivity associated with the isolated nucleoids, it was possible to estimate the relative protein and RNA contents of the bodies (after making minor corrections for the small amount of measured cross-over of the label into DNA). The results of these experiments (22) and other experiments on the molecular composition of the nucleoid are summarized in Table 1. As expected from the data of Fig. 1, the membrane-associated nucleoid has more bound protein and RNA (13) than the membrane-free structure. We refer to the latter body as "membrane-free," since little or no ^{14}C-oleic acid-labeled moieties (Table 1) or membrane proteins can be detected in association with it (Fig. 2).

Previously we showed (14, 21) that nearly all of the DNA-independent RNA polymerase activity of the crude lysate cosediments with the nucleoid and that these polymerase molecules have nascent RNA chains associated with them. We have estimated the number of polymerase molecules bound per genome equivalent of DNA (14) and the average size of the associated nascent RNA (13, 14); from these numbers it can be calculated that the amount of nascent RNA accounts for nearly all of the RNA associated with the nucleoid.

Proteins of the nucleoid have been separated (21) from the nucleic acid and analyzed by polyacrylamide gel electrophoresis (Fig. 2). The major protein components are those of the α and ββ' subunits of RNA polymerase. No σ factor was detected and the only other significant band was that due to an unknown protein of slightly greater mobility than α. The proteins of the nucleoid, then, are predominantly RNA polymerase; the level of purity of the polymerase with respect to other protein contaminants is comparable to the final product of many procedures for RNA polymerase purification.

Proteins from the membrane-associated nucleoid have also been isolated. The results of polyacrylamide gel electrophoresis of the sodium dodecyl sulfate-dissociated proteins are shown in Fig. 3. For comparison, total membranes have been isolated from the same bacterial strain, dissociated in the same manner, and electrophoresed under identical conditions. All of the major protein bands from the nucleoid complex have an equivalent band in total membrane proteins; however, one should

Table 1. Properties of the isolated nucleoid and the membrane-associated nucleoid

	Membrane-associated nucleoid	Nucleoid
Approximate sedimentation rate	3200 S	1600 S
Fraction of OD_{260} of purified nucleoid due to DNA (not corrected for scattering)	0.2	0.7
Protein fraction (relative to DNA)	~1.0	<0.1
RNA fraction (relative to DNA)		0.4
Polyamine fraction (relative to DNA)	<0.008	
Fraction of total incorporated ^{14}C-oleic acid bound to nucleoid	0.18	<0.01
Fraction of DNA-independent RNA polymerase activity bound to nucloid	>0.75	>0.75
Fraction of total OD_{260} of crude lysate associated with nucleoid	~0.1	~0.03

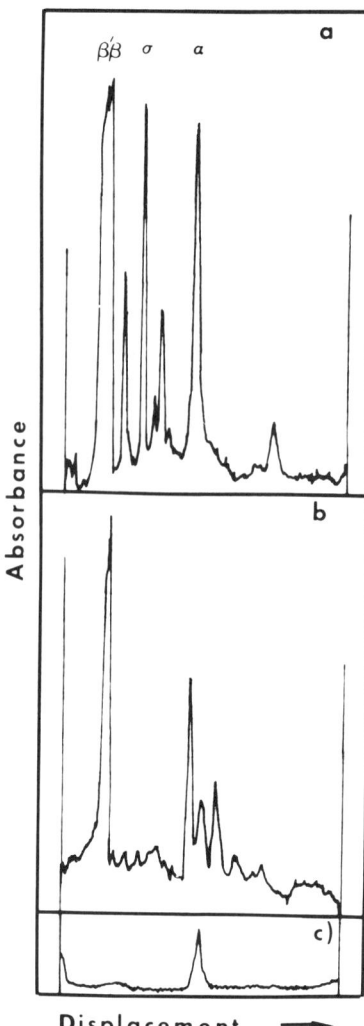

Fig. 2. Polyacrylamide gel electrophoresis of proteins isolated from purified nucleoid. A, RNA polymerase purified by the double gycerol gradient method of Burgess (2). Minor bands between β and σ and between σ and α are contaminants. B, Proteins purified from nucleoids which had previously been isolated as in Fig. 1B. The minor band, immediately to the right of α, is pancreatic DNase added to degrade the DNA of the nucleoid. C, Pancreatic DNase marker. The three different gels containing 0.1% SDS were prepared and run at the same time under identical conditions (21), stained with Coomassie blue, destained with trichloroacetic acid, and scanned with a recording spectrophotometer. Traces of the actual scans are shown above.

note that the relative proportions of the proteins differ in the membrane-associated nucleoid and in total membrane protein. Also some of the components present in total membrane may be absent in the membrane associated with nucleoid body. The last two observations suggest that the membrane fraction bound to the nucleoid body is not a random sample of the total bacterial membrane but rather a specific membrane fraction. This could be of significance since it is believed that the site of DNA replication is on the bacterial membrane (6, 18, 20). The amounts of RNA polymerase activity bound to the nucleoid and the membrane-associated nucleoid are roughly the same; however, bands corresponding to the subunits of RNA polymerase are not resolved in Fig. 3 since they are obscured by backgrounds due to other more abundant protein species. In visual observation of the stained gels, bands corresponding to the ββ' subunits of RNA polymerase can be observed.

The evidence that the molecules bound to the "membrane-associated nucleoid" and not found with the slower sedimenting "nucleoid" are membrane in properties is as follows: (a) a significant fraction of ^{14}C-oleic acid incorporated into bacteria cosediments with the membrane-associated nucleoid but is absent from the more purified nucleoid preparations (Table 1); (b) most of the additional proteins bound to the membrane-associated nucleoid but absent from the nucleoid have electrophoretic mobilities indistinguishable from total E. coli membrane proteins (Fig. 3); (c) ^{32}P-phospholipid is found in the membrane-associated nucleoid (22).

The following clarification is submitted for readers who will compare the protein analyses of Figs. 2 and 3 with our previously published work (13, 21). In our earlier studies we were unaware of the importance

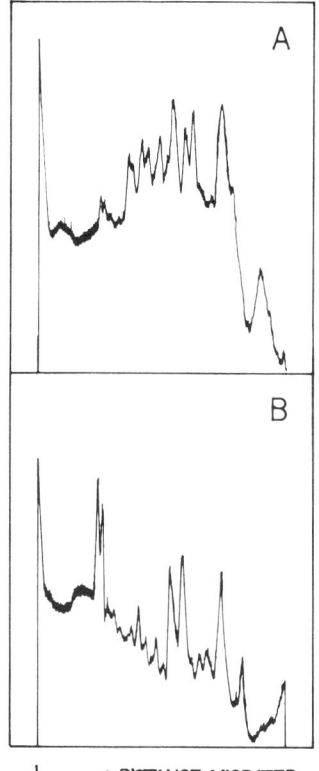

Fig. 3. Polyacrylamide gel electrophoresis of proteins isolated from the purified membrane-associated nucleoid. The nucleoid complexes, isolated by a scaled-up version of the procedure described in Fig. 1A, were degraded with pancreatic DNase and RNase. The degradation products were separated from proteins by passage through a Sephadex G-25 column; fractions containing proteins were pooled, lypholized, and redissolved in buffer containing 0.1% sodium dodecyl sulfate-0.14 M β-mercaptoethanol-0.01 M sodium phosphate, pH = 7.2. After heating for 10 min at 65°C and cooling to 20°C, about 30 μg of the reduced protein was electrophoresed for 20 volt-hrs/cm on 5% polyacrylamide gels containing sodium dodecyl sulfate as described by Shapiro et al. (19). A, Total membrane proteins. Membranes were isolated from strain D-10 by the procedure of Evans (4), then treated with Brij and deoxycholate under conditions described in Fig. 1A and collected by centrifugation. They were then lyophilized and treated in parallel with proteins from the nucleoid complex. B, Proteins isolated from the membrane-associated nucleoid. The two major bands at high molecular weight are not the ββ$^-$ bands of RNA polymerase; their mobility is greater than these polymerase subunits and they are also present in relatively lower proportions in total membrane protein.

of temperature regulation during lysis. Some of our prior reports have therefore described a procedure which isolates the membrane-associated nucleoid while describing proteins isolated from nucleoids which were prepared at higher lysis temperatures.

DNA of the Nucleoid Is Folded and the Conformation Is Stabilized by Other Molecules

Previously we reported that solutions containing moderate concentrations of isolated nucleoids have viscosities no greater than the solvent alone (21). After incubation with RNase or ionic detergents (sodium dodecyl sulfate) or a brief thermal treatment there was a large increase in viscosity. Similarly a decrease in sedimentation rate was noted (more then 10-fold) after exposure to the reagents. These changes occurred without apparent reduction of DNA molecular weights; thus they are attributed to conformation changes. Calculations based on the sedimentation rate of the nucleoid suggest that the DNA is packaged in particles of about the size of a nucleoid body (21) and that the density of the DNA is very high, although less than the density of DNA inside phage heads. Similarly, electron micrographs of the isolated structures (Fig. 4) show electron-dense particles of about the size of nucleoids. Frequently, preparations of the membrane-free nucleoids have fibers and bundles of fibers protruding from them which are probably DNA. All of our studies to date support the conclusion that the DNA of the isolated nucleoid is tightly folded into a conformation grossly resembling its *in vivo* condition.

Reagents that act to unfold the DNA are agents that are not known to have direct effects on DNA structure, but rather are more specific to the RNA or protein components of the nucleoid. Thus it seems that other macromolecules are involved in the stabilization of DNA conformation. It is almost a truism in polymer chemistry that interactions between macromolecules or between parts of a macromolecule that stabilize structure are also the interactions which *organize* structure. One might therefore speculate that the interactions which stabilize folding of DNA and perhaps also organize the folding come into play as DNA is replicated and are closely coupled to replication.

Both forms of the isolated nucleoid can be fixed by a brief incubation at 20°C with low concentrations of formaldehyde. In this state the DNA can no longer be opened with sodium dodecyl sulfate (SDS); however, as judged by viscosity and sedimentation properties, the sensitivity to ribonuclease remains. As shown in Fig. 5, the sedimentation of fixed nucleoids is unaffected by SDS treatment, but the DNA of the fixed nucleoid does not sediment under these conditions after the nucleoids

FOLDED BACTERIAL DNA 153

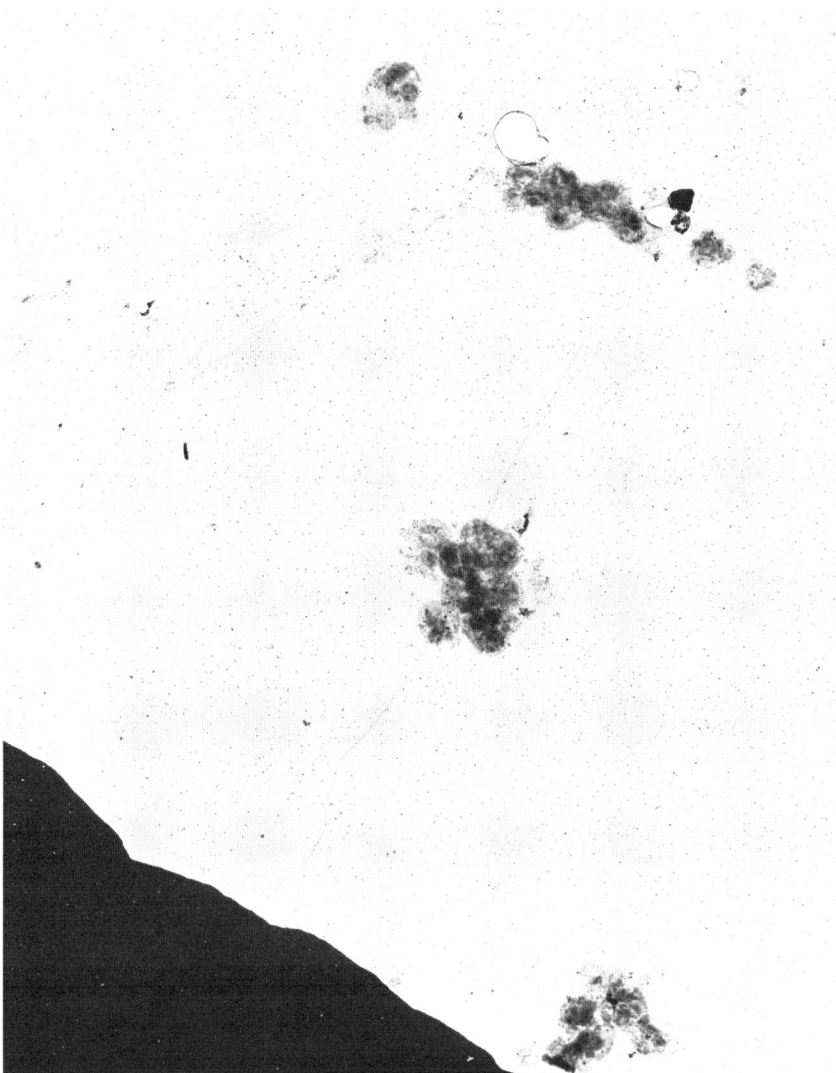

Fig. 4. Electron micrograph of isolated nucleoid. A nucleoid preparation made as described in Fig. 1B was prepared for electron microscopy as described by Miller *et al.* (12). We are grateful to Barbara Hamkalo for her cooperation in obtaining this electron micrograph.

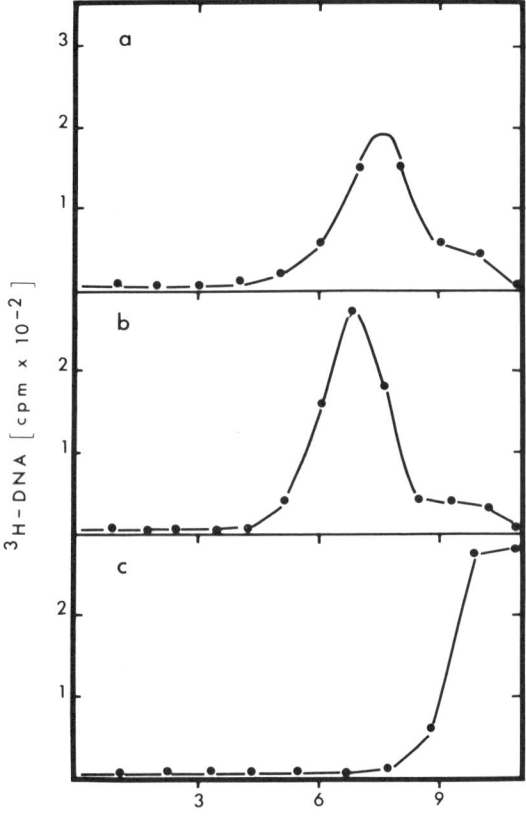

Fig. 5. Sedimentation on sucrose gradients of nucleoid after fixation by formaldehyde. Membrane-free nucleoids were isolated as described in Fig. 1 except that 0.01 M phosphate buffer, pH = 7.8, and 1 mM sodium citrate were substituted for the Tris buffer and EDTA in the sucrose gradient. The preparation was then incubated for 5 min at 20°C in the presence of 0.3% formaldehyde and the formaldehyde was removed by dialysis at 2°C against a solution containing 0.01 M sodium phosphate, pH 7.8, 0.1 M NaCl. Three identical aliquots of the fixed nucleoids were incubated for 20 min at 34°C in the presence of (A) no additions, (B) 0.5% sodium dodecyl sulfate, (C) 20 μg/ml of pancreatic RNase. These were then layered on three sucrose gradients made up as described in Fig. 1 except that the NaCl concentration was 0.1 M, the gradient was 5 to 20% sucrose at 22°C and, centrifugation was at 20,000 rpm, 22°C for 9 min.

are incubated with RNase. The change is not attributable to reductions in the size of the DNA; in control experiments it was shown that DNA from unfixed nucleoids sediments (after opening by SDS treatment) at the same rate as DNA from fixed nucleoids opened by RNase treatment (R. Hecht, unpublished result). This experimental result, when combined with previously published results as reviewed above, leads to the follow-

ing tentative conclusions concerning the properties of the nucleoid components which stabilize DNA folding.
1. More than one type of molecule bound to the nucleoid is essential for stabilizing folded DNA.
2. The binding of one kind of essential molecule to the folded DNA is disrupted by interaction with SDS; however, this molecule can be fixed to the nucleoid by formaldehyde. (The properties of the molecule are consistent with it being a protein although this is not required.) The binding of this molecule by itself is insufficient to stabilize the DNA folding.
3. The other kind of essential molecule is sensitive to RNase. To maintain DNA folding, its integrity, as well as that of the molecule mentioned above, is required.

The above statements rest upon the assumption that SDS in low concentrations does not interfere directly with nucleic acid-nucleic acid binding. Support for this assumption comes from observations that SDS at these concentrations does not reduce rates of RNA-DNA hybridization or significantly destabilize the DNA double helix.

Folded DNA Is an Efficient Template for RNA synthesis

If the folded DNA structure we have isolated bears any relation to the natural conformation of DNA, one might expect the isolated nucleoid to be an active template for RNA synthesis *in vitro*. It has been recognized for some time that *E. coli* DNA, as it is usually isolated, is a poor template for *E. coli* RNA polymerase compared to phage DNAs or even DNA isolated from mammalian viruses (7, 23). Template efficiency is defined here as the amount of RNA synthesized per RNA polymerase molecule per unit time per unit weight of DNA under constant conditions of synthesis.

Figure 6 records a comparison of T_4 DNA with folded and unfolded *E. coli* DNA as templates for purified RNA polymerase in RNA synthesis. The unfolded DNA, having a molecular weight larger than T_4 DNA (21), does not support as much RNA synthesis as similar amounts of folded DNA or T_4 DNA using constant amounts of RNA polymerase. The folded DNA, however, seems to be comparable in template efficienty to T_4 DNA over a range of RNA polymerase to DNA weight ratios. It is difficult to determine with certainty how much of the improved template efficiency of folded DNA is attributable to DNA structure and how much is attributable to the presence of unknown proteins which may be bound to the DNA. RNA synthesis due to endogenous polymerase bound to the folded DNA has been corrected for in the data of Fig. 6; however, other proteins bound to the DNA may have been inactivated by the brief heat treatment used to open the folded DNA. For this reason there could be differences in the folded and unfolded DNA other than a DNA conformation change. The experiment of Fig. 6 has been repeated using membrane-free nucleoids as template with no significant change in results. Since there is very little protein associated with the DNA in this case other than RNA polymerase (see Fig. 2), it seems that the improved template efficiency of the folded DNA may be, at least in part, attributable to DNA structure.

Whatever the basis for the improved template efficiency of folded DNA, it seems clear from the data of Fig. 6 that the template capacity (determined at near-saturating polymerase to DNA ratios) of folded DNA is not reduced from unfolded DNAs. This implies that most, if not

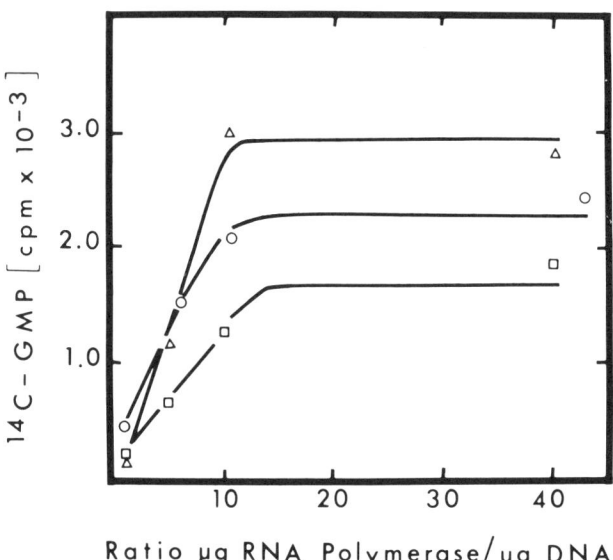

Fig. 6. Folded DNA as a template for RNA synthesis *in vitro*. Membrane-associated nucleoid was purified as described in Fig. 1A. The concentration of DNA in this preparation was determined by purifying the DNA from an aliquot; radioactive DNA was added to the aliquot before purification to control DNA losses internally. A series of RNA polymerase assays were made, each containing constant amounts of DNA but with varied amounts of added purified RNA polymerase. The assays containing ^{14}C-GTP were made as previously described (15); they were incubated 10 min at 37°C, the reaction was stopped and the amount of ^{14}C-RNA product was determined (15). Assays containing nucleoid were corrected for the RNA synthesis which occurred in the absence of added polymerase due to endogenous enzyme; at the maximum polymerase-to-DNA ratio this correction amounted to less than 10%. △——△, Nucleoid, folded DNA template; ○——○, T_4 DNA template; □——□, nucleoid, DNA template unfolded by heating 60°C for 5 min prior to the RNA synthesis.

all, of the tightly condensed DNA in the nucleoid is freely accessible to RNA polymerase. That is, the tightly packaged DNA structure appears to be organized in such a way that even the DNA sequences removed from its surface (which must include most of the linear extent of the genome) can interact freely with large macromolecules which were initially external to the nucleoid. This characteristic of the folded DNA is compatible with the *in vivo* situation, where studies of the rates at which different operons can be regulated and transcribed have also suggested that these operons are freely available to other macromolecules (9, 11).

RNA Fractions Bound to DNA

As described above, the effect of RNase on fixed and unfixed folded DNAs has suggested that RNA bound to the nucleoid stabilizes the folded conformation. Further support for this view has come from recent experiments where it was found that folded DNA could not be isolated from cells grown with rifampicin; at least part of the DNA unfolds during lysis or during the first purification step (Pettijohn, unpublished results). Details of these experiments will be described elsewhere.

There is at the onset a basic question. Do the total nascent RNA chains of the nucleoid somehow stabilize DNA folding or is a special class of RNA exclusively involved? In either case one might anticipate a novel interaction between RNA and DNA. For this reason we have begun a search for RNA components of the nucleoid bound to DNA through mechanisms not directly involving RNA polymerase.

Membrane-free nucleoids with ^{3}H-labeled RNA components and ^{14}C-labeled DNA have been purified; the DNA was unfolded by SDS treatment and sedimented on sucrose gradients to separate the high molecular weight DNA from the released RNA (Fig. 7). Most of the RNA sedimented free of the DNA at a slower rate; however, about 4% of the labeled RNA cosedimented with the DNA. Another aliquot of the unfolded nucleoid preparation was centrifuged to equilibrium in a CsCl gradient (Fig. 8); again a small fraction of the RNA banded with the DNA. The DNA:RNA ratios at the peaks of these two DNA-RNA bands were approximately the same, as indicated by the ^{3}H:^{14}C ratio; 1.5 for the sucrose gradient band and 1.6 for the CsCl gradient band. The ^{3}H radioactivity associated with the DNA is sensitive to RNase (see Fig. 7*b* for example, ^{3}H:^{14}C ratio reduced to 0.2) and is 70 to 90% degraded by alkali (R. Hecht, unpublished result). The residual resistance to RNase and alkali may be attributable to a small amount of cross-over label in the DNA.

The conditions of incubation with SDS should be sufficient to disrupt the ternary complexes, nascent RNA-RNA polymerase-DNA. This has

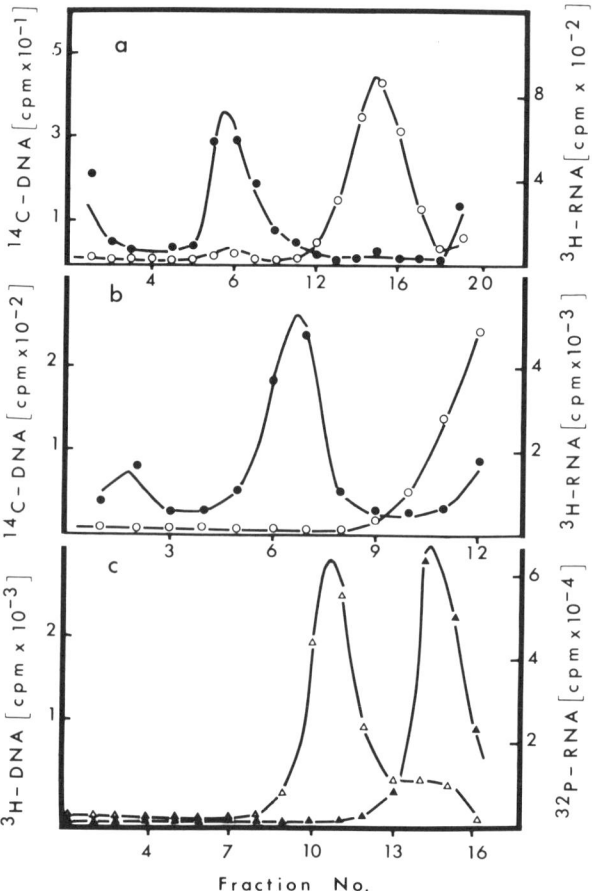

Fig. 7. Sedimentation in sucrose gradients of unfolded DNA with a bound RNA fraction. *A*, Strain D-10 was grown for one generation time in the presence of [^{14}C]thymidine and then labeled for 1 min with [5 – ^{3}H]uridine (2 μCi/ml, 18 Ci/mM). The cells were chilled to 0°C and nucleoids were isolated as described in Fig. 1B. An aliquot of the preparation was diluted into a solution containing 0.1 M NaCl, 0.01 M Tris (pH = 7.8), 1 mM EDTA, and the DNA concentration was reduced to less than 0.5 μg/ml; SDS was added to a final concentration of 0.5%; the mixture was incubated 1 hr at 34°C and layered with a large bore pipette onto a 5 to 30% sucrose gradient containing 0.1 M NaCl, 0.01 M Tris (pH = 7.8), 1 mM EDTA, and 0.5% SDS. The unfolded DNA was centrifuged 6.5 hr at 20,000 rpm in an SW 25.3 rotor at 22°C. Radioactivity in the gradient fractions was counted under conditions which discriminate between ^{3}H and ^{14}C. *B*, An aliquot of the above nucleoid preparation was diluted similarly and incubated 20 min at 34°C with 50 μg/ml of pancreatic RNase before adding SDS as in *A* above. This was layered on another sucrose gradient containing 0.5% SDS and centrifuged in a different rotor (SW 50) to separate the DNA from degraded RNA. The amount of radioactivity measured in *B* is greater than *A*, since complete gradient fractions were counted in *B* but not in *A*. ●——●, ^{14}C-DNA; ○——○, ^{3}H-RNA. *C*, Another nucleoid preparation made as *A* above except labeled only with [^{3}H]thymidine was diluted as in *A* except the final DNA concentration was about 10-fold greater. To this solution 1.5 × 10^5 cpm of ^{32}P-labeled RNA were mixed and allowed to equilibrate for 30 min at 0°C. The ^{32}P-RNA was total *E. coli* RNA purified from D-10 cells which had been pulse-labeled with ^{32}P-orthophosphate. The nucleoids were then treated with SDS as above and the unfolded DNA was sedimented as above on a sucrose gradient. △——△, ^{3}H-DNA; ▲——▲, ^{32}P-RNA.

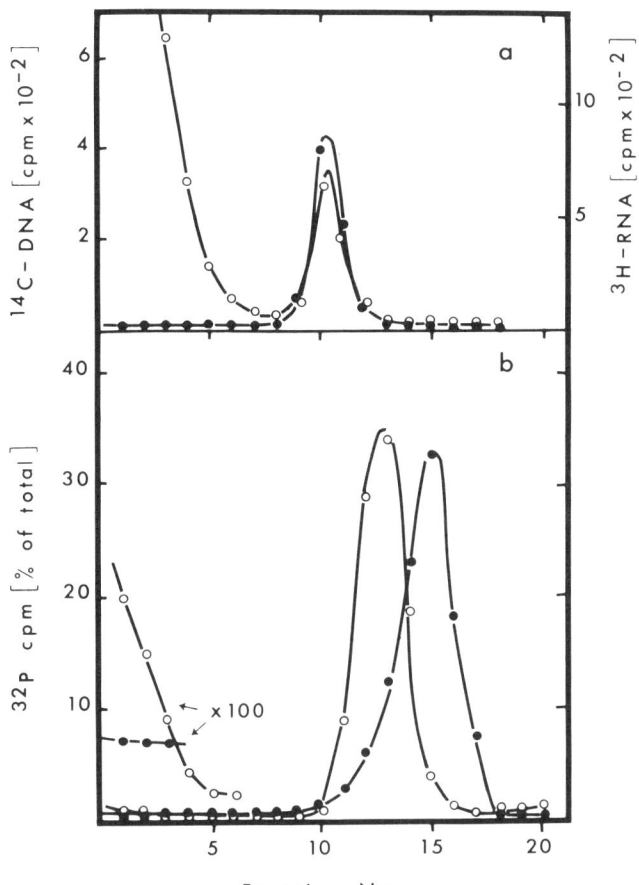

Fig. 8. Equilibrium centrifugation in CsCl of DNA-RNA complexes from the nucleoid. A, Another aliquot of the nucleoid preparation described in Fig. 7A was diluted and treated with SDS as in Fig. 7. CsCl was added to the solution to 1.710 g/cm^3 and the mixture was centrifuged 50 hr at 36,000 rpm in an SW 50 rotor at 22°C. ○——○, ^{3}H-RNA; ●——●, ^{14}C-DNA. B, Nucleoids isolated as in Fig. 1B from cells labeled 1 hr with ^{32}P-orthophosphate were treated with SDS and the opened DNA was sedimented on sucrose gradients to separate it from the released RNA as in Fig. 7A. Peak fractions from the DNA band were added to CsCl and centrifuged to equilibrium as above. Fractions from this CsCl gradient were counted by Cherenkov counting, ●——●. CsCl fractions from the DNA peak were pooled, heated to 95°C for 5 min, chilled, additional CsCl solution was added, and the solution was again centrifuged 50 hr, as above. The gradient was fractionated and counted, ○——○. In the latter CsCl gradient, 20 μg of nonradioactive tRNA or poly U carrier were added to minimize losses of the radioactive RNA and DNA.

been demonstrated by two different methods. (a) Nucleoids with ^{14}C-labeled DNA and ^{3}H-labeled proteins have been isolated, opened with SDS and sedimented on sucrose gradients as in Fig. 7; less than 0.7% (the detection limit) of the ^{3}H-protein initially bound to the nucleoid cosedimented with the DNA. Since most of the nucleoid-associated protein is RNA polymerase (Fig. 2), we conclude that the RNA polymerase, to the limits of detection, is removed from the DNA by this procedure (R. Hecht, unpublished result). (b) Ternary complexes have been made *in vitro* using purified bacteriophage DNA, purified RNA polymerase, and radioactive nucleoside triphosphates. We have observed, as have others (see for example reference 16), that SDS treatment is sufficient to separate, to the limit of deterction, nascent RNA initially bound in the ternary complex from the DNA.

Is it possible that the high molecular weight DNA physically traps (10) RNA and that the apparent RNA-DNA complex has no specific association? The experiments of Figs. 7 and 8 were done at very low DNA concentrations to reduce this possibility. Furthermore, no evidence for such trapping could be found when ^{32}P-pulse-labeled *E. coli* RNA was mixed with closed nucleoids before opening the DNA and sedimenting on sucrose gradients (Fig. 7c). The amount of ^{32}P-RNA cosedimenting with the DNA was <0.03% of the recovered RNA.

The DNA-bound RNA is released from the DNA (at least in part) by brief heating at 95°C (Fig. 8b). The released ^{32}P-RNA which sediments free of the DNA to the bottom of the CsCl gradient is alkaline degradable (>90%) and has an RNase resistant core no larger than purified *E. coli* rRNA (R. Hecht, unpublished data). The amount of ^{32}P-RNA recovered relative ^{32}P-DNA is about 0.5%. Although this DNA-bound RNA fraction is an attractive candidate for the DNA-packaging function, there is as yet no conclusive support for this possibility. Research in progress should clarify its significance.

Acknowledgments

We wish to thank Eileen Miles for technical assistance. This work was supported by research grants from the United States Public Health Service (GM18243) and the National Science Foundation (B025761), and by a United States Public Health Service Training Grant GM00781. This is Number 502 from the Department of Biophysics and Genetics, University of Colorado Medical Center.

References

1. Andrews, P. 1965. The gel-filtration behavior of proteins related to their molecular weight over a wide range. *Biochem. J.* 96: 595.

2. Burgess, R. 1969. A new method for the large scale purification of *Escherichia coli* deoxyribonucleic acid-dependent ribonucleic acid polymerase. *J. Biol. Chem.* 244: 6160.

3. Burgi, E., and Hershey, A. 1963. Sedimentation rate as a measure of molecular weight of DNA. *Biophys. J.* 3: 309.

4. Evans, D. J. 1969. Membrane adenosine triphosphatase of *Escherichia coli*. *J. Bacteriol.* 100: 914.

5. Fuhs, G. 1965. Symposium on the fine structure and replication of bacteria and their parts. I. Fine structure and replication of bacterial nucleoids. *Bacteriol. Rev.* 29: 277.

6. Ganesan, A., and Lederberg, J. 1965. A cell-membrane bound fraction of bacterial DNA. *Biochem. Biophys. Res. Commun.* 18: 824.

7. Hurwitz, J., Evans, A., Babinet, C., and Skalka, A. 1963. On the copying of DNA in the RNA polymerase reaction. *Cold Spring Harbor Symp. Quant. Biol.* 28: 59.

8. Kellenberger, E., Ryten, A., and Sechaud, J. 1958. Electron microscope study of DNA-containing plasms. II. Vegatative and mature phage DNA as compared with normal bacterial nucleoids in different physiological states. *J. Biophys. Biochem. Cytol.* 4: 671.

9. Kepes, A. 1963. Kinetics of induced enzyme synthesis. *Biochim. Biophys. Acta* 76: 293.

10. Konrad, M., and Stent, G. 1964. On "natural" DNA-RNA complexes in phage-infected cells. *Proc. Nat. Acad. Sci. U. S. A.* 51: 647.

11. Maaloe, O., and Kjeldgaard, N. 1966. The bacterial nucleus. *In* Maaloe, O., and Kjeldgaard, N. (eds.) *Control of macromolecular synthesis*, pp. 188–197, W. A. Benjamin, Inc., New York.

12. Miller, O. L., Beatty, B., Hamkalo, B., and Thomas, C. 1970. Electron microscopic visualization of transcription. *Cold Spring Harbor Symp. Quant. Biol.* 35: 505.

13. Pettijohn, D., Stonington, O., and Kossman, C. 1970. Chain termination of ribosomal RNA synthesis *in vitro*. *Nature* 228: 235.

14. Pettijohn, D., Clarkson, K., Kossman, C., and Stonington, O. 1970. Synthesis of ribosomal RNA on a protein-DNA complex isolated from bacteria: A comparison of rRNA synthesis *in vitro* and *in vivo*. *J. Mol. Biol.* 52: 281.

15. Pettijohn, D. 1972. Ordered and preferential initiation of ribosomal RNA synthesis *in vitro*. *Nature New Biol.* 235: 204.

16. Richardson, J. P. 1966. Enzymatic synthesis of RNA from T_7 DNA. *J. Mol. Biol.* 21: 115.

17. Robinow, C. 1962. Morphology of the bacterial nucleus. *Brit. Med. Bull.* 18: 31.

18. Ryter, A., Hirota, Y., and Jacob, F. 1968. DNA-membrane complex and nuclear segregation in bacteria. *Cold Spring Harbor Symp. Quant. Biol.* 33: 669.

19. Shapiro, A., Vinuela, E., and Maizel, J. 1967. Molecular weight estimation of polypeptide chains by electrophoresis in SDS-polyacrylamide gels. *Biochem. Biophys. Res. Commun.* 28: 815.

20. Smith, D., and Hanawalt, P. 1967. Properties of the growing point region in the bacterial chromosome. *Biochim. Biophys. Acta* 149: 519.
21. Stonington, O. G., and Pettijohn, D. E. 1971. The folded genome of *Escherichia coli* isolated in a protein-DNA-RNA complex. *Proc. Nat. Acad. Sci. U. S. A.* 68: 6.
22. Stonington, O. G. 1972. Bacterial nucleoid: Isolation and properties of folded DNA. Ph.D. dissertation. University of Colorado.
23. Travers, A., Kamen, R., and Cashel, M. 1970. The *in vitro* synthesis of ribosomal RNA. *Cold Spring Harbor Symp. Quant. Biol.* 35: 415.

Discussion

Battacharjee. Do you see your RNA associated with the DNA in short ^{32}P-labeled pulse experiments and/or under amino acid starvation conditions?

Pettijohn. We haven't done short pulse labeling with ^{32}P. We've done it with tritiated uridine, and yes, even in short pulses of a minute, you get about the same percentage of the RNA involved as we showed here. We haven't done a systematic study with starved cells, so I can't answer the other question.

Oeschger. How do you maintain your high ionic strength when you assay your RNA polymerase?

Pettijohn. There are two assays for the RNA polymerase; the one for the endogenous enzyme, we've previously shown, is not at all inhibited by the high salt. Its enzyme has already initiated the synthesis. It goes on as well in high salt, as well in the presence of rifampicin and all the other reagents. But naturally at that high salt concentration, you can't initiate RNA synthesis, and you therefore can't do the kind of studies I talked about earlier. As we previously reported, once having been isolated in the presence of high salt concentrations, the nucleoid still requires high salt to remain in a folded conformation, but the salt can be reduced somewhat. In fact, you can reduce it down to levels where you can get RNA polymerase initiation very efficiently.

Oeschger. What levels of salt? $M/10$, or $M/20$?

Pettijohn. The experiments were done in 0.1 M NaCl. The nucleoid is stable for long enough to do the experiment at that low salt. If you go much lower than that, it literally explodes and turns into a gel at high concentrations.

Gabbay. In your introductory remark, you said that in order to pack the DNA you need the counter ions, or positively charged proteins, and you find that there is an RNA, which you claim to actually pack this nucleoid; how can a polyanion pack other polyanions?

Pettijohn. I'm not saying that the RNA is packing the DNA. All I'm saying is that it's involved in the stability; the high salt and the RNA are both required. Take away the high salt, it explodes; take away the RNA, it explodes. The RNA could be involved as a linker, although our data so far do not rule on this. Just as single strands of DNA can associate in a double helix even though both have similar charges, there may be mechanisms for RNA binding to DNA directly or through intermediary complexes.

Protein ω: A DNA Swivelase from *Escherichia Coli?*

James C. Wang

Department of Chemistry
University of California
Berkeley, California 94720

The protein ω isolated from *Escherichia coli* is capable of converting a superhelical DNA to a much less twisted but covalently closed form. All evidence suggests that it is a DNA swivelase, *i.e.* an enzyme which is capable of introducing a swivel reversibly into a DNA helix. The possibility that it serves as a transient and movable swivel *in vivo* is discussed.

It is well known that the separation of two intertwined parental DNA strands accompanying semiconservative replication requires the strands to untwist (2-4, 7, 11, 22, 23). Consider the case in which a linear DNA is replicating from one end. The partially replicated molecule can be represented by a Y, with the stem of the Y representing the unreplicated portion of the molecule. For simplicity, only the untwisting of the parental strands is considered. While this could be achieved by a rotational motion of the stem around the helix axis, for a long DNA molecule replicating at a rate as fast as 1.5×10^5 base pairs per min, this rotational motion involving the bulk of the molecule poses some difficult problems, especially if there are membrane attachment sites on the DNA or there is more than one replicating fork per DNA molecule. Therefore it is generally believed that there is a "swivel" in the stem region, likely to be close to the fork, so that the rotation of a short segment rather than the complete stem accompanies the untwisting of the parental strands (7). That a swivel must exist at certain if not all times during the replication cycle is best illustrated by a covalently closed DNA;

topology alone forbids such a molecule to replicate semiconservatively without a swivel (2, 3, 19).

The molecular nature of such a swivel is yet to be elucidated. It is well recognized that the simplest swivel is a single chain scission in the double helix (7). Free rotation around one of the single bonds on the strand opposite to the nick serves the purpose of a swivel. The existence of a nicked replicative species (RF II) is best documented for the small DNA phage $\phi\chi$-174, although there is no evidence that the nick serves as *the* swivel (1, 12). When Tomizawa and Ogawa (17) examined the parental strands of partially replicated λ DNA for single chain scissions, they found that the number of scissions per strand was less than one and probably zero. Therefore they suggested that the swivel might be movable and transient, *i.e.* the swivel could be reversibly introduced into the double helix, perhaps at strategic positions close to the replication fork. Such a transient swivel is strongly suggested by some recent results. Both parental strands of a partially replicated circular DNA species were found to be covalently closed (9, 15). It is unlikely that the covalent closure of these strands occurred after lysis of the cells, since these molecules contain a substantial number of superhelical turns while postlysis closure of a nicked species would yield a species with very few superhelical turns (20).

It should be noted that efficient transcription *in vivo* might also need a DNA "swivel" so that the DNA molecule can rotate while the RNA polymerase molecules, with their associated nascent messenger RNA and ribosomes, can stay stationary (13).

Enzymes capable of introducing a temporary nick are known. An endonuclease which generates a nick with 5'-phosphoryl and 3'-hydroxyl end groups plus a ligase which joins such a nick would be an acceptable team. Recently it has been shown that ligase itself, in the presence of AMP, can perform both the nicking and rejoining, although the efficiency is rather low (14). The possibility that the *coli* protein ω might introduce a swivel reversibly into a DNA helix has been suggested previously (21). An activity somewhat similar to ω has been detected in extract from mouse cells (6). The evidence that protein ω serves as a swivel *in vitro* is reviewed here. Whether ω serves as a swivel *in vivo* remains to be elucidated.

Topological Labeling

If the function of an enzyme or enzymes is to introduce a transient swivel into a DNA helix, no chemical change in the DNA molecule is expected, and the detection of such a reaction by conventional methods would be difficult. For a superhelical DNA however, a swivel would by definition remove the topological constraint on the winding number of the molecule, and a reduction in the number of twists would result. Therefore a superhelical DNA is in this sense a "topologically labeled" DNA, and a swivel, either permanent or transient, would cause a change which can be measured experimentally.

A Macromolecular Species ω

The first hint that a transient swivel might be present in cell extracts came when the degree of superhelicity of λb2b5 DNA covalently closed *in vivo* was examined (20). Figure 1 depicts results for a sample of covalently closed λb2b5 DNA isolated from infected cells lysed by the Brig lysis procedure. The surprising observation was that there were several species different only in the degree of superhelicity. It was found that the most twisted species I_c was the *in vivo* species and the other species were derived from I_c by a macromolecular factor in the Brig lysate (20). It was postulated then that the macromolecular factor was an endonuclease, and the less twisted species were a result of the nicking by this endonuclease followed by the joining of the nick by ligase, which was shown to be active in the Brig lysate (20).

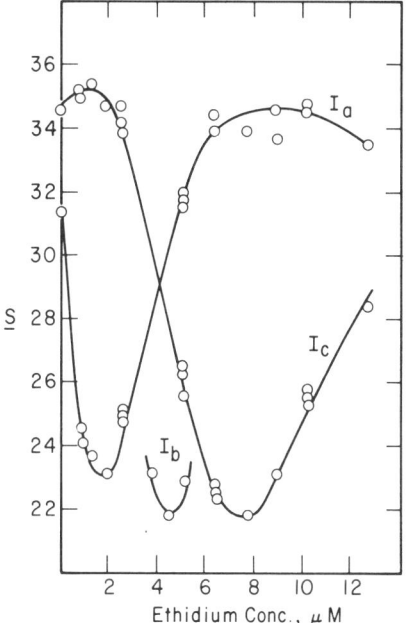

Fig. 1. Dependence of the sedimentation coefficients of covalently closed λb2b5 DNA species (isolated from cells infected at 30°C) on the ethidium concentration in 3 M CsCl, 0.01 M Na_3EDTA at 20°C. For a detailed discussion on the three species I_a, I_b, and I_c, cf. Wang (20).

It was soon clear that the activity which caused the reduction in the number of superhelical turns could not be a conventional endonuclease, since this activity could easily be separated from ligase on a DEAE-column. Even in the absence of ligase, this activity converts a highly twisted DNA to a less twisted, but covalently closed, form. This activity was designated ω (21). The purification procedures for ω have been described previously (21).

ω Is a Protein

The thermal lability, pronase sensitivity, and general chromatographic properties of ω all indicate that it is a protein (21). This is supported by its buoyant density in CsCl. Two milliliters of 2.8 M CsCl-0.01 M Tris, pH 7.9-0.001 M EDTA containing ω were layered with 3 ml of a light silicon oil and banded at 60,000 rpm for 24 hr at 4°C in an SW 65 rotor. Ten fractions with densities ranging from 1.2 to 1.5 g/cm^3 were collected and ω activity for each fraction was assayed after dialysis against 0.01 M Tris, pH 7.9, 0.01 M MgCl$_2$, 0.05 M KCl, and 10^{-4} M EDTA. The fractions with densities 1.246 and 1.290 g/cm^3 contained ω activity and none of the other fractions had ω activity.

Zone sedimentation on a 5-20% sucrose gradient (containing 0.1 M potassium phosphate, pH 6.5) with DNA polymerase as a marker showed that ω sedimented about the same as DNA polymerase, which has a molecular weight of 109,000 (10). A total of 40 fractions were collected from a 5-ml gradient, after spinning at 50,000 rpm and 4°C for 14 hr. The peak polymerase activity was found half-way down the gradient and so was the ω activity. The same patterns were observed when DNA polymerase and ω were sedimented in separate tubes, showing that the cosedimentation of the two is not due to intermolecular association. Due to the difficulty in assaying ω quantitatively, the error in locating the peak of the ω activity is relatively high. It can be stated that the sedimentation coefficient of ω is the same as that of DNA polymerase within an error of ±10%.

Recently, J. Carlson and W. Baase (personal communication) were able to assay ω activity after electrophoresis on acrylamide gel. The acrylamide gel was sliced and the protein extracted from each slice was denatured with sodium dodecyl sulfate (SDS) and electrophoresed again on SDS-acrylamide gel (16). It was found that the ω activity corresponds to a protein band with a molecular weight of 110,000 as evidenced from its mobility on SDS gel. Therefore both the sedimentation coefficient and the mobility of the denatured protein indicate that its molecular weight is about 110,000.

Reduction of the Number of Negative Superhelical Turns by ω

It has been firmly established by electron microscopy and sedimentation analysis in the presence of varying amounts of ethidium that ω reduces the number of negative superhelical turns (21). Figure 2 shows sedimentation results obtained with the 1.45-megadalton DNA from *coli* 15. The similarity between these results and the results previously published for λb2b5c DNA is obvious and the interpretation is the same. (Similar results were also obtained with PM-2 DNA as the substrate.) With λ DNA as the substrate, typically 15% of the DNA was converted to the nicked form after incubation with ω. With the much smaller DNA from *coli* 15, no detectable amount of nicked species was observed. This suggested that the ~15% nicked species observed for λb2b5 DNA did not result from ω itself but was due to a chemical or enzymological contaminant.

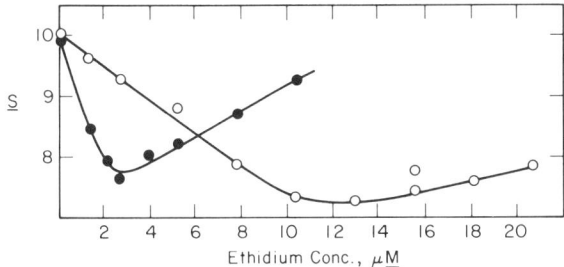

Fig. 2. Sedimentation coefficient of the 1.45-megadalton DNA from *coli* 15 as a function of ethidium concentration in 3 M CsCl, 0.01 M Na$_3$EDTA at 20°C. Open circles: untreated DNA. Most of the data points taken from Wang (20). Closed circles: after treatment with ω.

Is ω a Swivelase *in Vitro*?

There are only two ways a superhelical DNA can lose its superhelical turns. (*a*) No transient swivel is involved, the superhelical turns are removed by a concomitant change in the secondary structure of the DNA helix (and therefore the number of turns in the double helix). (*b*) A transient swivel is involved.

Since a direct demonstration of a transient swivel is difficult, all experiments carried out so far centered on the question of whether there was a concomitant change in the secondary structure of the DNA helix.

If the reduction in the number of negative superhelical turns is a result of a change in the secondary structure of the DNA helix, then there must be a reduction in the number of right-handed turns of the double helix: either a segment of the double helix is disrupted or the helix rotation angle per base pair is reduced. In either case, the structure change must result from either permanent chemical modification of the DNA, or irreversible binding of the protein (since the negative superhelical turns lost cannot be restored by standard deproteination procedures such as phenol extraction or pronase digestion).

Some of the experiments performed previously with λ DNA as the substrate were repeated with PM-2 DNA, since there is some advantage in using a superhelical DNA of lower molecular weight as the substrate. Figure 3 shows the circular dichroism (CD) of a sample of PM-2 DNA treated with ω. As a control, the CD of a sample treated with *Escherichia coli* ligase in the presence of AMP to remove the superhelical turns (14) was also measured. Both samples had been phenol-extracted to

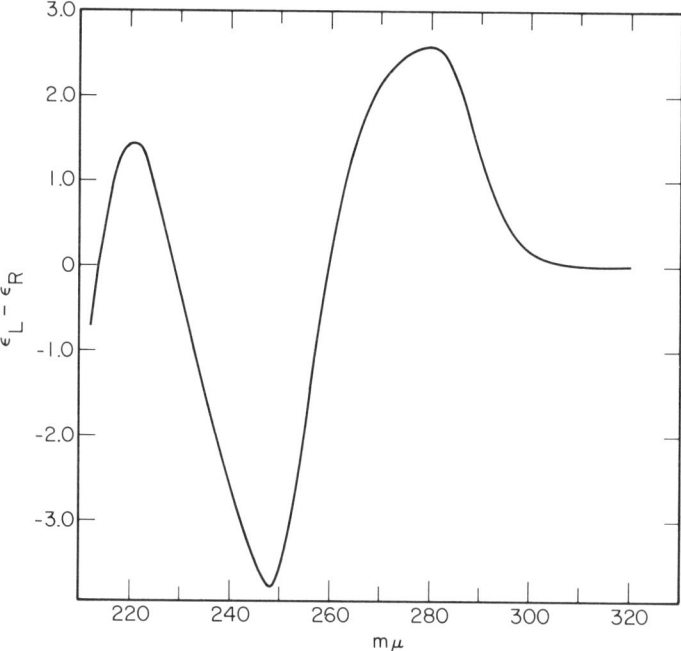

Fig. 3. The circular dichroism of ω-treated PM-2 DNA. The DNA was phenol-extracted after incubation with ω and dialyzed against 0.1 M NaCl, 0.01 M Tris, pH 8, and 10^{-4} M Na$_3$EDTA. The circular dichroism of PM-2 DNA relaxed by ligase in the presence of AMP was also measured in the same medium. The two curves are virtually superimposable. A Cary 60-CD dichrograph with an on-line PDP8/S computer for noise-averaging and data-smoothing was used in these measurements. (I thank Mrs. B. Dengler for her assistance in these measurements.)

remove protein. The CD of the two samples are virtually superimposable in the complete wave length region measured (205 to 360 mµ). This suggests that the secondary structure of PM-2 DNA after ω treatment is the same as PM-2 DNA relaxed by nicking and rejoining by ligase.

The buoyant densities in CsCl were also measured for the two PM-2 samples with *Micrococcus lysodeikticus* DNA as a marker. As previously observed with λ DNA, treatment of PM-2 with ω does not change the buoyant density of the DNA. It should be noted that the binding of a single protein molecule with a molecular weight of 10^5 to a PM-2 DNA molecule with a molecular weight of 6×10^6 would reduce the buoyant density by several milligrams per ml, which would be easily detectable. Therefore this experiment shows that ω does not bind irreversibly to DNA.

It has been shown previously, from the susceptibility of ω-relaxed λ DNA to the single strand specific *Neurospora* endonuclease, that no significant length of single stranded segments was generated by treatment with ω.

Furthermore, since ω does not require a cofactor, it is extremely unlikely that it could cause chemical modification to cause a 3% change in the average helix-winding angle, which would be necessary to account for the loss of the negative superhelical turns. The CD and buoyant density results are also against the possibility of chemical modification. As pointed out previously, even the methylation of 1% of the bases would have caused a buoyant density shift of 1 mg/ml.

All the evidence against a change in the secondary structure of the helix upon treatment with ω adds up to one conclusion: ω is a *swivelase*. It introduces a swivel reversibly into a DNA and its departure leaves no mark on the DNA, with the exception that the superhelical turns are lost during its transient presence.

ω Is Active Only on a Negatively Twisted DNA

As reported previously, the *coli* ω protein is only active on negatively twisted DNA (21). Champoux and Dulbecco (6) reported that an activity found in extract of mouse cell nuclei could relax both negatively and positively twisted DNA. To test whether a factor might have been lost during the purification of the *coli* ω so that it is no longer active on positively twisted DNA, experiments on positively twisted DNA were repeated with the ammonium sulfate fraction of ω and with crude Brig lysate. No activity capable of relaxing positively twisted DNA was found. (Ethidium was present in all experiments with positively twisted DNA; see Wang (21) for details.) As discussed previously (21), the specificity for negatively twisted DNA is probably linked to the destabilizing effect on the double helix by such twists (18).

The Gradual Reduction in the Number of Superhelical Turns

Two modes are usually observed for the reduction of superhelical turns by ω. (*a*) If the ω concentration is low but the reaction conditions are favorable, two major covalently closed species are present after incubation: molecules which have lost most of the superhelical turns and molecules which have hardly lost any superhelical turns. (*b*) If the ω concentration is high but the reaction conditions are unfavorable (low temperature of high ionic strength), the number of superhelical turns of all of the molecules is gradually reduced. The dividing line between the two modes is not sharp, as exemplified by results shown in Fig. 4. Here the kinetic analysis was carried out at 0°C. Samples were drawn at various time intervals, and the reaction was stopped by the addition of a stock solution of CsCl-containing ethidium. The resulting solution was banded in an analytical ultracentrifuge and photoelectric scans were taken at 350 mμ, a wave length at which bound ethidium absorbs more than free ethidium. (Therefore the traces gave the positions of the DNA-ethidium complexes.) The buoyant density of a covalently closed DNA

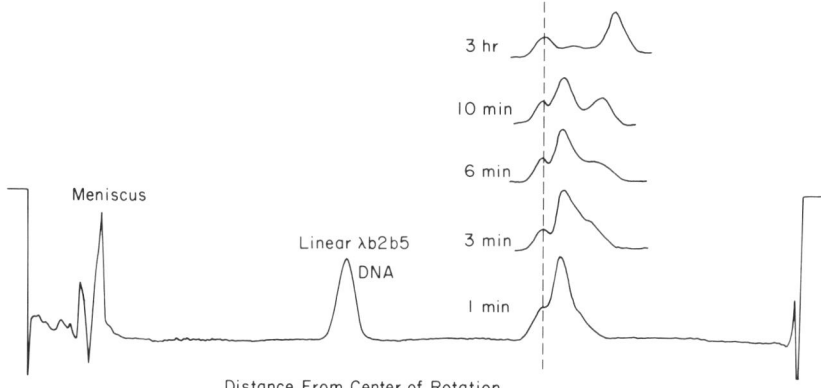

Fig. 4. CsCl-ethidium density gradient centrifugation of superhelical λb2b5 DNA incubated with ω at 0°C. Linear λb2b5 DNA was added to each sample as a marker (shown only in the 1-min tracing). A second superhelical λb2b5 DNA marker with a superhelical density of −0.044 was also added to each cell. The position of this marker was indicated by the dashed line shown in the figure. The superhelical λb2b5 DNA sample used as the substrate had an initial superhelical density of −0.039. The time of incubation is shown in the figure for each tracing. Centrifugation was done at 39,460 rpm for 48 hr at 20°C. the initial density of the CsCl solution was ⁻1.56 g/cm³, and the amount of ethidium bromide added was 100 μg/ml. Double sector cells were used with ethidium present in both sectors. Scans were taken at 350 or 546 mμ. A reflective mirror assembly (Spinco) was used instead of the usual refractive camera lens so that change of wave length had no effect on the focusing and magnification factor.

in the presence of ethidium is a measure of its degree of superhelicity (8). The gradual increase in buoyant density therefore shows a gradual reduction in the number of negative superhelical turns, in agreement with previous results obtained by measurements of sedimentation coefficients. The more or less partition of molecules into two major species can also be seen. For example, the 10-min sample contained roughly one-half of the molecules with little loss of superhelical turns; the other half showed a very significant reduction in the degree of superhelicity. In a parallel experiment, the reaction was stopped at time intervals by the injection of 1 M KOH to give a final KOH concentration of 0.1 M. Alkaline sedimentation runs were performed for the samples stopped at different times. In each case the majority of the DNA molecules remained covalently closed. These experiments suggest that at least under these unfavorable assay conditions, the conversion of the twisted DNA to the final product involves many nicking and rejoining events. In other words, during the time course of the relaxation, the swivel is put into and taken out of the helix many times. The bimodal distribution of twists sometimes observed is probably a result of slow exchange of bound protein molecules between DNA molecules. The initial distribution of the ω molecules among the DNA molecules is likely to be uneven due to the high local concentration when a small amount of ω stock (typically a few microliters) is added to the assay mixture (typically 30 to 50 μl).

The *in Vivo* Function of ω Is Unknown

The *in vitro* functions of ω suggest that it might function as a DNA swivelase *in vivo*. The swivel is transient and movable. Furthermore, while a transient and movable nick produced by a conventional endonuclease and ligase pair is subject to the difficulty that the nick might be attacked by other enzymes, the mechanism postulated for ω requires that the protein be present at the swivel point and therefore the transient ends produced are much less susceptible to attack by other enzymes (21). (It was observed that excess DNA polymerase I or exonuclease III present during incubation with ω had no effect on the ω-catalyzed reaction.) Experimental attempts to link ω to DNA replication, however, have so far been unsuccessful. A number of *E. coli* thermal-sensitive DNA replication mutants have been tested for the possibility of a thermal-sensitive ω *in vitro*. The ω from each strain grown at a permissible temperature was partially fractionated to the ammonium sulfate step (21), and the temperature sensitivity of the protein was compared with the parent strain. The mutants which have been tested so far include PC 1, PC 2, PC 3, PC 5, PC 6, PC 7, PC 8, 1026 (Bonhoeffer) and

E 101 (of loci dna C, dna C, dna G, dna A, dna B, dna D, dna B, dna E, and dna F, respectively (5, 24)). No abnormal thermal sensitivity of ω was found for any of the mutants. Many more mutants might have to be screened before any conclusion can be drawn. It should be noted that DNA replication mutants are usually selected for temperature sensitivity in DNA synthesis only and not in the syntheses of other macromolecules. If ω is a swivelase *in vivo*, it is probably also needed in RNA synthesis (13). Therefore we might have been looking for the ω tree in the wrong forest. This interesting possibility was pointed out to me by F. Bonhoeffer.

Conclusion

The experimental results accumulated so far suggest that ω is a DNA swivelase. The intriguing reaction it promotes *in vitro* and its possible functions in both replication and transcription *in vivo* demand further physicochemical as well as biochemical studies.

Acknowledgments

This work has been supported by grants from the United States Public Health Service, the National Science Foundations, and a fellowship from the Alfred P. Sloan Foundation.

References

1. Burton, A., and Sinsheimer, R. L. 1965. The process of infection with bacteriophage φχ-174. VII. Ultracentrifugal analysis of the replicative form. *J. Mol. Biol.* 14: 327.

2. Cairns, J. 1963. a. The bacterial chromosome and its manner of replication as seen by autoradiography. *J. Mol. Biol.* 6: 208.

3. Cairns, J. 1963. b. The chromosome of *Escherichia coli*. *Cold Spring Harbor Symp. Quant. Biol.* 28: 43.

4. Cairns, J., and Davern, C. I. 1967. The mechanics of DNA replication in bacteria. *J. Cell. Physiol.* 70 (suppl. 1): 65.

5. Carl, P. L. 1970. *Escherichia coli* mutants with temperature-sensitive synthesis of DNA. *Mol. Gen. Genet.* 109: 107.

6. Champoux, J. J., and Dulbecco, R. 1972. An activity from mammalian cells that untwists superhelical DNA. A possible swivel for DNA replication. *Proc. Nat. Acad. Sci. U.S.A.* 69: 143.

7. Delbrück, M., and Stent, G. S. 1957. On the mechanism of DNA replication. *In* McElroy, W. D., and Glass, B. (eds.), *The chemical basis of heredity*, p. 699, The Johns Hopkins Press, Baltimore.

8. Gray, H. B. Upholt, W. B., and Vinograd, J. 1971. A buoyant method for the determination of the superhelix density of closed circular DNA. *J. Mol. Biol.* 62: 1.

9. Jaenisch, R., Mayer, A., and Levine, A. 1971. Replicating SV40 molecules containing closed circular template DNA strands. *Nature New Biol.* 233: 72.

10. Jovin, T. M., Englund, P. T., and Bertsch, L. L. 1969. Enzymatic synthesis of deoxyribonucleic acid. XXVI. Physical and chemical studies of a homogeneous deoxyribonucleic acid polymerase. *J. Biol. Chem.* 244: 2996.

11. Levinthal, C., and Crane, H. R. 1956. On the unwinding of DNA. *Proc. Nat. Acad. Sci. U.S.A.* 42: 436.

12. Linqvist, B. H., and Sinsheimer, R. L. 1968. The process of infection with bacteriophage $\phi\chi$-174. XVI. Synthesis of the replicative form and its relationship to viral single-stranded DNA synthesis. *J. Mol. Biol.* 32: 285.

13. Maaloe, O., and Kjeldgaard, N. O. 1966. *Control of macromolecular synthesis*, Benjamin Co., Inc., New York.

14. Modrich, P., Lehman, I. R., and Wang, J. C. 1972. Enzymatic joining of polynucleotides. XI. Reversal of *Escherichia coli* deoxyribonucleic acid ligase reaction. *J. Biol. Chem.* 247: 6370.

15. Sebring, E. D., Kelly, T. J., Jr., Thoren, M. M., and Salzman, N. P. 1971. Structure of replicating simian virus 40 deoxyribonucleic acid molecules. *J. Virol.* 8: 478.

16. Shapiro, A. L., Viñuela, E., and Maizel, J. V., Jr. 1967. Molecular weight estimation of polypeptide chains by electrophoresis in SDS-polyacrylamide gels. *Biochem. Biophys. Res. Commun.* 28: 815.

17. Tomizawa, J.-I., and Ogawa, T. 1968. Replication of phage lambda DNA. *Cold Spring Harbor Symp. Quant. Biol.* 33: 533.

18. Vinograd, J., Lebowitz, J., and Watson, R. 1968. Early and late helix-coil transitions in closed circular DNA. The number of superhelical turns in polyoma DNA. *J. Mol. Biol.* 33: 173.

19. Vinograd, J., and Lebowitz, J. 1966. Physical and topological properties of circular DNA. *J. Gen. Physiol.* 49 (suppl.): 103.

20. Wang, J. C. 1969. Degree of superhelicity of covalently closed cyclic DNA's from *Escherichia coli*. *J. Mol. Biol.* 43: 263.

21. Wang, J. C. 1971. Interaction between DNA and an *Escherichia coli* protein ω .*J. Mol. Biol.* 55: 523.

22. Watson, J. D., and Crick, F. H. C. 1953. a. Genetic implications of the structure of deoxyribonucleic acid. *Nature* 171: 964.

23. Watson, J. D., and Crick, F. H. C. 1953. b. The structure of DNA. *Cold Spring Harbor Symp. Quant. Biol.* 18: 123.

24. Wechsler, J. A., and Gross, J. D. 1971. *Escherichia coli* mutants temperature-sensitive for DNA synthesis. *Mol. Gen. Genet.* 113: 273.

Discussion

Cairns. Have you considered doing an ^{18}O exchange experiment?

Wang. Yes, Dr. Boyer at UCLA talked to me about it and we decided that it was probably not feasible. Firstly, an ^{18}O in the middle of a DNA is awfully hard to detect; secondly, the mechanism I have written may or may not involve ^{18}O exchange.

Alberts. Do you know how many molecules of the ω protein you need to unravel the DNA molecule, the minimum number?

Wang. The minimum number is in the order of one or less, based on the assumption that our estimate of the purity is correct. Furthermore, we have shown that under conditions unfavorable for the reaction, the reaction can be quenched with the loss of only a fraction of the superhelical turns. That seems to say that going from a highly twisted DNA to a relaxed form involves many nicking and sealing events. So I kind of think that the number of enzyme molecules needed per nicking-and-sealing event is less than one. In other words, one ω molecule can catalyze more than one such event.

Initiation of DNA Synthesis

Randy Schekman, William
Wickner, Ole Westergaard, Douglas Brutlag,
Klaus Geider, Leroy Bertsch, and Arthur Kornberg

Department of Biochemistry
Stanford University School of Medicine
Stanford, California 94305

A soluble enzyme extract from uninfected *Escherichia coli coli* catalyzes the conversion of M13 and $\phi\chi$-174 ($\phi\chi$) single stranded DNAs to their double stranded replicative forms (RF). The M13 and $\phi\chi$ synthetic systems are distinct from each other, but in both cases, DNA synthesis is initiated by RNA priming as judged by: inhibition by RNA synthesis inhibitors, requirement for ribonucleoside triphosphates, and the presence of a phosphodiester bond linkage between DNA and RNA in the isolated RF product. Initiation of DNA synthesis by RNA priming may prove to be a mechanism of wide significance. The host enzymes which are required for M13 replicative form synthesis include RNA polymerase, a DNA polymerase, an unwinding protein, and additional factors. For $\phi\chi$ synthesis, host proteins involved in initiation and elongation of the host chromosome (i.e. dnaA and B) are required in addition to spermidine. The phage replicative systems provide convenient assays for purification of these host proteins and opportunities for clarifying their roles in replication.

DNA polymerases from *Escherichia coli* and phage-infected cells extend DNA chains, but as yet they have not been shown to initiate a chain. How is new DNA synthesis initiated? One possibility is that all DNA synthesis takes place by covalent extension of pre-existing DNA chains. Thus, oligonucleotide fragments of a DNA chain or ends produced by endonucleolytic scissions of a chain might serve as primers. Experience with available enzymes favors this scheme. An alternative to this possibility is that a new enzyme initiates chains by itself, or in conjunction with one of the known DNA polymerases. Studies with intact cells favor this suggestion, but no enzyme has yet been found to do this job.

Another alternative occurred to us. Since RNA polymerase starts new RNA chains, and DNA polymerase is known to extend a ribonucleotide terminus during DNA synthesis covalently (15), a brief transcriptional operation by RNA polymerase might provide an RNA primer for DNA synthesis. This priming piece of RNA could later be recognized and excised by nuclease action. Thus, a *de novo* initiation event catalyzed by RNA polymerase would be an essential step for the start of DNA synthesis.

We found that the conversion of the single stranded DNA circle (SS) of M13 phage into a double stranded circle (replicative form, RF) in *E. coli* was prevented by rifampicin (4). Studies with enzyme fractions have shown that RNA synthesis by RNA polymerase is required (4). Our findings indicate that the RNA synthesis provides a primer to initiate DNA synthesis.

There are strong indications that RNA polymerase action is also involved in the replication of double stranded DNA. Multiplication of M13 RF is immediately and profoundly inhibited by rifampicin (4); so is the replication of the RF-like colicinogenic factor $ColE_1$ (3). Furthermore, Blair *et al.* have found that alkaline or RNase digestion introduces breaks in the supercoiled $ColE_1$ (3), suggesting that a segment of RNA is present in the supercoiled DNA. A transcriptional event in the initiation of phage λ DNA replication has been described (6) and rifampicin sensitivity of F factor replication (1) and the start of *E. coli* chromosome replication have been attributed to an RNA synthetic action (12).

DNA synthesis which is unaffected by rifampicin implies a mechanism of strand initiation without RNA involvement or else RNA synthesis catalyzed by an enzyme system with properties different from RNA polymerase. The ongoing synthesis of an *E. coli* chromosome (18) and the conversion of φχ-174 (φχ) SS to RF *in vivo* (18) are examples of such rifampicin-resistant processes.

Olivera and Bonhoeffer showed that an *E. coli* lysate supported on a cellophane disc converted φχ SS to RF (13). We developed a high speed supernatant enzyme fraction (Fraction I, Fig. 1) capable of carrying out the conversion of an M13 or φχ-174 template to its RF (4, 22). With both M13 and φχ the product sediments in a neutral sucrose velocity gradient as RF II ((phage) RF with discontinuity in at least one strand). The product was analyzed further by subjecting denatured RF II to velocity sedimentation in alkaline sucrose or equilibrium sedimentation in alkaline CsCl. The labeled product in each case appeared as a full length, linear, complementary strand, and the template as an intact, circular, viral strand.

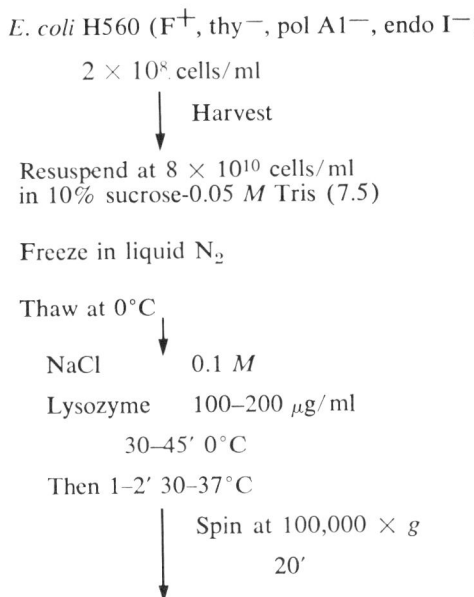

Fig. 1. Preparation of Fraction I enzyme.

With Fraction I, from which much of the cellular DNA has been removed, assay of the conversion of SS to RF could be measured directly by incorporation of labeled nucleotide into an acid-insoluble material. Total nucleotide incorporation into the acid-insoluble fraction was identical with that identified in the RF region of sucrose velocity gradients.

Rifampicin, which prevents RNA polymerase from initiating synthesis (19), inhibits the replication of M13 but not φχ (Table 1). The same result was found with streptolydigin, which prevents RNA polymerase from propagating a chain (17). From these results, one might conclude that RNA synthesis is not required for conversion of φχ SS to RF or that RNA is synthesized by a system distinguishable from RNA polymerase.

Table 1. Effect of RNA synthesis inhibitors on conversion of M13 and φχ-174 SS to RF in vitro

Inhibitor	Single strand template (% of control)	
	M13	φχ
Rifampicin, 5 µg/ml	10	80
Streptolydigin, 600 µg/ml	10	90
Actinomycin D, 5 µg/ml	7	11

Reproduced with permission from reference 15.

Three lines of evidence implicate RNA synthesis in both the φχ and the M13 reactions: (a) inhibition by low levels of actinomycin D, (b) requirement for all four ribonucleoside triphosphates, and (c) covalent linkage of DNA to RNA in the product.

Actinomycin D inhibits RNA synthesis by virtually all RNA polymerases (bacterial and animal) by intercalating into a duplex DNA template and binding to deoxyguanosine (20). φχ synthesis was inhibited by actinomycin D almost completely (as was M13) at levels which do not significantly affect DNA replication (Table 1). These results indicate that RNA synthesis is catalyzed by a new or modified form of RNA polymerase and point to a duplex template region within the single stranded, circular φχ DNA. Such duplex or hairpin structures have been demonstrated in φχ and M13 single strands (7, 14). Some years ago, Sekiguchi and Iida (16) examined an E. coli mutant whose altered permeability rendered it susceptible to actinomycin. One of the interesting characteristics was the profound inhibition of DNA as well as RNA synthesis by low levels of the antibiotic. A possible explanation now for the actinomycin effect on DNA synthesis is the interruption of RNA-primed DNA initiations.

Fraction I enzyme had a strong requirement for ATP for maximal DNA synthesis. Although a dependence on added CTP, GTP, and UTP was not apparent in this fraction, it was clearly demonstrable with an enzyme fraction treated with DEAE-cellulose and precipitated with ammonium sulfate. With both M13 and φχ as templates, all four ribonucleoside triphosphates were required for maximal synthesis; 2- to 5-fold lower DNA synthesis was observed when any one was omitted.

A phosphodiester bridge between a deoxyribonucleotide and a ribonucleotide exists in the isolated RF product. The experiment entails alkaline hydrolysis of the RF II synthesized in the presence of four [α-^{32}P]deoxyribonucleoside triphosphates and isolation of the ribonucleotides from the alkaline digest. With both M13 and φχ, approximately 1 mole of [^{32}P]ribonucleoside monophosphate was isolated for each mole of RF produced (Table 2). The mixtures of 2'- and 3'-ribonucleotides to which ^{32}P was transferred contained largely AMP in the case of M13, but included significant amounts of all four ribonucleotides (with AMP and GMP predominating) in the case of φχ.

In order to determine the relationship between chromosome replication and the φχ and M13 SS to RF reactions, the properties of several host mutants thermosensitive in chromosome duplication were analyzed. Two classes of these mutants have previously been isolated (9). dnaA and C mutants continue DNA synthesis at 42°C but fail to initiate a new chromosome, and dnaB, D, E, F, and G mutants stop DNA synthesis immediately at 42°C. In extracts (Fraction I) of the thermosensitive dnaA and dnaB mutants there was little or no synthesis of φχ DNA at 37°C

Table 2. The *in vitro* product contains RNA covalently attached to DNA

Labeled ribonucleotide released by alkaline hydrolysis	M13 RF* (moles nucleotide/mole RF)		φχ RF*	
	Exp. 1	Exp. 2	Exp. 1	Exp. 2
Ap	1.2	1.2	0.40	0.40
Gp	<0.1	<0.2	0.40	0.35
Cp	<0.1	<0.2	0.10	0.20
Up	<0.1	<0.1	0.20	0.10

Reproduced with permission from reference 15.
*<0.1 mole of nucleotide per mole of RF was found in otherwise identical, unhydrolyzed samples.

although the rates at 25°C were at or near those of extracts of wild type or temperature-resistant revertants. When equal amounts of enzyme fractions from the dnaA and dnaB mutants were combined, synthesis was at the wild type level (Fig. 2). The rate of M13 DNA synthesis was the same in extracts of these mutant cells as in those of wild type cells whether measured at 25 or 37°C.

Complementation at 37°C of Fraction I of the dnaB mutant provided a linear assay to determine levels of dnaB gene product in extracts

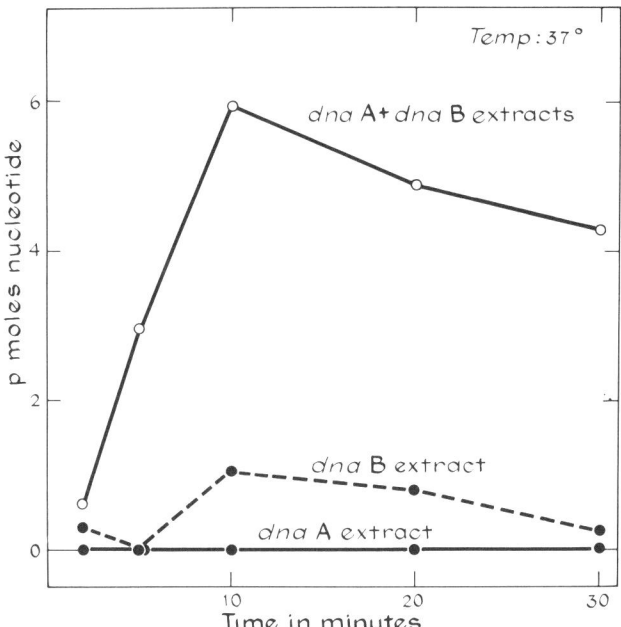

Fig. 2. Thermolabile dnaA and dnaB mutant extracts complement at the nonpermissive temperature. Equal amounts of Fraction I from dnaA and dnaB mutant were assayed together or separately at 37°C. (Reproduced with permission from reference 15.)

of wild type cells (Fig. 2). A similar complementation assay for levels of the dnaA gene product was feasible using Fraction I of the dnaA mutant. Complementation reactions were rifampicin-resistant and the product was RF II. The dnaA and dnaB proteins have been partially purified in our laboratory. In addition to our own work on the dna mutants, current studies by R. Wickner *et al.* (personal communication) indicate involvement of proteins specified by dnaC, dnaD, and dnaG genes. DNA polymerase III may be involved in the φχ and M13 reactions. Under certain conditions, extracts of dnaE mutants (defective in DNA polymerase III, references 8, 21) appear to be defective in the φχ and M13 SS to RF reactions. Addition of purified DNA polymerase I or II did not restore activity, whereas partially purified DNA polymerase III did.

Several outstanding questions remain regarding the M13 and φχ SS to RF conversions. One is why φχ and M13, so similar in DNA structure, exploit two different synthetic systems and, beyond that, what roles these two systems serve in the replication of *E. coli* DNA.

With regard to the φχ reaction, its relation to *E. coli* DNA synthesis is obvious. The two share requirements for some of the same gene products. Mechanistically, the conversion of φχ SS to RF resembles the ongoing replication of the *E. coli* chromosome. Recently, Sugino *et al.* (21) made the important discovery that the newly synthesized fragments of DNA in the nascent part of the chromosome contain covalently linked RNA sections which are subsequently removed when the fragments become linked to the main body of the chromosome. This finding strengthens the validity of Okazaki's hypothesis that DNA growth is discontinuous and the generality of our proposal for RNA priming as the mechanism of DNA initiation.

There are many unanswered questions regarding the initiation event: the specificity and number of starts, the size of the primer, the specificity of termination, and the nature of excision. For an accurate answer to all these questions, purification of the synthetic system is absolutely essential. Purification efforts have shown that the M13 reaction requires RNA polymerase, a DNA-binding (unwinding) protein (isolated by N. Sigal and B. Alberts, manuscript in preparation), a fraction which con-

Table 3. Synthetic systems for M13 and φχ

	M13	φχ
Rifampicin, streptolydigin inhibition	Yes	No
Actinomycin inhibition	Yes	Yes
Ribonucleoside triphosphate required	Yes	Yes
Covalent linkage of DNA to RNA	Yes	Yes
Host chromosome replication proteins: dnaA, B, C, G	No	Yes

Reproduced with permission from reference 15.

tains DNA polymerase III and other factors. In addition to the requirements for dnaA and B proteins in the φχ reaction, spermidine stimulates φχ DNA synthesis 5- to 20-fold. Although the stimulation was not always observed with different Fraction I preparations, more purified fractions showed an invariable and near absolute requirement for spermidine. A comparison of the M13 and φχ reactions is shown in Table 3.

Acknowledgments

This work was supported in part by grants from the National Institutes of Health and the National Science Foundation. W. Wickner is a Basic Science Fellow of the National Cystic Fibrosis Research Foundation. O. Westergaard is an American Cancer Society-Eleanor Roosevelt-International Cancer Fellow. D. Brutlag is a Predoctoral Fellow of the National Science Foundation; his present address is Division of Plant Industries, C.S.I.R.O., Canberra, Australia. K. Geider is a Fellow of the National Institutes of Health.

References

1. Bazzicalupo, P., and Tocchini-Valentini, G. P. 1972. Curing of an *Escherichia coli* episome by rifampicin. *Proc. Nat. Acad. Sci. U. S. A.* 69: 298.

2. Berg, P., Fancher, H., and Chamberlain, M. 1963. The synthesis of mixed polynucleotides containing ribo- and deoxyribonucleotides by purified preparations of DNA polymerase from *Escherichia coli*. In Vogel, H. J., Bryson, V., and Lampen, J. O. (eds.), *Informational macromolecules*, pp. 467-483, Academic Press, New York.

3. Blair, D. G., Sherratt, D. J., Clewell, D. B., and Helinski, D. R. 1972. RNase and alkali-sensitive supercoiled $ColE_1$ DNA synthesized in *E. coli* in the presence of chloramphenicol. *Fed. Proc.* 31: 1269.

4. Brutlag, D., Schekman, R., and Kornberg, A. 1971. A possible role for RNA polymerase in the initiation of M13 DNA synthesis. *Proc. Nat. Acad. Sci. U. S. A.* 68: 2826.

5. Clewell, D. B., Evenchik, B., and Cranston, J. W. 1972. Direct inhibition of $ColE_1$ plasmid DNA replication in *Escherichia coli* by rifampicin. *Nature* 237: 29.

6. Dove, W. F., Inokuchi, H., and Stevens, W. F. 1971. Replication control in phage lambda. In Hershey, A. D. (ed.) *The bacteriophage lambda*, pp. 747-771, Cold Spring Harbor Laboratory, New York.

7. Forsheit, A. B., and Ray, D. S. 1970. Conformations of the single-stranded DNA of bacteriophage M13. *Proc. Nat. Acad. Sci. U. S. A.* 67: 1534.

8. Gefter, M. L., Hirota, Y., Kornberg, T., Wechsler, J. A., and Barnoux, C. 1971. Analysis of DNA polymerases II and III in DNA thermosensitive mutants of *Escherichia coli*. *Proc. Nat. Acad. Sci. U. S. A.* 68: 3150.

9. Gross, J. D. 1971. DNA replication in bacteria. *Current topics in microbiology and immunology*, Springer Verlag, Berlin.

10. Kornberg, T., and Gefter, M. 1971. DNA synthesis in cell-free extracts: purification and properties of DNA polymerase II. *Proc. Nat. Acad. Sci. U. S. A.* 68: 761.

11. Lancini, G., Pallanza, R., and Silvestri, L. G. 1969. Relationships between bacteriocidal effect and inhibition of ribonucleic acid nucleotidyl-transferase by rifampicin in *Escherichia coli*. *J. Bacteriol.* 97: 761.

12. Lark, K. G. 1972. Evidence for the direct involvement of RNA in the initiation of DNA replication in *Escherichia coli* 15 T⁻. *J. Mol. Biol.* 64: 47.

13. Olivera, B. M., and Bonhoeffer, F. 1972. Replication of $\phi\chi 174$ DNA *in vitro* by *Escherichia coli* polA⁻ extracts. *Proc. Nat. Acad. Sci. U. S. A.* 69: 25.

14. Schaller, H., Voss, H., and Gucker, S. 1969. Structure of the DNA of bacteriophage fd. II. Isolation and characterization of a DNA fraction with double strand-like properties. *J. Mol. Biol.* 44: 445.

15. Schekman, R., Wickner, W., Westergaard, O., Brutlag, D., Geider, K., Bertsch, L., and Kornberg, A. 1972. Initiation of DNA synthesis. III. Synthesis of $\phi\chi 174$ replicative form requires RNA synthesis resistant to rifampicin. *Proc. Nat. Acad. Sci. U. S. A.* 69: 2691.

16. Sekiguchi, M., and Iida, S. 1967. Mutants of *Escherichia coli* permeable to actinomycin. *Proc. Nat. Acad. Sci. U. S. A.* 58: 2315.

17. Siddhikol, C., Erbstoeszer, J. W., and Weisblum, B. 1969. Mode of action of streptolydigin. *J. Bacteriol.* 99: 151.

18. Silverstein, S., and Billen, D. 1971. Transcription: Role in the initiation and replication of DNA synthesis in *Escherichia coli* and $\phi\chi 174$. *Biochim. Biophys. Acta* 247: 383.

19. Sippel, A., and Hartmann, G. 1968. Mode of rifamycin on the RNA polymerase reaction. *Biochim. Biophys. Acta.* 157: 218.

20. Sobell, H. M., Jain, S. C., and Sakore, T. D. 1971. Stereochemistry of actinomycin-DNA binding. *Nature* 231: 200.

21. Sugino, A., Hirose, S., and Okazaki, R. 1972. RNA-linked nascent DNA fragments of *Escherichia coli*. *Proc. Nat. Acad. Sci. U. S. A.* 69: 1863.

22. Wickner, W., Brutlag, D., Schekman, R., and Kornberg, A. 1972. RNA synthesis initiates *in vitro* conversion of M13 DNA to its replicative form. *Proc. Nat. Acad. Sci. U. S. A.* 69: 965.

Discussion

Alberts. Have you checked to see which labeled deoxyribonucleotide transfers its α-^{32}P to the ribonucleotide? Is there any specificity?

Schekman. In the conversion of the M13 single strand to replicative form, any of the four deoxyribonucleotides can transfer α-^{32}P to ribosomal AMP;

however, deoxyguanosine seems to predominate. The equivalent experiment has not been done with ϕχ.

Fidanian. It would appear from the work that you have presented and the work of Doctors Okazaki and Gefter presented earlier, that the length of the RNA primer rather than its sequence is important in determining DNA synthesis. The RNA segments of Okazaki pieces end in rU, and the ϕχ RNA primer ends in either rA or rG.

Schekman. We have no data regarding the specificity of DNA synthesis with RNA primers of different size or sequence using the enzymes of this reaction.

Fidanian. The second question is, aren't you surprised that streptolydigin does not block ϕχ replicative form synthesis?

Schekman. No.

Finadian. Why not?

Schekman. If RNA polymerase is not involved in the reaction, then it is not surprising that a drug specific for RNA polymerase would not inhibit the reaction.

Ryder. Randy, do you think that the stimulating of DNA synthesis by ATP can be accounted for by the requirement of ATP in RNA primer formation?

Schekman. Possibly. We have found that in a partially purified enzyme fraction the requirement for ATP is much higher than for the other ribotriphosphates. Perhaps, as in the case of RNA polymerase, the K_m for the initiating nucleotide is higher than the K_m for nucleotides involved in propagation.

Smith. Is there any evidence that a preformed ribooligonucleotide could substitute for the four ribonucleoside triphosphates and RNA polymerase in the system?

Schekman. We haven't tried that experiment. The only experiments that we have done on model systems are similar to those reported by Dr. Gefter yesterday with RNA and DNA polymerase.

Fidanian. I have one more question about the complex that you have purified. Does this complex replicate ϕχ or M13?

Schekman. The crude extract shows absolute discrimination between ϕχ and M13 templates. However, a partially purified fraction which uses ϕχ as a template will also use M13 as a template, with both reactions resistant to rifampicin.

Ray. In the two-stage reaction where the DNA template is isolated after RNA primer formation, does subsequent DNA synthesis require ATP?

Schekman. We don't know.

In Vitro Studies on Escherichia coli DNA Replication Factors and on the Initiation of Phage λ DNA Replication

A. Klein, V. Nüsslein, B. Otto, and A. Powling

Friedrich-Miescher-Laboratorium der Max-Planck-Gesellschaft
74 Tübingen, Germany

Dna ts mutants of *Escherichia coli* have been characterized biochemically by *in vitro* complementation. Three complementation groups have been found which coincide with the genetically defined groups dna B, dna E, and dna G. The purification of dna E gene product (polymerase III) and of dna G gene product (necessary for the formation of small Okazaki pieces) is described.

In vitro initiation of λ DNA replication has been achieved. It depends on the products of λ genes O and P which are shown to be complementable *in vitro*. In addition a functional RNA polymerase is required for the initiation reaction. Besides the first initiation event, reinitiation of λ DNA replication can be demonstrated by density shift experiments.

DNA replication comprises a number of sequential processes, only some of which have been investigated experimentally (initiation, synthesis, joining), some of which have only been inferred from theory (unwinding), and conceivably some of which have never been thought of.

For the analysis of these processes *in vitro* systems are desirable, in which the macromolecules participating in replication can be exchanged. Two such systems have been developed recently (13, 14). Synthesis in these systems depends on the proper functioning of several different proteins which are required for *in vivo* replication. This can be shown by the use of dna mutants, which are defective in essential replication functions at elevated temperatures. When the *in vitro* systems are prepared from such mutants they fail to synthesize DNA at nonpermissive temperatures. Addition of wild type crude extracts can restore the activity of such *in vitro* systems. This complementation of *in vitro* systems which show impaired synthesis due to mutational defects can be used as an assay for proteins which are required for replication.

Using this complementation assay we have purified two essential proteins, the dna E and the dna G gene products. The dna E protein is DNA polymerase III (3, 8). The data dna G protein seems to be required for the initiation of Okazaki pieces (9); its exact function is unknown. Complementation experiments which have led to the isolation and characterization of polymerase III and dna G proteins are described briefly at the beginning of this paper.

While in the first part of this paper we deal with proteins which are required for the fork movement, we will show in the second part that systems which are impaired in initiation of chromosomal replication can also be complemented *in vitro*. In these investigations we studied the replication of bacteriophage λ DNA, taking advantage of the initiation mutants λO and λP.

In Vitro Complementation of *Escherichia coli* Mutants Defective in DNA Replication

Dna mutants have been isolated in different laboratories (for review see reference 4) and genetically characterized (15). Not included in this review was a set of approximately 50 mutants which we have isolated recently from a polymerase I-negative strain (16). Twenty-six of these mutants show reduced DNA-synthesizing activity when tested *in vitro*. The *in vitro* nonpermissive temperature varies from strain to strain and is in some mutants drastically lower than the one observed *in vivo*. As characterized by complementation tests (see below), five of the mutants belong to the dna E group, eight to the dna B group, and one to the dna G group.

In order to classify the mutants according to their complementation groups we mixed two strains to be tested and compared the *in vitro* activity of the mixed lysates with the activity of the pure lysates. If the strains complement (see Fig. 1), they belong to different groups; if not, they belong to one group (or are noncomplementable). So far we have not found any noncomplementable strains. As shown in Fig. 1c the B strain is complemented by a mitomycin C-inhibited wild type lysate. For the E and G mutants complementability is shown by a second type of complementation test. A crude protein preparation of *E. coli* wild type stimulates the DNA-synthesizing activity in these strains (Fig. 2).

Since unspecific stimulation can be observed with the system (*e.g.* pancreatic nuclease stimulates all dna E^+ strains) the specificity of each

INITIATION AND REPLICATION FACTORS 187

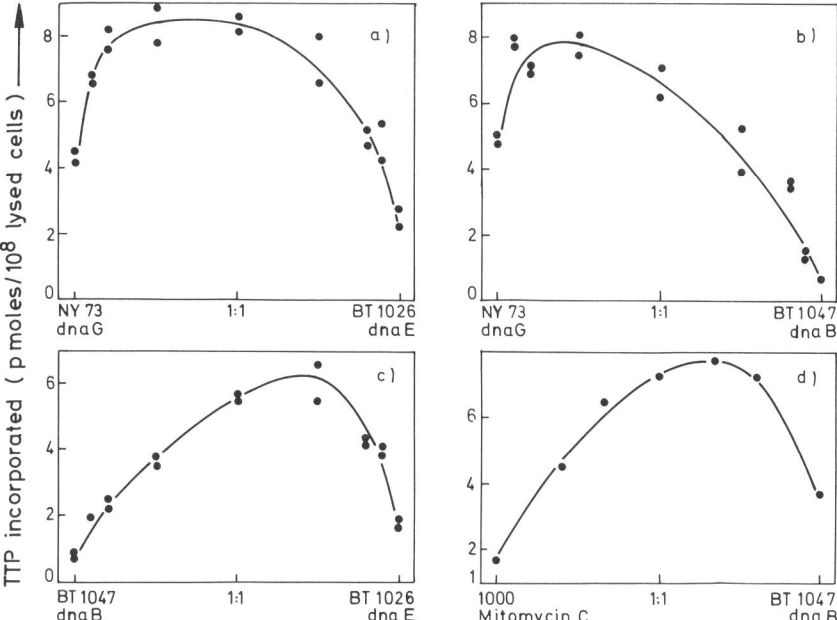

Fig. 1. Complementation of dna ts mutants in mixed lysates. Strains NY73 (a gift from Dr. J. A. Wechsler), BT1026, BT1047, and strain 1000, the dna⁺ parent of BT1026 and BT1047, were grown in penassay broth, supplemented with 2 µg/ml ^{14}C-thymidine (640 mCi/mole) at 30°C to 2×10^8 cells/ml. Strain 1000 cells were treated with mitomycin C (50 µg/ml) at 30°C (10). Cells were cooled, mixed in the desired ratios, washed with buffer A (40 mM Tris-HCl, pH 7.8, 10 mM EGTA, 20% w/v sucrose), and resuspended in the same buffer at 5×10^{10} cells/ml at 0°C. One microliter of the cell suspension together with 1 µl of lysozyme solution (1 mg/ml of lysozyme, 1% Brij-58 in buffer A) was spread on a cellophane membrane disc (1.2 cm ϕ) sitting on an agar plate A (2% Bacto agar, 40% w/v glycerol in buffer A) and the plate stored at −20°C. For lysis the discs were transferred onto agar plate B (2% Bacto agar in buffer B: 20 mM Tris-HCl, pH 7.8, 5 mM MgCl$_2$, 0.1 mM EDTA) and were air-dried on the plate in the cold. The discs were incubated for 15 min on 50 µl of incorporation mixture at 37°C (100 mM KCl, 1 mM ATP, 20 µM each of dATP, dCTP, dGTP, ^{3}H-TTP (500 Ci/mole), 170 µM thymidine in buffer B). For stopping and determination of acid-precipitable radioactivity, discs were placed upside down on glass filters soaked with 0.05 N NaOH, 0.05% sodium dodecyl sulfate, 1% saturated sodium pyrophosphate, 0.01 mg/ml of calf thymus DNA. The discs were taken off and the filters washed in 0.25 M trichloroacetic acid, 0.05 M trichloroacetic acid, and methanol, dried, and counted.

complementation has to be checked. We regard the stimulation to be specific if at least one other mutant exists which does not complement the system under the test conditions, and if in the case of complementation with extracts the amount of cell extract necessary for complementation is reasonably low. In the case of E and G mutants further tests of specificity have been used (see below). Both complementation assays give rough estimates of the ratios of the activities present in

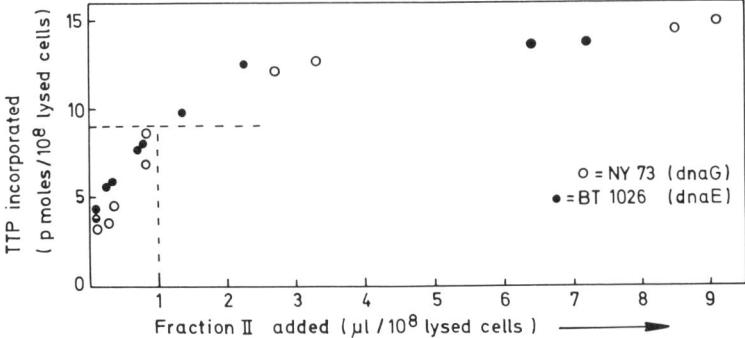

Fig. 2. *In vitro* complementation of dna ts mutants by addition of cell extracts. Preparation of Fraction II. Strain 1000 cells were grown in penassay broth complemented with thymine to 4×10^8 cells/ml. Cells were harvested, suspended in standard buffer (20 mM Tris-HCl, pH 7.5, 0.1 mM EDTA, 0.1 mM DTE, 10% glycerol), and sonicated. After removal of cells debris (30-min centrifugation at 15,000 × g) DNA was precipitated by addition of 2% streptomycin sulfate and removed by centrifugation. One milliliter of Fraction II contains the extract of 7×10^{10} cells. Lysate-carrying discs were prepared as described in the legend to Fig. 1 and stored (up to 1 week) at −20°C. The discs were transferred to an agar plate B, the desired amount of the protein fraction, diluted with buffer B to give 5 μl, was added, and the discs were air-dried in the cold. Dna E lysate-carrying discs were incubated floating on incubation mixture for 15 min at 37°C. Dna G lysates were incubated for 10 min at 37°C following a preincubation period of 10 min on nonradioactive incubation mixture at the same temperature.

the cell to the activities necessary for DNA replication. From Fig. 2 it can be seen that cell extracts prepared from 7×10^7 cells cause half-maximal stimulation of 10^8 mutant cells. We have been unable to find much higher activities in various kinds of cell preparations. This might indicate that there is no surplus of active E and G gene products present in the cells. The B function seems to be more limiting than E or G (see Fig. 1, b and c), which indicates that the B function may be either limiting or poorly complementable in the system.

Purification of E and G Gene Products

With the help of the complementation test we have purified the E and G proteins. Dna E activity was compared with DNA polymerase III activity (3, 8) throughout the purification procedure.

No activity is lost in the DNA precipitation step (see legend to Fig. 2). Both activities are precipitated with 42% ammonium sulfate. They bind to single stranded DNA agarose columns (13) when loaded at an ionic strength corresponding to 50 mM KCl. E protein is eluted from these columns with 200 mM KCl; G protein is eluted over a wide range of ionic strength up to 500 mM KCl. E protein was further purified by gel filtration on a 1.5 M agarose column. The main activity and also polymerase III activity are found at a position which corresponds to 150,000 daltons molecular weight. In some cases we have observed a trailing or a second peak (between 15 and 50% of dna E activity) at about 60,000 daltons molecular weight. This material is relatively low in polymerase III activity and has not yet been investigated any further. Polymerase III was further purified on a phosphocellulose column (pH 7.5 loaded with 40 mM KCl and eluted between 120 and 140 mM KCl). With the exception of the extra E activity found by gel filtration, the ratio of dna E activity to polymerase III activity stayed quite constant throughout the purification procedure. The enzyme has been purified 20,000-fold but nevertheless is not electrophoretically pure. Analyzing the four major protein bands found in sodium dodecyl sulfate gels and the specific activities of purified enzyme and crude cell extracts, we calculated the number of enzyme molecules extractable from one cell to be between 5 and 15.

G protein eluted from the DNA agarose column did not bind to phosphocellulose at pH 7.5 and 40 mM KCl. The unadsorbed proteins were further purified and concentrated on a second DNA agarose column. At this stage the protein is purified 10,000-fold. The specificity of the purified activity has been tested on the basis of Lark's (9, 11) observations that the dna G mutant is unable to synthesize small Okazaki pieces at nonpermissive temperature. Thus, addition of wild type G protein should not only raise DNA-synthesizing activity but also specifically stimulate the synthesis of small pieces. A sedimentation analysis of DNA intermediates synthesized in the mutant after maximal stimulation with the purified activity (Fig. 3) shows a pattern similar to the one obtained with wild type lysates, whereas the unstimulated or half-maximally stimulated mutant is unable to synthesize short Okazaki pieces. Wild type cells can also be influenced by G factor in that G factor added to lysed wild type cells results in a shift of Okazaki pieces to smaller molecular weight (data not shown). This last result we take as further indication that there is no high surplus of G factor in the cells.

Initiation of Phage λ DNA Replication *in Vitro* (7)

Two functions coded for by phage λ genes O and P are known to play a role in the initiation of λ DNA replication *in vivo* (5). Besides these initiation proteins transcription near the origin of replication is required for initiation (2). In the *in vitro* system described here we have obtained initiation of λ DNA replication which also depends on these three functions (Table 1).

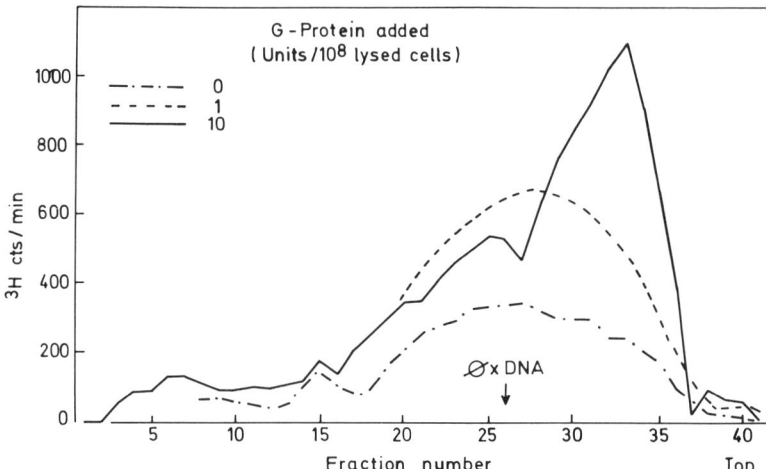

Fig. 3. Sedimentation profile of denatured DNA synthesized *in vitro* by the dna G mutant with and without addition of G factor. Lysate-carrying discs (2.4 cm φ; approximately 1.5 × 10⁸ cells per disc), with and without additional purified G activity, were prepared as described in the legends to Figs. 1 and 2. Incubation at 37°C was for 5 min on nonradioactive and for 20 min on radioactive incorporation mixture supplemented with 1 mM NMN for ligase inhibition and 0.3 mM each of rGTP, rUTP, and rCTP. The discs were heated to 60°C for 5 min in 10 mM Tris-HCl, pH 8.0, 5 mM EDTA, 0.5% sarkosyl. DNA was carefully removed from the discs, layered on top of a sucrose gradient (5 to 20% sucrose in 0.5 M NaCl, 10 mM Tris, pH 8.0, 1 mM EDTA, 0.1% Sarkosyl) and spun at 37,000 rpm at 20°C for 4.5 hr in a Spinco SW 41 rotor. Acid-precipitable radioactivity was determined.

Table 1. *In vivo* and *in vitro* complementation of O and P products: Effect of rifampicin (RIF) on the *in vitro* λ DNA synthesis

	Radioactivity bound to λ DNA filter (cpm/10^4 cpm input)	
Cells used for system preparation	− RIF	+ RIF (200 μg/ml)
Uninfected cells	7	
Cells infected with λCI 0125	9	
Cells infected with λCI P80	17	
Cells infected with λCI 0125 and cells infected with λCI P80 mixed 1 : 1 before lysis	166	33
Cells mixedly infected with λCI 0125 and CI P80	785	580

The system was prepared as described in Fig. 4 with the following modification: Mitomycin C treatment was omitted. Multiplicity of infection was 8 for single and 4 + 4 for mixed infections. Incubation of infected cells was for 7 min. When indicated rifampicin (200 μg/ml) was added and the culture incubated for another 2 min before chilling, and the drug was present throughout the preparation and incubation of the system at the same concentration. *In vitro* incubation (8 × 10⁸ cells/4.7 cm φ disc) was for 15 min at 34°C. The discs then were incubated in 0.1 M EDTA, 20 μg/RNase for 1 hr at 37°C and with 400 μg/ml of Pronase for another 12 hr at 60°C. DNA was dialyzed, phenol-extracted, and hybridized (17) in the presence of excess nonradioactive *Escherichia coli* DNA on 2.4-cm nitrocellulose filters loaded with 40 μg of λ DNA. Values are averages of three experiments.

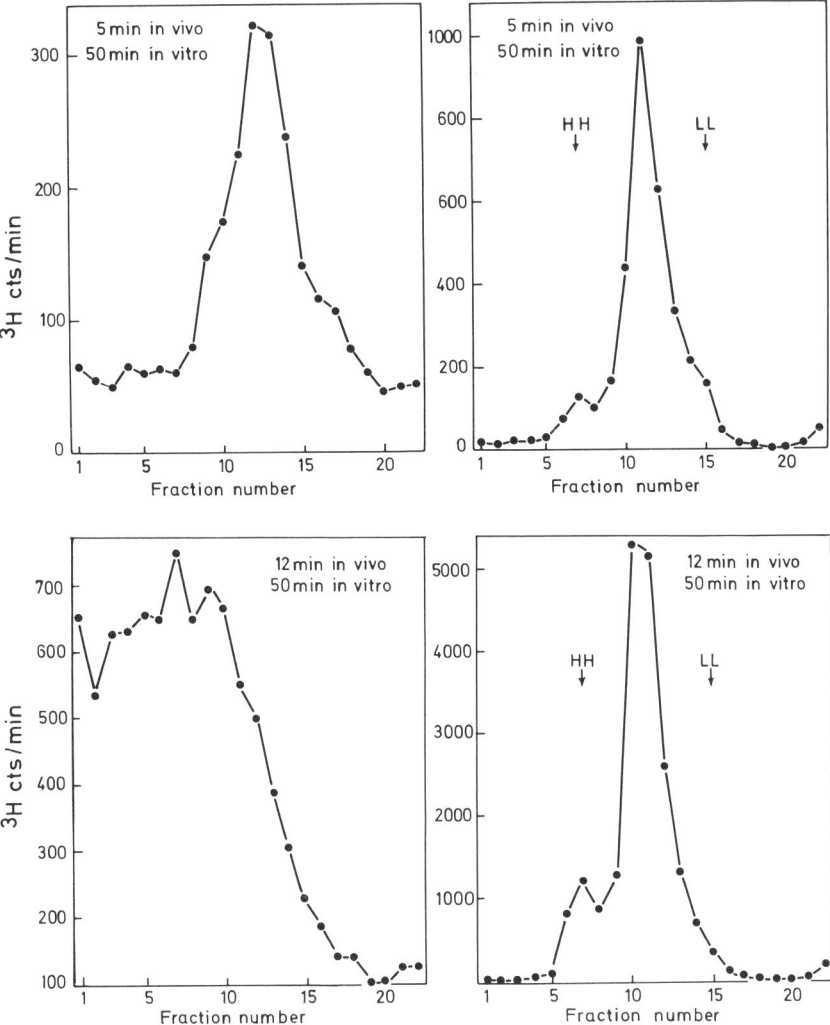

Fig. 4. Sedimentation (left) and density (right) analysis of λ DNA synthesized *in vitro*. *Escherichia coli* H502 was grown in penassay broth supplemented with 2 μg/ml of thymine at 37°C to 2×10^8 cells/ml. Cells were suspended in 10 mM Tris-HCl, pH 7.5, 10 mM $MgCl_2$ at 2×10^9 cells/ml and treated with mitomycin C (10). KCN, 3 mM, was added and the cells were infected at a multiplicity of 8. After 15-min adsorption at 37°C the cells were suspended in penassay broth + 20 μg/ml of thymine at 2.5×10^8 cells/ml at 37°C and aerated for 5 or 12 min. Cells were chilled and the *in vitro* system prepared as described in the legend to Fig. 1. The discs were incubated for 50 min at 26°C on incorporation mixture (500 Ci/mole of ^{3}H-dCTP. TTP was replaced by dBUTP) complemented with rCTP, rGTP, rUTP, 1 mM each and 0.1 mM NAD. For sedimentation analysis the reaction was stopped by putting the discs into 0.1 N NaOH-0.1 M EDTA. DNA was carefully removed from the discs, layered on top of a 5 to 20% alkaline sucrose gradient and spun at 38,000 rpm at 15°C for 3.5 hr in a Spinco SW 41 rotor. For density analysis the reaction was stopped in 10 mM Tris-HCl, pH 8.0, 5 mM EDTA, 0.5% Sarkosyl, 0.5 mg/ml of Pronase. The mixture was incubated for 1 hr at 37°C, the disc was removed and DNA was sheared by passing the solution repeatedly through an 18-gauge syringe. CsCl was added to a final density of 1.73 g/cm^3 and the solution was spun in a Spinco SW 56 rotor for 24 hr each at 35,000 and 25,000 rpm. Radioactivity was determined after acid precipitation. Sedimentation in the left panel is from right to left. Marker λ phage DNA sedimented to fraction 12.

Lysates from cells infected with λO^- or λP^- mutants do not synthesize λ DNA *in vitro*. However, a mixed lysate from cells infected with either λO^-P^+ or λO^+P^- phage does synthesize λ DNA. This shows that the two gene products complement *in vitro* as they do *in vivo*. We can thus study initiation of replication *in vitro*. This initiation needs a functional RNA polymerase. Rifampicin, present during preparation and incubation of the system, inhibits λ DNA synthesis to a much larger extent when initiation has to occur *in vitro* than when initiation has already occurred in mixedly infected cells prior to the system preparation (Table 1, line 5). Rifampicin sensitivity of *in vitro* initiation is not observed when rifampicin-resistant bacteria are used which contain a modified RNA polymerase (data not shown). As pointed out to us by W. F. Dove, there is a remote possibility that the O or P products, or both, are extremely unstable *in vivo*, so that they have already decayed during the 2-min period during which rifampicin was administered to the cells prior to the system preparation. This point is presently under investigation.

Replication and Reinitiation of λ DNA *in Vitro*

Replication of λ DNA is known to proceed according to different mechanisms at different stages of infection (1). In the early phase rings are replicated and in a later phase long concatomers are synthesized on a rolling circle. Both phases can be observed *in vitro*. λ-Infected cells lysed in early phase produce λ-specific DNA with a molecular weight of λ DNA. *In vitro* incorporation of precursors into λ-specific DNA of higher molecular weight is found only if the phage was allowed to start late-phase replication *in vivo* (Fig. 4).

In both cases a small but reproducible amount of native DNA is of the heavy-heavy type when heavy precursors (dBUTP instead of dTTP) are incorporated *in vitro* for more than 10 min (Fig. 4). This result shows that *in vitro* synthesized λ DNA acts as a template for DNA synthesis, which means that reinitiation occurs *in vitro*.

Acknowledgments

We thank Dr. F. Bonhoeffer for his encouragement and helpful criticism and Miss J. Huf for her diligent assistance.

References

1. Carter, B. J., Shaw, B. D., and Smith, M. G. 1969. Two stages in the replication of bacteriophage λ DNA. *Biochim. Biophys. Acta* 195: 494.
2. Dove, W. F., Inokuchi, H., and Stevens, W. F. 1971. Replication control in phage Lambda. *In* Hershey, A. D. (ed.) *The bacteriophage Lambda*, p. 747, Cold Spring Harbor Monograph Series.
3. Gefter, M. L., Hirota, Y., Kornberg, T., Wechsler, J. A., and Barnoux, C. 1971. Analysis of DNA polymerase II and III in mutants of *E. coli* thermosensitive for DNA synthesis. *Proc. Nat. Acad. Sci. U. S. A.* 68: 3150.
4. Gross, J. 1972. DNA replication in bacteria. *Curr. Top. Microbiol. Immunol.* 57:40.
5. Joyner A., Isaacs, L. N., Echols, H., and Sly, W. S. 1966. DNA replication and messenger RNA production after induction of wild type bacteriophage λ and λ mutants. *J. Mol. Biol.* 19: 174.
6. Kaiser, D. 1971. Lambda DNA replication. *In* Hershey, A. D. (ed.) *The bacteriophage Lambda*, p. 195, Cold Spring Harbor Monograph Series.
7. Klein, A., and Powling, A. 1972. *In vitro* initiation of lambda DNA replication. *Nature New Biol.* 239: 71.
8. Kornberg, T., and Gefter, M. L. 1971. Properties of DNA polymerase II. *Proc. Nat. Acad. Sci. U. S. A.* 68: 762.
9. Lark, K. G. 1972. Genetic control over the initiation of the synthesis of short deoxynucleotide chains in *E. coli*. *Nature New Biol.* 240: 237.
10. Lindqvist, B. H., and Sinsheimer, R. L. 1967. The process of infection with bacteriophage ϕχ174. XV. Bacteriophage DNA synthesis in abortive infections with a set of conditional lethal mutants. *J. Mol. Biol.* 30: 69.
11. Olivera, B., Lark, K. G., Herrmann, R., and Bonhoeffer, F. Discontinuous DNA synthesis *in vitro:* A method for defining the role of factors in replication. *DNA Synthesis in vitro*. The Second Annual Harry Steenbock Symposium, Madison, Wisc., July 10-12, 1972.
12. Schaller, H., Nüsslein, C., Bonhoeffer, F. J., Kurz, C., and Nietzschmann, I. 1972. Affinity chromatography of DNA binding enzymes on single-stranded DNA-agarose columns. *Eur. J. Biochem.* 26: 474.
13. Schaller, H., Otto, B., Nüsslein, V., Huf, J., Herrmann, R., and Bonhoeffer, F. 1972. DNA replication *in vitro*. *J. Mol. Biol.* 63: 183.
14. Scheckman R., Wickner, W., Westergaard, O., Brutlag, D., Geider, K., Bertsch, L. L., and Kornberg, A. Initiation of DNA synthesis. *DNA synthesis in vitro*. The Second Annual Harry Steenbock Symposium, Madison, Wisc., July 10-12, 1972.
15. Wechsler, J. A., and Gross, J. D. 1971. *E. coli* mutants temperature-sensitive for DNA synthesis. *Mol. Gen. Genet.* 113: 273.

16. Wechsler, J., Nüsslein, V., Otto, B., Klein, A., Bonhoeffer, F., Herrmann, R., Gloger, L., and Schaller, H. 1973. Isolation and characterization of thermosensitive *E. coli* mutants defective in DNA replication. *J. Bacteriol.* In press.

17. Winocour, E. 1968. Further studies on the incorporation of cell DNA into polyoma related particles. *Virology* 34: 571.

Discussion

Alberts. You said that two parts purified together; was that G and E or B?

Nusslein. E and G don't purify together. The second E activity has the same molecular weight as G so they purify together on the agarose column.

Alberts. What about B?

Nusslein. We didn't find B in the supernatant. From the experiments I've shown in the first slide, we believe that B is poorly complemented in our system and I don't know why we didn't find it in the supernatant. We should use one of the methods developed here in the States.

Alberts. Do you know whether the G binds the DNA agarose?

Nusslein. Yes, it binds when applied at very low salt, but it comes off (varying from experiment to experiment) at different salt concentrations.

Alberts. Gene 32 protein also does not bind to phosphocellulose, but binds to DNA.

Szybalski. In the λ experiments, was there always endogenous λ DNA template?

Nusslein. We have tried to shift added DNA to heavy density, but have not succeeded so far.

Szybalski. It also would help to have some λ DNA mutants to be able to have the same origin but different DNA.

Proteins of the T4 Bacteriophage Replication Apparatus

J. Barry, H. Hama-Inaba, L. Moran, and B. Alberts

Department of Biochemical Sciences
Princeton University
Princeton, New Jersey 08540

J. Wilberg

Department of Radiation Biology and Biophysics
University of Rochester School of Medicine
Rochester, New York 14642

An *in vitro* complementation assay for T4 DNA synthesis in cell lysates has been developed. Use of this assay has enabled both the T4 gene 45 protein and a complex of gene 44 and 62 proteins to be purified to homogeneity in active form. These proteins are required for T4 DNA replication *in vivo*, but their exact function is unknown. It is postulated that they function in a replication apparatus containing at least three other phage-specified proteins: the products of genes 43 (T4 DNA polymerase), 32 (DNA-unwinding protein), and 41. This last protein has not yet been purified in active form.

Escherichia coli cells infected with bacteriophage T4 amber mutants in genes 32, 41, 43, 45, 44, or 62 synthesize little or no DNA, even though all four deoxyribonucleoside triphosphates are present (12, 35). For the first four of these gene products, temperature-sensitive variants have been isolated which cause T4 DNA replication to cease so abruptly after a shift from 25 to 42°C (normally a permissive temperature) that replication forks appear to stop completely before moving even one-fifth the length of one mature T4 genome (26; M. Curtis and B. Alberts manuscript in preparation). These gene products are therefore likely to be *directly* involved in building the T4 DNA replication apparatus (4). As illustrated in Fig. 1, five of the above "replication genes" occupy

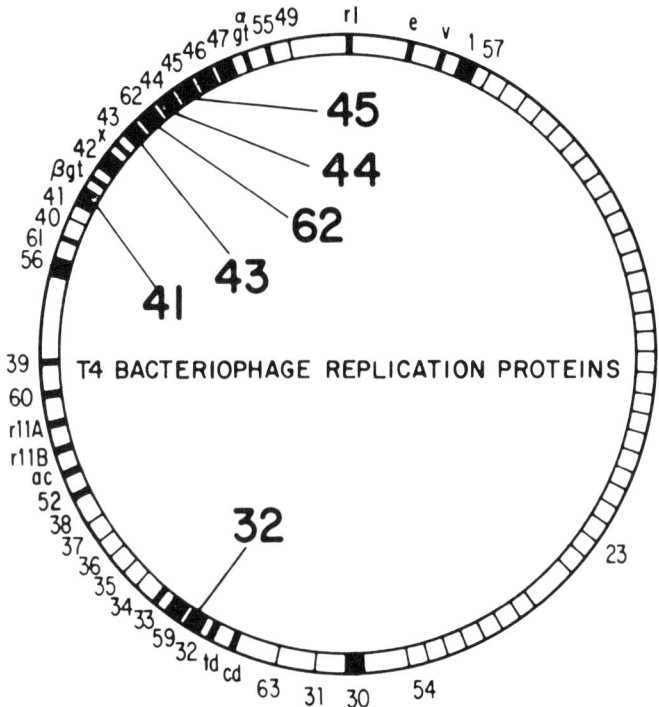

Fig. 1. A genetic map of T4 bacteriophage with DNA replication genes emphasized. The broadest black segments indicate the relative map locations of mutationally identified genes with a dramatic effect on DNA replication. These genes are assumed to be only indirectly involved in the polymerization process if their functional absence allows normal initial DNA synthesis, with "DNA arrest" only at later times (genes 30, 46, 47, and 59; references 17, 18, 21, 29, 38), or if they are necessary for providing the proper deoxyribonucleotide triphosphates (genes 1, 42, and 56). However, Chiu and Greenberg (8) have suggested that the gene 42 protein (dCMP hydroxymethylase) may have an additional, more direct role in replication; and this could be true for the products of genes 56 and 1 as well. The narrowest lines represent genes affecting phage morphogenesis, as defined by Wood et al. (37); on the right half of the map most of these gene numbers have been omitted for clarity. Segments of intermediate width are used to denote genes not included in either of the first two classes. Among these are mutants in genes 39, 52, 60, and 61, which have a "DNA delay" phenotype (35, 39) (adapted from reference 37).

adjacent regions of the T4 genetic map; this might suggest that their proteins mutually interact to form some larger complex (33).

The products of T4 genes 43 and 32 have been shown to be a DNA polymerase (11, 34) and a DNA-unwinding protein (1), respectively; both of these proteins have been previously purified and characterized in detail (3, 5, 9, 14). Besides their separate activities, the DNA-unwinding protein and the DNA polymerase appear to form a loose complex with each other; in addition, the presence of the DNA-unwinding protein can increase the *in vitro* rate of polymerization by the polymerase on an exonuclease III-degraded, double stranded DNA template by 5 to 10-fold (19).

Recently, it has been possible to demonstrate that this polymerase stimulation is specific: whereas 32 protein is without effect on the three known *E. coli* DNA polymerases, an unwinding protein isolated from uninfected *E. coli* (N. Sigal and B. Alberts, in preparation) strongly stimulates its homologous DNA polymerase (*E. coli* polymerase II), but not the T4 polymerase (Gefter *et al.*, this symposium). This specificity implies that polymerase and unwinding protein interact together *in vivo*, as well as *in vitro*.

In order to decipher their roles in T4 DNA replication, the products of genes 41, 44, 45, and 62 must also be identified and purified. For this purpose we have developed an *in vitro* DNA-synthesizing system which shows a requirement for these gene products, and which can be used to provide an assay for their purification, despite the fact that their function in replication is unknown (6). This system consists of a concentrated T4-infected cell lysate which, when supplied with deoxyribonucleoside triphosphates, supports a brief period of rapid synthesis using the endogenous DNA as template (25, 32). We have found that when such lysates are prepared from replication-defective cells, the DNA synthesis they support is specifically stimulated by addition of extracts which contain the missing gene product. Using this complementation test as an assay, it has been possible to purify both the gene 45 protein and a complex of gene 44 and 62 proteins to apparent homogeneity in an active form.

Materials and Methods

Bacteria, Bacteriophage, and Enzymes

The host strain for all experiments was *E. coli* D110 (Pol A1, endI$^-$, thy$^-$, su$^-$) obtained from Dr. C. C. Richardson (23). The following T4 mutants were obtained from the Cal Tech collection: *am* HL618 (gene 32$^-$), *am* B22 (gene 43$^-$), *am* N81 (gene 41$^-$), *am* N82 (gene 44$^-$), *am* E10 (gene 45$^-$), *am* E1140 (gene 62$^-$), and *am* BL292 (gene 55$^-$). In addition, T4 phage SP62-*am*N55 (gene ?, gene 42$^-$), whose SP62 genotype causes it to overproduce several early gene products when DNA replication is blocked, was used (Wiberg, Mendelsohn, Warner, Hercules, Aldrich and Munro, manuscript submitted). Purified T4 DNA polymerase was a gift from Dr. Wai Mun Wang; T4 gene 32 protein was prepared according to the method of Alberts and Frey (3).

Preparation of Infected Cells

E. coli D110 was grown in log phase to a concentration of 4×10^8 cells/ml in M-9 minimal medium supplemented with 0.3% casein hydrolysate, 1 µg/ml of thiamine, and 0.02 mg/ml of thymidine. The cells were infected by adding the appropriate mutant T4 bacteriophage at a multiplicity of infection (m.o.i.) of 5, followed by incubation for 20 min at 37°C (for protein purification from T4 SP62-*am* N55-infected cells, a 60-min incubation was used). The infected cells were harvested by centrifugation at 4°C, washed twice in 20% sucrose containing 0.05 M Tris-HCl, pH 7.4, plus 1 mM Na$_3$EDTA, and stored as a frozen pellet at $-70°C$.

Preparation of Receptor Cell Lysates

A pellet of frozen cells was thawed and evenly suspended in 4 volumes of 25% sucrose, 0.05 M Tris-HCl, pH 7.4. After addition of one more volume of 10 mM Na$_3$EDTA containing 2 mg/ml of egg white lysozyme (Worthington Biochemical), the mixture was incubated for 30 to 60 min in an ice bath. Five more volumes of buffer containing 0.05 M Tris-HCl, pH 7.4, 0.03 M MgSO$_4$, and 1% Brij-58 (Atlas Chemical Industries) were then added, and the incubation was continued for 15 to 30 min to complete lysis (13, 25). The final concentration of the lysed cells was about 4×10^{10} per ml; little clearing on lysis is noticeable at this high cell concentration.

Preparation of Donor Extracts

A lysate prepared as described above was centrifuged at 30,000 × g for 15 min to pellet cell membranes and associated DNA. The supernatant was then centrifuged at 165,000 × g for 35 min to pellet ribosomes. This second supernatant is called the "extract."

Complementation Assay

Fifty microliters of mutant-infected, receptor cell lysate are mixed with 50 µl of donor extract or buffer in an ice bath. A 25-µl aliquot is added to 25 µl of a mixture containing 0.05 M Tris-HCl, pH 7.4, 0.2 mM deoxyadenosine, 2 mM ATP, and 0.04 mM each: dGTP, dCTP, and TTP, and ^{3}H-dATP (250 µCi/µmole). After incubation at 37°C for 20 min, DNA synthesis is stopped by adding 50 µl of 0.2 M Na$_3$EDTA and chilling. The entire mixture (100 µl) is then spotted on a glass fiber filter (Whatman GF-A), batch-washed in 5% trichloroacetic acid and ethanol (7), and dried and counted by standard techniques. In order to quantitate activity, donor extracts are serially diluted into a buffer containing 0.05 M Tris-HCl, pH 7.4, 5 mM MgSO$_4$, 1 mM β-mercaptoethanol, 10% glycerol and 100 μg/ml of bovine serum albumin (CalBiochem); each dilution is then tested by mixing duplicate aliquots with receptor cell lysate as described above.

Results

DNA synthesis in whole cell lysates of T4-infected cells can be quantitated by measuring incorporation of added radioactively labeled, deoxyribonucleoside triphosphates into acid-insoluble material. As shown in Fig. 2A, such synthesis is of brief duration, lasting only a few minutes at 37°C. The product made is stable for at least 30 min of further incubation at 37°C; it can be degraded by added pancreatic DNase but not by pancreatic RNase. Fig. 2B demonstrates that the rate of DNA synthesis measured in lysates decreases below uninfected values when cells are

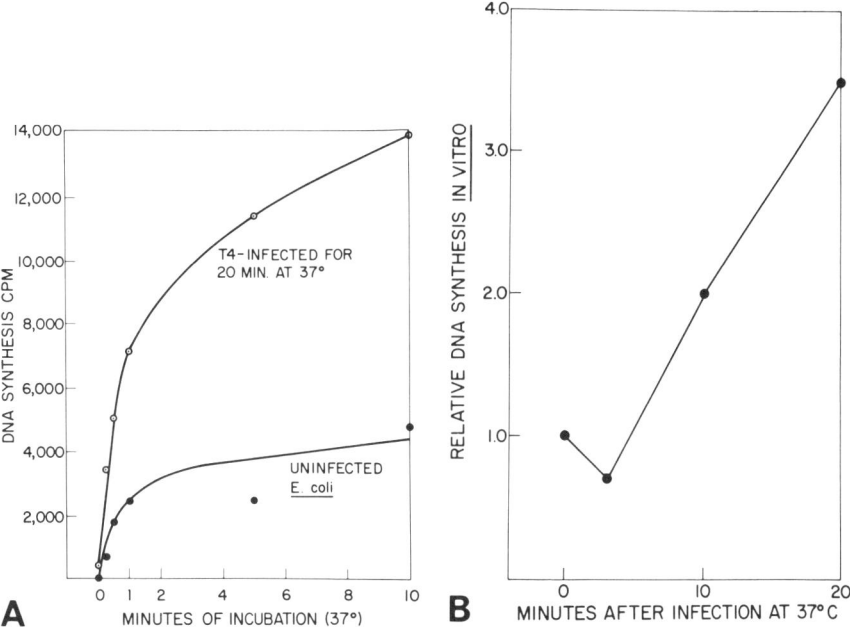

Fig. 2. A, time course of DNA synthesis in concentrated cell lysates (10^{10} cell equivalents/ml). For assay conditions, see "Materials and Methods." The T4 phage used was mutant am BL292 (gene 55$^-$), chosen to inhibit lysis; its DNA replication is normal. B, DNA synthesis $in\ vitro$ with cell lysates prepared from uninfected cells and cells infected with T4-am BL292 for the indicated times at 37°C. In all cases, the lysates are incubated for 10 min at 37°C.

Table 1. Extent of in vitro DNA synthesis in cell lysates

Cell lysate*	cpm	Wild type (%)
Wild type T4†	8750	(100)
T4 Gene 32⁻	1980	22
T4 Gene 41⁻	1700	19
T4 Gene 43⁻	120	1
T4 Gene 44⁻	450	5
T4 Gene 45⁻	810	9
T4 Gene 62⁻	670	8
Uninfected cells	2200	

*For the particular mutants used and details of the assay, see "Materials and Methods." A filter blank of 50 cpm has been subtracted.

†The level of activity in this lysate may be misleadingly enhanced relative to other lysates, since more T4 DNA template is present due to intracellular DNA synthesis.

harvested shortly after infection, and then increases dramatically at later times of infection. This time course thus resembles that for *in vivo* DNA synthesis. We have routinely used cells infected for 20 min at 37°C; for these cells the initial rate of synthesis is very rapid, corresponding to more than 10^3 molecules of ^{3}H-dATP incorporated into DNA per cell per sec. Thus, reminiscent of the *in vitro* system described previously by Smith et al. for uninfected *E. coli* (32), a transient period of DNA synthesis is observed at close to *in vivo* replication rates.

T4 DNA replication *in vivo* exhibits a nearly absolute requirement for the products of T4 genes 32, 41, 43, 44, 45, and 62 (12, 35). *In vitro* DNA synthesis in our whole cell lysates is likewise reduced if cells have been infected with these bacteriophage mutants. As listed in Table 1, the DNA synthesis in these mutant lysates is reduced to anywhere from 20 to 1% of the wild type level, depending on the particular gene product missing.

In Vitro Complementation for DNA Synthesis

In preliminary tests, purified 32 protein stimulated the amount of DNA synthesis in its deficient lysate 1.7-fold, while purified 43 protein (T4 DNA polymerase) stimulated DNA synthesis in its deficient lysate 1.5-fold. Neither protein stimulated DNA synthesis in the opposite lysate. Individual lysates deficient in the products of T4 genes 41, 44, 45, or 62 were then tested for their ability to be complemented by extracts containing these proteins. Extracts were prepared and used for complementation as described under "Materials and Methods." Figure 3 (top) illustrates the level of DNA synthesis in a 62-deficient lysate after addition of varying amounts of an extract made from cells containing the product of gene 62. As shown, the donor extract could be diluted 27-fold and still evoke a 2-fold stimulation of DNA synthesis. However, an extract prepared in exactly the same manner from 62-deficient cells

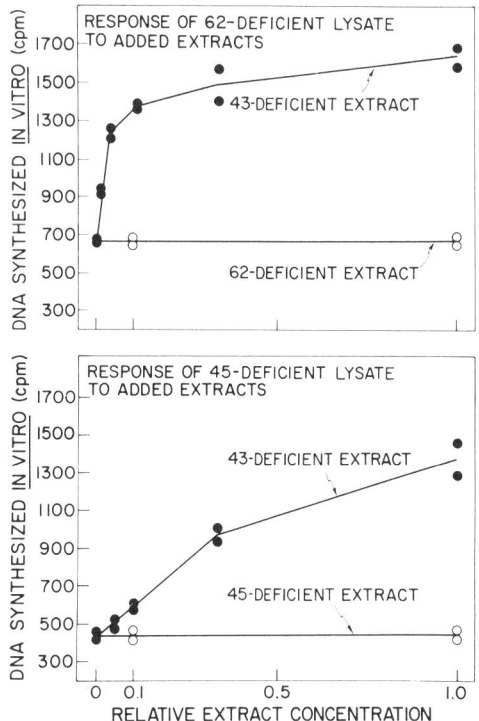

Fig. 3. Complementation of 62- and 45-deficient receptor cell lysates by added donor extracts. For details of the complementation assay and phage mutants used, see "Materials and Methods."

is seen to have no stimulatory effect. Similar complementation of a 45-deficient lysate by the gene 45 product is shown in Fig. 3 (bottom). Analogous complementations have been obtained with receptor cell lysates deficient in the products of genes 44 and 41, and with donor extracts prepared from any of the six mutant-infected cells listed in Table 1. Results of these cross-complementations, expressed as DNA synthesis relative to a buffer control, are presented in Table 2. Note that in all cases addition of a homologous extract is without effect, while heterologous extracts stimulate synthesis from 1.5- to 2.7-fold. This demonstrates that the stimulations observed are due to addition of the missing gene product. An apparent exception to the general pattern in Table 2 is that very little complementation is observed when 44-deficient donor extracts are mixed with 62-deficient lysates, or when 62-deficient donor extracts are mixed with 44-deficient lysates. This initially puzzling observation will be discussed in detail below.

Reproducible response to these gene products requires a high concentration of receptor cell lysate. For example, in the experiment shown

Table 2. Complementation of T4 mutant lysates by various mutant extracts: DNA synthesis (relative to buffer control at 1.0)*

T4 donor extract	T4 receptor cell lysate			
	Gene 62$^-$	Gene 45$^-$	Gene 41$^-$	Gene 44$^-$
Gene 32$^-$	2.7	1.6	2.1	1.7
Gene 41$^-$	2.1	2.2	1.0	2.3
Gene 43$^-$	2.3	1.6	1.9	1.5
Gene 44$^-$	1.2	1.6	1.8	1.0
Gene 45$^-$	2.1	1.0	1.5	1.9
Gene 62$^-$	1.1	2.0	1.6	1.2
Buffer control	(1.0 = 1800 cpm)	(1.0 = 500 cpm)	(1.0 = 2000 cpm)	(1.0 = 600 cpm)

*The incubations were performed with undiluted donor extracts as described under "Materials and Methods," except that 3×10^9 rather that 10^{10} receptor cell equivalents per milliliter were used in the final assay mix. This lower cell concentration probably accounts for the rather poor complementations obtained (see Table 3).

in Table 3, DNA synthesis in a gene 62-deficient cell lysate was stimulated 2.4-fold by two extracts containing 62 protein. However, when this receptor cell lysate was diluted 4-fold to 2.5×10^9 cell equivalents/ml, its DNA synthesis was barely stimulated by the same extracts. This is at least in part due to the relatively high backgrounds ("buffer controls") seen after dilution, but it might also suggest that at least some of the factors necessary for DNA synthesis must be present at very high concentration if efficient complementation is to occur (see also reference 27).

Purification of Gene 62 and Gene 44 Proteins

Using the complementation response as an assay, 62 activity was originally purified from cells infected with T4 am B22 (gene 43$^-$) and later from cells infected with T4 SP62-am N55. The latter cells appear to produce 10 to 20 times as much 62 activity as the former, as estimated by complementation. This is illustrated in Fig. 4, where diluted aliquots

Table 3. Effect of receptor lysate concentration on the extent of complementation observed: DNA synthesis in 62-deficient receptor cell lysate (cpm)

Donor extract	Undiluted receptor lysate*	Receptor lysate diluted 4-fold
Buffer control	1100	870
Gene 62$^-$	1150	700
Gene 41$^-$	2640	1090
Gene 43$^-$	2740	1030

*10^{10} cell equivalents/ml in final assay mix.

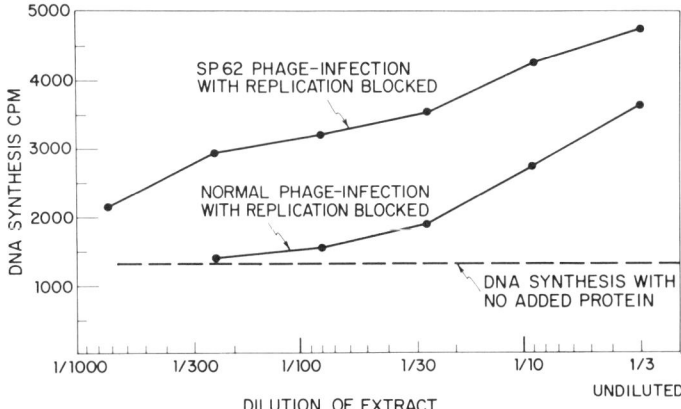

Fig. 4. Response of DNA synthesis in 62-deficient receptor cell lysates to extracts from cells infected with normal and SP62 mutant phage (both cultures infected for 40 min at 37°C). The SP62 mutation is known to cause overproduction of several T4 gene products only in the absence of DNA synthesis (Wiberg et al., manuscript submitted). To stop DNA synthesis, the "normal" phage contained the am B22 mutation and the SP62 phage contained the am N55 mutation.

of donor extracts from the SP62-am N55-infected cells and the am B22-infected cells (both of which are wild type for gene 62 and blocked in DNA replication) have been compared in their ability to stimulate 62-deficient lysates. Overproduction of the replication gene products by phage containing the SP62 mutation was expected from the earlier results of Wiberg (Wiberg et al., manuscript submitted for publication).

The quantitation of gene 62 activity required the assay of several dilutions of the fractions obtained at each stage of purification, with a unit of activity defined as that giving half-maximal stimulation. The purification method developed consists of passing a Brij-lysed extract through a large DEAE-cellulose column; this binds all of the nucleic acids and most of the protein. The gene 62 activity, appearing in the column breakthrough, is further purified by adsorption and elution from a hydroxylapatite column. Preparative isoelectric focusing then yields the final Fraction IV, isoelectric at pH 8.2. Details of this procedure have been published elsewhere (6).

Electrophoresis on sodium dodecyl sulfate-containing polyacrylamide gels (28) followed by protein staining (Fig. 5) shows that Fraction IV consists almost entirely of two proteins, having apparent molecular weights of 34,000 and 20,000 daltons, respectively. As shown in Table 4, this fraction does not stimulate DNA synthesis in lysates deficient in the products of genes 32, 41, 43, or 45. However, both 44- and 62-deficient lysates are complemented. Most significantly, as shown in Fig. 6, the percentage of stimulation of these two lysates is similar

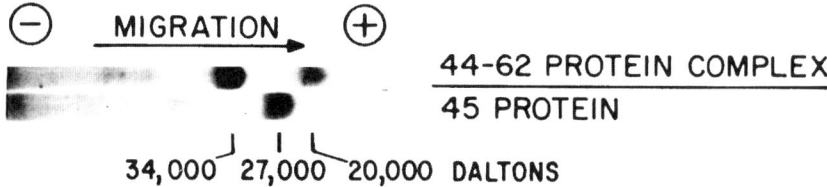

Fig. 5. Electrophoresis of purified gene 45 protein (20 μg) and the gene 44-62 protein complex (60 μg) on sodium dodecyl sulfate (SDS)-containing polyacrylamide gels. Gels were run as previously described (2), stained in 0.25% Coomassie blue in methanol-glacial acetic acid-H_2O (5:1:5 by volume) for 60 min at 50°C, and destained at 50°C overnight in several changes of a 7.5% acetic acid, 5% methanol solution. At this stage of purification, the 45-protein represents 0.3%, and the 44-62 complex 0.5%, of the total extract protein.

Table 4. Complementation of different mutant lysates by the fraction IV protein complex: DNA synthesis *in vitro* (cpm)*

Receptor cell lysate	Concentration of Fraction IV protein in the reaction mixture		Stimulation of DNA synthesis
	(0 μg/ml)	(4 μg/ml)	
Gene 32⁻	2000	2200	1.1
Gene 41⁻	1700	1770	1.0
Gene 43⁻	170	120	0.8
Gene 45⁻	370	350	0.9
Gene 44⁻	450	920	2.0
Gene 62⁻	660	1740	2.5

*For assay conditions, see text.

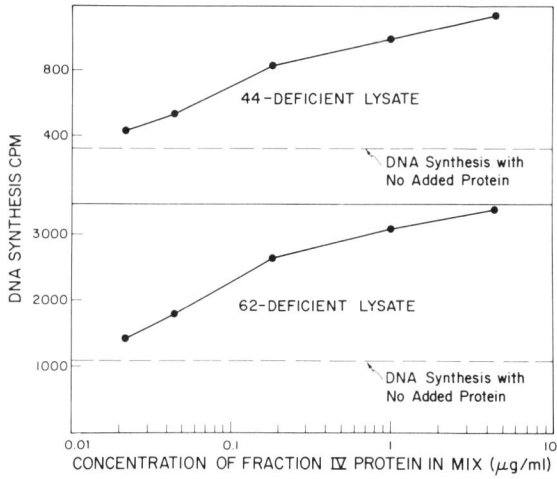

Fig. 6. Response of DNA synthesis in 44- and 62-deficient receptor lysates to dilutions of Fraction IV protein. For assay conditions, see "Materials and Methods."

over a 250-fold concentration range of purified complex. It appears therefore that one of the two proteins in Fraction IV is the gene 62 protein and that the other is the gene 44 protein.

Attempts were made to separate the two proteins in Fraction IV by taking advantage of the difference in their size. However, on a 5 to 20% sucrose gradient, the proteins sedimented together as a homogeneous complex at 7.1 S. (Catalase (11.3 S) and *E. coli* alkaline phosphatase (6.2 S) were used as internal markers, and the gradient buffer contained 0.15 M potassium phosphate, pH 7.0, 5 mM MgSO$_4$, 1 mM β-mercaptoethanol, and 10% glycerol.) With the same buffer on a Bio-Gel A-0.5 M column (Bio-Rad Laboratories), the two proteins eluted together at an apparent (spherical) molecular weight of about 300,000 daltons (in this case, the column was standardized with catalase, alkaline phosphatase, and myoglobin markers). Across the peak, the molar ratio of 34,000-dalton protein to 20,000-dalton protein was constant at about 2 : 1, as estimated by elution of bands stained with Coomassie blue from sodium dodecyl sulfate-polyacrylamide gels (20). Combining the sedimentation and gel filtration data (30) suggests that the complex is actually quite asymmetrical (axial ratio of about 5 : 1 for a prolate ellipsoid), with a molecular weight of about 164,000 daltons.* The ratio of chains in this complex should therefore be 4 : 2 (*i.e.* expected molecular weight of 176,000 daltons).

The radioactive proteins in partially fractionated crude extracts have been analyzed by double label techniques after pulse labeling and frac-

---

*Comparing two proteins, the Svedberg equation gives, assuming $\bar{v}_1 = \bar{v}_2$:

$$\frac{M_1}{M_2} = \frac{S_1}{S_2} \cdot \frac{f_1}{f_2}.$$

Our standard, beef liver catalase, is a roughly spherical protein of 244,000 daltons which sediments at 11.3 S (10, 22). Since frictional coefficients for spherical molecules are proportional to the cube root of their molecular weights, the gel filtration result implies that

$$\frac{f_{44-62}}{f_{catalase}} = 1.07.$$

[$1.07 = (300,000/244,000)^{1/3}$.] Thus

$$M_{44-62} = 244,000 \left(\frac{7.1}{11.3}\right)(1.07) = 164,000 \text{ daltons.}$$

The same S value was obtained for the 44-62 protein complex in the buffer indicated in the text, and in a buffer containing 0.02 M Tris-HCl, pH 8.1, 5 mM MgSO$_4$, 1 mM β-mercaptoethanol, and 10% glycerol.

tionation on sodium dodecyl sulfate-polyacrylamide gels. Comparing different mutant-infected cells (all of which have the SP 62 genotype), we find that a protein with a molecular weight of about 34,000 daltons is missing in a 44-deficient extract and that a protein with a molecular weight of about 20,000 daltons is missing in a 62-deficient extract. It appears then that the heavier component in Fraction IV is the product of gene 44 and the lighter component the product of gene 62.

Attempts to prove this directly by using double label techniques to investigate the isolated 44-62 complex have been frustrated by the fact that this complex does not appear to form from its separate components with any reasonable efficiency *in vitro*. This is clear from the results presented in Table 5, where an extract prepared from a 1 : 1 mixture of SP62-*am* N82-infected cells (44 protein-deficient) and SP62-*am* E1140-infected cells (62 protein-deficient) has been tested for its ability to complement both 62- and 44-deficient receptor lysates. For comparison, results obtained with extracts made from each of these cells separately, and from SP62-*am* N55-infected cells (wild type for both 44 and 62 proteins), are also shown. It can be seen that very little 44-62 protein complex has been generated *in vitro*, since the level of complementation observed with the 1 : 1 mixed extract is only about one one-hundredth the level obtained with the *in vivo* generated complex. This could be due to instability or degradation of one of the components of the complex

Table 5. Attempt to form the 44-62 complex *in vitro* from its separated components: DNA synthesis in the complementation assay (relative to buffer control at 1.0)

Genotype of donor extract	Extract dilution	62-Deficient receptor cell lysate	44-Deficient receptor cell lysate
44^+62^+	1 : 10	3.2	3.2
	1 : 100	2.2	2.9
	1 : 1000	2.0	2.7
44^+62^-	1 : 10	1.1	2.5
	1 : 100	1.1	1.0
	1 : 1000	1.2	0.8
44^-62^+	1 : 10	1.7	0.8
	1 : 100	1.1	0.8
	1 : 1000	1.0	0.9
$44^+62^- + 44^-62^+$ Mixture	1 : 10	2.0	2.7
	1 : 100	1.3	1.5
	1 : 1000	1.0	0.9
Buffer control		(830 cpm = 1.0)	(470 cpm = 1.0)

The mutant phage used contained the SP62 genotype, as described in the text, in order to increase the sensitivity of the assay. For one extract, equal amounts of 44^-62^+- and 44^+62^--infected cells were mixed just prior to lysis. To allow time for complex formation, all extracts were preincubated for 2 hr at 4°C before being assayed.

when the other is absent, or to actual difficulty in complex formation. In any case, the fact that complementation activity of 44-deficient extracts for 62-deficient lysates, and of 62-deficient extracts for 44-deficient lysates (as seen in Table 5, and earlier in Table 2).

Even though we could not demonstrate formation of the complex from its separated components *in vitro*, we hoped that it might be possible to prove that the 20,000-dalton protein in the complex is the product of gene 62 by an appropriate double label experiment. Thus, T4 *am* E1140 (gene 62⁻) infected cells labeled with ^{14}C-leucine after infection were mixed with ^{3}H-leucine-labeled cells infected with a wild type revertant of this phage, and Fraction IV was prepared. As expected, this fraction was highly enriched for the ^{3}H isotope. However, we had anticipated that the 62-deficient lysate would exchange its ^{14}C-labeled 44 protein into the ^{3}H-labeled wild type complex, in which case only the 20,000-dalton protein in the complex should be ^{3}H-enriched. The actual result obtained when the complex was analyzed by sodium dodecyl sulfate polyacrylamide gel electrophoresis is shown in Fig. 7. It can be

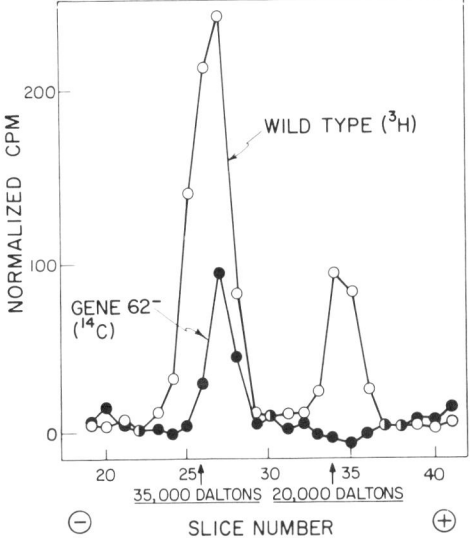

Fig. 7. Electrophoretic analysis of radioactive Fraction IV proteins on a polyacrylamide gel containing sodium dodecyl sulfate. T4 *am* E1140 (gene 62-) infected cells were labeled with ^{14}C-leucine and T4 wild type-infected cells were labeled with ^{3}H-leucine; in both cases labeling was from 5 to 15 min post-infection at 25°. To insure that the two phages differed only by a single mutation, the wild type was selected on *Escherichia coli* B as a spontaneous revertant from the *am* E1140 stock. After isoelectric focusing, the proteins were precipitated with 5% trichloracetic acid after adding lysozyme carrier, and redissolved in gel sample buffer containing 1% sodium dodecyl sulfate. Subsequent sample treatment, electrophoresis, and gel counting techniques were as previously described (2). The counts shown have been corrected for overlap and normalized to a ^{14}C:^{3}H ratio of 1.0 in the original extract.

seen that there are two major peaks of radioactive protein, one at 34,000 daltons and one at 20,000 daltons, and that *both* of these peaks are greatly enriched for ^{3}H. It appears therefore that the complex formed *in vivo* is sufficiently tight that it does not readily exchange components *in vitro*. Our assignment of gene numbers to its two polypeptide chains must therefore remain tentative.

Purification of Gene 45 Protein

Gene 45 protein was originally identified on polyacrylamide gels of crude extracts by Hosoda and Levinthal (16), and is thus a relatively major species. Since it constitutes about 2% of the protein made in a normal infected cell labeled between 5 and 7 min postinfection at 37°C, it was convenient to utilize double label techniques during its purification. However, in order to insure that biological activity was being preserved, the 45 protein obtained at each step was also monitored by the standard complementation assay. A satisfactory purification method developed consists of removal of nucleic acids from a Brij-lysed extract on DEAE-cellulose, followed by gradient elution from a hydroxylapatite column, selective ammonium sulfate precipitation, and chromatography on a DEAE-Sephadex column (Hama-Inaba *et al.*, manuscript in preparation). When T4 SP62-*am* N55-infected cells were used as the starting material, a 300- to 400-fold purification was sufficient to yield a preparation for which sodium dodecyl sulfate polyacrylamide gel electrophoresis and

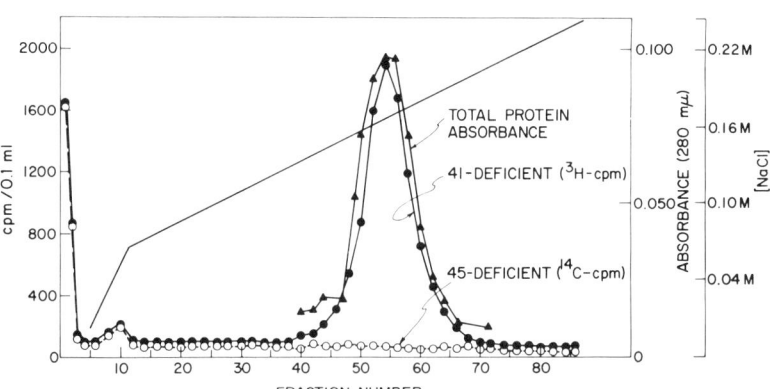

Fig. 8. Final step in the purification of T4 gene 45 protein: salt elution from a DEAE-Sephadex column. As indicated in the text, 41-deficient cells labeled with ^{3}H-leucine and 45-deficient cells labeled with ^{14}C-leucine had been mixed prior to purification with a large excess of unlabeled SP62-am N55-infected cells, from which the majority of the 45-protein was derived. The counts have been normalized to a ^{14}C : ^{3}H ratio in the extract of 1.0; the absence of ^{14}C in the purified protein demonstrates that it was not synthesized in 45-deficient cells. On sodium dodecyl sulfate polyacrylamide gel electrophoresis, the ^{3}H-radioactivity migrated with the major protein band at 27,000 daltons.

protein staining reveal a single polypeptide chain at 27,000 daltons (see Fig. 5, above). A small amount of ^{14}C-leucine-labeled cells carrying an amber mutation in gene 45 and ^{3}H-leucine-labeled cells wild type for gene 45 had been added at the start of the purification; since the purified 27,000-dalton protein contained ^{3}H, but no detectable ^{14}C (Fig. 8), it is clearly a product of T4 gene 45.

As would be expected for a protein consisting of a single polypeptide chain, the purified 45 protein complements only 45-deficient cell lysates of all those tested (for mutants used, see Table 4). Figure 9 shows the stimulation of DNA synthesis observed in a 45-deficient cell lysate with varying concentrations of purified 45 protein. Comparison with Fig. 6 (above) reveals that the 45 protein is active in this complementation assay only at approximately 10-fold higher concentrations than those required for the 44-62 protein complex. This observation is consistent with the relative activities of the two proteins in crude extracts seen earlier (Fig. 3), and might suggest either that 45 protein has a relatively low affinity for its substrate site, or that it is difficult to insert into the abortive replication complex present in a 45-deficient lysate.

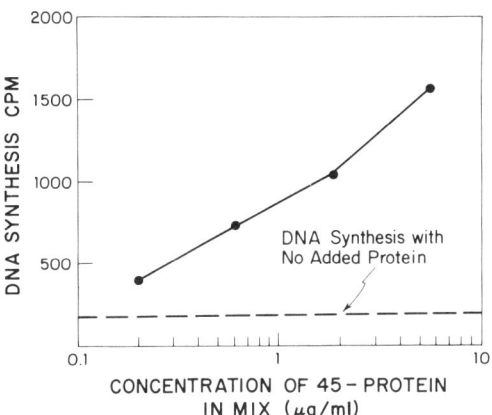

Fig. 9. Response of DNA synthesis in a 45-deficient receptor cell lysate to dilutions of purified 45 protein. For details of assay, see "Materials and Methods."

Discussion

The *in vitro* DNA synthesis described here is of brief duration, and its relation to true DNA replication is not known. In particular, we do not understand the wide variation in the background levels of (uncomplemented) DNA synthesis in different mutant lysates (Table 1). Perhaps several different partial reactions are possible *in vitro;* clearly, true replication forks need not be operative in any of our complemented lysates. Nevertheless, this system exhibits a requirement for proteins whose precise functions are unknown, but whose involvement in DNA replication has been established from the isolation of conditional lethal mutants. Using this system to provide an assay for isolation of these proteins, gene 45 protein and a complex of gene 44 and 62 proteins have been successfully purified in active form. In both cases, double label techniques have been used to confirm the identity of the isolated proteins. Thus, of the six proteins known to be essential for T4 DNA replication yet not needed for deoxyribonucleoside triphosphate formation, only the product of gene 41 remains to be isolated. Although a protein missing in T4 *am* N81 (gene 41$^-$)-infected cells can be identified and purified extensively by double label techniques, we have not yet been able to preserve the 41 complementation activity during this purification (L. Moran, unpublished results).

In searching for functions for these new replication proteins, it is perhaps useful to consider the possible nature of the T4 replication apparatus, as might be inferred from the properties of its two previously characterized proteins: T4 DNA polymerase and gene 32 protein. The requirement of T4 DNA polymerase for a single stranded DNA template and the ability of 32 protein to destabilize the double helix suggest that 32 protein precedes the polymerase during replication, exposing the template strands and aligning the bases in a conformation optimal for rapid polymerase action (19). Any model proposed must account both for the discontinuous synthesis of the 5'-terminated daughter DNA strand (see Okazaki *et al.* and Olivera *et al.*, this symposium), and for the apparent stoichiometric requirement for 32 protein in T4 DNA replication forks (31). This latter observation has suggested to us that a replication unit capable of synthesizing DNA has an unique three dimensional structure, containing a fixed number of every protein molecule which functions in it (4). In this view, some of the additional proteins required for T4 DNA replication might have purely structural roles, while others could be directly involved in synthesizing the 5'-terminated strand.

The tight binding of gene 44 and 62 proteins found here, as well as the weaker interaction of gene 32 and 43 proteins observed previously (19), may be remnants of a much larger structure which constitutes the DNA replication apparatus *in vivo* (see also Schekman *et al.*, this symposium). It is our hope that, like the ribosome (24), this entire apparatus can be induced to self-assemble *in vitro,* given the proper conditions and suitable concentrations of the appropriate pure proteins and DNA. Whether gene products in addition to those tested in Table 1 will be necessary for such a reconstruction remains an open question: although studies with known *E. coli* DNA mutants have thus far failed to implicate host components in T4 DNA replication (15), a requirement for host RNA polymerase (36) or T4, "DNA delay" gene products (39), or both, cannot be ruled out.

Acknowledgments

Contributions to this work by Mrs. Linda Frey are gratefully ackowledged. Supported by grants from the National Institutes of Health (GM-14927) and the American Cancer Society (E-510) at Princeton. J. W. was supported under contract with the Atomic Energy Commission at the University of Rochester Atomic Energy Project; this report therefore has been assigned report No. UR-3490-293.

References

1. Alberts, B., Amodio, F., Jenkins, M., Gutmann, E., and Ferris, F. 1968. Studies with DNA-cellulose chromatography. I. DNA-binding proteins from *Escherichia coli*. *Cold Spring Harbor Symp. Quant. Biol.* 33: 289.
2. Alberts, B. M. 1970. Function of gene 32-protein, a new protein essential for the genetic recombination and replication of T4 bacteriophage DNA. *Fed. Proc.* 29: 1154.
3. Alberts, B., and Frey, L. 1970. T4 bacteriophage gene 32: a structural protein in the replication and recombination of DNA. *Nature* 227: 1313.
4. Alberts, B. 1971. On the structure of the replication apparatus. *In* Ribbons, D. W., Woessner, J. F., and Schultz, J. (eds.) *Nucleic acid-protein interactions and nucleic acid synthesis in viral infection,* North-Holland, Amsterdam.
5. Aposhian, H. V., and Kornberg, A. 1962. Enzymatic synthesis of deoxyribonucleic acid. IX. The polymerase formed after T2 bacteriophage infection of *E. coli:* a new enzyme. *J. Biol. Chem.* 237: 519.
6. Barry, J. and Alberts, B. 1972. *In vitro* complementation as an assay for new proteins required for T4 bacteriophage DNA replication: purification of the complex specified by T4 genes 44 and 62. *Proc. Nat. Acad. Sci. U. S. A.* 69: 2117.
7. Bollum, F. J. 1966. Filter paper disk techniques for assaying radioactive macromolecules. *In* Cantoni, G. L., and Davies, D. R. (eds.) *Procedures in nucleic acid research*, pp. 296–300, Harper and Row, New York.
8. Chiu, C-S., and Greenberg, G. R. 1968. Evidence for a possible direct role of dCMP hydroxymethylase in T4 phage DNA synthesis. *Cold Spring Harbor Symp. Quant. Biol.* 33: 351.
9. Delius, H., Mantell, N. J., and Alberts, B. 1972. Characterization by electron microscopy of the complex formed between T4 bacteriophage gene 32-protein and DNA. *J. Mol. Biol.* 67: 341.

10. DeRosier, D. J. 1971. Three dimensional image reconstruction of helical structures. *Phil. Trans. Roy. Soc. London, Ser. B, Biol. Sci.* 261: 209.
11. DeWaard, A., Paul, A. V., and Lehman, I. R. 1965. The structural gene for DNA polymerase in bacteriophages T4 and T5. *Proc. Nat. Acad. Sci. U. S. A.* 54: 1241.
12. Epstein, R. H., Bolle, A., Steinberg, C. M., Kellenberger, E., Boy de la Tour, E., Chevallez, R., Edgar, R. S., Susman, M., Denhardt, G. H., and Lielausis, A. 1963. Physiological studies of conditional lethal mutants of bacteriophage T4D. *Cold Spring Harbor Symp. Quant. Biol.* 28: 375.
13. Godson, G. N. and Sinsheimer, R. L. 1967. Lysis of *Escherichia coli* with a neutral detergent. *Biochim. Biophys. Acta* 149: 476.
14. Goulian, M., Lucas, Z. J., and Kornberg, A. 1968. Enzymatic synthesis of DNA. XXV. Purification and properties of DNA polymerase induced by infection with phage T4. *J. Biol. Chem.* 243: 627.
15. Gross, J. D. 1972. DNA replication in bacteria. *Curr. Top. Microbiol. Immunol.* 57: 39.
16. Hosoda, J. and Levinthal, C. 1968. Protein synthesis by *Escherichia coli* infected with bacteriophage T4D. *Virology* 34: 709.
17. Hosoda, J. and Mathews, E. 1971. DNA replication *in vivo* by polynucleotide-ligase defective mutants of T4. *J. Mol. Biol.* 55: 155.
18. Hosoda, J., Mathews, E., and Jansen, B. 1971. Role of genes 46 and 47 in bacteriophage T4 reproduction. I. *In vivo* DNA replication. *J. Virol.* 8: 372.
19. Huberman, J., Kornberg, A., and Alberts, B. 1971. Stimulation of T4 bacteriophage DNA polymerase by the protein product of T4 gene 32. *J. Mol. Biol.* 62: 39.
20. Johns, E. W. 1967. The electrophoresis of histones in polyacrylamide gel and their quantitative determination. *Biochem. J.* 104: 78.
21. Kozinski, A. W. 1968. Molecular recombination in the ligase negative T4 amber mutant. *Cold Spring Harbor Symp. Quant. Biol.* 33: 375.
22. Martin, R. C., and Ames, B. N. 1961. A method for determining the sedimentation behavior of enzymes: application to protein mixtures. *J. Biol. Chem.* 236: 1372.
23. Moses, R. E., and Richardson, C. C. 1970. Replication and repair of DNA in cells of *Escherichia coli* treated with toluene. *Proc. Nat. Acad. Sci. U. S. A.* 67: 674.
24. Nomura, M. 1970. Bacterial ribosome. *Bacteriol. Rev.* 34: 228.
25. Okazaki, R., Sugimoto, K., Okazaki, T., Imae, Y., and Sugino, A. 1970. DNA chain growth: *In vivo* and *in vitro* synthesis in a DNA polymerase-negative mutant of *E. coli*. *Nature* 228: 223.
26. Riva, S., Cascino, A., and Geiduschek, E. P. 1970. Coupling of late transcription to viral replication in bacteriophage T4 development. *J. Mol. Biol.* 54: 85.
27. Schaller, H., Otto, B., Nüsslein, V., Huf, J., Herrman, R., and Bonhoeffer, F. 1972. DNA replication *in vitro*. *J. Mol. Biol.* 63: 183.

28. Shapiro, A. L., Vinuela, E., and Maizel, J. V. 1967. Molecular weight estimation of polypeptide chains by electrophoresis in SDS-polyacrylamide gels. *Biochem. Biophys. Res. Commun.* 28: 815.

29. Shah, D. B. and Berger, H. 1971. Replication of gene 46-47 amber mutants of bacteriophage T4D. *J. Mol. Biol.* 57: 17.

30. Siegel, L. M. and Monty, K. J. 1966. Determination of molecular weights and frictional ratios of proteins in impure systems by use of gel filtration and density gradient centrifugation. *Biochim. Biophys. Acta* 112: 346.

31. Sinha, N. K. and Snustad, D. P. 1971. DNA synthesis in bacteriophage T4-infected *E. coli:* evidence supporting a stoichiometric role for gene 32-product. *J. Mol. Biol.* 62: 267.

32. Smith, D. W., Schaller, H., and Bonhoeffer, F. 1970. DNA synthesis *in vitro. Nature* 226: 711.

33. Stahl, F. W. 1967. Circular genetic maps. *J. Cell Physiol.* 70 (suppl. 1): 1.

34. Warner, H. R., and Barnes, J. E. 1966. DNA synthesis in *E. coli* infected with some DNA polymerase-less mutants of bacteriophage T4. *Virology* 28: 100.

35. Warner, H. R., and Hobbs, M. D. 1967. Incorporation of uracil-C^{14} into nucleic acids in *Escherichia coli* infected with bacteriophage T4 and T4 amber mutants. *Virology* 33: 376.

36. Wickner, W., Brutlag, D., Schekman, R., and Kornberg, A. 1972. RNA synthesis initiates *in vitro* conversion of M13 DNA to its replicative form. *Proc. Nat. Acad. Sci. U. S. A.* 69: 965.

37. Wood, W. B., Edgar, R. S., King, F., Lielausis, I., and Henninger, M. 1968. Bacteriophage assembly. *Fed. Proc.* 27: 1160.

38. Wu, R., Ma, F., and Yeh, Y. 1972. Suppression of DNA-arrested synthesis in mutants defective in gene 59 of bacteriophage T4. *Virology* 47: 147.

39. Yegian, C. D., Mueller, M., Selzer, G., Russo, V., and Stahl, F. W. 1971. Properties of the DNA-delay mutants of bacteriophage T4. *Virology* 46: 900.

Discussion

Greenberg. Che-Shen Chiu and I have suggested that gene 42, the structural gene of dCMP hydroxymethylase, may have a second function, perhaps in DNA synthesis. The first suggestion that this might be so was made in the infection by *ts*L13, a mutant of gene 42. Even though relatively unstable, the enzyme was formed at nonpermissive temperatures in more than sufficient quantity to form DNA, but DNA was not synthesized. The enzyme inactivation *in vitro* at 42°C is not reversed at lower temperatures. However, brief low temperature pulses *in vivo* under conditions requiring protein synthesis allow the native enzyme to be formed and DNA synthesis to then occur at 42°C for at least 30 min. At permissive temperature, in our laboratory and that of John Drake, gene 42 *ts* mutants increase the reversion rate of rII mutants, though the effect is about one-tenth or one-twentieth of that shown by gene 43 (polymerase) mutants in the studies of Speyer and coworkers. M. Wovcha, P. Tomich, and C. S. Chiu have studied sucrose-plasmolyzed pol A$^-$ cultures infected by T4 phage, measuring ^{3}H-dTTp incorporation into DNA. The system forms DNA for about an hour. With DNA zero *ts* and *am* mutants in such genes as 44, 45 and 43, DNA synthesis does not occur under nonpermissive conditions. In T4-infected preparations washed through sucrose layers, the deoxynucleoside triphosphates are necessary for maximal synthesis. However, all of the dCTP is very rapidly converted to dCMP by dCTPase and must first be converted to 5-HMdCMP dCMP hydroxymethylase and then to HMdCTP. This was supported by the fact that the label from 5-^{3}H-dCTP was not incorporated into the DNA and by measurement of ^{3}H released as a result of dCMP HMase activity. As expected, HMdCMP replaces dCTP in DNA synthesis in this system with T4 wild type phage. Am122 (gene 42)-infected plasmolyzed systems do not form DNA. However, 5-HMdCMP does not bypass *am*122. At nonpermissive temperature *ts*L13 does not stop DNA synthesis in these preparations. We feel it may be dissociated from its *in vivo* function. The system is being used to determine whether gene 42 mutants act at the level of the DNA-synthesizing complex or perhaps at the level of regulation of synthesis of deoxyribonucleoside triphosphates.

Discontinuous DNA Synthesis in Vitro: A Method for Defining the Role of Factors in Replication*

Baldomero M. Olivera and K. G. Lark

Department of Biology
University of Utah
Salt Lake City, Utah 84112

Richard Herrmann and Friedrich Bonhoeffer

Friedrich-Miescher Laboratorium der
  Max-Planck-Gesellschaft and
Max-Planck-Institute fur
  Virusforschung
Tubingen, Germany

The accumulation of discontinuously synthesized DNA intermediates was observed when DNA joining was inhibited in the cellophane disc system. The size of these intermediates was apparently determined by the relative rates of chain initiation and chain elongation.

These findings provide a basis for defining the function of various factors in DNA replication. It appears that lowering the deoxynucleoside triphosphate concentration preferentially slows down chain elongation, while freezing the cells before use at $-80°C$ preferentially inactivates chain initiation. The DNA G locus appears to be involved in chain initiation, while at least some DNA B mutations affect chain elongation. This method can also be used to elucidate factors which may not be genetically defined; preliminary evidence is presented for two functions which are preferentially inhibited in the presence of single stranded $\phi\chi$-174 DNA.

---

*Much of the work described in this talk is detailed in a series of three papers (2, 3, 5).

The *in vivo* replication of double stranded DNA requires a number of different factors. A similar *in vitro* replication of double stranded DNA using a purified enzyme, or a combination of purified enzymes, has not been demonstrated. In addition to multiple macromolecular factors, semiconservative replication may require the integrity of a structural complex which could be broken by homogenization and purification procedures, as well as high concentrations of some macromolecular components which in conventional enzyme assays are greatly diluted from the natural concentrations found within the cell. Recently, a number of *in vitro* systems capable of semiconservative replication have been developed, in which all of the acid-soluble pools within the cell are released but the macromolecular components remain close to their natural concentrations, and structural integrity is hopefully not perturbed.

The cellophane disc system developed by Schaller *et al.* (7) has one feature which most of the other *in vitro* systems do not possess. In addition to keeping the macromolecular components concentrated, and structural complexes presumably intact, it permits easy mixing of macromolecules from two different cells. Thus, this system lends itself to complementation assays *in vitro*, as described by Klein *et al.*, in this symposium. Such assays allow the purification of active factors but do not guarantee that the mode of action of the factor will be elucidated.

In this paper we describe experiments in which the cellophane disc system has been used to provide a basis for defining the role of various factors in the initiation or elongation of discontinuously synthesized DNA chains.

Experimental Plan

In all of our experiments the basic principle is to inhibit DNA joining so that the discontinuously synthesized DNA chains can be examined and analyzed. DNA joining is inhibited by adding nicotinamide mononucleotide to the incubation mixture: at sufficiently high concentrations this compound completely inhibits DNA ligase (6, 9). Under these conditions DNA synthesis proceeds in the absence of DNA joining. The fact that the newly synthesized DNA is unsealed does not seem to inhibit the rate of DNA synthesis measurably.

The typical experimental protocol is to grow a culture of *Escherichia coli* in a medium containing ^{14}C-thymine: the bulk DNA synthesized *in vivo* is therefore labeled with ^{14}C. Generally, a polymerase I$^-$ endonuclease I$^-$ strain is used to minimize repair replication. The cells are harvested, spread at a high concentration on a cellophane disc which is sitting on an agar plate, and lysed gently. Upon lysis, all of the small molecules inside the cells (such as the deoxynucleoside triphosphates) dialyze through the cellophane into the agar, leaving a film on top of the disc containing only the macromolecular components of the cell. The cellophane disc is then placed on top of an incubation mixture containing deoxynucleoside triphosphates (labeled with ^{3}H) and ATP (as well as NMN). These molecules dialyze through the cellophane into the macromolecular film, providing precursors for DNA synthesis *in vitro*. The DNA made on the disc is therefore labeled with tritium, and is readily distinguished from the ^{14}C-DNA synthesized *in vivo*.

Two Modes of Discontinuous DNA Synthesis *in Vitro*

Results with this system suggest the mechanism in Fig. 1. Discontinuous synthesis occurs on both strands. However, on one strand, DNA is synthesized discontinuously with regular Okazaki pieces as an intermediate, while on the other strand, the intermediates are much larger than regular Okazaki pieces.

Figure 2 shows an alkaline sucrose gradient analysis of the products of this *in vitro* system. It is seen that the prelabeled ^{14}C-DNA is rather large and sediments to the bottom of the tube if some care is taken to avoid shear. However, the DNA made *in vitro* in the absence of DNA joining is smaller than the bulk DNA and is found as two discrete classes: a peak of material sedimenting at about 9 S, and a second class of material sedimenting at about the rate of T7 DNA with an average sedimentation coefficient of about 38 S under these conditions. We have done numerous experiments of this general type with several different strains and always find only the two size classes of DNA

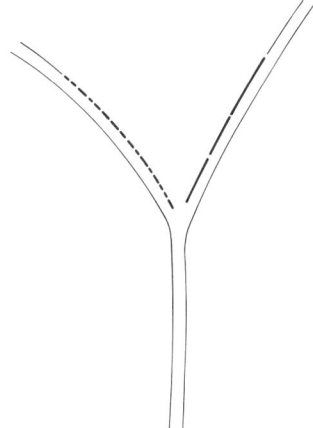

Fig. 1. A postulated representation of the replication fork in the cellophane disc system in the presence of NMN. The bold lines represent the DNA synthesized *in vitro*, the thinner lines the prelabeled DNA made *in vivo*.

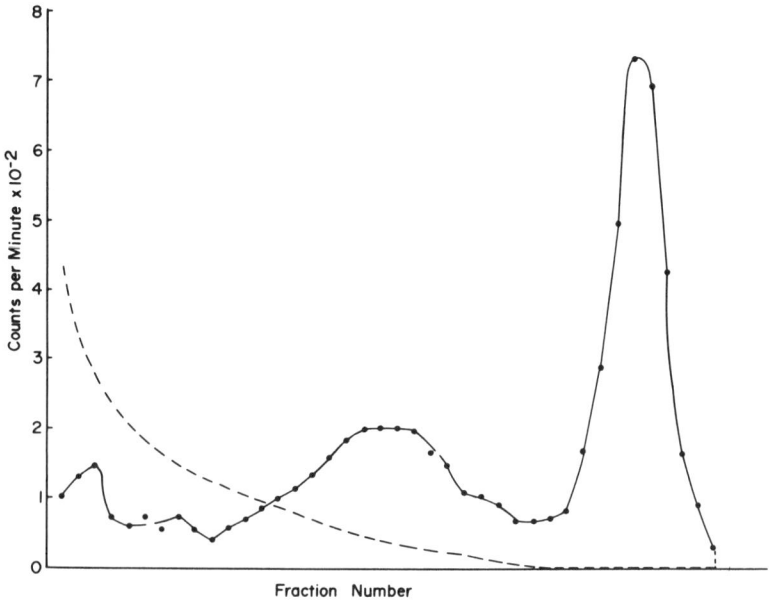

Fig. 2. Alkaline sucrose gradient analysis of DNA synthesis in the absence of DNA joining. A lysate of *Escherichia coli* H560, a Pol A1⁻ Endo 1⁻ strain which was grown in ^{14}C-thymine (2 μg/ml, 320 mCi/mole) was prepared on a cellophane disc (2 cm diameter) as described by Shaller *et al.* (7). The disc containing the lysate was incubated for 15 min at 37°C on top of an incubation mixture (0.1 ml) containing 20 mM morpholino propane sulfonic acid, pH 7.2, 100 mM KCl, 5 mM $MgCl_2$, 1 mM ATP, 20 μM each of dATP, dCTP, and dGTP, 170 μM thymidine, and 20 μM ^{3}H-TTP (specific activity 500 Ci/mole). The incubation mixture also contained 1 mg/ml of NMN. At the end of the incubation period, the discs were added to 0.3 ml of a solution containing 0.1 M Tris, pH 8.0, 0.005 M EDTA, and 0.5% Sarkosyl. The solutions were then heated to 60°C for 10 min and any remaining lysate taken off the cellophane disc with the aid of a thin glass rod. The eluted mixture was then analyzed by sedimentation on an alkaline sucrose gradient (5 to 20% sucrose, containing 0.2 M NaOH, 0.5 M NaCl, 0.05 M Tris, and 0.003 M EDTA), which was centrifuged for 60 min at 50,000 rpm on the SB 283 rotor of the B60 International Centrifuge. The direction of sedimentation is from right to left. ——, ^{3}H radioactivity (DNA synthesized *in vitro*); - - -, ^{14}C radioactivity (prelabeled DNA).

present in approximately equal amounts.

We believe that these pieces are a consequence of discontinuous synthesis and not of continuous synthesis followed by endonuclease action. A series of *in vitro* experiments utilizing extremely short incubations were performed, and the chains could be observed to grow under these conditions. Using density shift experiments in which dBUTP is substituted for dTTP in the incubation mixture we have determined that it takes about 30 sec to synthesize a fully heavy 9 S Okazaki piece and about 8 min to synthesize a fully heavy 38 S piece.

If Fig. 1 were correct, definite predictions could be made with regard to the complementarity of sequences. If the small Okazaki pieces come from a unique strand, and the larger pieces come from the opposite strand, then neither the small 9 S pieces nor the larger size class should contain self-complementary sequences. However, the small pieces should contain base sequences which are complementary to the large pieces. We have tested these predictions by doing hybridization experiments. Neither purified 9 S pieces nor the bigger 38 S pieces are self-hybridizable, but if the two classes are mixed together efficient hybridization occurs (2). This is strong evidence that the two size classes come from different strands.

It might be added that both of these two classes of pieces appear to be normal intermediates in DNA synthesis in this system since they can be chased into larger DNA if the disc is incubated with a mixture containing DPN, the cofactor for the DNA-joining enzyme of E. coli.

A Rationale for Two Modes of Discontinuous Synthesis

Why should the two different strands of the E. coli chromosome have two different sizes of unsealed DNA intermediates? We would like to suggest a model to explain these results.

Our hypothesis is based on the lack of equivalence of the two daughter strands in replication. Although the polymerization reaction in both strands probably involves the addition of nucleotide units at the 3' end, the fact that the two strands are antiparallel makes them basically nonequivalent. One strand is growing in the direction of replication while the other strand is in fact growing biochemically away from this direction. Thus, while the strand growing backward (the 5' strand) must obligatorily be synthesized discontinuously, the other strand could in principle be synthesized continuously.

The essence of the model which we propose is that there are many potential sites on both template strands for initiating the discontinuously synthesized chains but that it is not obligatory that these initiation sites be used. For example, if in the 3' strand a growing daughter strand were being elongated very rapidly, this growing strand might grow past a newly exposed site on the template strand. If this should happen, then no initiation would take place at that site. Thus, the length of the chains which accumulate could result from a balance between the relative rates of initiation and of elongation of previously initiated DNA chains. If no initiations occurred at all, the 3' strand would be synthesized continuously. We suggest that *in vivo* every initiation site is used, and thus the 3' strand is synthesized as Okazaki pieces. *In vitro*, however, initiations are less frequent, leading to the accumulation of intermediate sizes of DNA chains.

On the 5' strand, previously initiated DNA chains which are growing away from the over-all direction of growth cannot cover a newly exposed potential initiation site. How long will a potential initiation site remain exposed (*i.e.* single stranded) if an initiation does not take place? After the parental strands are unwound, a potential site on the 3' strand only will be exposed until the previously initiated chain has grown to cover it. On the other hand, a potential initiation site on the 5' strand will remain exposed for a much longer period of time since the previously initiated chain will grow away from it. The only way such a site will be covered is if a subsequently exposed potential site becomes the site for an initiation; polymerization would then take place in the direction of the previously exposed initiation site. This situation will only occur if initiations become exceedingly improbable. Since on the average a potential site on the 5' strand is exposed and available for initiation for a much longer period of time than a potential site on a 3' strand, it will be expected that if the initiation process becomes somewhat inefficient, more initiations will have occurred on the 5' than on the 3' strand. The consequence of this is that the unsealed DNA intermediates which accumulate should be different in size in the two different strands; this is consistent with the experimental observations.

Changes in Size of DNA Intermediates

The explanation above leads directly to several experimentally useful predictions. Since the size of the DNA intermediates which accumulate is dependent on the relative rates of initiation and polymerization, the size of these intermediates should change if the rate of one process is altered relative to the other process. We have observed several situations in which the size of the discontinuously synthesized intermediates is reproducibly altered.

If the deoxynucleoside triphosphate concentration is lowered 5-fold, two size classes of intermediates accumulate but the molecular weight of the larger pieces is considerably less (Fig. 3). The size of these intermediates is independent of incubation times greater than 10 min, but more chains of the same size accumulate, indicating that these are the size of completed but unsealed chains. Our interpretation of this experiment is that by lowering the deoxynucleoside triphosphate concentrations, the replication polymerase (or polymerases) elongates the growing chains more slowly, since the concentration of DNA precursors is now less than optimal. If the DNA chains are polymerized more slowly, then relatively more initiations can occur and a decrease in the average chain length should be observed.

An increase in the length of Okazaki pieces is also possible. We

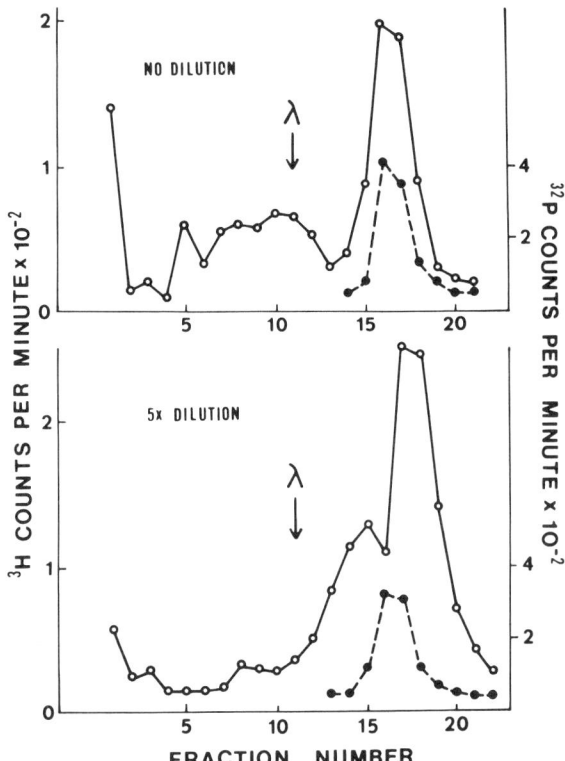

Fig. 3. Change in the size distribution of DNA intermediates with a change in deoxynucleoside triphosphate concentration. A lysate of *Escherichia coli* H560 which had been stored frozen at −80°C was incubated on a cellophane disc for 10 min at 37°C on top of an incubation mixture containing the regular levels of the dexoynucleoside triphosphates (20 μM) and over an incubation mixture containing a 5-fold diluted level of deoxynucleoside triphosphates (4.0 μM each dATP, dGTP, dCTP, and 4.6 μM dTTP); each incubation mixture also contained 1 mg/ml of NMN. The DNA was then taken off the cellophane discs and sedimented on an alkaline sucrose gradient using the SB 405 rotor of the B60 International centrifuge for 1 hr at 50,000 rpm. ^{32}P-ϕχ-174 and λDNA were added as markers; the position of the λDNA is indicated by an arrow. ○──○, ^{3}H radioactivity (DNA synthesized *in vitro*); ●──●, ^{32}P radioactivity (ϕχ-174 DNA as marker).

have observed both a physiological and a genetic situation which appear to fulfill this prediction. The physiological situation is observed if cells are frozen at −80°C before they are spread on the cellophane discs. Under these conditions, the Okazaki pieces increase in size (Fig. 4). We believe that an initiation factor is preferentially inactivated under these conditions.

It has been found that temperature-sensitive mutants exist in the DNA G locus which fail to form Okazaki pieces when incubated *in vivo* at the nonpermissive temperature (3). In Fig. 5, an experiment

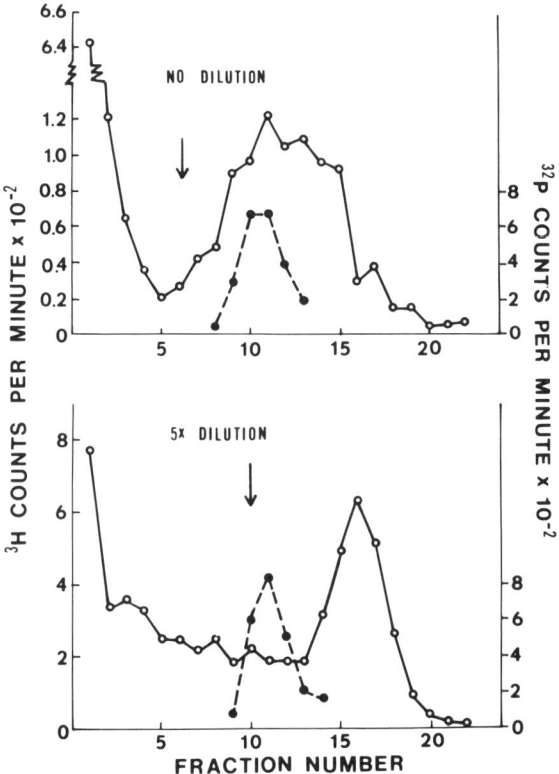

Fig. 4. Changes in the size of Okazaki pieces. The experiment figured here is essentially identical with the experiment described in the legend to Fig. 3 except that the discs were incubated for 30 min instead of 5 (which does not cause any change in the size of the DNA made, although it changes the total amount synthesized), and the sedimentation runs were performed for 3 hr at 50,000 rpm (instead of for 1 hr). ^{32}P-ϕx-174 DNA was added as a marker. The arrows indicate the point to which 50% of the radioactivity has sedimented. The profile of the Okazaki pieces in the lower figure is identical with the profile exhibited by freshly grown cells in a Pol A1$^-$ strain in the presence of regular levels of triphosphates and NMN. ○──○, ^{3}H radioactivity (DNA synthesized *in vitro*); ●──●, ^{32}P radioactivity (ϕx-174 DNA as marker).

is shown in which NY 73, a DNA G temperature-sensitive, polA$^-$ strain (8) is incubated in the NMN cellophane disc system at the permissive and nonpermissive temperatures. Even at permissive temperatures this mutant seems to make abnormally long pieces. As is seen in the figure the smaller size class of pieces sediments at about the same rate as ϕx-174 DNA (18 S), which is significantly larger than the regular size of Okazaki pieces (9 S). As the temperature is raised, this mutant gradually fails to make recognizable Okazaki pieces. It is not clear from the data whether synthesis stops altogether in the strand in which Okazaki pieces are normally made, or whether a few residual initiations can

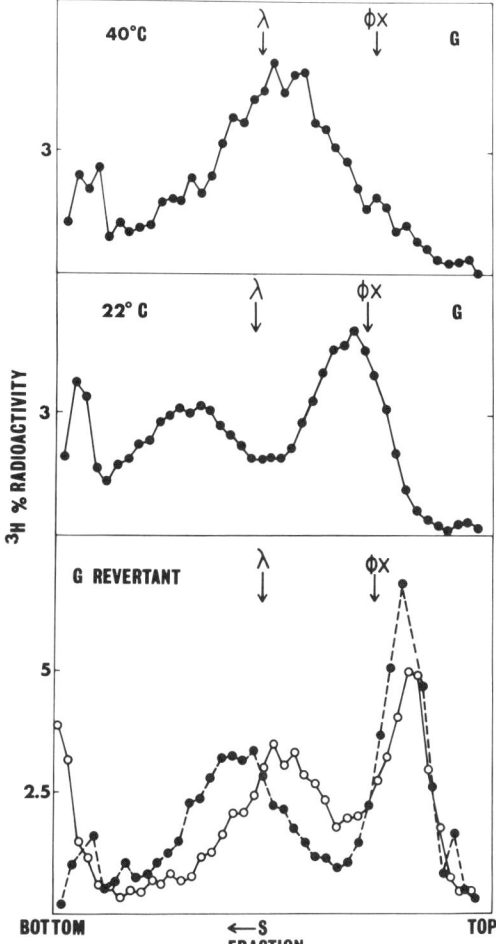

Fig. 5. Sedimentation of DNA synthesized at different temperatures by lysates of NY 73 and by lysates of NY 73 rev 5 in the presence of NMN. Cells were grown in penassay broth supplemented with ^{14}C-thymine (0.05 µCi/2 µg/ml). Lysates of 10^9 cells on large (3.3 cm) discs were pre-equilibrated on drops of nonradioactive reaction mixture, containing NMN, at the temperatures indicated below. These discs were then transferred to radioactive reaction mixtures at the same temperature and incubated for 30 min. The mixture was identical with that used in the experiment in Fig. 3 except that the specific activity of the ^{3}H-TTP was raised to 1 mCi/µM for the experiments with NY 73, and to 0.5 mCi/µM for NY 73 rev 5. After incubation, the discs were placed in glass Petri dishes on top of ice and covered with 0.3 ml of 0.1 N NaOH, 0.02 M EDTA. After lysates were dissolved, they were poured onto alkaline sucrose gradients, centrifuged for 4 hr at 39,000 rpm in the SB 283 rotor of an International centrifuge. The temperatures of incubation for the NY 73 strain (G) are indicated in the figure. The solid line is for an incubation at 40°C for the revertant, and the dotted line for an incubation at 22°C. Both λ and φχ DNAs were used as reference markers. The total ^{3}H radioactivity per gradient was G: 40°C, 10,900; 22°C, 65,000; and for the revertant: 40°C, 24,900; 22°C, 50,900. The ^{14}C-DNA (not shown) sedimented toward the bottom of the gradient in all cases.

continue to occur. It is evident, however, that initiations are certainly much less frequent. Revertants of this DNA G strain make normal Okazaki pieces *in vitro* at both permissive and nonpermissive temperatures.

This evidence therefore suggests that the DNA G product is probably involved in the initiation of the discontinuously synthesized chains.

Strong evidence in support of the role of the DNA G product in DNA chain initiation is also provided by the fact that purified DNA G product causes the larger class of pieces which accumulate in the NMN cellophane disc experiment to become smaller, suggesting that if more DNA G product is present, more chain initiations take place. This is shown most strikingly when the purified DNA G product is added to a cellophane disc containing lysates of DNA G mutant, and the mixture is incubated at the nonpermissive temperature. In the absence of added DNA G product, lysates of the mutant do not synthesize Okazaki pieces. If the DNA G product is added, not only does the amount of DNA synthesized *in vitro* increase dramatically, but the two size classes of intermediates accumulate, and a very clear peak of Okazaki

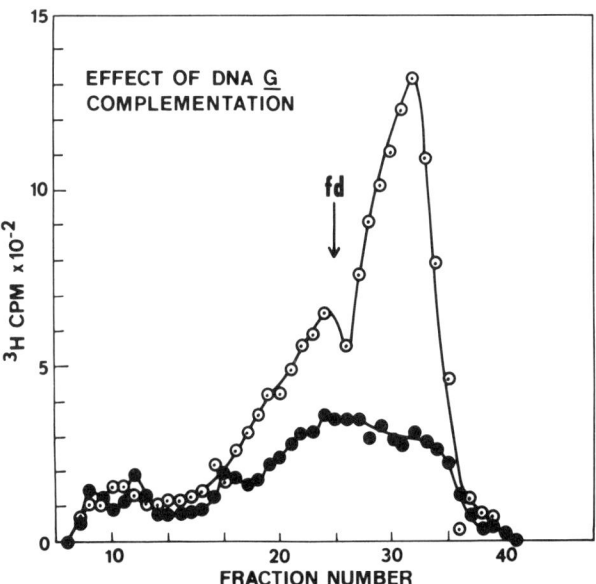

Fig. 6. Effect of DNA G complementation. A lysate of NY 73 was incubated on cellophane discs at the nonpermissive temperature (37°C, 5 min in nonradioactive, and 20 min in radioactive incorporation mixture) as described in Figs. 2 and 5 in the absence (closed circles) and presence (open circles) of purified DNA G protein. The experimental details regarding the purification of the DNA G product, as well as the complementation assay, are described by Klein et al. (this symposium). At the end of the incubation, the discs were analyzed by alkaline sucrose gradient sedimentation (SW 41 rotor, 37K, 20°C) using fd DNA as a marker.

pieces can be distinguished (Fig. 6). Thus complementation of DNA G mutants with the purified DNA G product results in apparent chain initiation.

These observations are a particular property of mutants in the DNA G locus. Temperature-sensitive strains in the DNA B and DNA E loci apparently continue to initiate discontinuous pieces at the nonpermissive temperature *in vitro*, although these strains exhibit aberrant discontinuous DNA synthesis. For DNA E mutants, and at least some DNA B mutants, the best interpretation appears to be that the rate of polymerization or the elongation of chains is affected by the mutation. This is shown in the experiment in Fig. 7 in which only short chains accumulate *in vitro* when a DNA B mutant is incubated at the nonpermissive temperature.

Effect of Exogenous DNA on DNA Intermediates

Some preliminary experiments with the system described above show potential for uncovering additional factors involved in DNA replication which have not yet been genetically defined. The basic strategy in these experiments is to introduce a perturbation into the system which causes abnormal discontinuous DNA synthesis.

There are many proteins in cell extracts which bind DNA. These DNA-binding proteins are presumably involved in DNA metabolism, and perhaps some are involved in replication (1). We have therefore added exogenous DNA to cellophane discs in an attempt to bind factors involved in DNA replication; the effect on discontinuous replication in the endogenous replication fork is then observed. The DNA we have added is $\phi\chi$-174 DNA, a single stranded circular DNA. Under these conditions, we know that only about one molecule of the single stranded DNA is replicated per lysed bacterial cell (4): this means that essentially all of the ^{3}H incorporation is due to the endogenous replication fork on the *E. coli* chromosome.

The effect of adding different levels of exogenous DNA is 2-fold: there is an effect on the sealing of the pieces and an inhibition of the synthesis of the larger size class of pieces (Figs. 8 and 9). It is seen in Fig. 8 that even though this is a polA$^-$ strain in which sealing of the discontinuously synthesized pieces is slow, there is an obvious difference in the size of the DNA obtained after an incubation in the presence of NMN, which causes complete inhibition of DNA ligase, and in DPN which is the cofactor for DNA ligase. A significant fraction of the DNA is sealed in the presence of DPN, and is as large as the prelabeled DNA. However, when $\phi\chi$-174 DNA is added, small pieces accumulate exclusively whether the incubation is in the presence of NMN

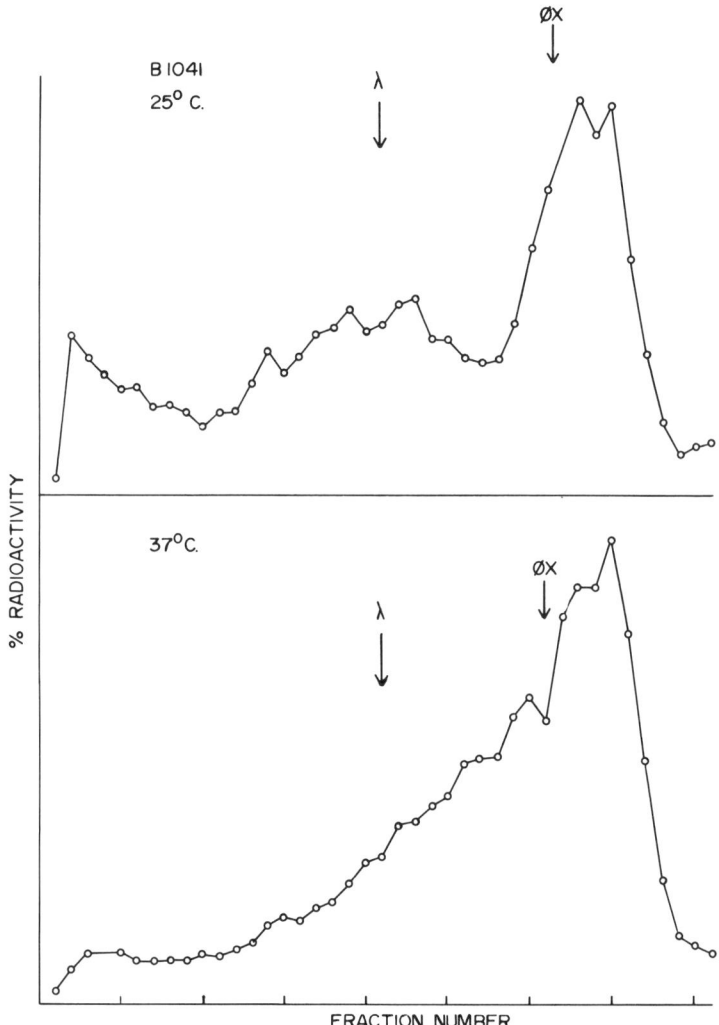

Fig. 7. Sedimentation of DNA synthesized at the permissive and nonpermissive temperatures by lysates of *Eschericihia coli* H560 DNA B ts 1041. The experimental details are identical with Fig. 5 except that lysates were incubated for 15 min on the cellophane discs. The alkaline sedimentations of DNA at the permissive temperature, 25°C, and at the temperature of transition to nonpermissive growth, 37°C, are shown. The sedimentation profiles of DNA synthesized at higher temperatures (40 or 42°C) were similar to the 37°C profile except that an even smaller fraction of the DNA sedimented at a rate greater than that of ϕx-174 DNA. The counts per min incorporated per 10^9 cells were: 25°C, 32,980; 37°C, 22,660; 40°C, 8,520; 42°C, 5,640.

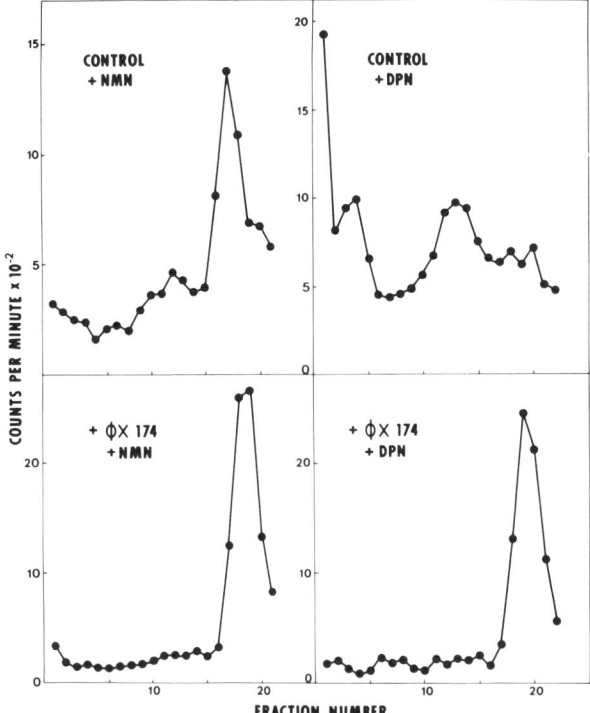

Fig. 8. Effect of φχ-174 DNA on sealing of the discontinuously synthesized chains. A lysate of *Escherichia coli* H560 was incubated for 15 min at 37°C on a cellophane disc as described in Fig. 2, except that 0.2 ml of incubation mixture and large discs (3.3 cm) containing 3×10^8 bacteria were used, and DPN (0.3 mg/ml) was added instead of NMN in the experiments indicated. In the experiments on the bottom panels, purified φχ-174 DNA (from 7.5×10^{10} phage) was also added. The cellophane discs were analyzed by alkaline sucrose gradient centrifugation as described in the legend to Fig. 2. Only the ^{3}H-DNA synthesized *in vitro* is shown. The prelabeled ^{14}C-DNA sedimented to the bottom fraction of the tube in all cases. The direction of sedimentation is from right to left.

or DPN. Since all studies with purified DNA ligase indicate that the presence of large amounts of single stranded DNA is not inhibitory to the ligase-catalyzed reaction, these results suggest that a factor necessary for a sealing step prior to the DNA ligase-joining step is being inhibited by the addition of exogenous single stranded DNA.

The second effect that is seen is the inhibition of synthesis of the larger pieces. When φχ-174 DNA is added, fewer large pieces are synthesized relative to small pieces, but the fraction left still sediments at about the same rate (Fig. 9). One explanation for these results currently being investigated is that φχ DNA binds some factor necessary for the replication of only one of the two DNA strands, and that in the absence of this factor, there is selective inhibition of synthesis on this

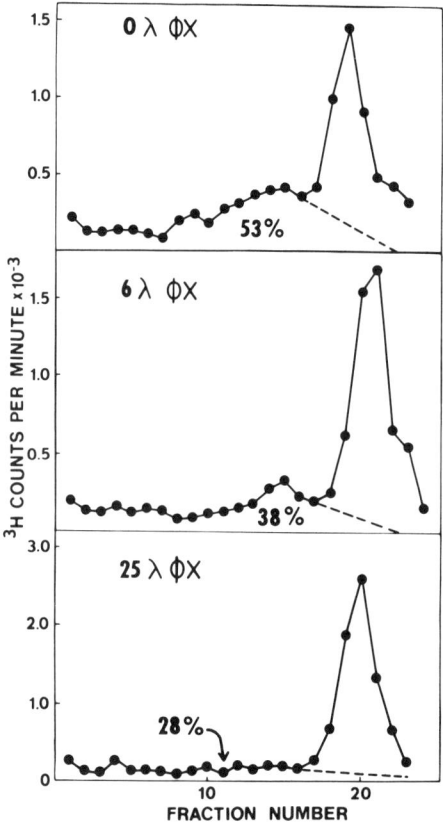

Fig. 9. Dosage effect of φχ-174 DNA on the synthesis of larger pieces. Lysates of *Escherichia coli* H560 (3 × 10⁸ cells per disc) were incubated for 5 min at 37°C and then subjected to alkaline sucrose gradient analysis as in the experiment in Fig. 8. The amount of φχ-174 DNA was varied as indicated. The experiment on the uppermost panel had no φχ DNA added; the middle panel, 1.8 + 10¹⁰ molecules; the bottom panel, 7.5 × 10¹⁰ molecules. The direction of sedimentation is from right to left.

strand. If this were correct it would mean that it is possible to uncouple synthesis on one of the template strands from synthesis on the other.

In summary, we have developed an *in vitro* system which permits us to examine the unsealed intermediates during discontinuous DNA replication. This system has potential for defining the role of different factors because aberrant synthesis of the discontinuously replicated chains immediately implies either the dysfunction or absence of a particular factor involved in DNA synthesis. We have identified some genetic and physiological factors which seem to be involved in DNA chain initiation and chain elongation. We believe further use of this system should permit a better definition of the role of other factors in discontinuous DNA synthesis.

Acknowledgments

Miss Julita Huf, Mrs. Purita Fernandez, Mr. Ron Lundquist, Mr. Edward Meenen, Miss Ning Chai, and Mrs. Cheri Dueber participated in some of the experiments described in this talk. This work was supported by National Science Foundation Grant GB 22869 and Public Health Service Grant 5R01-CA 12096 (to B. M. O.), and National Institutes of Health Grant 5R01-AII0056 and American Cancer Society Grant VC-3B (to K. G. L.). F. B. was a Visiting Senior Scientist of the National Science Foundation.

References

1. Alberts, B. M., Amodio, F. J., Jenkins, M., Gutman, E. D., and Ferris, F. L. 1968. Studies with DNA-cellulose chromatography. I. DNA-binding proteins from *Escherichia coli*. *Cold Spring Harbor Symp. Quant. Biol.* 33: 289.
2. Herrmann, R., Huf, J., and Bonhoeffer, F. 1972. II. Cross hybridization and rate of chain elongation of the two classes of DNA intermediates. *Nature* 240: 235.
3. Lark, K. G. 1972. Genetic control over the initiation of the synthesis of short deoxynucleotide chains in *E. coli*. *Nature* 240: 237.
4. Olivera, B. M., and Bonhoeffer, F. 1972a. Replication of $\phi\chi 174$ by *Escherichia coli polA*$^-$ *In vitro*. *Proc. Nat. Acad. Sci. U. S. A.* 69: 25.
5. Olivera, B. M., and Bonhoeffer, F. 1972b. Discontinuous DNA replication *in vitro* I: Two distinct size classes of intermediates. *Nature* 240: 233.
6. Olivera, B. M., and Lehman, I. R. 1967. Diphosphopyridine nucleotide: A cofactor for the polynucleotide-joining enzyme from *Escherichia coli*. *Proc. Nat. Acad. Sci. U. S. A.* 57: 1700.
7. Schaller, H., Otto, B., Nusslein, V., Huf, J., Herrmann, R., and Bonhoeffer, F. 1972. Deoxyribonucleic acid replication *in vitro*. *J. Mol. Biol.* 63: 183.
8. Wechsler, J. A., and Gross, J. D. 1971. *Escherichia coli* mutants temperature-sensitive for DNA synthesis. *Mol. Gen. Genet.* 113: 273.
9. Zimmerman, S. G., Little, J. W., Oshinsky, C. K., and Gellert, M. 1967. Enzymatic joining of DNA strands: A novel reaction of diphosphopyridine nucleotide. *Proc. Nat. Acad. Sci. U. S. A.* 57: 1841.

Discussion

Caro. Dr. Louarn in my laboratory has looked at the formation of Okazaki pieces specific for λ in λ lysogenic cells by hybridizing the pieces to the two strands of λ. In a wild type cell, the pieces hybridize only to the strand which corresponds to the direction of synthesis $3' \rightarrow 5'$. In polymerase I$^-$ strains we get results which are quite similar to what you get. There are long pieces which are about the size of λ and a little shorter, and short pieces which are about 1 μ long. The short pieces hybridize to the 5' direction and the long pieces hybridize to the opposite direction so that is exactly what your work predicts.

Olivera. We are certainly happy to hear that.

Schekman. You have previously shown that this system is capable of converting φχ single stranded DNA to its replicative form and I was wondering two things. (a) In your last experiment where you were adding φχ DNA, were you synthesizing *Escherichia coli* DNA or φχ DNA, and (b) if it was *E. coli* DNA, how come φχ DNA wasn't synthesized?

Olivera. In this system, as we had shown previously, the lysates have a very limited capacity to replicate φχ DNA, and on the average they replicate only one φχ DNA molecule per lysed cell. In these experiments we are adding phage at a multiplicity of 60 or more and the incorporation into φχ is negligible under these conditions so that essentially all of the incorporation is at the *E. coli* fork and there is a negligible incorporation into φχ.

Krieger. Have you looked at the difference in the amount of ^{3}H label triphosphates incorporated into the very long DNA in polA$^-$ and polA$^+$ cells? I'm interested in the 3' end, the long 3' end of the rest of the chromosome.

Olivera. No, we have not examined polA$^+$ strains at all because one gets a repair kind of incorporation which sort of messes up the picture.

Szybalski. You said that in DNA G the initiation seems to be repressed. Does it mean that mainly the 5' → 3' chain is being replicated and the 3' → 5' not, or did you check that?

Olivera. Well, these are Lark's experiments and his experiments are consistent with two possible interpretations. One interpretation is that one is indeed inhibiting synthesis on the 3' → 5' strand and simply getting some residual synthesis on the 5' → 3' strand. The other interpretation is that one is getting residual incorporation very, very weakly on both strands and one is getting pieces that are very much longer on the 3' → 5' strand (because this strain actually shuts off rather slowly and it takes a few minutes before it shuts off.)

Szybalski. But did he hybridize?

Olivera. That experiment is being done at the moment.

Ray. I would just like to point out that it's rather interesting to note that the size of your big piece of DNA is about λ size and corresponds fairly well with the size you would predict for units of the *coli* chromosome from the kind of data that Worcel presented. From the number of nicks required to convert the entire *coli* chromosome to a slower sedimenting form, you could imagine individual units of the chromosome of about λ size each being involved there. It's interesting that both of these are comparable within an order of magnitude or so.

Kushner. In the last slide that you showed, it seems to me that the disappearance of the larger fragments seems to indicate that the presence of single stranded DNA leads to more efficient initiation on the strand where you are making a larger piece. It seems to me that all your large pieces were incorporated into that smaller peak or was I mistaken?

Olivera. The data are not good enough to completely eliminate that, but we don't think that's the case. If initiation is becoming more frequent, we would expect that, as the amount of φχ DNA was increased, one would get a gradual shift in size of the large pieces to smaller pieces and then finally all the way into Okazaki size pieces. This is, in fact, what we see if we keep on lowering triphosphate concentrations. However, what we in fact see in the last slide is that these large pieces seem to be about the same size except they just seem to decrease in amount. In other words, as we add φχ, these pieces don't shift down in molecular weight, they just seem to decrease in amount.

Kushner. On the bottom of the slide, you seem to have increased the scale, so that your peak is actually much bigger than the top one.

Olivera. In these experiments, one should always correct for the amount of prelabeled DNA present, and it just so happens that in this particular experiment, more cells were spread on the discs and if we correct for the amount of ^{14}C-DNA, it normalizes to give the same amount of incorporation.

Kushner. Where does that DNA go? You don't get initiation on that other strand? You're only synthesizing one strand then?

Olivera. That's very preliminary, as I said, and we'd certainly like that interpretation but we have no definitive proof for it.

Brown. What happens to these larger pieces if they don't inhibit ligase? They should be rather continuously synthesized. Are they?

Olivera. No, they are discontinuously synthesized but they are sealed together.

Brown. So you don't see them. They disappear if you don't inhibit ligase?

Olivera. Right. One just sees much larger DNA under these conditions.

Alberts. From what you said, you seem to be assuming one large piece per growing fork? That model had essentially one large piece per growing fork.

Olivera. Oh no, under these conditions it depends on the time of incubation. Normally we incubate for 10 min, which should make about two large pieces, but if we incubate longer then presumably we will be getting more.

Alberts. Normally only one is growing at any one time.

Olivera. It's hard to tell in the *in vitro* system.

Caro. *In vivo* it's about one long piece for polA$^-$ and you don't see that in polA$^+$.

Self-association of Gene 32 Protein

Robert B. Carroll,
Kenneth E. Neet, and David A. Goldthwait

Case Western Reserve University
Department of Biochemistry
Cleveland, Ohio 44106

Physical studies show that the gene 32 protein associates in a monomer-dimer-further aggregate mode.

Gene 32 protein of coliphage T4 has been isolated and extensively characterized by Alberts *et al.* (2-4, 7, 9). They have shown that gene 32 protein binds in long aggregates along the DNA backbone, and that protein-protein interactions are important to this cooperative binding.

With the idea that a knowledge of these protein-protein interactions might be important to an understanding of the role of gene 32 protein in replication and recombination, we have undertaken biophysical studies of gene 32 protein self-association.

We studied its molecular weight by sedimentation equilibrium centrifugation using a model E analytical ultracentrifuge. Figure 1 shows the effect of three treatments upon the molecular weight of gene 32 protein as a function of protein concentration. In the left-hand panel the protein was treated with 6 M guanidine hydrochloride. A molecular weight of 38,000 was obtained using a partial specific volume determined by the method of Edelstein and Schachman (8). We repeated the determination of the monomer molecular weight in sodium dodecyl sulfate as done by Alberts *et al.* (2-4) and obtained a value of 34,000. Our monomer molecular weights are therefore in essential agreement with those of Alberts.

The molecular weights obtained under a variety of other conditions are twice the monomer molecular weight. The center and right panels show the results obtained in buffer pH 10 (center) and in 2.0 M KCl (right). The dimer may also be obtained in ethanol and also in urea

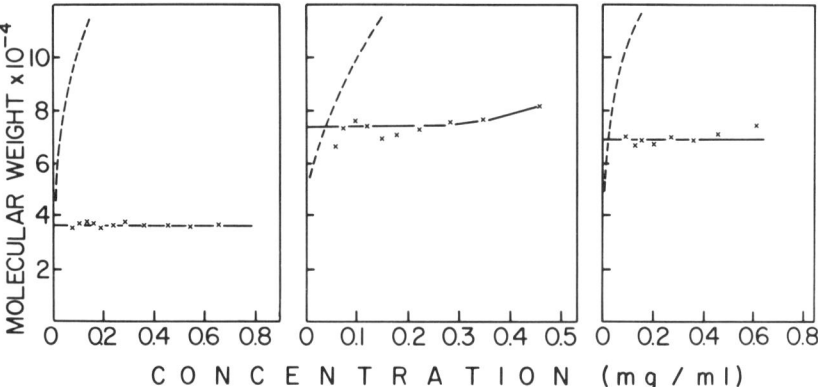

Fig. 1. Monomer and dimer molecular weights of gene 32 protein. Gene 32 protein was prepared by a modification of the method of Alberts and Frey (3). The molecular weight of gene 32 protein was determined by sedimentation equilibrium centrifugation by the method of Yphantis (10) in the presence of: (left) 6.0 M guanidine hydrochloride; (center) 0.02 M Tris-HCl, pH 10.0, 0.001 M β-mercaptoethanol, 0.10 M KCl; (right) 2.0 M potassium chloride. The control, without perturbant, is represented by dashed lines (- - -, see Fig. 2), the experimental sample by solid lines.

gels. These conditions evidently do not break one of the bonds which is broken by sodium dodecyl sulfate and also by guanidine hydrochloride. These results were substantiated with polyacrylamide gels (5).

Figure 2 illustrates the ability of gene 32 protein to self-associate. At optimum conditions of Tris-HCl, pH 8, and 0.1 M KCl the protein aggregates in a concentration-dependent manner; in other words, the molecular weight of the complex increases with protein concentration. Preliminary analysis indicates that this is an indefinite association as determined by the method of Adams (1). This finding agrees well with the presence of long aggregates of 32 seen on the DNA by Delius et al. (7) and the requirement for 170 32 protein monomers at the replication fork postulated by Alberts and Frey (3).

We can also see aggregates up to 12-mers on plain polyacrylamide gels (5). We conclude from our ability to isolate stable dimers that gene 32 protein aggregates by two different interactions, necessitating two different protein interaction sites designated by aa and bb interactions. At least one DNA binding site is necessary, giving us a monomer like that in Fig. 3A.

We propose that these may associate to give either molecules illustrated in Fig. 3, B or C. If gene 32 protein binding to DNA has polarity, and no data exist either for or against polarity, then this polarity is illustrated by the arrows above the DNA binding site. Polarity of binding

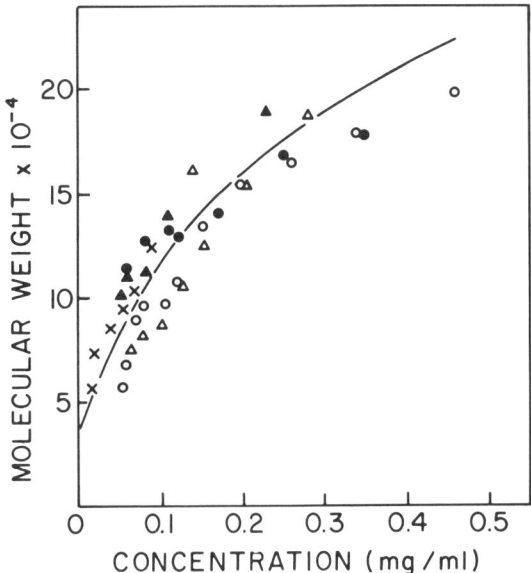

Fig. 2. The self-association of gene 32 protein. Gene 32 protein at an initial concentration of 1.06 mg/ml was centrifuged at 18,000 rpm in a 30-mm light path double sector cell (×). Gene 32 protein at initial concentrations of 1.0 mg/ml (○), 0.75 mg/ml (●), and 0.50 mg/ml (△) were centrifuged in a 12-mm three-channel Yphantis cell at 22,000 rpm. Gene 32 protein at an initial concentration of 1.00 mg/ml was centrifuged in a 12-mm double sector cell at 26,000 rpm (▲). The partial specific volume (as in Fig. 1, center and right panels) was calculated from the amino acid composition (6).

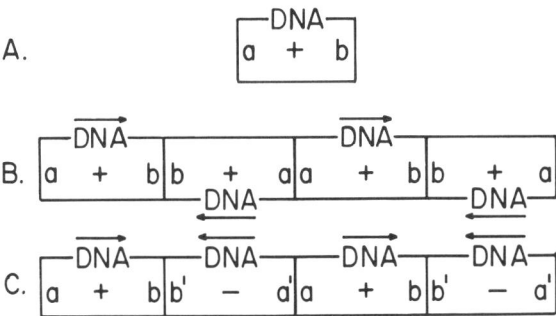

Fig. 3. The conformation of the gene 32 protein aggregate. A, A model of the gene 32 protein monomer with a and b protein-protein interaction sites and one DNA-Protein monomer with a and b protein-protein interaction sites and one DNA-protein interaction site. B and C, Two possible conformations of the aggregate, aa and bb are the two types of protein-protein bonds. + and − indicate the two sides of the monomer. The arrows indicate possible polarity of binding between DNA and protein.

might indicate 3B as a more probable model of the conformation of the gene 32 protein aggregate.

Acknowledgment

We acknowledge support from the National Institutes of Health (CA-11322 and 5-KO6GM 21 444) and from the American Cancer Society (NP-13A, In-57-J, and PRA-55).

References

1. Adams, E. T. 1967. Analysis of self-associating systems by sedimentation equilibrium experiments. *Fractions* 3: 1.
2. Alberts, B. M., Amodio, F. J., Jenkins, M., Gutmann, E. D., and Ferris, F. L. 1968. Studies with DNA-cellulose chromatography. I. DNA-binding proteins from *E. coli. Cold Spring Harbor Symp. Quant. Biol.* 33: 289.
3. Alberts, B. M., and Frey, L. 1970. T4 Bactreiophage gene 32. A structural protein in the replication and recombination of DNA. *Nature* 227: 1313.
4. Alberts, B. M. 1970. Function of gene 32-protein, a new protein essential for the genetic recombination and replication of T4 bacteriophage DNA. *Fed. Proc.* 29: 1154.
5. Carroll, R. B., Neet, K. E., and Goldthwait, D. A. 1972. Self-association of gene 32 protein of bacteriophage T4. *Proc. Nat. Acad. Sci. U. S. A.* 69: 2741.
6. Cohn, E. J., and Edsall, J. T. (eds.). 1943. *Proteins, amino acids and peptides*, p. 370, American Chemical Society.
7. Delius, H., Mantell, N. J., and Alberts, B. 1972. Characterization by electron microscopy of the complex formed between T4 bacteriophage gene 32-protein and DNA. *J. Mol. Biol.* 67: 341.
8. Edelstein, S. J., and Schachman, H. K. 1967. The simultaneous determination of partial specific volumes and molecular weights with microgram quantities. *J. Biol. Chem.* 242: 306.
9. Huberman, J. A., Kornberg, A., and Alberts, B. M. 1971. Stimulation of T4 bacteriophage DNA polymerase by the protein product of T4 gene 32. *J. Mol. Biol.* 62: 39.
10. Yphantis, D. A. 1964. Equilibrium ultracentrifugation of dilute solutions. *Biochemistry* 3: 297.

RNA Tumor Virus DNA Polymerase

The DNA Polymerases of RNA Tumor Viruses and Their Relationship to Cellular DNA Polymerases

Howard M. Temin and Satoshi Mizutani

McArdle Laboratory
University of Wisconsin
Madison, Wisconsin 53706

RNA tumor viruses replicate through a DNA intermediate, the DNA provirus. All infectious virions of RNA tumor viruses have endogenous RNA-directed DNA polymerase activity, which is believed responsible for the synthesis of the DNA provirus. Uninfected chicken cells contain endogenous RNA-directed DNA polymerase activity. The DNA polymerase solubilized from this activity is not neutralized by antibody which neutralizes viral DNA polymerase or by antibody which neutralizes large chicken DNA polymerases; viral polymerase is not neutralized by antibody to soluble chicken DNA polymerases; and soluble chicken polymerases are not neutralized by antibody to viral polymerase. There is, therefore, no serological relationship found between the viral DNA polymerase, the large chicken DNA polymerases, and the polymerase from the chicken endogenous RNA-directed DNA polymerase activity. The rate of DNA synthesis with different template-primers is different for the viral and both serological types of chicken DNA polymerases.

RNA tumor viruses were first isolated from tumors, but have since been isolated from a variety of normal animals and cells. RNA tumor viruses are characterized by virions about 100 nm in diameter with a lipid-containing envelope and a ribonucleoprotein core of no clearly observable symmetry. The virion contains in its core a single stranded RNA of 70 S, which dissociates into smaller pieces upon heating, and RNA-directed DNA polymerase activity. RNA tumor viruses do not kill the cells that they infect. They may be produced by infected cells or they may exist in a nonproductive state. In both productive and nonproductive states, the virus information is regularly passed to daughter cells at mitosis (see 15 for references).

Two other types of viruses, Visna and the syncytium-forming viruses, have virions and replication similar to those of the RNA tumor viruses. However, Visna and syncytium-forming viruses kill infected cells in their normal hosts (10). The types of RNA viruses with DNA polymerases in their virion are listed in Table 1 (see 17).

Table 1. RNA viruses with a virion DNA polymerase

Non-cell killing
  C-type
    Strongly transforming (Rous sarcoma virus)
    Weakly transforming (Rous-associated virus-1)
    Nontransforming (erythroblastosis-associated virus)
  B-type (murine mammary tumor virus)
  Unclassified
Cell killing
  Visna virus
  Syncytium-forming ("foamy") viruses

The DNA Provirus Hypothesis

In 1964, Temin proposed in the DNA provirus hypothesis that RNA tumor viruses replicate through a DNA intermediate; that is, that in the replication of RNA tumor viruses there is transfer of information from virion RNA to DNA. The present evidence for this hypothesis is as follows (see 15 for references to early work): (*a*) There is regular inheritance of viral genomes (proviruses). In multiply infected virus-producing cells, there are one or two copies of each provirus. These proviruses are regularly passed to the daughter cells. In nonvirus-producing cells, the virus genome is regularly passed to all progeny cells. (*b*) Virus production is sensitive to treatment with actinomycin D. This treatment directly blocks virus RNA synthesis, and this blockage is not a secondary effect of the cytotoxicity of actinomycin D. (*c*) No reproducible evidence exists for RNA to RNA replication by RNA tumor viruses. No double stranded RNA, no RNA that is complementary to virus RNA, and no RNA replicase activity has been reproducibly found in RNA tumor virus-infected cells. (*d*) Early virus-specific DNA synthesis is required after infection of cells by RNA tumor viruses. This requirement for early virus-specific DNA synthesis is in addition to the necessity for cellular DNA synthesis and normal progression through the cell cycle required for activation of virus RNA synthesis and progeny virus production (6). (*e*) The early virus-specific DNA can be specifically labeled with 5-bromodeoxyuridine and then inactivated by exposure to visible light. (*f*) Increased hybridization between virus RNA and DNA of transformed cells has been found (2, 3, 12). (*g*) Infectious DNA with virus information can be extracted from Rous sarcoma virus (RSV)-infected rat and chicken cells (5, 8, 14). (*h*) All infectious virions contain DNA polymerase activity.

Virion DNA Polymerase

The evidence for the virion DNA polymerase activity having a role in virus infection by forming the DNA provirus is as follows (see 15, 17). (*a*) Virus infection takes place in stationary cells in the presence of cycloheximide, an inhibitor of protein synthesis. This result suggests that, if a DNA provirus exists, the enzyme responsible for its synthesis is not itself synthesized after infection. Enzymes which use heterogeneous RNA as a template for DNA synthesis do not appear to be widely distributed in uninfected cells (see below). Therefore, a virion DNA polymerase appears necessary for formation of the DNA provirus in the absence of protein synthesis. (*b*) RSV*a*, a noninfectious variant of Bryan high titer Rous sarcoma virus, does not contain DNA polymerase activity. (*c*) S^+H^- murine sarcoma virus particles, noninfectious particles produced by certain murine sarcoma virus-infected cells, contain very low or no DNA polymerase activity (11). (*d*) The virions of RNA tumor viruses can be irreversibly inactivated by preincubation with certain derivatives of rifamycin which inhibit virion DNA polymerase activity (19).

This evidence is consistent with the hypothesis that the virion DNA polymerase is necessary for provirus formation. Unfortunately, none of the evidence is conclusive. No temperature-sensitive mutants of the virion DNA polymerase or RNA-DNA intermediates in infection have been reported.

The RNA tumor virus virion DNA polymerase has been much studied since its discovery by Baltimore (1) and Temin and Mizutani (16). The work up until the end of 1971 has been recently reviewed (17). It has been found that all infectious virions of RNA tumor viruses contain endogenous DNA polymerase activity. For full activty, the virions must be disrupted and incubated at a temperature of about 40°C, pH 8.0, in the presence of a reducing agent, all four deoxyribonucleoside triphosphates, and a divalent cation, either magnesium or manganese. The activity is sensitive to pretreatment with ribonuclease, but is resistant to pretreatment with deoxyribonuclease and is partially resistant to treatment with actinomycin D. The initial product of the endogenous virion DNA polymerase activity is an RNA-DNA hybrid containing the DNA product and both an RNA template and an RNA primer. The final product of the endogenous reaction is double stranded DNA which specifically hybridizes with the 70 S RNA of the virion.

DNA polymerases have been purified from virions of avian and murine RNA tumor viruses. There appear to be only a small number of molecules of DNA polymerase per virion. The avian virus DNA polymerase appears to have a molecular weight of 160,000 by glycerol gradient centrifugation. The murine virus DNA polymerase appears to have a molecular weight of 90,000. The purified polymerases have an absolute requirement for template and primer. The primer can be either RNA or DNA and incorporation takes place at the 3'-OH end of the primer. The purified polymerases have little template specificity, but can use a variety of natural and synthetic RNAs and DNAs as templates. The polymerases appear not to copy efficiently long stretches of single stranded DNA even in the presence of a primer.

The areas of greatest biochemical research at the present time will be discussed in this symposium. They are: What is the nature of the primer in the virion 70 S RNA for the endogenous reaction? Is it one of the 35 S RNA subunits or a specific smaller 4 or 5 S primer molecule? What controls the formation of the DNA product? Especially, what controls the extent of copying of the RNA template and the size of the DNA product? Are other activities present in purified RNA tumor virus virions important in endogenous DNA synthesis? Is the virion DNA polymerase coded for by the virus or by the cell? Is there a relationship between the virion DNA polymerase and the usual cellular DNA polymerases or cellular DNA polymerases involved in cellular RNA-directed DNA synthesis?

Cellular Endogenous RNA-directed DNA Polymerase Activity

In our laboratory, we have been concentrating on the last two questions. Our work has been based on the general hypothesis, the protovirus hypothesis, that RNA tumor viruses arose from normal cellular elements which transfer information from DNA to RNA to DNA (15). We have followed two approaches to testing this hypothesis. One has been to isolate endogenous RNA-directed DNA polymerase activity from non-virous-infected cells; the second has been to isolate as many DNA polymerases as possible from uninfected chicken cells to see if some of them are related, primarily by serological tests, to the avian RNA tumor virus virion DNA polymerase.

Our initial work demonstrated endogenous ribonuclease-sensitive DNA polymerase activity in both Rous sarcoma virus-infected and uninfected rat cells (4). We have shifted our present work to chicken cells because we have available more different types of chicken cells, and we know more about possible contaminating viruses in chicken cells than in rat cells. From tissue culture cells or 5-day-old chicken embryos, we have isolated endogenous RNA-directed DNA polymerase activity (7).

This endogenous DNA polymerase activity has been shown to be RNA-directed by its sensitivity to ribonuclease, its resistance to deoxyribonuclease, and its partial resistance to actinomycin D. The product DNA of the chicken endogenous DNA polymerase activity hybridizes to the RNA present in the same fraction from chicken cells. Early product DNA is attached by noncovalent bonds to RNA (18). The early product DNA is released from this RNA by ribonuclease or heating, as shown by its decrease in sedimentation rate.

The RNA template of this chicken endogenous RNA-directed DNA polymerase activity is not related to the RNAs of known chicken RNA viruses containing DNA polymerase activity (7). The DNA product of the chicken endogenous RNA-directed DNA polymerase activity does not hybridize to RNA from Rous sarcoma or reticuloendotheliosis viruses. (Of course, the same DNA product hybridizes to the RNA from the same chicken fraction.)

To determine whether the polymerase of the chicken endogenous RNA-directed DNA polymerase activity is related to the avian RNA tumor virus DNA polymerase, neutralization tests were carried out. An antibody, kindly supplied by Dr. R. Nowinski of the McArdle Laboratory (9), which neutralized the activity of the avian myeloblastosis or other avian leukemia-sarcoma virus DNA polymerases was used. This antibody did not neutralize the activity of the endogenous RNA-directed DNA polymerase activity from chicken cells (7). This antibody also did not neutralize exogenous DNA-directed polymerase activity of the 3 to 4 S polymerase isolated from the chicken fraction with endogenous RNA-directed DNA polymerase activity (Fig. 1). The antibody also did not neutralize the activity of a 10 S soluble chicken DNA polymerase (Fig. 1).

To determine whether the DNA polymerase of the chicken endogenous DNA polymerase activity was related to other chicken DNA polymerases, further neutralization tests were performed. Antibody was prepared against a partially purified 10 S DNA polymerase isolated from the soluble fraction of 7-day-old chicken embryos. This antibody neutralized the activity of the homologous DNA polymerase and several other DNA polymerase activities isolated either from the soluble fraction or the nuclear fraction of 7-day-old chicken embryos (Table 2). However, this antibody to a 10 S chicken DNA polymerase did not neutralize either the endogenous RNA-directed or the exogenous calf thymus DNA-

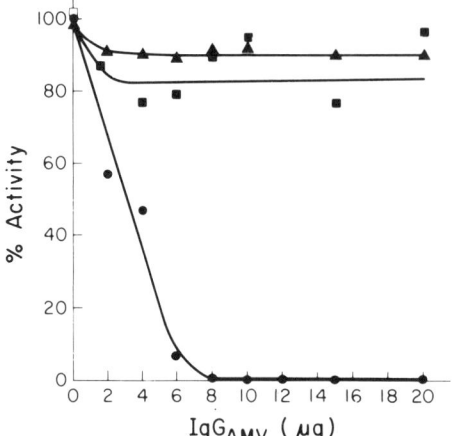

Fig. 1. Neutralization of DNA polymerases by antibody to avian myeloblastosis virus DNA polymerase. DNA polymerases were purified from virions of RSV and from 7-day-old chicken embryos. Polymerase, 3 to 4 S, of the chicken endogenous DNA polymerase activity, and a 10 S soluble chicken DNA polymerase were used. Antibody, kindly supplied by Dr. Robert Nowinski, McArdle Laboratory (9), was against the purified DNA polymerase of avian myeloblastosis virus. IgG was prepared from the antibody, different amounts of IgG were incubated with polymerases (about 20 µg/ml) for 1 hr at room temperature, and the rate of residual calf thymus DNA-directed polymerase activity was measured for 20 min at 40°C. One hundred per cent activity was 2600 cpm incorporated per hr for RSV polymerase (●); 18,000, for chicken fraction 3 to 4 S polymerase (■); and 7,400, for chicken 10 S soluble polymerase (▲).

Table 2. Neutralization of soluble chicken DNA polymerases

DNA polymerase		
S-value	Purification	Activity remaining (%)
10*	DEAE-B	0.5
7	DEAE-B	5
10	DEAE-A	10
5	DEAE-A	10
10	Unabsorbed to PC	15
10	Nuclear	20

DNA polymerases were purified 100- to 200-fold from the soluble fraction of 7-day chicken embryos by successive chromatography on phosphocellulose and DEAE-cellulose and sedimentation in a glycerol gradient. Antibody was prepared in a rat to the 10 S polymerse purified on phosphocellulose and eluted from DEAE-cellulose at 0.3 M NaCl. IgG was prepared from immunized and control rats, incubated with polymerase (about 20 µg/ml of IgG and of DNA polymerase) for 1 hr at room temperature, and the rate of residual calf thymus DNA-directed polymerase activity measured for 20 min at 40°C. One hundred per cent activity was from 15,000 to 20,000 cpm incorporated per hr.

*Used for immunization.

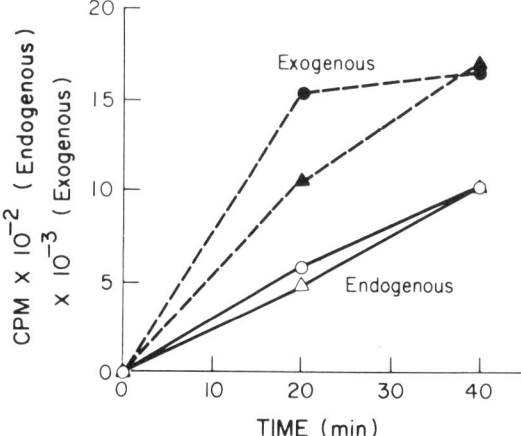

Fig. 2. Neutralization of exogenous and endogenous DNA polymerase activity of chicken fraction by antibody to soluble chicken DNA polymerase. Antibody was prepared in rats against a partially purified 10 S polymerase isolated from the soluble fraction of 7-day-old chicken embryos. IgG was prepared from this antibody and incubated for 1 hr with a fraction from 7-day-old chicken embryos containing endogenous DNA polymerase activity with no added template-primer or added cell thymus DNA, and the rate of residual endogenous or calf thymus DNA-directed polymerase activity measured. Endogenous activity without IgG (△), with IgG (○); exogenous activity without IgG (▲) and with IgG (●).

directed polymerase activity of the chicken fraction (Fig. 2). In further tests at a variety of antibody concentrations, no neutralization was found of the partially purified 3 to 4 S polymerase solubilized from the chicken endogenous DNA polymerase activity or the DNA polymerase from Rous sarcoma virus (Fig. 3). Therefore, these two antibodies provided no evidence by neutralization for serological relationships between the DNA polymerase of the chicken endogenous RNA-directed DNA polymerase activity, the Rous sarcoma virus virion DNA polymerase, and the other chicken DNA polymerases.

The Rous sarcoma virus DNA polymerase then is neither simply one of the common DNA polymerases of the cell (see also 13), nor the DNA polymerase of the cellular endogenous RNA-directed DNA polymerase activity. Furthermore, use of these two antibodies gives no evidence that the DNA polymerase of Rous sarcoma virus is related to DNA polymerases of the cell. These results strongly suggest that the Rous sarcoma virus DNA polymerase is coded by the virus genome and that if the Rous sarcoma virus DNA polymerase originated from a cellular DNA polymerase, the origin was long enough ago so that serological relatedness, as seen by these two antibodies, has been lost.

Although these antibody experiments establish no close relationship between the Rous sarcoma virus DNA polymerase and the cellular DNA

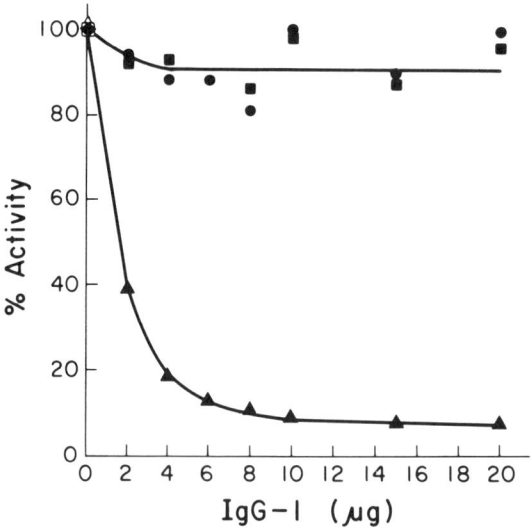

Fig. 3. Neutralization of DNA polymerase activity with antibody to soluble chicken 10 S DNA polymerase. The same three polymerases used in the experiment of Fig. 1 were incubated at the same concentrations with different concentrations of the antibody described in Table 2 and Fig. 2. RSV polymerase, ●; chicken 10 S polymerase, ▲; and chicken fraction 3 to 4 S polymerase, ■.

polymerases, there might be common structural features between the polymerase of the virus RNA-directed DNA polymerase activity and the cell endogenous RNA-directed DNA polymerase activity. To test this hypothesis, comparison was made of the activity of these polymerases with different template-primers. In each case, preliminary experiments were carried out to find the optimum concentration of magnesium or manganese for the polymerase and template-primer combination. Then the rate of DNA synthesis was measured at the optimum divalent cation concentration (Table 3). The Rous sarcoma virus DNA polymerase and the soluble cellular DNA polymerases were found to behave as previously reported by others (see references in 17). That is, the soluble cellular DNA polymerases could use DNA and poly(rA) · poly(dT) as template-primers, but did not use poly(rA) · oligo(dT) or 70 S viral RNA as template-primers. The tumor virus DNA polymerase was able to use all of the template-primers tested. Surprisingly, the 3 to 4 S DNA polymerase isolated from the chicken fraction with endogenous RNA-directed DNA polymerase activity behaved like the soluble chicken DNA polymerases in its template-primer preferences. The failure of this poly-

Table 3. Relative activity of different template-primers and different DNA polymerases

	DNA polymerase		
Template-primer	RSV	Chicken fraction, 3-4 S	Chicken soluble, 10 S
Native calf thymus DNA	100	100	100
RSV-RNA	55	0.1	0.1
poly (dA · dT)	200	75	23
poly (dA) · oligo (dT)	20	1.3	42
poly (rA) · poly (dT)	320	4	8.5
poly (rA) · oligo (dT)	100	0.7	3

DNA polymerases were purified from virions of RSV and from 7-day-old chicken embryos. The 3 to 4 S polymerase of the chicken endogenous DNA polymerase activity and a 10 S soluble chicken DNA polymerase were used. The relative rate of incorporation of ^{3}H-TTP was measured during a period of linear rate of incorporation. One hundred was 13,800 pmoles incorporated per hr per mg of protein of RSV polymerase; 1,500 pmoles, for chicken fraction 3 to 4 S polymerase; and 6,400 pmoles, for chicken soluble 10 S polymerase.

merase from the chicken endogenous RNA-directed DNA polymerase activity to use RNA or poly(rA) · oligo(dT) as template-primers may relate to nuclease contamination of the polymerase. To test this hypothesis a mixing experiment was performed with RSV polymerase and the chicken fraction 3 to 4 S polymerase (Table 4). It was found that the chicken fraction 3 to 4 S polymerase inhibited the RSV RNA-directed and the poly(rA) · oligo(dT)-directed DNA polymerase activity of the RSV polymerase. Therefore, until nuclease is removed a true picture of the activity of the chicken fraction 3 to 4 S polymerase with different template-primers cannot be obtained.

DNA polymerases of different types appear to exist in Rous sarcoma virus and in chicken cells. Further studies are necessary to determine whether there is any relationship between them or if they have independent origins.

Table 4. Mixing experiment

	DNA polymerase(s)		
Template · primer	RSV	Chicken fraction, 3 to 4 S + RSV	Chicken fraction, 3 to 4 S
RSV-RNA	1100*	225	100
poly (rA) · oligo (dT)	8100	100	0

RSV polymerase and chicken fraction 3 to 4 S polymerase were incubated separately or together with the indicated template-primers as in the experiment of Table 3.
*Counts per min incorporated per 30 min.

Acknowledgments

The work under discussion from our laboratory was supported by Public Health Service Research Grant CA-07175 from the National Cancer Institute and Grant VC-7 from the American Cancer Society. H. M. T. holds Research Career Development Award 10K3-CA-8182 from the National Cancer Institute.

Noted added in proof: Some but not all preparations of the 3 to 4 S DNA polymerase solubilized from the chicken fraction had, after removal of ribonuclease activity, template-primer preferences similar to RSV DNA polymerase.

A 3 to 4 S DNA polymerase was also partially purified from the soluble fraction of chicken embryo homogenates. The activity of this DNA polymerase was not neutralized either by antibody to AMV DNA polymerase or by antibody to large soluble chicken DNA polymerase. Antibody to the small soluble chicken DNA polymerase neutralized the activity of the 3 to 4 S DNA polymerase solubilized from the chicken fraction, partially neutralized the chicken endogenous RNA-directed DNA polymerase activity, but did not neutralize the activity of RSV DNA polymerase or large chicken DNA polymerases.

References

1. Baltimore, D. 1970. RNA-dependent DNA polymerase in virions of RNA tumor viruses. *Nature (London)* 266: 1209.

2. Baluda, M. A. 1972. Widespread presence, in chickens, of DNA complementary to the RNA genome of avian leukosis viruses. *Proc. Nat. Acad. Sci. U. S. A.* 69: 576.

3. Bishop, J. M., Faras, A. J., Garapin, A. C., Hansen, C., Jackson, N., Levinson, W., Taylor, J. M., and Varmus, H. E. 1972. RNA directed DNA polymerase in the replication of Rous sarcoma virus. *In* Bowen, J. (ed.) *Molecular studies in viral neoplasia*, 25th M. D. Anderson Symposium, University of Texas Press, Houston. In press.

4. Coffin, J. M., and Temin, H. M. 1971. Ribonuclease-sensitive deoxyribonucleic acid polymerase activity in uninfected rat cells and rat cells infected with Rous sarcoma virus. *J. virol.* 8: 630.

5. Hill, N., and Hillova, J. 1971. Production virale dans les fibroblastes de poule traités par l'acide desoxyribonucleique de cellules XC de rat transformées par le virus de Rous. *C. R. Hebd. Seances Acad. Sci. Ser. D, Sci. Natur. (Paris)* 272: 3094.

6. Humphries, E. H., and Temin, H. M. 1972. Cell cycle-dependent activation of Rous sarcoma virus-infected stationary chicken cells: avian leukosis group-specific antigens and ribonucleic acid. *J. Virol.* 9: 82.

7. Kang, C.-Y., and Temin, H. M. 1971. Endogenous RNA-directed DNA polymerase activity in uninfected chicken embroyos. *Proc. Nat. Acad. Sci. U. S. A.* 69: 1550.

8. Montaıgnier, L., and Vigier, P. 1972. Production virale dans les fibroblastes de poule traités par l'acide desoxyribonucleique des cellules XC de rat transformées par le virus de Rous. *C. R. Hebd. Seances Acad. Sci. Ser. D, Sci. Natur. (Paris)* 272: 3094.

9. Nowinski, R., Watson, K., Yaniv, A., and Spiegelman, S. 1972. Serological analysis of the deoxyribonucleic acid polymerase of avian oncornavirus. II. Comparison of avian deoxyribonucleic acid polymerases. *J. Virol.* 10: 959.

10. Parks, W. P., and Todaro, G. J. 1972. Biological properties of syncytium forming ("foamy") viruses. *Virology* 47: 673.

11. Peebles, P. T., Haapala, D. K., and Gadzar, A. F. 1972. Deficiency of viral ribonucleic acid-dependent deoxyribonucleic acid polymerase in noninfectious virus-like particles released from murine sarcoma virus-transformed hamster cells. *J. Virol.* 9: 488.

12. Rosenthal, P., Robinson, H. L., Robinson, W. S., Hanafusa, T., and Hanafusa, H. 1971. DNA in noninfected cells and virus-infected cells complementary to avian tumor virus RNA. *Proc. Nat. Acad. Sci. U. S. A.* 68: 2336.

13. Ross, J., Scolnick, E. N., Todaro, G. J., and Aaronson, S. A. 1971. Separation of murine cellular and murine leukemia virus DNA polymerases. *Nature New Biol.* 231: 163.

14. Svoboda, J., Holozanek, I., and Mach, O. 1972. Detection of chicken sarcoma virus after transfection of chicken fibroblasts with DNA isolated from mammalian cells transformed with Rous virus. *Folio Biol. (Praha)* 18: 149.

15. Temin, H. M. 1971. Mechanism of cell transformation by RNA tumor viruses. *Annu. Rev. Microbiol.* 25: 610.

16. Temin, H. M., and Mizutani, S. RNA-dependent DNA polymerase in virions of Rous sarcoma virus. *Nature* 226: 1211.

17. Temin, H. M., and Baltimore, D. 1972. RNA-directed DNA synthesis and RNA tumor viruses. *Advan. Virus Res.* 17: 129.

18. Temin, H. M., Kang, C.-Y., and Mizutani, S. 1972. Endogenous RNA-directed DNA polymerase activity in normal cells. In Beers, R. A. (ed.) *Cellular modification and genetic transformation by exogenous nucleic acid*, 6th Miles International Symposium on Molecular Biology, Johns Hopkins Press, Baltimore. In press.

19. Ting, R. C., Yang, S. S., and Gallo, R. C. 1972. Reverse transcriptase, RNA tumor virus transformation and derivatives of rifamycin SV. *Nature New Biol.* 236: 163.

RNA-directed and -primed DNA Polymerase Activities in Tumor Viruses and Human Lymphocytes

Robert C. Gallo, Prem S. Sarin,
R. Graham Smith, Samuel N. Bobrow,
Mangalasseril G. Sarngadharan,*
Marvin S. Reitz, Jr.,* and John W. Abrell*

Laboratory of Tumor Cell Biology
National Cancer Institute
National Institutes of Health
Bethesda, Maryland 20014

1. Purified DNA polymerases of RNA tumor viruses from different species and even within the same species show significant biochemical heterogeneity, e.g. the relative capacity to transcribe exogenous viral 70 S RNA and the requirement for divalent cation. Yet all the tumor virus DNA polymerases appear to have the following properties in common: they (a) catalyze endogenous RNase-sensitive DNA synthesis in crude disrupted virions, (b) transcribe heteropolymeric regions of RNA, (c) have a greater response to dT_{12-18} · poly rA than to dT_{12-18} · poly dA (when the enzymes are sufficiently purified), and (d) transcribe poly rC with oligo dG primer. These characteristics are extremely useful in distinguishing the viral enzyme from the DNA polymerases of normal cells.
2. We find that all purified DNA polymerases we have tested (bacterial, mammalian, and viral) can transcribe *homopolymeric* RNA stretches, e.g. poly A in the presence of appropriate primers, whether with synthetic hybrid templates such as poly dT · poly rA or natural RNAs with poly rA "tracts." Therefore, this property clearly is not specific for the DNA polymerase of RNA tumor viruses.
3. Two DNA polymerases have been purified from human blood lymphocytes. These appear to have properties identical with the main mammalian DNA polymerases described in other

*Bionetics Research Laboratories, Bethesda, Md.

laboratories. The high molecular weight enzyme (~160,000 daltons) is cytoplasmic while the lower molecular weight enzyme (~30,000 daltons) is found in the cytoplasm and nucleus. Both enzymes prefer $dT_{12-18} \cdot $ poly dA to $dT_{12-18} \cdot $ poly rA, will not utilize $dG_{12-18} \cdot $ poly rC as a template-primer, and fail to transcribe heteropolymeric regions of natural RNAs.

4. A DNA polymerase was isolated from a cytoplasmic 60,000 × g "pellet" from fresh human leukemic blood lymphocytes and from phytohemagglutinin-*stimulated* normal blood lymphocytes (but not found in unstimulated normal cells). This polymerase carries out an *endogenous* RNase-sensitive synthesis of DNA. Part of the DNA products (with the leukemic cells) move in the RNA region of a Cs_2SO_4 equilibirum density gradient. The polymerase purified from the *leukemic* pellet prefers $dT_{12-18} \cdot $ poly rA to $dT_{12-18} \cdot $ poly dA, accepts $dG_{12-18} \cdot $ poly rC, and transcribes the heteropolymeric region of viral 70 S RNA. The DNA product of the reaction with 70 S RNA is about 4 to 8 S in size and it hybridizes back to the template RNA but not to heterologous RNA. The enzyme purified from the "pellet" of normal lymphocytes will accept neither exogenous RNA nor $dG_{12-18} \cdot $ poly rC, and shows marked preference for $dT_{12-18} \cdot $ poly dA to $dT_{12-18} \cdot $ poly rA, *i.e.* in these analyses the purified leukemic "pellet" polymerase behaves like the purified tumor virus enzyme, while the purified normal lymphocyte polymerase behaves like the major two cellular DNA polymerases. It appears that either the endogenous RNase-sensitive DNA polymerase reaction of normal cells is RNA-*primed* and DNA-directed, or RNA-directed, this enzyme, unlike the viral or leukemic "pellet" polymerases, has template specificity, requiring an RNA present in the "pellet."

Oncornavirus DNA Polymerase

Before discussing our studies on cellular DNA polymerases, a brief description of some of the important properties of the oncornavirus (RNA tumor virus) DNA polymerase is appropriate. No attempt has been made to supply complete references to some established points since the properties of the viral enzyme were reviewed last year (7) and a detailed review is now in press (30) by the discoverers of the enzyme (1, 31).

The main common features of the DNA polymerases from various oncornaviruses are the following. (*a*) The enzyme is located in a particulate fraction, *i.e.* in the virion (and probably in the core of the virion), requiring high salt, or non-ionic detergent, or both, to be solubilized. (*b*) It catalyzes endogenous, partially actinomycin D-resistant, RNase-sensitive DNA synthesis in disrupted particles when supplied with appropriate buffers and deoxynucleoside triphosphates. The DNA product of the endogenous reaction (*c*) is about 4 to 6 S in size, (*d*) is covalently attached to RNA, and (e) hybridizes to the virion 70 S RNA. (f) Synthesis of DNA catalyzed by the solubilized and partially purified enzyme as well as the endogenous reaction are inhibited by a variety of rifampicin derivatives and streptovaricins. (g) The response of a partially purified enzyme to particular template-primers is useful in distinguishing this enzyme from other DNA polymerases; e.g. dT_{12-18} · poly rA and dG_{12-18} · poly rC are utilized while dT_{12-18} · dA is a very poor template-primer. The property of greatest interest and usefulness is the ability of this polymerase to transcribe certain heterologous (or homologous) predominantly single stranded natural RNAs such as viral 70 S RNA or cellular mRNA. However, it should be emphasized that this has been achieved to date *only* for the DNA polymerases from avian RNA tumor viruses, i.e. avian myeloblastosis virus (AMV) and Rous sarcoma virus (RSV), and for a DNA polymerase we have purified (26) from fresh human leukemic blood cells which will be described in detail in a later section. To date, no other DNA polymerases have been shown to transcribe heteropolymeric regions of exogenous, primarily single stranded natural RNAs. It could possibly be that the main reason for lack of evidence that the purfied polymerase from mammalian RNA tumor viruses transcribes heteropolymeric portions of added RNA is that not enough virus is available. Alternatively, since the endogenous reaction of the crude disrupted virion is clearly RNA-directed, some component of the mammalian virion DNA polymerase is probably lost or inactivated during purification of the enzyme, a component which must be either more stable, more firmly bound, present in greater amount, or not required in avian systems. This leads to a discussion of some dissimilarities in the properties of DNA polymerases from different virions.

Heterogeneity of RNA tumor viruses from different species and even within the same species is well known by morphological (3) and immunological (11) criteria. The extent of the heterogeneity of the DNA polymerase, however, was somewhat surprising and perhaps is not yet appreciated. It is clear from immunological (22) and biochemical studies that the DNA polymerases of different types of RNA tumor viruses differ even within the same species and between C-type viruses from different species. For example, we find that with dT_{12-18} · poly rA as the template-primer the DNA polymerase from the Mason-Pfizer agent (M-PMV) (an RNA virus isolated from a monkey breast tumor which is morphologically different from the C-type RNA tumor viruses) requires Mg^{++}, while the polymerase from a monkey C-type virus, the simian sarcoma virus, shows a marked preference for Mn^{++}. Reports on molecular weight of the enzyme also indicate species variation. We (J. Abrell and R. Gallo, unpublished results) finds that the molecular weight of the Mason-Pfizer agent is about 120,000 daltons, a value similar to the 110,000 daltons reported for AMV polymerase (17) but higher than the

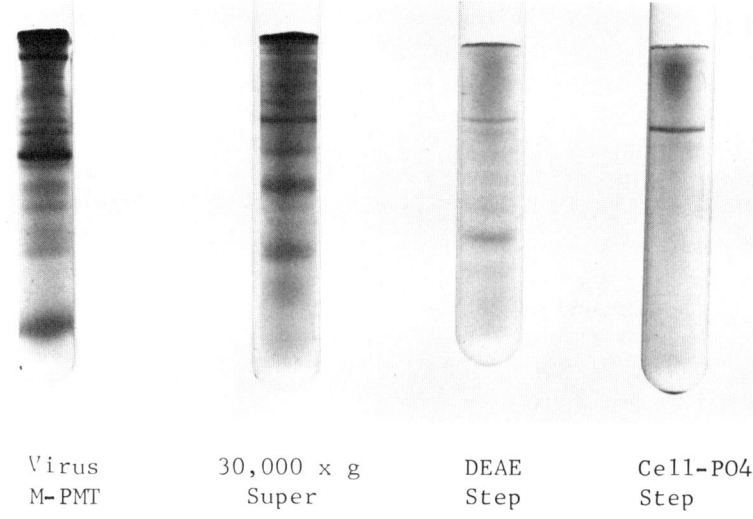

|Virus|30,000 x g|DEAE|Cell-PO4|
|M-PMT|Super|Step|Step|

Fig. 1. Disc gel electrophoresis in the presence of sodium dodecyl sulfate according to the procedure of Weber and Osborn (32). Virus M-PMT is Mason-Pfizer monkey tumor virus disrupted with Triton X-100. The supernatant from the disrupted virus after centrifugation at 30,000 × g is 30,000 × g Super. DEAE Step is the activity pool after chromatography on DEAE-cellulose. Cell-PO4 Step is the activity pool after chromatography on phosphocellulose.

value (70,000 to 90,000 daltons) found in different laboratories (16, 25) for the mouse virus polymerase (Rauscher leukemia virus). Further, our evidence suggests that there is only one polypeptide for the polymerase of the Mason-Pfizer agent (Fig. 1), while Kacian *et al.* reported evidence for two subunits for the AMV polymerase (17).

DNA Polymerases of Human Blood Lymphocytes

We became interested in characterizing the DNA polymerases of human peripheral blood lymphocytes for several reasons. First, these cells were readily available to us, and there was no detailed study of DNA polymerases that was carried out with *fresh* human cells. Second, this cell system offers a unique opportunity to study the biochemical changes which occur in conversion of "resting" cells to cells entering the mitotic cycle. This is easily achieved by stimulation of these cells with phytohemagglutinin (PHA). Some of the early biochemical events that occur following stimulation with PHA have been extensively studied. For instance, it is known that both DNA and RNA polymerase activities of crude extracts increase prior to the onset of DNA synthesis and reach a maximum about 3 days after the addition of PHA to the cells (10,

15, 20). This system may be extremely useful in learning which DNA polymerase catalyzes replication of DNA. Third, in characterizing DNA polymerases from neoplastic cells, an appropriate normal cell control is often extremely difficult, if not impossible, to obtain. In this system, we are able to study the polymerases of normal and leukemic cells which appear to be as comparable as possible. Lymphoblasts from blood of patients with lymphocytic leukemia are similar in cell type and metabolic state to lymphoblasts derived from the conversion of resting normal blood lymphocytes to "blast" cells with PHA. Fourth, in looking for DNA polymerases in cells with characteristics of the "reverse transcriptase" of RNA tumor viruses, it is of obvious interest to look at human neoplastic cells, particularly the leukemic lymphocytes which are available in gram quantities. Furthermore, in this case the "control" normal cells might be just as interesting, because the state of differentiation of these cells can be changed with addition of antigen as well as PHA. Reversal of information flow (RNA → DNA) could be involved in cell differentiation and in the immunological response as postulated in the protovirus theory (29).

Two Approaches

Peripheral blood lymphocytes were obtained from healthy donors and from patients with acute leukemia, and separated from other cell components of the blood as previously described (10, 23). Two approaches were used for isolation and purification of DNA polymerases. (a) The first approach is a modification of the procedure used by Chang and Bollum in other mammalian cells (6), and involved rigorous extraction of the disrupted cells with high concentrations of salt and with non-ionic detergent. The objective is to "solubilize" and identify every DNA polymerase which is present in the cell, providing that irreversible inactivation of one or more enzyme (or enzymes) does not occur from this treatment. No attempt is made at subcellular fractionation. The details of this procedure are described elsewhere (27). (b) A second approach was based on an attempt to isolate a cytoplasmic particulate fraction in which a DNA polymerase might be complexed with its template nucleic acid, in other words, a method that might isolate virus particles, components of virus, or other "organelle"-associated DNA polymerases. We, of course, were looking for an RNA-directed (or RNA-primed, or both) endogenous DNA polymerase system. Instead of vigorous extraction techniques, cells were manually gently homogenized so that intact nuclei could be removed. An outline of the methods and the results is presented later. With the first approach we isolated two major DNA polymerases from both normal and leukemic human lymphocytes, but the experiments described below are with PHA-stimulated normal lymphocytes.

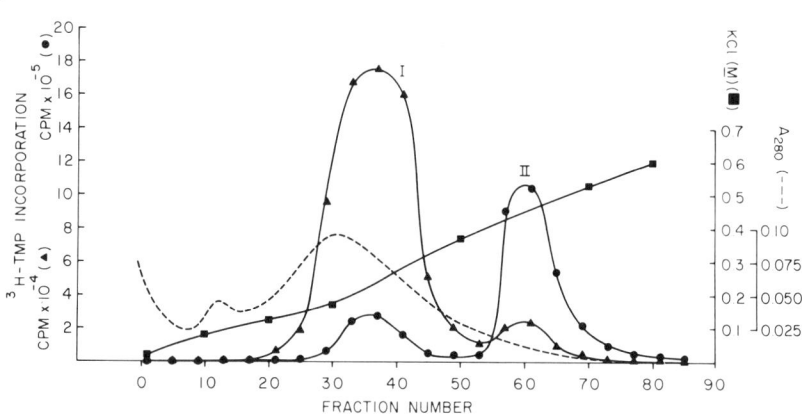

Fig. 2. Separation on phosphocellulose of normal lymphocyte DNA polymerases I and II. Fraction 2 (156 ml) was absorbed to a column, 2.0 × 6.0 cm, of phosphocellulose (Whatman P-11) equilibrated with TDEG buffer (see legends to Tables 1 and 2 for buffer definitions). The flow rate was 18 ml/hr. Following a 15-ml wash with TDEG buffer + 0.1 M KCl, 160 ml of a linear gradient of KCl in buffer, extending between extremes of 0.1 and 0.7 M, was attached. Fractions of 1.6 ml were collected and assayed (System A, ▲——▲; System B, ●——●) (see Table 2). Concentration of KCl (■——■) was measured with a conductivity meter.

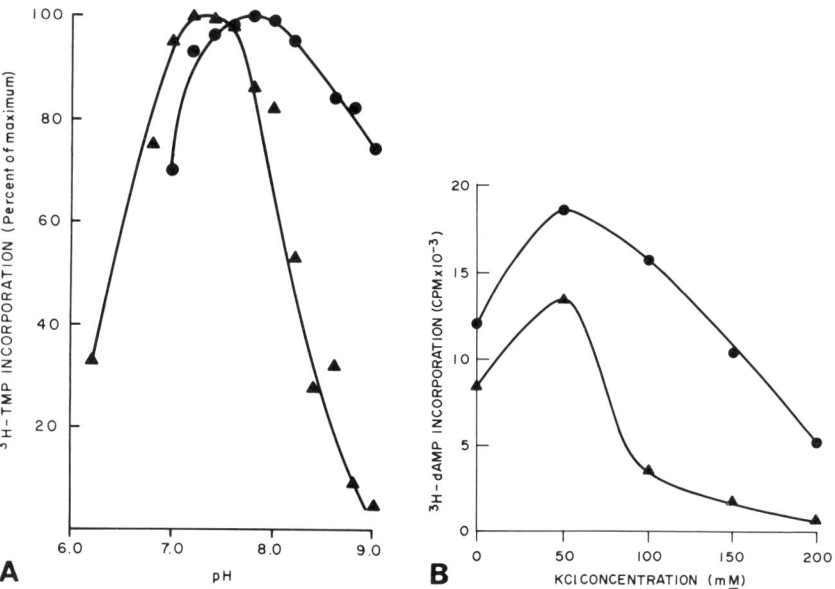

Fig. 3. A, pH optima of normal lymphocyte DNA polymerases I (▲——▲) and II (●——●). The assay ixtures (50 μl) included: activated salmon sperm DNA, 50 μg/ml; $MgCl_2$, 10 mM; KCl, 50 mM; dithiothreitol, 5 mM; dATP, dGTP, and dCTP, 8×10^{-5} M each; 5×10^{-6} M ^{3}H-TTP (7400 cpm/pmole); 0.2 μg of DNA polymerase I; 0.1 μg of DNA polymerase II; and Tris-HCl buffers of various pH values (determined at 25°C) as indicated. B, Effect of KCl concentration on activity of normal lymphocyte DNA polymerases I (▲——▲) and II (●——●). Reaction conditions were as described in Fig. 3A, except that various concentrations of KCl were added as indicated, and the pH (7.8) was constant.

Table 1. Purification of human lymphocyte DNA polymerase I

Fraction	Total protein (mg)	Specific activity (pmole/hr/μg)	Total activity (pmole/hr)
1. S-100	1600	0.35*	5.6×10^{5}*
2. DEAE-eluate	330	4.8	1.6×10^{6}
3. Phosphocellulose (0.23 M peak)	27.5	51	1.4×10^{6}
4. "Sephadex G-200"	0.63	79	5.0×10^{4}

Approximately 20 g of packed human blood lymphocytes (stimulated for 72 hr with phytohemagglutinin) were thawed and suspended in 50 ml of TDE buffer (0.05 M Tris-HCl, pH 7.8; 0.001 M dithiothreitol, 0.001 M EDTA). Following homogenization in a Dounce homogenizer, the suspension was adjusted to 10% (v/v) glycerol, and centrifuged at 8,500 × g for 15 min, yielding a pellet (P-8.5) and a supernatant (S-8.5). The pellet (P-8.5) was suspended in 40 ml of TDEG buffer (TDE buffer + 10% (v/v) glycerol + 0.5% Triton X-100) and sonicated until all nuclei were ruptured. The supernatant (S-8.5) was centrifuged at 41,000 × g for 30 min, to remove lysosomes. The supernatant (S-41) was stirred with slow addition of Triton X-100 to a final concentration of 0.5% (v/v) and combined with the sonicate of P-8.5. This mixture was centrifuged at 100,000 × g for 1 hr. The supernatant (S-100; fraction 1) was absorbed to a column, 2.5 × 35, of DEAE-cellulose equilibrated with TDEG buffer. The column was washed with 400 ml of TDEG buffer, and 800 ml of a linear gradient of NaCl in buffer, from 0 to 0.3 M, was attached. Fractions of 8 ml were collected and assayed as described in the legend to Table 2. The bulk of DNA polymerase activity (fraction 2, DEAE-eluate), eluting as a broad peak centering around 0.12 M NaCl, was collected and absorbed to a column, 3.0 × 15 cm, of phosphocellulose (Whatman P11) equilibrated with TDEG buffer + 0.1 M NaCl. The flow rate was 45 ml/hr. Following a 200-ml wash with TDEG buffer + 0.1 M NaCl, 800 ml of a linear gradient of NaCl in buffer, extending between extremes of 0.1 and 0.7 M, was attached. Fractions of 7 ml were collected and assayed for activity in System A (see legend to Table 2). A single symmetrical peak of activity eluting at 0.23 M KCl (fraction 3) was collected and concentrated on a phosphocellulose column, 2.0 × 0.9 cm. The concentrate was chromatographed on "Sephadex G-200" as described in the legend to Fig 4. Tubes containing the bulk of DNA polymerase activity were pooled as fraction 4 of DNA polymerase I. For estimation of purification, activity was measured in System A (see legend to Table 2). An asterisk (*) denotes that activity was not proportional to input of protein.

If care isn't taken to remove all nucleic acids from the earliest fractions, artifactual "multiple forms" of DNA polymerases elute from the various chromatographic columns. The two distinct DNA polymerases are easily separated by phosphocellulose chromatography (Fig. 2). We arbitrarily refer to the early eluting species as polymerase I and the late eluting peak as polymerase II. A summary of the purification is shown in Tables 1 and 2 and some of the conditions for optimal activity in Fig. 3.

The evidence that these are indeed two distinct DNA polymerases is summarized below.
1. Separation by phosphocellulose chromatography (Fig. 2).
2. DNA template preferences. The most sensitive template for detection of polymerase I is an "activated" DNA in the presence of Mg^{++}, while polymerase II shows preference for dT_{12-18} · poly dA in the presence of Mn^{++} (Fig. 2).

Table 2. Purification of human lymphocyte DNA polymerase II

Fraction	Total protein (mg)	Specific activity		Total activity	
		Activated DNA (System A) (pmole/hr/μg)	poly dA · dT$_{10}$ (System B) (pmole/hr/μg)	Activated DNA (pmole/hr)	poly dA · dT$_{10}$ (pmole/hr)
1. S-100	208	2.3	*3.8	4.8×10^5	7.9×10^5
2. DEAE-batch	148	2.9		4.3×10^5	
3. Phosphocellulose (0.45 M peak)	2.54	11.4	280	2.9×10^4	7.2×10^5
4. "Sephadex G-200"	0.083	30	1100	2.5×10^3	9.2×10^4

Approximately 5 g of packed human blood lymphocytes (see legend to Table 1) were lysed as described in the legend to Table 1. Following homogenization, the suspension was adjusted to 0.8 M KC1 and 10% glycerol stirred at 4°C for 8 hr, and centrifuged at 48,200 × g for 15 min. The pellet was re-extracted once with 10 ml of TDEK buffer (TDE buffer + 0.8 M KC1 + 10% glycerol). Supernatants were combined, centrifuged at 105,000 × g for 1 hr, and dialyzed for 8 hr against TDEG buffer (TDE buffer + 10% glycerol). This material (fraction 1; S-100) was passed over a column, 3.5 × 10 cm, of DEAE-cellulose equilibrated with TDEG buffer and washed through batchwise with TDEG buffer + 0.3 M KC1. The protein-containing eluant was collected and dialyzed for 4 hr against 20 volumes of TDEG buffer. This material (fraction 2; DEAE-batch) was chromatographed on a phosphocellulose column as described in the legend to Fig. 2. Tubes 31 to 43 (Fig. 2) were pooled as DNA polymerse I. Tubes 55 to 60 (fig. 2), containing most of the activity of DNA polymerase II, were pooled and concentrated as described in the legend to Table 1. The concentrate (fraction 3; phosphocellulose eluate) was chromatographed on "Sephadex G-200" as described in the legend to Fig. 4. Tubes 34 to 40 were pooled as fraction 4 of DNA polymerase II. Fractions were assayed in 50-μl volumes containing the following ingredients. System A, activated salmon sperm DNA, 50 μg/ml; 50 mM Tris-HCl, pH 7.8; 10 mM MgCl$_2$; 50 mM KC1; 5 mM dithiothreitol; 8×10^{-5} M each of dATP, dGTP, and dCTP; 7×10^{-6} M ^{3}H-TTP (7400 cpm/pmole) and 10 μl of enzyme solution. System B, dT$_{10}$ · dA, 50 μg/ml; 50 mM Tris-HCl, pH 7.8; 0.5 mM MnCl$_2$; 50 mM KC1; 5 mM dithiothreitol; 7×10^{-6} M ^{3}H-TTP (7400 cpm/pmole), and 10 μl of enzyme solution. System A was incubated at 37°C for 10 min, while System B was incubated at 30°C for 10 min. Specific activities were calculated from initial reaction rates; incorporation was a linear function of time for at least 15 min.

*Activity was not proportional to input of protein.

3. Molecular weight estimates. DNA polymerase I is a high molecular weight enzyme of about 150,000 to 160,000 daltons, which can apparently dimerize to about 300,000 daltons. An estimate of the molecular weight of DNA polymerase II is 25,000 to 35,000 daltons. These estimates are based on determinations made from Sephadex G-200 gel filtration with appropriate marker proteins (Figs. 4 and 5) and by sucrose gradient sedimentation and conversion of the S values to molecular weights as described by Martin and Ames (21).

4. Sensitivity to N-ethylmaleimide (NEM). As shown in Fig. 6, polymerase I is very sensitive to NEM, while polymerase II is not affected.

5. Sensitivity to arabinosyl cytosine triphosphate (ara-CTP). A summary of some of the properties of these two polymerases from normal lymphocytes is presented in Table 3. Included in the table is the differential sensitivity of the two enzymes to ara-CTP, polymerase I being much more sensitive (K_i about 1.1 μM while K_i with polymerase II is > 10 μM).

6. Sensitivity to rifamycin derivatives. We recently reported our results of a "screen" of the effects of 201 derivatives of rifamycin SV on both these DNA polymerases and compared these results to inhibition of viral reverse transcriptase (33). It is of interest that some derivatives can preferentially inhibit one cellular polymerase more than the other, but in general inhibitions were comparable.

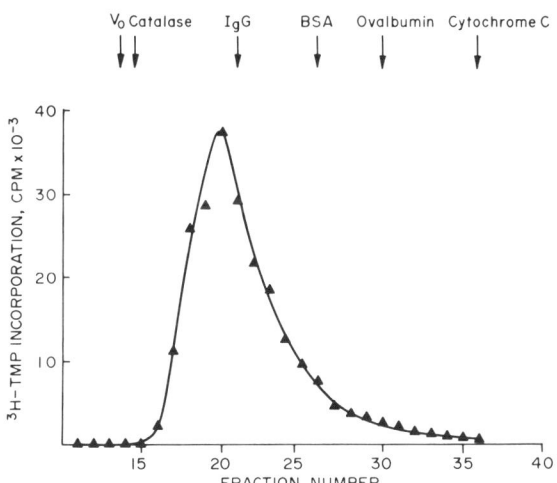

Fig. 4. Sephadex G-200 chromatography of normal lymphocyte DNA polymerase I. Aliquots of fraction 3 (phosphocellulose eluate) of DNA polymerase I were loaded onto a column, 1.5 × 28 cm, of "Sephadex G-200" equilibrated with TDEG buffer (see legends to Tables 1 and 2 for bufferdeinition). Fractions, 1.05 ml, were collected at a flow rate of 7 ml/hr. Fractions were assayed for activity as described in the legend to Fig. 3A. BSA, bovine serum albumin.

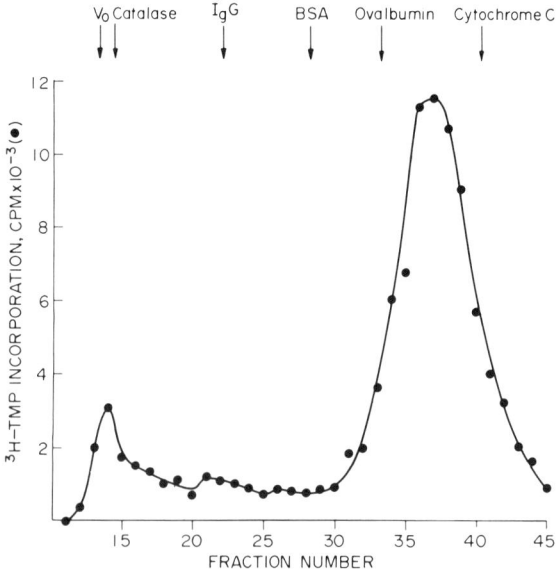

Fig. 5. Sephadex G-200 chromatography of normal lymphocyte DNA polymerase II. Aliquots of fraction 3 (phosphocellulose eluate) of DNA polymerase II were loaded onto a column, 1.5 × 28 cm, of "Sephadex G-200" equilibrated with TDEG buffer. Fractions, 1.05 ml, were collected at a flow rate of 7 ml/hr. Fractions were assayed for activity as described in the legend to Fig. 3A except that poly d(A-T) was used as the template. Marker proteins were dissolved in TDEG buffer and chromatographed individually on the same column. V_0 is void volume measured with blue dextran. BSA, bovine serum albumin.

7. Location. In agreement with Chang and Bollum (6), we find that polymerase I is predominantly found in the cytoplasm. Polymerase II is found in both nucleus and cytoplasm. Of great interest, but indeed perplexing, is the finding that polymerase I, although apparently a cytoplasmic enzyme, is the polymerase which predominantly increases on stimulus to DNA synthesis and as stated above is the enzyme chiefly inhibited by ara-CTP, a compound apparently cytotoxic for these cells only during DNA synthesis. Similar results have been found by M. Goulian (personal communication).

Changes on Stimulation with PHA

We have already noted above that the total DNA polymerase activities of crude extracts increase strikingly following stimulation of a resting normal lymphocyte with PHA. We find *no* DNA polymerase I in resting cells and low levels of DNA polymerase II. Following stimulation with

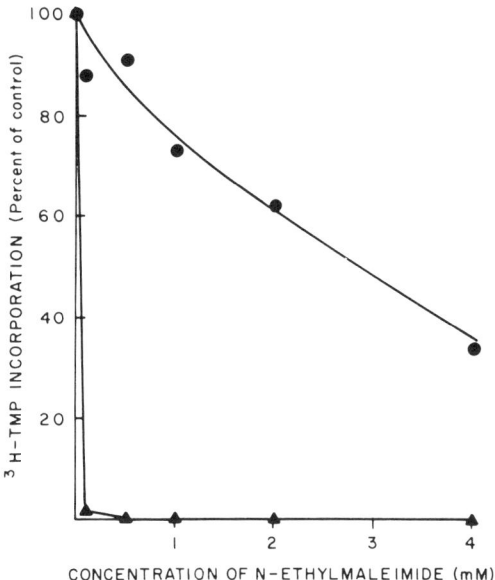

Fig. 6. Sensitivity of normal lymphocyte DNA polymerases to N-ethylmaleimide. Activity of DNA polymerase I (▲——▲) and DNA polymerase II (●——●) was determined as described in the legend to Fig. 3A, except that enzymes were preincubated for 10 min at 37°C with the indicated amounts of N-ethylmaleimide. The control was a preincubation mixture with solvent (ethanol) alone.

Table 3. Summary of characteristics of DNA polymerase I and II of normal human blood lymphocytes

Characteristic	DNA polymerase I	DNA polymerase II
Size	7–10 S	3.3 S
	(150,000–300,000 daltons)	(30,000 daltons)
pI	4.5–6.5	9.4
Template acceptance		
Activated DNA (Mg^{++})	28 pmoles/hr/μg	30 pmoles/hr/μg
Initiated homopolymeric DNA (Mn^{++}), e.g.		
poly dA · dT_{10}	28 pmoles/hr/μg	880 pmoles/hr/μg
Inhibition by 0.05 mM NEM	+++	0
Inhibition by ara-CTP*	+++	+
	($K_i = 1.1$ μM)	($K_i > 10$ μM)
Subcellular localization	Cytoplasm	Nucleus and cytoplasm
Response to proliferation	Marked increase	Slight increase
Presence in resting cells	Absent	Present

*These experiments were performed in collaboration with Dr. A. Schrecker, National Cancer Institute.

PHA both activities increase, reaching values at about the 3rd day after addition of PHA (R. G. Smith and R. C. Gallo, unpublished results).

Transcription of Appropriately Primed Ribohomopolymers by Normal Lymphocyte DNA Polymerase I and II

Both enzymes can copy some ribohomopolymers when accompanied by appropriate primers. For example, both synthesize poly dT from poly dT · poly rA (Fig. 7 and Table 4) or even from dT_{12-18} · poly rA, although much less efficiently (Table 5). This activity is markedly increased when Mn^{++} is the divalent cation. Thus, these templates are

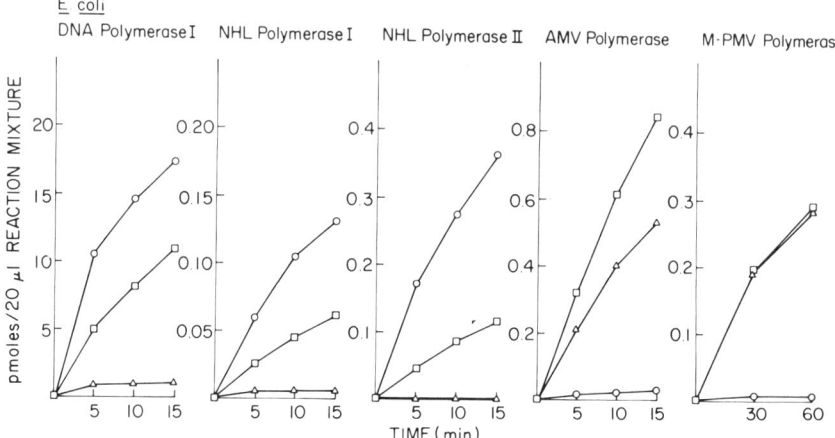

Fig. 7. Kinetics of DNA polymerase activities with synthetic templates. DNA polymerase assays were carried out at 37°C in 100-μl standard reaction mixtures composed of 50 mM Tris-HCl buffer, pH 8.3; 30 mM KCl; 10 mM $MgCl_2$; 5 mM dithiothreitol; 80 μM dATP; 5.6 μM ^{3}H-TTP (15,000 cpm/pmole); and 50 μg/ml of template. Aliquots (20 μl) were taken at the indicated times, precipitated, and counted. The M-PMV was assayed as above, except that the template concentrations were 20 μg/ml, and standard assay contained 5 mM $MgCl_2$. Results are expressed on the basis of 20 μl of reaction mixture, as the protein concentrations of the purified viral enzymes were not measured. Five microliters of AMV polymerase and 20 μl of M-PMV polymerase were routinely used in their respective reaction mixtures. The *Escherichia coli* DNA polymerase was assayed at 2.8 ng of protein per 20 μl of reaction mixture. The NHL polymerases I and II had protein concentrations of 144 ng/20 μl and 36 ng/20 μl, respectively. The dT_{12-18} · poly dA and dT_{12-18} · poly rA were prepared in ratios of 1:1 by annealing equimolar amounts of the polymer and oligomer in 0.01 M Tris-HCl, pH 7.2; 0.1 M NaCl, at 70°C for 5 min, followed by slow cooling to room temperature over a period of 8 hr. The poly (rA) · poly (dT) was a product of Miles Laboratories. The other polymers and oligomers were obtained from Collaborative Research Inc. □——□, activity with poly (rA) · poly (dT); ○——○, activity with dT_{12-18} · poly dA; △——△, activity with poly(rA) · oligo(dT)$_{12-18}$. This figure was taken from an article by Robert et al. (24) and is reproduced here with the permission of the publishers.

Table 4. Template-primer activity of various RNA and DNA-RNA hybrid species

Template-primer	Specific activity (pmole/hr/μg)			
	Lymphocyte DNA Polymerse I		Lymphocyte DNA Polymerase II	
	System A	System B	System A	System B
Poly dT · rA	1.10	8.72	7.97	241
Poly rA · rU	0	0.37	0	0
70 S RNA (avian myeloblastosis virus)	0	0	0.31	0
70 S RNA plus ribonuclease A (10 μg/ml)			0.31	
70 S RNA plus dT$_{12-18}$				
^{3}H-TTP plus dATP, dCTP, dGTP	0.17	0.55	0.22	9.78
^{3}H-dATP plus dCTP, dGTP, TTP	0.03	0	NT*	NT
^{3}H-dCTP plus dGTP, TTP, dATP	0	0	NT	NT
^{3}H-dGTP plus dCTP, dATP, TTP	0	0	NT	NT
Messenger RNA (rabbit globin)	0	0	0.58	0
mRNA plus ribonuclease (10 μg/ml)			0.58	
mRNA plus dT$_{12-18}$	0.032	0.043	0.38	2.04

Reaction conditions were as described for Systems A and B in Table 2, except that various template-primers and substrates were used as indicated. Concentrations and specific activities of labeled substates were: 5.3×10^{-6} M ^{3}H-dATP (10,300 cpm/pmole); 4.4×10^{-6} M ^{3}H-dCTP (12,500 cpm/pmole); 8.0×10^{-6} M ^{3}H-dGTP (6,900 cpm/pmole); and 7.0×10^{-6} M ^{3}H-TTP (7400 cpm/pmole). Rabbit globin messenger RNA, a kind gift of Dr. F. Anderson, was purified by phenol extraction of reticulocytes followed by centrifugation on a sucrose gradient; 6 to 14 S material was collected, ethanol-precipitated, and redissolved in distilled water. Preincubation of nucleic acid with boiled ribonuclease A (10 μg/ml) was for 30 min at 37°C.
*NT = not tested.

not specific for reverse transcriptase of RNA tumor viruses (especially in the presence of Mn^{++}), a point which has been emphasized elsewhere (2, 8, 9, 13, 24). Therefore, activity *per se* with these synthetic templates means little more than that some DNA polyermase activity is present.

Since poly A is present in the 70 S RNA of RNA tumor viruses (12, 14, 19) as well as in mRNA, activity with these templates also is not to be taken as specific for a "reverse transcriptase" unless it is shown that heteropolymeric regions of the RNA are copied. As shown in Table 4, both DNA polymerase I and II of normal lymphocytes synthesize only poly dT from mRNA and AMV 70 S RNA. This is indicated by the incorporation of ^{3}H-TTP, lack of incorporation of ^{3}H-dCTP or ^{3}H-dGTP, and insensitivity of the ^{3}H-TTP incorporation to pretreatment

Table 5. Template-primer activity of synthetic oligomer-homopolymeric duplexes

	Specific activity (pmole/μg/hr)			
	Lymphocyte DNA polymerase I		Lymphocyte DNA polymerase II	
Template-primer	System A	System B	System A	System B
$dT_{10} \cdot dA$	2.43	14.2	8.33	389
$dT_{12-18} \cdot rA$	0.215	2.53	0	2.25
Poly dT $\cdot$ rA	1.10	8.72	7.97	241
$dG_{12-18} \cdot rC$	0		0	

Reaction conditions were described in the legend to Table 2, except that various template-primers were used as indicated. In reactions directed by AT-containing template-primers, 7.0×10^{-6} M ^{3}H-TTP (7400 cpm/pmole) was the only substrate present. When $dG_{12-18} \cdot rC$ was used, 8.0×10^{-6} M ^{3}H-dGTP (5900 cpm/pmole) was the only substrate present.

of the RNAs with pancreatic RNase (10 μg/ml of RNase in 0.2 M NaCl for 30 min at 37°C). The details of these experiments are reported elsewhere (27).

Templates which Discriminate between Oncornavirus DNA Polymerase and the Main Cellular DNA Polymerases

As shown in Fig. 7, the relative response to $dT_{12-18} \cdot$ poly rA *versus* $dT_{12-18} \cdot$ poly dA (in presence of Mg^{++} and incorporating ^{3}H-TTP) is useful in differentiating between the enzyme from RNA tumor viruses and cellular DNA polymerases. Both DNA polymerase I and II of lymphocytes and *Escherichia coli* DNA polymerase I show marked preference for $dT_{12-18} \cdot$ poly dA (see Table 5 as well as Fig. 7). In contrast, the enzymes from AMV and from the Mason-Pfizer agent show an opposite response, and in our experience this has been true of the polymerase from viruses derived from all species tested (avian, mouse, cat, and primate).

A *second* and even more useful synthetic template-primer was first utilized by Baltimore (2). He showed that the AMV polymerase transcribed a poly C template with primer oligo dG while the *E. coli* DNA polymerase I showed no activity. In agreement with these results, we find that normal lymphocyte DNA polymerases *have no* activity with this template-primer (4, 27) (Table 5), and each of the viral enzymes we have purified (AMV, RLV, and the Mason-Pfizer agent) will transcribe the poly C strand (J. Abrell and R. Gallo, unpublished results).

The *third* distinguishing feature is the transcription of heteropolymeric portions of predominantly single stranded natural RNAs. As stated before, this has not been shown with any purified DNA polymerase except that from *avian* RNA tumor viruses and a DNA polymerase we found in human acute leukemic cells (see section below).

RNA-directed and -primed DNA Polymerase System from a Cytoplasmic "Pellet" of Human Leukemic Cells

Endogenous RNase-sensitive DNA Polymerase in "Pellet"

With the second approach for isolating cellular DNA polymerases (outlined in Fig. 8), we find a cytoplasmic particulate fraction in human leukemic blood cells which contains an endogenous and completely RNase-sensitive DNA polymerase "system" (Fig. 9 and references 4 and 26). The polymerase activity in the particulate fraction bands at a density of 1.09 g/ml (Fig. 9) in a glycerol velocity gradient centrifugation and at a density of 1.16 g/ml after equilibrium density centrifugation in sucrose. This activity has been found in the cells of most, but not all, leukemic patients (over 30 have been examined). (It should be emphasized that these cells are *fresh* blood cells and are not known to be producing RNA tumor viruses. Furthermore, examination of this particulate fraction by electron microscopy does not reveal intact virions.) The nucleic acid of the peak containing the polymerase is more than 95% RNA. (The nature of the RNA is not yet resolved.) The DNA products of early time points of these endogenous reactions migrate in the RNA, RNA-DNA hybrid, and DNA regions of a Cs_2SO_4 equilibrium density gradient (Fig. 10). Treatment of the product with alkali removes the counts in the RNA and hybrid regions (Fig. 10), but heating the product at 95°C for 10 min did not change its position in the gradient. The DNA product hybridizes back to RNA isolated from the pellet (M. Reitz, P. Savin, M. Sarngadhavan, and R. Gallo, unpublished results). These results taken together show than an RNA species is required for this reaction and suggest that the DNA is covalently attached to RNA. Thus, a native RNA serves as a primer and *possibly* as template for DNA synthesis. However, the data relating to a template function of the RNA are equivocal. It was now necessary to purify the polymerase from the "pellet" and determine its response to various templates, especially its ability to transcribe heteropolymeric portions of viral 70 S RNA.

Purification of the Leukemic "Pellet" DNA Polymerase

A detailed report of the methods for purification of the enzyme is published elsewhere (26). In summary, the peak DNA polymerase fractions from the glycerol gradient (Fig. 9) are combined, concentrated, and

Procedure for Preparation of 60K Pellet

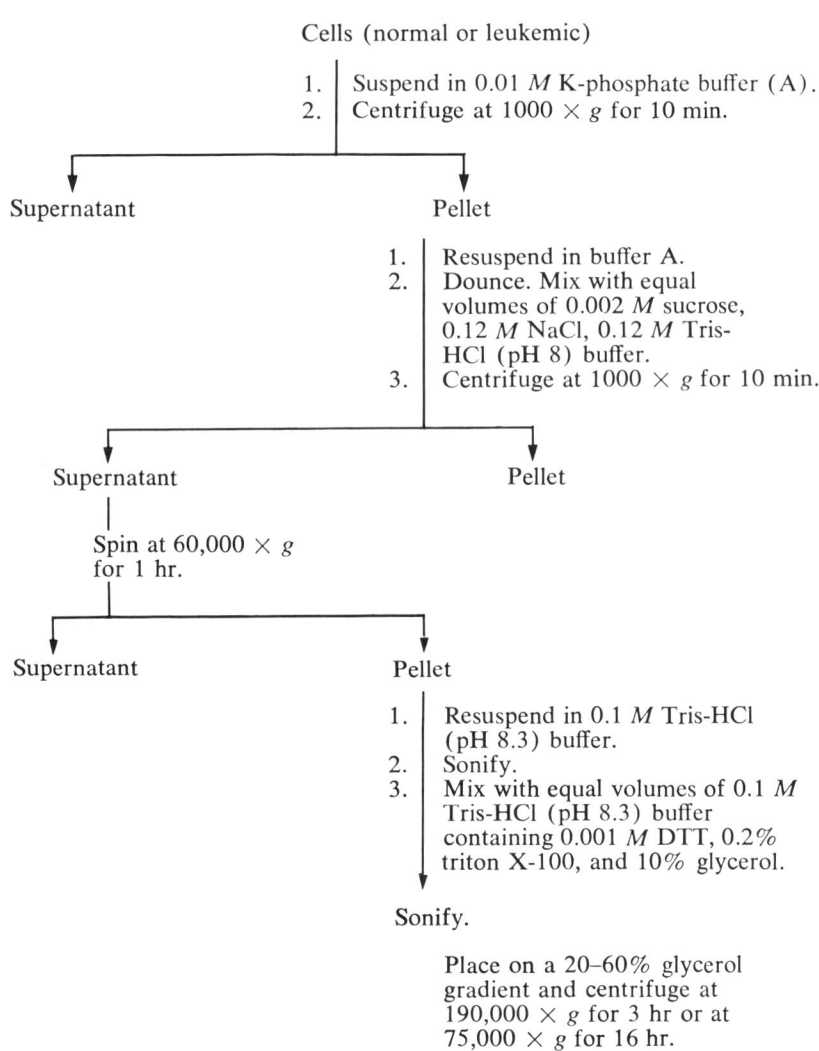

Fig. 8. Scheme for isolation of "cytoplasmic pellet" from human leukemic cells. DTT, dithiothreitol.

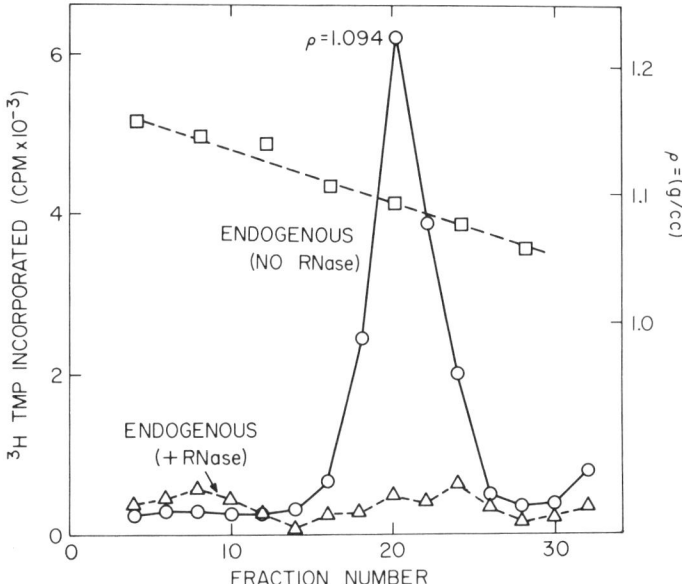

Fig. 9. Endogenous DNA polymerase activities of the leukemic 60,000 × g "pellet" through a glycerol gradient centrifugation. The pellet was centrifuged on a 20 to 60% glycerol gradient in 8 mM Tris-HCl (pH 7.9), containing 0.8 mM EDTA, and 0.8 mM dithiothreitol (DTT) for 3 hr at 40,000 rpm in an SW 41 rotor. Fractions (0.3 ml) were collected and the endogenous DNA polymerase activities measured on 15-μl aliquots which were preincubated at room temperature for 1 hr with 0.2 M NaCl plus (△——△) or minus (○——○) RNase (20 μg/ml). Samples were incubated at 37°C for 1 hr in the standard reaction mixture (0.05) ml which contained: 50 mM Tris-HCl (pH 8); 5 mM MgCl$_2$; 10 mM NaCl; 5 mM DTT; 80 μM each of dATP, dCTP, and dGTP; 0.4 μM ^{3}H-dTTP (18,900 cpm/pmole). The reaction was stopped by the addition of 50 μg of yeast tRNA and 2 ml of 10% trichloroacetic acid containing 0.02 M sodium pyrophosphate, and the precipitate collected on Millipore filters and counted.

nucleic acids are removed by DEAE-cellulose chromatography. The "nucleic acid-free" proteins are applied to phosphocellulose and Sephadex G-200 chromatographic columns. An outline of the purification scheme, starting with the glycerol gradient "pellet" fraction, is shown in Fig. 11. The enzyme elutes as one sharp peak in both columns (Fig. 12 and references 8, 9, 26). The final preparation is free of detectable RNase (assayed by incubating the enzyme with ^{3}H-labeled AMV 70 S RNA overnight and determining change in percentage recovery as 70 S RNA by gel electrophoresis) and DNase (assayed by determining the appearance of acid-soluble counts from labeled poly d(AT) and labeled λ DNA). The enzyme was purified more than 200-fold from the pellet. (No attempt is made to estimate purification from earlier fractions because of the presence of other polymerases and very high activities

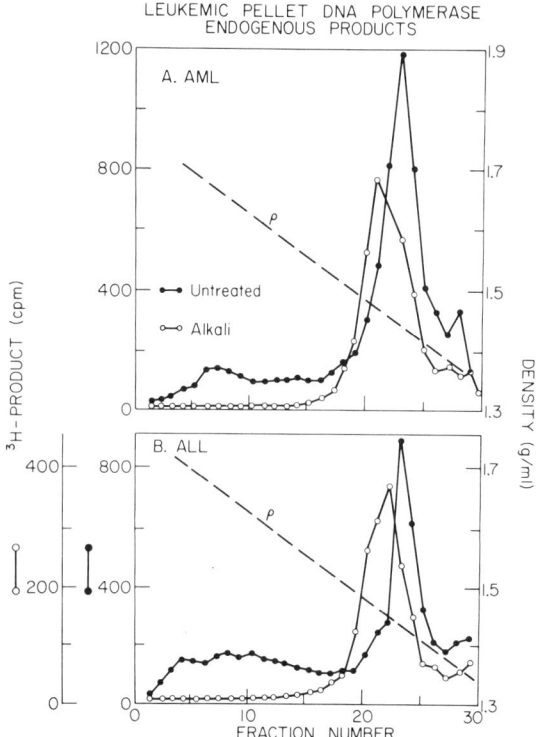

Fig. 10. Analysis of endogenous DNA product of leukemic "pellet" DNA polymerase by Cs_2SO_4 equilibrium density gradient centrifugation. DNA product was prepared in 0.2 ml of a standard endogenous reaction mixture of DNA polymerase (purified to the 60,000 × g pellet step) with the following alterations: 0.5 μCi each of ^{3}H-dATP (18 Ci/mmole), ^{3}H-dCTP (22.6 Ci/mmole), ^{3}H-dGTP (12.4 Ci/mmole), and ^{3}H-TTP (18.6 Ci/mmole) were used in place of unlabeled nucleoside triphosphates, and dimethylbenzyl N-demethyl rifampicin was added to a concentration of 50 μg/ml. The reactions were terminated after 1 min by addition of sodium dodecyl sulfate to 1% and NaCl to a final concentration of 0.12 M. This was extracted twice with phenol-m-cresol (9:1), 0.1% in 8-quinilinol, and twice with ether. These extracts were diluted to 1 ml with 10 mM Tris, 0.1 M NaCl, and 1 mM EDTA, pH 7.24. Stripped yeast tRNA, (500 μg) 100 μg of calf thymus DNA, and 25 μl of 0.1 M cetyltrimethylammonium bromide (CTAB) were added to the final extracts. The CTA-nucleic acid salts were precipitated on ice for 30 min and collected by centrifugation (15,000 × g; 10 min). The pellet was dissolved in 1 ml of 1 M NaCl and 3 ml of EtOH were added. The nucleic acids were precipitated for 2 hr at −15°C and collected by centrifugation as above. The precipitate was then dissolved in TNE (0.01 M Tris, pH 7; 0.1 M NaCl; 0.001 M EDTA) buffer and the NaCl concentration adjusted to 0.4 M. Samples were run either without further treatment or after alkali digestion (10 min, 95°C 0.3 N KOH). DNA product was diluted to 1 ml with TNE buffer and mixed in a 5-ml polyallomer tube with 4 ml of a solution of 1 ml of TNE per g of cesium sulfate (Schwarz-Mann optical grade). The tubes were centrifuged in an SW 50.1 rotor at 36,000 rpm for 65 hr at 20°C. Density was determined by refractive index with a Bausch and Lomb Abbe refractometer. Fractions were precipitated with 2 ml of 10% trichloroacetic acid containing 0.02 M pyrophosphate and filtered on a Millipore filter and counted in Liquiflour-toluene scintillation fluid. A, DNA product from acute myelogenous leukemic lymphoblast "pellet" DNA polymerase; B, DNA product from acute lymphocytic leukemic lymphoblast "pellet" DNA polymerase.

Purification of DNA Polymerase from Human Leukemic Pellet

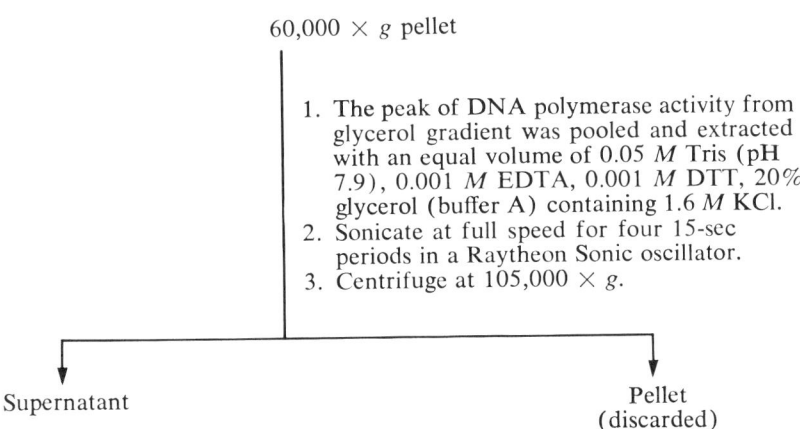

1. Dialyzed *vs.* 100 volumes buffer A.
2. Chromatograph on DEAE cellulose.

↓

1. DEAE cellulose peak activity pooled.
2. Dialyzed and chromatographed on phosphocellulose column.

↓

1. Phosphocellulose peak activity pooled.
2. Dialyzed and chromatographed on Sephadex G-200 column.

↓

G-200 enzyme

Fig. 11. Scheme for purification of DNA polymerase from leukemic pellet.

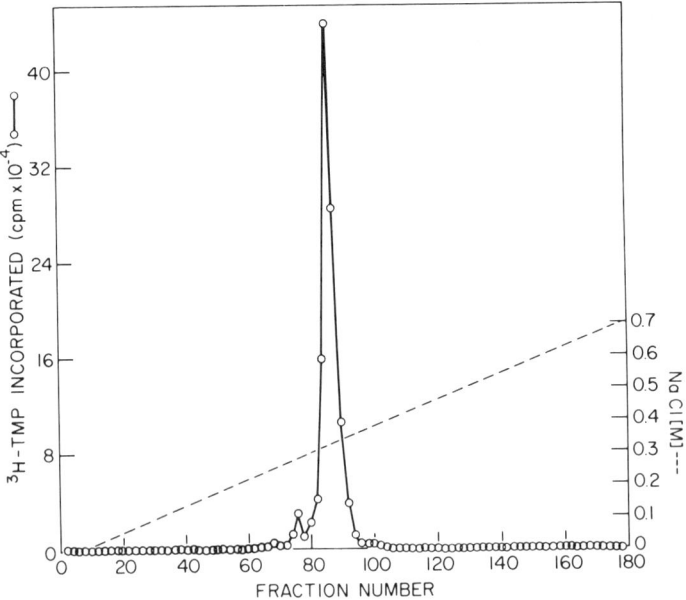

Fig. 12. Phosphocellulose chromatography of human leukemic DNA polymerase purified from the high speed "pellet." The pooled DNA polymerase activity (43.4 mg) from DEAE-cellulose chromatography was applied to a phosphocellulose column (12 × 1 cm) previously equilibrated with buffer B. After extensive washing of the column with buffer B, which removed most of the nonadsorbed protein, the DNA polymerase activity was eluted with 200 ml of a linear 0 to 0.7 M NaCl gradient in the same buffer. Fractions (1.1 ml) were collected and 25-μl aliquots were incubated at 37°C for 1 hr in th standard assay system described in Fig. 9 with poly d(AT) (Miles Laboratories) (50 μg/ml) as a template. Concentration of ^{3}H-dTTP was 0.4 μM (25,000 cpm/pmole). O——O, cpm; ---, salt concentration. See reference 26 for details of buffer.

of phosphatases and nucleases.) A summary of the purification (after isolation of the pellet) is given in Table 6.

Response to DNA and DNA-RNA Hybrid Template-Primers

As illustrated in Table 7, the DNA polymerase purified from three different leukemic subjects showed activity with dT_{12-18} · poly rA. Further, it prefers this template-primer over dT_{12-18} · poly dA (in Mg^{++}). This characteristic is different from that of the main cellular DNA polymerases and similar to the oncornavirus enzyme, as previously noted. In addition, this polymerase will transcribe poly C with dG_{12-18} as primer (Table 7). As mentioned above, no response at all to this template-primer has been seen in our laboratory with any normal cellular DNA polymerase. There was no activity with dT_{12-18} or dG_{12-18} (with Mg^{++} or Mn^{++}), indicating that there was no terminal addition activity associated with this DNA polymerase.

Table 6. Purification of DNA polymerase from human leukemic pellet

Purification step	Total protein (mg)	Total activity (cpm)	Specific activity cpm/mg protein	Specific activity pmoles/mg protein
Dialyzed pellet extract	56.7	2.0×10^7	3.59×10^5	337
DEAE-cellulose peak	53.3	4.8×10^7	9.19×10^5	862
Phosphocellulose peak	0.3	2.3×10^7	7.87×10^7	73,827
Sephadex G-200	0.23	6.1×10^6	2.62×10^7	24,578

The sonicated high speed pellet was mixed with an equal volume of 1.6 M KCl in 50 mM Tris-HCl (pH 7.9) containing 1 mM dithiothreitol (DTT), 1 mM EDTA, and 20% ethylene glycol (buffer B). The mixture was dispersed by sonic treatment and centrifuged at 105,000 $\times$ g for 2 hr. The supernatant solution was dialyzed against buffer B and then chromatographed on a DEAE-cellulose column (12 $\times$ 1.8 cm) equilibrated with buffer B. The DNA polymerase activity which eluted with 0.25 M NaCl in buffer B was pooled, dialyzed, and concentrated against 30% polyethylene glycol (PEG) in buffer B. This was further purified by chromatography on a phosphocellulose column as described in Fig. 12. The concentrated phosphocellulose enzyme was next purified by gel filtration on a Sephadex G-200 column equilibrated with 10 mM potassium phosphate buffer (pH 7.5) containing 20% ethylene glycol and 1 mM DDT. The peak fractions were pooled and stored at $-20°C$. The enzyme was assayed as described in Fig. 9 using poly d(AT) (Miles Laboratory) (50 µg/ml) as template. Concentration of H^3-dTTP was 5.9 µM (1066 cpm/pmole) and of dATP, 80 µM. Incubations were for 30 min. Protein concentrations were determined using the procedure of Lowry et al.

Table 7. Template characteristics of the human leukemic DNA polymerase

Template*	^{3}H-TMP incorporated (pmole/10 µg enzyme†) Patient 1	Patient 2	Patient 3
dT$_{12-18}$ · poly rA	71.9	22.0	446.0
dT$_{12-18}$ · poly dA	56.4	5.5	89.0
poly DT · poly rA	45.7	104.8	1219.0
Activated salmon sperm DNA	23.1	37.4	79.0
poly d(AT)	265.7	N.T.‡	N.T.
dG$_{12-18}$ · poly rC	6.7 (^{3}H-dGMP incorporated)	N.T.	N.T.

Activated salmon sperm DNA used in this experiment is an excellent template for the human normal blood lymphocyte DNA polymerases. Activity with normal lymphocyte DNA polymerase I, a high molecular weight enzyme, was 257 pmoles/10 µg of protein and 98 pmoles/10 µg of protein/hr with the low molecular weight enzyme (DNA polymerase II). (27). Poly d(AT), poly dT · poly rA, and poly dA were obtained from Miles Laboratory; dT$_{12-18}$ · poly rA, dG$_{12-18}$ · poly rC, and dT$_{12-18}$ from Collaborative Research. The conditions for preparation of dT$_{12-18}$ · from Collaborative Research. The conditions for preparation of dT$_{12-18}$ · poly dA have been described elsewhere (24). In all these assays Mg^{++} was the divalent cation and other details of the assay are as described in the legend to Fig. 9.

*Purified enzyme was incubated for 30 min in the standard assay system with the indicated templates (50 µg/ml). In the case of dG$_{12-18}$ · poly rC, the reaction mixture contained ^{3}H-dGTP in place of ^{3}H-dTTP.

†These values were obtained with enzymes purified from the cells obtained from three different leukemic patients. Note that each enzyme shows preference for dT$_{12-18}$ · poly rA over dT$_{12-18}$ · poly dA. There was no activity with dT$_{12-18}$ or dG$_{12-18}$ and dNTP substrates alone (i.e. without template).

‡N. T. = not tested.

Transcription of Heteropolymeric Portions of AMV 70 S RNA

Enzymes purified from cells of five different leukemic patients have been shown to transcribe AMV, or RLV 70 S RNA, or both. In all cases, the reaction has been inhibited by 60 to 100% by pretreatment of the RNA with RNase. A typical example of kinetics and RNase sensitivity is shown in Fig. 13. The RNase sensitivity suggested that transcription was not limited to the poly A tracts of the RNA (12, 14, 19). Further evidence that the product is heteropolymeric is the fact that all four deoxynucleoside triphosphates are incorporated (Table 8) and the results of product analysis (see below).

The enzyme has an absolute requirement for Mg^{++} and four deoxynucleoside triphosphates (Table 8). The molecular weight of the enzyme is estimated as 130,000 daltons by elution from Sephadex G-200 chromatography with several marker proteins and by converting a sedimentation value obtained by glycerol gradient centrifugation to molecular weight (21) (Fig. 14).

Product Analysis of the DNA Synthesized with AMV RNA as Template

We find that the size of the DNA product varies between approximately 4 to 7 S. These estimates were obtained by glycerol gradient centrifu-

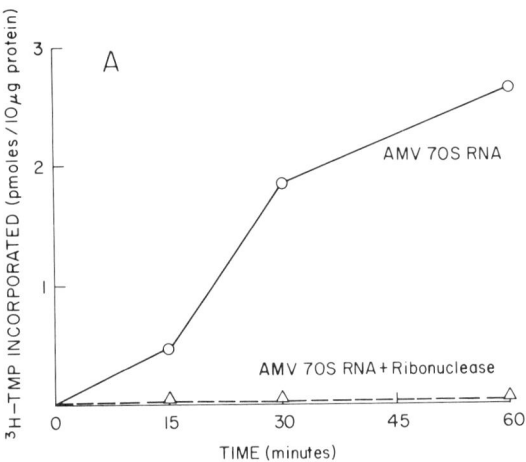

Fig. 13. Kinetics and ribonuclease sensitivity of the AMV 70 S RNA-directed DNA synthesis with purified enzyme from leukemic pellet. For the ribonuclease reaction, the AMV 70 S RNA was first incubated with pancreatic ribonuclease (20 µg/ml) at 37°C for 1 hr before initiating the polymerase. AMV 70 S RNA (O——O); AMV 70 S RNA plus ribonuclease (△- - -△).

Table 8. Characteristics of the AMV RNA-directed DNA synthesis by purified human leukemic DNA polymerases

Additions	Deoxynucleotide incorporation (pmoles/10 μg protein)*		
	Patient 1	Patient 2	Patient 3
Complete (^{3}H-dTTP)	0.45	1.5	1.04
Complete (^{3}H-dTTP) + dT$_{12-18}$	2.25	7.3	5.1
Plus RNase	0.07	0.15	0.1
Minus AMV RNA	<0.05		
Minus Mg^{++}	<0.05		
Minus dATP, dCTP, dGTP	<0.05		
Complete (^{3}H-dTTP)	0.52		
Complete (^{3}H-dATP)	0.76		
Complete (^{3}H-dCTP)	0.49		
Complete (^{3}H-dGTP)	0.66		

DNA polymerase assays were carried out with purified DNA polymerase from "pellet" using RNA (2.5 μg) as a template. Other conditions were the same as described in Fig. 9. RNase treatment was carried out as described in Fig. 13. The concentrations of the labeled deoxynucleoside triphosphates were 5 μM and the unlabeled deoxynucleoside triphosphates were 80 μM. The specific activities of the labeled deoxynucleoside triphosphates were: ^{3}H-dTTP, 23,000 cpm/pmole; ^{3}H-dATP, 8,400 cpm/pmole; ^{3}H-dCTP, 10,100 cpm/pmole; and ^{3}H-dGTP, 5,600 cpm/pmole. Deoxynucleotide incorporations were for 60 min, but are not maximum values.

*These values, were obtained with enzymes purified from the cells obtained from different leukemic patients.

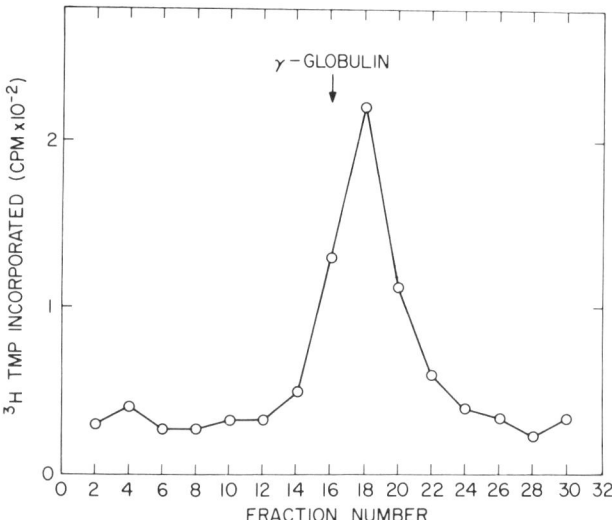

Fig. 14. Molecular weight estimation of human leukemic DNA polymerase by glycerol gradient centrifugation. One milliliter of the Sephadex G-200 enzyme was applied to a 10 to 60% glycerol gradient in 8 mM Tris-HCl (pH 7.9) containing 0.8 mM EDTA, 0.8 mM DTT, and centrifuged at 40,000 rpm in an SW 41 rotor for 14 hr. Fractions (0.3 ml) were collected from the bottom of the tube and 20-μl samples were assayed for activity by incubating for 1 hr with dT$_{12}$ · poly rA as a template. γ-Globulin was run as a marker on a parallel gradient. The position of the marker is indicated by the arrow.

gation of labeled DNA product using internal unlabeled RNA markers (Fig. 15).

Analysis of a 60-min DNA product of an AMV 70 S RNA-directed reaction (with ^{3}H-TTP as labeled substrate) by Cs_2SO_4 equilibrium density gradient shows that counts are in the RNA and RNA-DNA hybrid region of the gradient. The counts are removed by treatment of the product with alkali (Fig. 16). With heat treatment (95°C for 10 min), all the labeled product bands in the DNA region of the gradient if dT_{12-18} is used with the 70 S RNA, and ^{3}H-TTP is the labeled substrate. In this case dT_{12-18} serves as primer and the product is covalently attached to dT_{12-18} (Fig. 16). With the same experiment (oligo dT stimulation, AMV RNA template) but with other deoxynucleoside triphosphates labeled instead of ^{3}H-TTP, some of the counts do remain in the RNA and in the hybrid region after heat (Fig. 17). This suggests that more than one primer is utilized, i.e. the dT_{12-18} and some portion of the native viral RNA.

When dT_{12-18} is *not* added, and ^{3}H-TTP is again a labeled substrate, our results indicate that the product is entirely covalently attached to

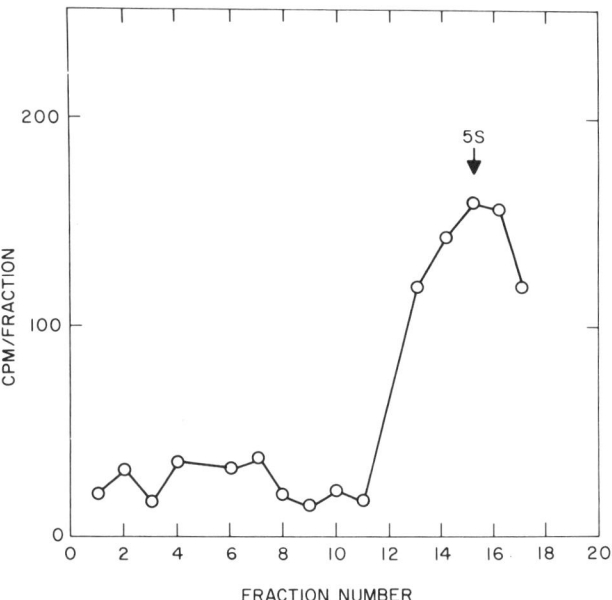

Fig. 15. Size of the DNA product by glycerol gradient centrifugation. DNA product of purified human leukemic pellet DNA polymerase with AMV 70 S RNA as a template in the presence of oligo dT was prepared and isolated as described in Fig. 10. The product was incubated with 0.3 N KOH at 37°C for 20 hr. This solution was neutralized and centrifuged with *Escherichia coli* 5 S, 16 S, and 23 S RNA markers on a 5 to 30% glycerol gradient in TNE buffer for 20 hr at 40,000 rpm. Samples were then collected and counted. The position of the 5 S marker is indicated by the arrow.

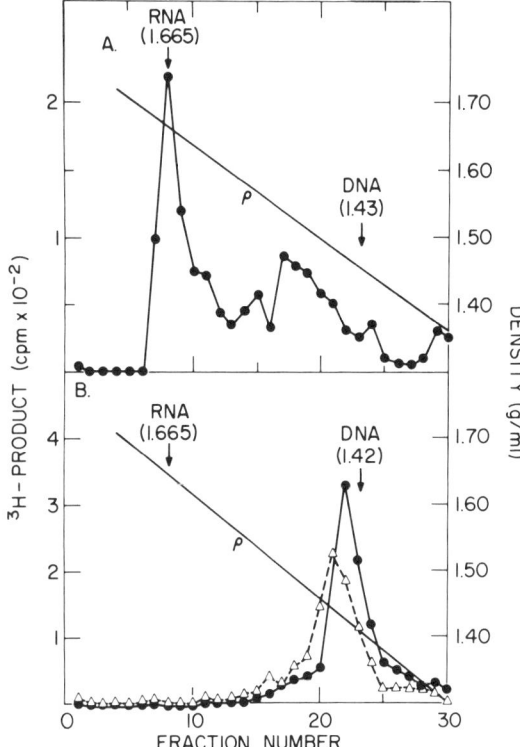

Fig. 16. Cs$_2$SO$_4$ equilibrium density gradient centrifugation of DNA product from purified human leukemic pellet DNA polymerase with AMV 70 S RNA as a template. A standard reaction mixture (0.4 ml), but containing 0.5 mCi of ^{3}H-TTP (18.6 Ci/mmole), was incubated with AMV 70 S RNA and oligo dT at 37°C for 60 min and then adjusted to 0.12 M NaCl and 1% sodium dodecyl sulfate. The DNA product was then purified and analyzed by Cs$_2$SO$_4$ equilibrium density gradient centrifugation as described for Fig. 10. A, Native DNA product (●——●); B, heated at 95°C for 10 min (●——●); treated with alkali (0.3 N KOH, 95°C 10 min) (△- - -△).

the viral RNA since heat treatment in this case does not move the counts to the DNA region of the gradients (Fig. 18).

That the DNA product is complementary to the template RNA is shown in Figs. 19 and 20. In Fig. 19 in the upper panel the DNA was annealed to the template RNA (AMV 70 S RNA) and the counts again move in the RNA and hybrid regions of the gradient. When incubated with heterologous RNA (Fig. 19, lower panel) there is no hybridization as indicated by the fact that the counts remain in the DNA region of the gradient.

The results of product analysis which indicate that *heteropolymeric* regions of the viral RNA are transcribed are the complementarity of the DNA product to the template RNA and the failure of the greater

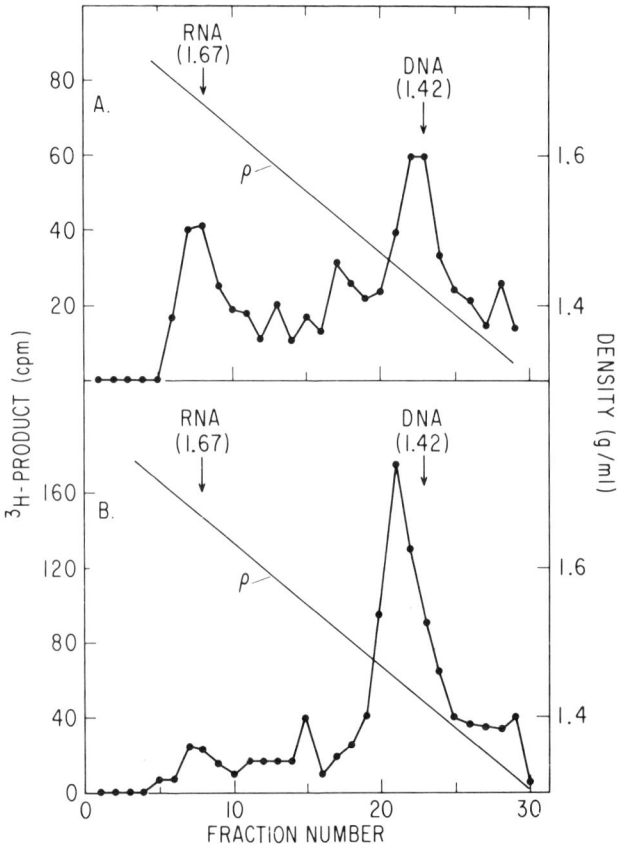

Fig. 17. Cs$_2$SO$_4$ equlibrium density gradient centrifugation of ^{3}H-dATP, ^{3}H-dCTP, and ^{3}H-dGTP labeled DNA product from purified human leukemic pellet DNA polymerase with AMV 70 S RNA as a template and with oligo dT. A standard reaction mixture containing 50 mM Tris-HCl (pH 8); 5 mM MgCl$_2$; 10 mM NaCl; 5 mM dithiothreitol; 80 μM TTP; and 10 μM each ^{3}H-dATP (14 Ci/mmole), ^{3}H-dCTP (23 Ci/mmole) and ^{3}H-dGTP (12 Ci/mmole) (New England Nuclear) was incubated with 2.5 μg of AMV 70 S RNA and 1 μg of oligo dT at 37°C for 60 min and then adjusted to 0.12 M NaCl and 1% sodium dodecyl sulfate. The DNA product was then purified and analyzed as described in Fig. 10. A, Native DNA product; B, heated at 95°C for 10 min.

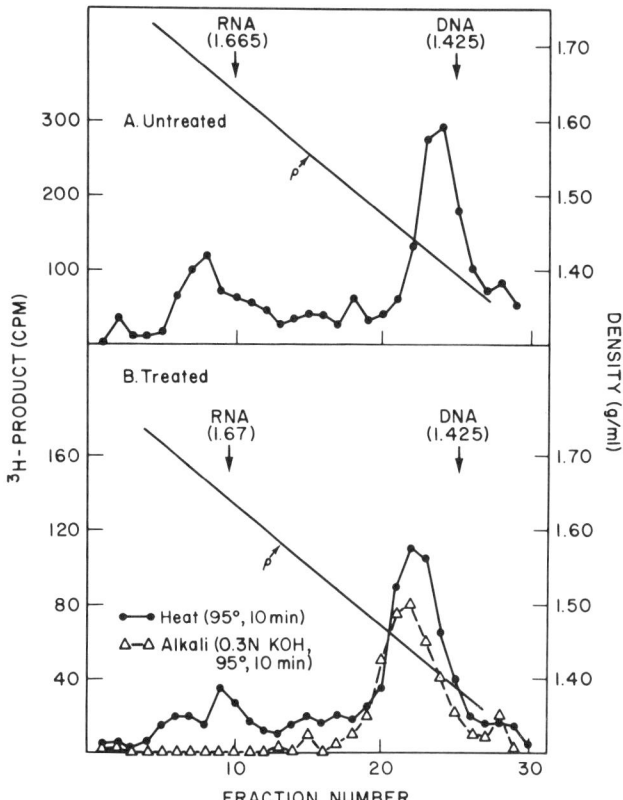

Fig. 18. Cs_2SO_4 equilibrium density gradient centrifugation of DNA product from purified human leukemic pellet DNA polymerase with AMV 70 S RNA as a template without oligo dT. A standard reaction mixture, but containing 0.5 mCi each of ^{3}H-dATP (14 Ci/mmole, ^{3}H-dCTP (23 Ci/mmole), ^{3}H-dGTP (14 Ci/mmole), and ^{3}H-TTP (18.6 Ci/mmole), was prepared. This was incubated with 2.5 μg of AMV 70 S RNA at 37°C for 60 min and then adjusted to 0.12 M NaCl and 1% sodium dodecyl sulfate. The product was then isolated and analyzed as described in Fig. 10. A, Native DNA product (●——●); B, heated at 95°C for 10 min (●——●); treated with 0.3 N KOH (95°C 10 min) (△- - -△).

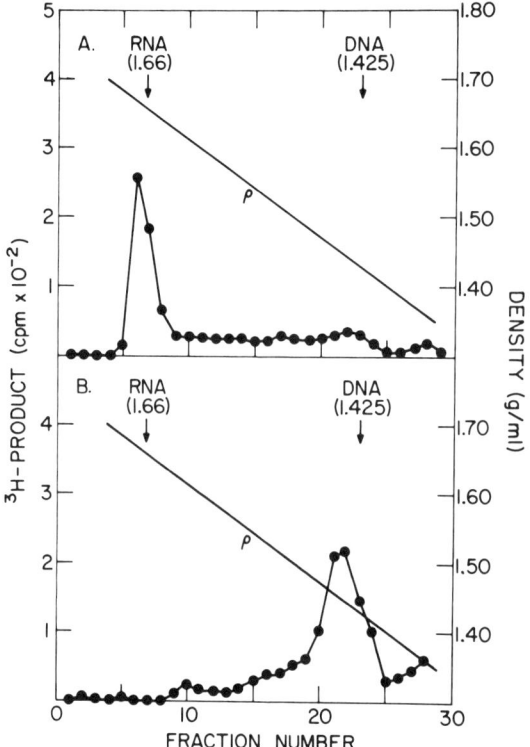

Fig. 19. Specific hybridization of DNA product of purified human leukemic pellet DNA polymerase (with AMV 70 S RNA as template) with AMV 70 S RNA. A, Upper panel, the product was annealed with AMV 70 S RNA; B, lower panel, with MS 2 RNA. Product was prepared as described in Fig. 16, then alkali-treated (0.3 N KOH, 95°C, 10 min, followed by room temperature, 20 hr), and neutralized. The product was then adjusted to 2 × SSC (2-fold concentrated 0.15 M sodium chloride-0.015 M sodium citrate) and an equal volume of formamide was added. Five micrograms of AMV 70 S RNA (A) or 5 μg of MS 2 RNA (B) were added to 0.5-ml portions of this solution, annealed for 72 hr at room temperature, and then the CS_2SO_4 density gradient centrifugation was carried out.

portion of the DNA product to hybridize to poly A (Fig. 20). This is true with either dT_{12-18} as a primer (if TTP is not the labeled substrate) or any labeled substrate without dT_{12-18} (data not shown).

As with normal lymphocytes, when conventional purification techniques are utilized two major DNA polymerases are found in leukemic cells. "DNA polymerase I" from leukemic cells elutes slightly after polymerase I of normal lymphocytes. At the present time we are not certain that this represents a difference in the two enzymes, nor are we certain whether the leukemic "pellet" DNA polymerase is a new enzyme or the same as one of the two major soluble *leukemic* polymerases.

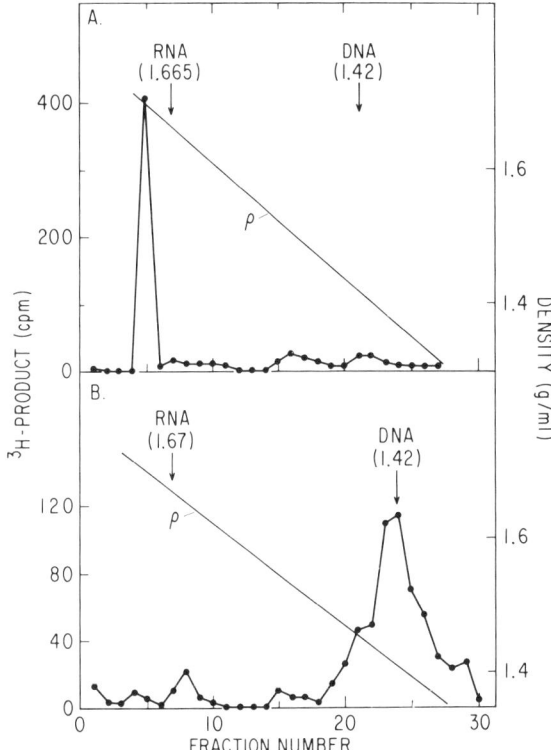

Fig. 20. Demonstration of complementarity between DNA product and RNA template (in an oligo dT-stimulated reaction) and lack of complementarity of most of the product to poly rA. Product was prepared as described in Fig. 17, then alkali-treated (0.3 N KOH, 95°C, 10 min, followed by room temperature, 20 hr) and neutralized. This solution was then adjusted to 2 × SSC and an equal volume of formamide was added. AMV 70 S RNA (2.5 µg) or 0.1 µg of poly rA was added to 0.2-ml portions of this solution and annealed for 72 hr at room temperature. A, The product annealed with AMV 70 S RNA; B, the product annealed with poly rA.

Do Normal Cells Contain an RNA-directed DNA Polymerase Activity?

As noted in a previous section, all mammalian cellular DNA polymerases which we have examined can transcribe appropriately primed poly A, but we have already shown (Table 4) that the two major DNA polymerases of cells do not transcribe heteropolymeric portions of exogenously supplied RNA. On the other hand, in three types of normal cells a cytoplasmic "pellet" fraction which contains an endogenous RNase-sensitive DNA polymerase system has been identified. This has been reported by Coffin and Temin in rat tissue culture cells (5), Kang and

Temin in chick embryo (18), and by our laboratory in PHA-stimulated normal blood lymphocytes (4, 9). Our observations with the pellet fraction in normal lymphocytes are summarized below.

1. The activity is not found in unstimulated normal lymphocytes.

2. The fraction is isolated with techniques identical with those used with the leukemic "pellet." The endogenous DNA polymerase activity is sensitive to RNase (Fig. 21, upper panel, and references 4, 9).

3. In most experiments analysis of the DNA product by equilibrium density gradient centrifugation in Cs_2SO_4 shows all the counts in the DNA region of the gradient (4, 9). However, in a few recent experiments

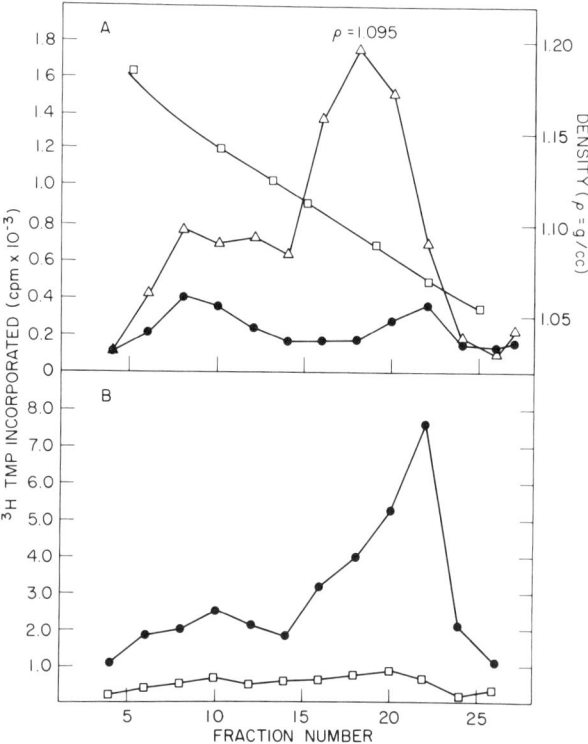

Fig. 21. Endogenous RNase-sensitive DNA polymerase activity of fractions of a high speed pellet from PHA-stimulated *normal* human blood lymphocytes. The partially purified "pellet" was layered on a 20 to 60% glycerol gradient. Upper panel, A, △——△, endogenous DNA polymerase activity; ●——●, endogenous DNA polymerase activity following preincubation of gradient fractions with RNase; □——□, density of fraction measured optically. Lower panel, B, ●——●, DNA polymerase activity in the presence of dT_{12-18} · poly dA; □——□, DNA polymerase activity with dT_{12-18} · poly rA, each at a concentration of 50 µg/ml. The much greater acceptance of dT_{12-18} · poly dA over dT_{12-18} · poly rA is unlike the viral reverse transcriptase.

a significant portion of the counts was found in the RNA and in the hybrid regions. Therefore, the product analysis of these endogenous reactions remains ambiguous and undoubtedly complicated by the presence of nucleases that are present in these fractions.

4. The enzyme from the normal cell pellet was purfied by DEAE-cellulose, phosphocellulose, and gel filtration column chromatography (4). Unlike the "pellet" DNA polymerase from leukemic cells and unlike the oncornavirus DNA polymerase, this polymerase shows marked preference for dT_{12-18} · poly dA compared to dT_{12-18} · poly rA (in Mg^{++}) (Fig. 21, lower panel, and Table 9). In short, the affinity for different templates by this enzyme is like the major two normal cellular DNA polymerases. This has apparently also been Temin's experience (28) with the chick embryo polymerase isolated from a similar pellet fraction which also catalyzes endogenous RNase-sensitive DNA synthesis (18).

Two explanations are immediately apparent to explain these observations. First, as Temin speculated (28), unlike the viral enzyme, the RNA requirement for this DNA polymerase may be highly specific. (It may require the RNA present in the pellet.) Second, it is possible that the pellet endogenous reaction in normal stimulated lymphocytes is RNA-primed but DNA-directed. These alternatives are now under experimental test. If the former interpretation proves correct, the observations will be of obvious interest in relationship to Temin's protovirus hypothesis (29).

A summary and comparison of some of the properties of all these lymphocyte DNA polymerases are shown in Table 10.

Table 9. Activity of "pellet" DNA polymerase from normal stimulated blood lymphocytes with templates useful in distinguishing viral reverse transcriptase from cellular DNA polymerases

Template	^{3}H-TMP incorporated (pmoles/10 μg enzyme)
dT_{12-18} · poly dA	155
dT_{12-18} · poly rA	32
dG_{12-18} · poly rC	0
70 S RNA (^{3}H-TTP plus dATP, dCTP, dGTP) (avian myeloblastosis virus)	<0.1 (and not RNase-sensitive)
70 S RNA plus dT_{12-18}	
^{3}H-TTP plus dATP, dCTP, dGTP	10.0
^{3}H-dCTP plus dGTP, TTP, dATP	<0.1

With dG_{12-18} · rC as template, the substrates were 8.0×10^{-5} M dCTP and 8.0×10^{-6} M ^{3}H-dGTP (3730 cpm/pmole). 70 S RNA and dT_{12-18} were used at a concentration of 1.25 μg per reaction mixture (50 μl). With ^{3}H-dCTP as the labeled substrate, a concentration of 4.4×10^{-6} M was used at a specific activity of 12,500 cpm/pmole. All assays contained Mg^{++} as the divalent cation and other details of the assay are as described in the legend to Fig. 9.

Table 10. Human leukemic pellet DNA polymerase versus normal lymphocyte DNA polymerases*

		DNA polymerase		
Property	Leukemic 60K pellet	Normal I	Normal II	Normal III (60K pellet)
Chromatographic behavior				
DEAE cellulose	Bound at low salt	Bound at low salt	Not bound	
Phosphocellulose	Bound up to 0.35 M KCl	Bound up to 0.2 M KCl	Bound up to 0.45 M KCl	Bound up to 0.35 M KCl
pH optimum	Nuetral	Neutral	Alkaline	Not determined
Template preference	Strongly favors poly d(AT)	Strongly favors activated salmon sperm DNA	Strongly favors activated salmon sperm DNA	Factors activated salmon sperm DNA
Template activity				
with AMV 70 S RNA	Active	Not active	Not active	Not active
$dT_{12-18} \cdot \text{poly rA} : dT_{12-18} \cdot \text{poly dA}$	1.40–5.0	0.008	0.022	0.11
$dG_{12-18} \cdot \text{poly rC}$	Active	Not active	Not active	Not active
Response to N-ethylmaleimide	Very sensitive	Very sensitive	Resistant	Moderately sensitive
Molecular weight	130,000	300,000 or 160,000	40,000	170,000

*The various templates were used at a concentration of 50 µg/ml. All assays were carried out with magnesium.

Acknowledgments

We wish to thank Dr. Joseph Beard (Duke University) for the generous supply of AMV virus; the Special Virus Cancer Program (National Institutes of Health) for the help in supply of all viruses; Dr. J. Freireich (M. D. Anderson Hospital and Tumor Institute), Dr. K. McCredie (M. D. Anderson Hospital and Tumor Institute), Dr. J. Burchenal (Memorial Hospital for Cancer and Allied Diseases), Dr. J. Holland (Roswell Park Memorial Institute), and Dr. E. Henderson (National Cancer Institute) for human leukemic cells; Dr. J. Hurwitz (Albert Einstein College of Medicine) for supply of labeled poly d(A-T) and λ DNA; Dr. R. Wells (University of Wisconsin) for supply of some of the synthetic template primers and for some helpful discussions; and Mrs. Cecelia Trainor, Miss Carla Davis, Mrs. Martha Feller, Miss Mary Gunn, Mrs. Irene Smith, and Miss Elizabeth Roseberry for excellent technical assistance. Parts of these studies were supported by the Special Virus Cancer Program, National Cancer Institute, National Institutes of Health (Contract No. NIH-71-2025).

References

1. Baltimore, D. 1970. RNA-dependent DNA polymerase in virions of RNA tumor viruses. *Nature* 226: 1209.

2. Baltimore, D., and Smoler, D. 1971. Primer requirement and template specificity of the DNA polymerase of RNA tumor viruses. *Proc. Nat. Acad. Sci. U. S. A.* 68: 1507.

3. Bernhard, W. 1958. Electron microscopy of tumor cells and tumor viruses: A review. *Cancer Res.* 18: 491.

4. Bobrow,, S. N., Smith, R. G., Reitz, M. S., and Gallo, R. C. 1972. Normal human lymphocytes contain a ribonuclease-sensitive DNA polymerase which is distinct from viral reverse transcriptase. *Proc. Nat. Acad. Sci. U. S. A.* 69: 3228.

5. Coffin, J. M., and Temin, H. M. 1971. Ribonuclease-sensitive deoxyribonucleic acid polymerase activity in uninfected rat cells and rat cells infected with Rous sarcoma virus. *J. Virol.* 8: 630.

6. Chang, L. M. S., and Bollum, F. J. 1971. Low molecular weight deoxyribonucleic acid polymerase in mammalian cells. *J. Biol. Chem.* 246: 5835.

7. Gallo, R. C. 1971. Reverse transcriptase, the DNA polymerase of oncongenic viruses. *Nature* 234: 194.

8. Gallo, R. C., Sarin, P. S., Sarngadharan, M. G., Reitz, M. S., Smith, R. G., and Bobrow, S. N. 1972. Viral 70 S RNA directed DNA synthesis with a purified DNA polymerase from human acute leukemic cells. *Molecular studies in viral neoplasia.* The University of Texas M. D. Anderson Hospital and Tumor Institute at Houston, 25th Annual Symposium on Fundamental Cancer Research, 1972. In press.

9. Gallo, R. C., Sarin, P. S., Sarngadharan, M. G., Smith, R. G., Bobrow, S. N., and Reitz, M. S. Biochemical properties of reverse transcriptase activities from human cells and RNA tumor viruses. *Proceedings of the Sixth International Miles Symposium on Molecular Biology.* In press.

10. Gallo, R. C., and Whang-Peng, J. 1971. Observations on the regulatory effects of the transfer RNA minor base, $N^6\text{-}\Delta^2$-isopentenyladenosine, on human lymphoctyes. In Beers, R. F., and Braun, W. (eds.) *Biological effects of polynucleotides*, pp. 303–304, Springer-Verlag, New York.

11. Gilden, R. V., and Oroszlan, S. 1971. Structure and immunologic relationships among mammalian C-type viruses. *J. Amer. Vet. Med. Ass.* 158: 1099.

12. Gillespie, D., Marshall, S., and Gallo, R. C. 1972. RNA of RNA tumour viruses contains poly A. *Nature New Biol.* 236: 227.

13. Goodman, N. C., and Spiegelman, S. 1971. Distinguishing reverse transcriptase of an RNA tumor virus from other known DNA polymerases. *Proc. Nat. Acad. Sci. U. S. A.* 68: 2203.

14. Green, M., and Cartas, M. 1972. The genome of RNA tumor viruses contains polyadenylic acid sequences. *Proc. Nat. Acad. Sci. U. S. A.* 69: 791.

15. Hausen, P., and Stein, H. 1968. On the synthesis of RNA in lymphocytes stimulated by phytohemagglutinin. I. Induction of uridine-kinase and the conversion of uridine to UTP. *Eur. J. Biochem.* 4: 401.

16. Hurwitz, J., and Leis, J. P. 1972. RNA-dependent DNA polymerase activity of RNA tumor viruses. I. Directing influence of DNA in the reaction. *J. Virol.* 9: 116.

17. Kacian, D. L., Spiegelman, S., Bank, A., Terada, M., Metafora, S., Dow, L., and Marks, P. A. 1972. In vitro synthesis of DNA components of human genes for globins. *Nature New Biol.* 235: 167.

18. Kang, C.-Y., and Temin, H. M. 1972. Endogenous RNA-directed DNA polymerase activity in uninfected chicken embryos. *Proc. Nat. Acad. Sci. U. S. A.* 69: 1550.

19. Lai, M. M. C., and Duesberg, P. H. 1972. Adenylic acid-rich sequence in RNAs of Rous sarcoma virus and Rauscher mouse leukaemia virus. *Nature* 235: 383.

20. Loeb, L. A., Agarwal, S. S., and Woodside, A. M. 1968. Induction of DNA polymerase in human lymphocytes by phytohemagglutinin. *Proc. Nat. Acad. Sci. U. S. A.* 61: 827.

21. Martin, R. G., and Ames, B. N. 1961. A method for determining the sedimentation behavior of enzymes: Application to protein mixtures. *J. Biol. Chem.* 236: 1372.

22. Parks, W. P., Scolnick, E. M., Ross, J., Todaro, G. J., and Aaronson, S. A. 1972. Immunological relationships of reverse transcriptases from RNA tumor virus. *J. Virol.* 9: 110.

23. Riddick, D. H., and Gallo, R. C. 1970. Correlation of transfer RNA methylase activity with growth and differentiation in normal and neoplastic tissues. *Cancer Res.* 30: 2484.

24. Robert, M. S., Smith, R. G., Gallo, R. C., Sarin, P. S., and Abrell, J. W. 1972. Viral and cellular DNA polymerase: Comparison of activities with synthetic and natural RNA templates. *Science* 176: 798.
25. Ross, J., Scolnick, E. M., Todaro, G. J., and Aaronson, S. A. 1971. Separation of murine cellular and murine leukemia virus DNA polymerase. *Nature New Biol.* 231: 163.
26. Sarngadharan, M. G., Sarin, P. S., Reitz, M. S., and Gallo, R. C. 1972. Reverse transcriptase activity of human acute leukemic cells: Purification of the enzyme, response to AMV 70S RNA, and characterization of the DNA product. *Nature New Biol.* 240: 67.
27. Smith, R. G., and Gallo, R. C. 1972. DNA-dependent DNA polymerases I and II from normal human blood lymphocytes. *Proc. Nat. Acad. Sci. U.S.A.* 69: 2879.
28. Temin, H. M. Presented at the 25th Annual Symposium on Fundamental Cancer Research at the University of Texas M. D. Anderson Hospital and Tumor Institute at Houston, March 1972.
29. Temin, H. M. 1971. The protovirus hypothesis: Speculations on the significance of RNA-directed DNA synthesis for normal development and for carcinogenesis. *J. Nat. Cancer Inst.* 46: 3.
30. Temin, H. M., and Baltimore, D. 1972. RNA-directed DNA synthesis and RNA tumor viruses. *Advan. Virus Res.* 17: 129.
31. Temin, H. M., and Mizutani, S. 1970. RNA-dependent DNA polymerase in virions of Rous sarcoma virus. *Nature* 226: 1211.
32. Weber, K., and Osborn, M. 1969. The relativity of molecular weight determinations by dodecyl sulfate polyacrylamide gel electrophoresis. *J. Biol. Chem.* 211: 4406.
33. Yang, S. S., Herrera, F., Smith, R. G., Reitz, M., Lancini, G., Ting, R., and Gallo, R. 1972. Rifamycin antibiotics: Inhibitors of Rauscher (murine) leukemia virus reverse transcriptase and of purified DNA polymerases from human normal and leukemic lymphoblasts. *J. Nat. Cancer Inst.* 49: 7.

Discussion

Dorson. In the product you sedimented that gave a 5 S value, was the template the endogenous reaction?

Gallo. No, that was with the purified enzyme and the viral (avian myeloblastosis virus) 70 S RNA as template.

Dorson. Why was it 5 S (*i.e.* so small)? Do you have any explanation for that?

Gallo. Perhaps for the same but as yet unidentified reason, that the viral enzyme gives a small product. We really don't know. It's exactly the same when you use a purified DNA polymerase from a known RNA tumor virus. I don't know if anyone has an explanation yet. I think perhaps Dr. Bishop might get into this later.

Bishop. No, we don't know why.

Stodolsky. Have any of these RNA-directed DNA polymerases been checked to see whether they have the propensity of the pol Al polymerase or pol A polymerase of *Escherichia coli* to form hairpins of newly synthesized material?

Temin. Is someone going to talk about that later? Probably someone will.

Krieger. Has anybody tried to use polymerase I from *E. coli* and see whether, with oligo dT primer and the 70 S RNA, it will copy or not?

Gallo. So far, the experiments we have done are negative. In other words, with *E. coli* DNA polymerase I the heteropolymeric region of viral 70 S RNA has not been transcribed. Of course, like all other cellular DNA polymerases we have studied, *E. coli* polymerase I will transcribe appropriately primed poly rA and we have shown that viral 70 S RNA contains relatively large tracts of poly rA. Therefore, this portion of the viral RNA is transcribed. What we can say is if one normalizes the activity of viral (or the leukemic) polymerase with a template (primer) which all these enzymes can utilize, *e.g.* poly d(AT) and poly(dT) · poly(rA), under these conditions, the *coli* polymerase will not transcribe viral 70 S RNA while the viral and leukemic enzymes do. On the other hand, we have not attempted to "force" the reaction with the *coli* enzyme. Therefore, I think it still may be an open question, but so far, the experiments that have been done in several laboratories indicate that the answer is "no."

RNA-dependent DNA Polymerase Activity of RNA Tumor Viruses

IV. Characterization of AMV Stimulatory Protein and RNase H-associated Activity

Jonathan Leis, Ira Berkower, and Jerard Hurwitz

Department of Developmental Biology and Cancer
Albert Einstein College of Medicine
Bronx, New York 10461

A protein has been isolated from avian myeloblastosis virus which stimulates the rate and yield of AMV RNA-primed DNA synthesis with purified AMV polymerase. It specifically affects AMV polymerase and does not stimulate other DNA polymerases under conditions tested. The AMV polymerase, in conjunction with this protein, transcribes extended single stranded regions of DNA and permits the enzyme to initiate synthesis from single strand breaks in DNA. RNase activity associated with the AMV reverse transcriptase degrades homoribopolymers in the presence of their complementary homodeoxyribopolymer, with the exception of poly(U) · poly(dA) as well as RNA · DNA hybrids formed with fd DNA and *Escherichia coli* RNA polymerase. The products of the reaction are oligonucleotides containing 3'-hydroxyl and 5'-phosphate termini and no detectable mononucleotides. The enzyme is a processive exonuclease which shows an absolute requirement for ends of RNA chains; it does not degrade circular poly(A) · poly(dT). The direction of attack is presently unknown.

RNA tumor viruses contain a DNA polymerase capable of transcribing viral RNA into DNA (1, 28). Studies in several laboratories (6, 8, 9, 18, 24, 25), with virions rendered permeable with non-ionic detergents, have established the following chronological sequence for DNA synthesis. The initial product is RNA · DNA hybrids in which the RNA and

DNA are covalently linked. This material is converted to nonconvalent hybrids and then eventually into single and double stranded DNA.

We and others (4, 12, 17, 23, 24) have previously shown that highly purified preparations of avian myeloblastosis virus or Rous sarcoma virus DNA polymerase catalyze repair-like reactions on RNA, DNA, or RNA · DNA hybrids. Deoxynucleotide incorporation occurs specifically from 3'-OH ends of primer strands attached to template strands. The product of the reaction with AMV RNA is an RNA · DNA hybrid in which the DNA is covalently attached to the RNA (17, 29). Only after prolonged incubation does material possessing characteristics of free DNA appear among the products (27). Evidence presented in this communication suggests that the latter observation is due to polymerase and an additional protein factor isolated from AMV referred to as stimulatory protein.

Ribonuclease H (RNase H), an enzyme capable of specifically degrading polyribonucleotides in RNA · DNA hybrids, was discovered in calf thymus by Hausen and Stein (10). A similar nuclease activity was reported by Mölling et al. (19) to copurify with the AMV reverse transcriptase. These authors proposed that the RNase H activity, in conjunction with the AMV polymerase, played an important role in the generation of free DNA from RNA · DNA transcript products. We have confirmed their observation and have detected the presence of RNase H in our purified AMV polymerase preparations. However, since stimulatory protein plays an important role in the production of "free DNA," we decided to characterize the RNase H activity and to determine its effect, if any, on transcription of RNA by the reverse transcriptase. A comparative study of the properties of the AMV · RNase H activity with RNase H activity purified 2000-fold from *Escherichia coli* was also carried out and is presented below.

Purification of AMV DNA Polymerase

The DNA polymerase of AMV was purified from 2 to 3 g, wet weight, of plasma-derived virus by a modification of the purification procedure described by Hurwitz and Leis (12). Polymerase activity was measured by d(AT) copolymer synthesis (12) and by AMV RNA-primed DNA synthesis (17). The crude extract prepared from detergent-treated virus was precipitated with solid ammonium sulfate (2.6 g added to 10 ml) and the insoluble material collected by centrifugation at 105,000 × g for 10 min. The precipitate was extracted by suspension in a solution (1 to 2 volumes/g, wet weight, of ammonium sulfate precipitate) containing ammonium sulfate (1.4 g added to 10 ml), 5% glycerol, 20 mM Tris · HCl, pH 7.5, 2 mM DTE, and 0.1 mM EDTA (buffer A). The insoluble material was collected by centrifugation and suspended in 22% saturated solution of ammonium sulfate (1.2 g added to 10 ml) prepared in buffer A and the insoluble residue was collected by centrifugation. More than 95% of the RNase activity measured by acid solubilization of [^{3}H]poly(U) was extracted from the ammonium sulfate precipitate with a loss of only 5 to 10% of the polymerase activity by this procedure. The AMV polymerase was recovered from the precipitate (95% yield) by suspension in 2% saturation ammonium sulfate (0.1 g added to 10 ml) in buffer A. After centrifugation, the residue was discarded and the supernatant fluid, referred to as 2% ammonium sulfate fraction, was adjusted to a final glycerol concentration of 40%. This fraction catalyzed the incorporation of 0.1 μmole of nucleotide/mg/30 min.

The 2% ammonium sulfate fraction was diluted with 10% glycerol, 20 mM Tris · HCl, pH 8.4, 0.1 mM EDTA, 2 mM DTE (buffer B) to a salt concentration less than 0.04 M ammonium sulfate (measured by conductivity) and applied to a 50-ml Whatman P-11 phosphocellulose column equilibrated with buffer B. The polymerase was eluted with a 220-ml linear ammonium sulfate gradient (0 to 0.4 M) in buffer B as previously described (12). Polymerase fractions eluted from the column were dialyzed against buffer B containing 65% glycerol for 3 hr at 4°C. The specific activity of the major polymerase fraction referred to as polymerase I was greater than 1 μmole/mg/30 min.

Isolation of AMV Stimulatory Protein

The AMV polymerase I fraction was diluted with 10% glycerol, 20 mM sodium phosphate, pH 7.2, 2 mM DTE, 0.1 mM EDTA (buffer C) to a salt concentration less than 0.04 M ammonium sulfate and applied to a 7-ml P-11 phosphocellulose column equilibrated with buffer C. The protein peak which did not adsorb to the phosphocellulose was collected and dialyzed against buffer B containing 65% glycerol. These dialyzed fractions contained the AMV stimulatory protein at a concentration of 0.57 μg/ml. The AMV polymerase was eluted from the column with a 20-ml linear gradient of ammonium sulfate (0 to 0.4 M) in buffer C and dialyzed against buffer B containing 65% glycerol for 3 hr at 4°C. The AMV polymerase, which elutes from the column at a concentration of 0.1 M ammonium sulfate, is referred to as AMV polymerase II and has a specific activity of greater than 1 μmole/mg/30 min. AMV polymerase II fractions are free of detectable RNase as previously described (12, 17). The stimulatory protein fraction possessed no detectable RNase activity toward ^{3}H-labeled AMV RNA or [^{3}H]poly(U) as measured either by gradient centrifugation or generation of acid-soluble radioactivity. The protein fraction did not produce acid-soluble material when tested with RNA · DNA hybrids. These studies for RNase H activity were carried out with labeled RNA polymerase products (2 nmoles, 100 cpm/pmole) prepared with fd DNA and with the synthetic homopolymers [^{3}H]poly(A) (14 cpm/pmole) · poly(dT) and [^{3}H]poly(U) (4 cpm/pmole) · poly(dA). There was no detectable DNase activity in AMV polymerase and stimulatory protein preparations as measured by the methods previously described (12, 17). Purified AMV polymerase preparations were found to contain detectable RNase activity as described below.

During the isolation of the reverse transcriptase from AMV, it was observed that the yield of DNA synthesized in reactions primed with AMV RNA decreased with purification. The amount of DNA synthesized relative to RNA added at 120 min was 20% with the 2% ammonium sulfate fraction, 14% with a polymerase I preparation, and only 3% with a polymerase II preparation (Fig. 1). These values were obtained with approximately identical amounts of AMV polymerase. In the latter case, this value is a yield and is not related to enzyme inactivation since addition of more enzyme after incorporation ceased was without effect; the addition of more RNA resulted in an immediate resumption of DNA synthesis which again reached a plateau at a value corresponding to a yield of 3%.

We have isolated a heat-labile, nondialyzable factor which separates from the AMV polymerase after the second phosphocellulose chromato-

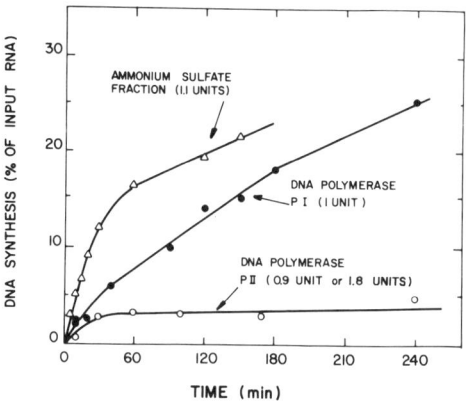

Fig. 1. Yield of DNA synthesis with AMV polymerase preparations. Assay conditions were as described in Table 1 with [^{3}H] dTTP (1300 cpm/pmole), 82 pmoles of AMV RNA, and polymerase preparations as indicated. Two per cent ammonium sulfate fraction, 1.1 units (△——△); AMV polymerase I, 1 unit (●——●); AMV polymerase II, 0.9 or 1.8 units (○——○).

graphic step. The factor restored the higher yield of DNA synthesis catalyzed by polymerase II preparations. The factor is quantitatively inactivated by heating at 100°C for 4 min or by Pronase treatment, but is not affected by incubation with nuclease SI, an enzyme which specifically digests single stranded RNA or DNA. A small amount of AMV polymerase which contaminates the stimulatory protein preparations can be inactivated by heating the fractions at 70°C for 2 min without concomitant loss in stimulatory protein activity.

Effect of Stimulatory Protein on the Rate of RNA-primed DNA Synthesis

Upon addition of stimulatory protein to AMV polymerase II fractions, the rate of AMV RNA-primed DNA synthesis increased as much as 8-fold (Fig. 2). The rate of DNA synthesis rose exponentially with increasing concentration of factor (see inset in Fig. 2). At the maximally stimulated rate, the DNA synthesized relative to RNA added increased from 3 to 40% after 3 hr of incubation. Preincubation of RNA with stimulatory protein for 10 min at 38°C did not alter the marked increase in rate of yield observed above. However, the addition of larger amounts of stimulatory protein (34.2 mµg) resulted in an inhibition of the rate of DNA synthesis.

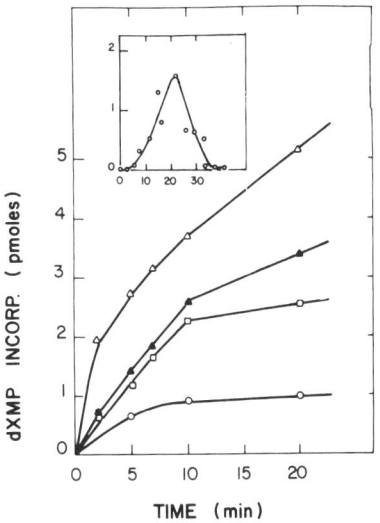

Fig. 2. Influence of stimulatory protein on rate of DNA synthesis. DNA synthesis was measured in 0.1-ml reaction mixtures as described in Table 1 with 1.84 nmoles of [^{3}H]dCTP (1370 cpm/pmole), 10 nmoles each of cold dATP, dGTP, and dTTP, 125 pmoles of AMV RNA, and 0.2 unit of purified AMV polymerase II. Varying amounts of stimulatory protein were added as indicated. Polymerase alone (O——O); addition of stimulatory protein, 11.4 mµg (□——□); 22.8 mµg (△——△); 34.2 mµg (▲——▲). In the case of the insert figure, the abscissa represents millimicrograms of stimulatory protein added and the ordinate represents the increase in the rate of DNA synthesis (picomoles/2 min) with the rate of DNA synthesis in the absence of stimulatory protein subtracted.

Repair of Extended Single Stranded Regions of DNA and Specificity of Stimulatory Protein for Polymerase

The stimulatory effect of the protein factor on deoxynucleotide incorporation is not limited to RNA but is also evident with suitably constructed DNA templates. ^{3}H-labeled λ DNA was briefly treated with exonuclease III to remove deoxynucleotides from the 3'-OH ends exposing single stranded regions of different lengths (33). λ DNA from which 50 nucleotides were excised was quantitatively repaired by polymerase II preparation in the absence of stimulatory protein. In contrast, λ DNA more extensively treated with exonuclease III so that 300 nucleotides were removed was only partially (15%) repaired by the polymerase (Fig. 3A). However, when both the AMV polymerase and the stimulatory protein were added, this λ DNA was almost fully repaired (80% after 90 min of incubation, Fig. 3A). In the presence of the stimulatory protein alone, no detectable nucleotide incorporation was detected. Thus, in the presence of stimulatory protein, the AMV polymerase can transcribe

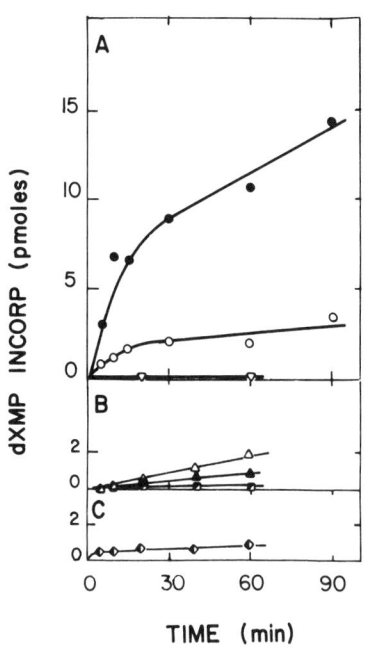

Fig. 3. Action of AMV stimulatory protein with various DNA polymerases. The extent of deoxynucleotide incorporation with exonuclease III-treated DNA was measured in 0.05-ml reaction mixtures as described in Table 1 except that they contained 10 nmoles each of dATP, dCTP and dGTP, 0.18 nmole of [α-^{32}P]dTTP (1060 cpm/pmole), 470 pmoles of sonically treated λ [^{3}H] DNA (9.3 cpm/pmole) treated with exonuclease III (33) (which removed 32% of the total nucleotides). Where indicated the following other additions were made: 0.006 unit of purified AMV polymerase II and 11.4 mμg of AMV stimulatory protein, 0.018 unit of Escherichia coli DNA polymerase II (32) or 0.023 unit of E. coli DNA polymerase III (16), 0.006 unit of HeLa cell DNA polymerase I (31) and 0.009 unit of HeLa cell R-DNA polymerase (7). A, AMV polymerase (○——○); AMV polymerase with stimulatory protein (●——●); stimulatory protein without polymerase (▽——▽). B, HeLa DNA polymerase I with or without stimulatory protein (◪——◪); HeLa R-DNA polymerase; without stimulatory protein (△——△): with stimulatory protein (▲——▲). C, E. coli polymerases II or III; with or without stimulatory protein (◆——◆).

longer single stranded regions of a DNA template than in its absence. The addition of the stimulatory protein to reaction mixtures containing λ DNA in which 50 nucleotides had been removed led to deoxynucleotide incorporation two to three times greater than the number of deoxynucleotides initially removed by exonuclease III treatment.

The AMV stimulatory protein appears to affect the AMV polymerase specifically under the conditions tested (Fig. 3). This factor did not affect the repair of the extensively exonuclease III-treated λ DNA when added to mixtures containing the following DNA polymerases: *E. coli* DNA polymerases II or III (Fig. 3C), DNA polymerase I and R-DNA polymerase isolated from HeLa cells by Weissbach and coworkers (7, 31) (Fig. 3B), and Rauscher leukemia viral DNA polymerase (not shown). *E. coli* DNA polymerase II, however, was stimulated by the *E. coli* unwinding protein, confirming the results reported by Alberts and coworkers (personal communication). The *E. coli* unwinding protein had no effect on the AMV polymerase, in agreement with its reported specificity for *E. coli* DNA polymerase II described by Alberts *et al.* (personal communication).

Activation by AMV Stimulatory Protein of Duplex DNA-containing Single Strand Breaks

In the presence of stimulatory protein the AMV polymerase initiates DNA synthesis from single strand breaks in native DNA (Fig. 4). Colicin El DNA, a double stranded covalently closed circular DNA of molecular weight 4×10^6, is inactive as a template with AMV polymerase regardless of the presence of the stimulatory protein. When a limited number of single strand breaks were introduced in colicin El DNA with pancreatic DNase, a small amount of DNA synthesis catalyzed by AMV polymerase was observed. The addition of the stimulatory protein, however, resulted in a marked increase in DNA synthesis. The stimulation was significantly reduced if the nicked DNA was repaired with T4 DNA ligase prior to the addition of AMV polymerase. Incubation of stimulatory protein and AMV polymerase II preparations with deoxynucleoside triphosphates and $5'$-^{32}P-labeled nicked salmon sperm DNA (1200 cpm/pmole) prepared with polynucleotide kinase did not lead to detectable acid-soluble radioactivity. Since we have not detected any $5' \rightarrow 3'$ exonuclease activity in either preparation, the above results suggest that AMV polymerase in conjunction with the stimulatory protein can synthesize DNA from single strand breaks by strand displacement. In the absence of this factor the purified enzyme is unable to catalyze DNA synthesis from single strand breaks as previously reported (12).

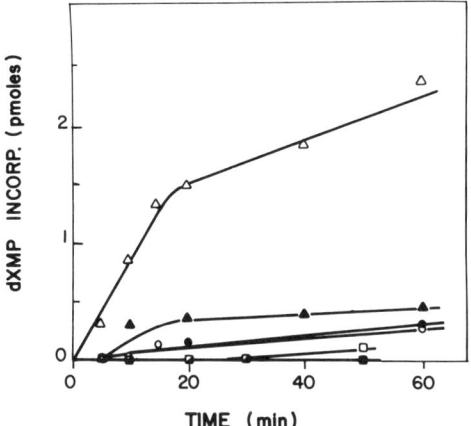

Fig. 4. Activation of duplex DNA containing single strand breaks by AMV stimulatory protein. Assay conditions were as described in Fig. 3 except that ^{3}H-labeled colicin E1 DNA (188 pmoles) (3) was added as template and the [^{32}P]dTTP was 910 cpm/pmole. Nicked colicin E1 DNA was prepared by incubation of this DNA with 26.6 pg of pancreatic DNase/nmole of DNA for 15 min at 38°C. DNase was inactivated by heating at 65°C for 10 min and the DNA utilized as described above. In reaction mixtures containing repaired DNA, the nicked DNA was incubated with 0.6 unit of T4 DNA ligase (30) in the presence of 5 nmoles of ATP for 15 min at 38°C and heated to 60°C for 15 min to inactivate DNA ligase. AMV polymerase and deoxynucleoside triphosphates were then added and incubation carried out as described above. Polymerase without stimulatory protein, native DNA (■——■); nicked DNA (●——●); repaired DNA (○——○); polymerase with stimulatory protein, native DNA (□——□); nicked DNA (△——△); repaired DNA (▲——▲).

RNA Template Specificity of Stimulatory Protein

DNA synthesis primed with AMV RNA is stimulated in the presence of stimulatory protein (Table 1). AMV RNA heated at 80°C for 2 min suffers a change in sedimentation coefficient from 60 to 35 S and smaller fragments and a concomitant loss of priming activity, as reported by several laboratories (5, 17). As much as 50% of this priming activity of heat-treated RNA was recovered in the presence of stimulatory protein. A similar, though not as pronounced, effect was observed with RSV RNA. Again, 50% of the activity of the heated RNA was regained in the reactions carried out in the presence of stimulatory protein.

The stimulatory protein exhibits specificity in reactions primed with RNA. For example, RNA preparations isolated from bacteriophages f2 and Qβ, in contrast to tumor viral RNA-primed reactions, were inhibited by the addition of the stimulatory protein (Table 1). Variation in the concentration of the phage RNA or in the time of incubation did not increase deoxynucleotide incorporation under these conditions.

Table 1. Specificity of action of stimulatory protein

RNA added	Amount of RNA added (pmoles)	Deoxynucleotide incorporation (pmoles/30 min)	
		−Factor	+Factor
AMV	125	1.43	5.14
AMV heated 100°C for 2 min	110	0.36	2.70
RSV	200	0.87	1.67
RSV heated 100°C for 2 min	200	0.16	0.88
f2	130	0.44	0.19
Qβ	190	0.20	<0.02

Purified AMV polymerase II (0.1 unit) was incubated with different RNA preparations in the presence and absence of 11.4 mμg of stimulatory protein. Reaction mixtures (0.05 ml) contained 1 μmole of Tris · HCl, pH 8.0, 0.5 μmole of $MgCl_2$, 0.25 μmole of KCl, 0.3 μmole of DTE, 5 nmoles each of dATP, dGTP, dTTP, 1 nmole of [^{3}H]dCTP (1370 cpm/pmole). The reaction was stopped after 30 min at 38°C by the addition of 0.1 ml of 0.1 M sodium pyrophosphate, 0.02 ml of denatured salmon sperm DNA (2.6 μmoles/ml), and 5% trichloroacetic acid. The precipitate was collected on Gelman type E glass fiber filters, dried, and counted in 10 ml of toluene scintillation fluid in a scintillation counter.

Strand Displacement in RNA-primed Reactions

We have previously reported that free DNA is not formed in reactions primed with AMV RNA in the presence of highly purified AMV polymerase (17). This observation was re-examined utilizing the stimulatory protein. DNA products labeled with [^{32}P]dTTP during a 60-min incubation in the presence of stimulatory protein were analyzed by isopycnic banding in Cs_2SO_4 (Fig. 5A). The DNA product banded not only at a density corresponding to that of RNA (1.68), indicating its hybrid structure, but also in the region corresponding to DNA. The latter material was isolated and rebanded in CsCl before and after treatment with alkali (Fig. 5B). Tritium-labeled T7 DNA was added as an internal marker. The DNA product banded at a higher density than the reference marker, while the alkali-treated DNA product banded at a density coincidental to the marker. These results suggest that the nucleic acid isolated from the DNA region after banding in Cs_2SO_4 contained small amounts of RNA linked to the DNA. Identical results were obtained after heat denaturation of the DNA product (100°C for 4 min), suggesting that the RNA component banding at the density expected of DNA was covalently attached.

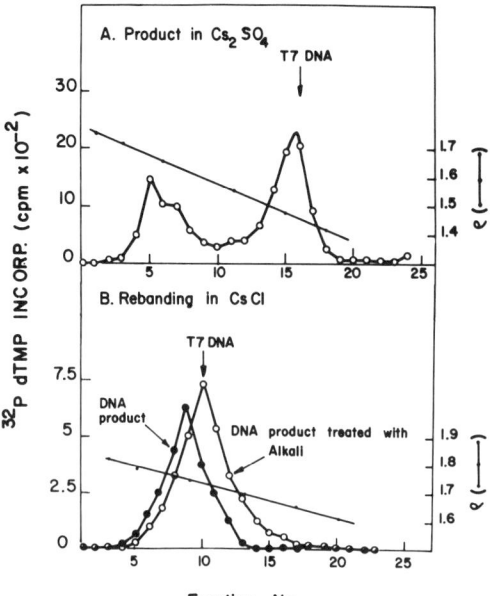

Fig. 5. Equilibrium sedimentation of DNA polymerase products in Cs_2SO_4 and CsCl. A, A polymerase product (97,000 cpm) labeled with [^{32}P]dTTP (10,000 cpm/pmole) was prepared with AMV RNA (140 pmoles), 1.1 units of purified AMV polymerase II, and 57 mµg of stimulatory protein in 0.2 ml volume as described in Table 1. After 60 min at 38°C, the incubation mixture was diluted to 4 ml with a solution containing 5 mM EDTA, 0.25% Sarkosyl, 10 mM Tris · HCl, pH 7.3, 0.3 M NaCl, and 0.3 M sodium citrate. Solid Cs_2SO_4 (2.7g) was added and the solution centrifuged for 65 hr at 33,000 rpm in a SW 50.1 rotor at 25°C in a polyallomer tube. Fractions were collected from a hole pierced in the bottom of the tube, refractive indices measured, 0.05-ml aliquots acid-precipitated, and radioactivity determined. The recovery of ^{32}P in this procedure was 60%. DNA product (O——O); density (—·—). B, Fractions 15 to 18 from the above Cs_2SO_4 gradients were combined and dialyzed against 4 liters of distilled water overnight at 4°C. Two aliquots of the dialyzed materials each containing 4100 cpm were utilized as follows: one aliquot was treated with 0.67 N NaOH for 30 hr at 38°C and then neutralized with HCl. Both samples were then diluted to 3 ml with a solution containing 8 mM EDTA, 0.33% Sarkosyl, 16 mM Tris · HCl, pH7.3, 0.4 M NaCl, and 0.4 M sodium citrate. T7 [^{3}H] DNA (10,000 cpm) was added to each as an internal marker. Solid CsCl (3.6 g) was added to each and the mixtures were spun in a centrifuge for 63 hr as described above. Fractions were collected, refractive indices measured, and the amount of acid-insoluble radioactivity was determined. The recovery of ^{32}P from both aliquots was 70 to 80%. DNA product (●——●); DNA product treated with alkali (O——O); density (—·—).

Purification and Properties of AMV RNase H

RNase H activity was measured during the course of purification of the AMV reverse transcriptase by the formation of acid-soluble radioactivity from [^{3}H]poly(A) in the presence of stoichiometric amounts of poly(dT), an assay communicated to us by D. Baltimore. DNA synthesis, measured by d(AT) copolymer- or 60 S AMV RNA-primed synthesis, and RNase H activity were found to be closely associated during purification, though the two activities did not coincide as indicated below. The ratio of the d(AT) synthesis to RNase H activity in the 2% ammonium sulfate fraction was 1.45. Upon adsorption and elution on two successive phosphocellulose columns (pH 8.4 and 7.2, respectively) RNase H activity was slightly displaced from DNA polymerase activity. The ratio of d(AT) synthesis to RNase H varied from 1 to 4 across the enzyme peak of the first phosphocellulose column (pH 8.4) and from 4 to 8 across the peak of the second phosphocellulose column (pH 7.2). The chromatographic separation of these two activities has been variable with different enzyme preparations. These results suggest that although the two enzyme activities have not been physically separated, they may reside on two different enzyme molecules. Another possible interpretation is that the two activities reside on the same enzyme but at different parts of the molecule similar to polymerase and $5' \rightarrow 3'$ exonuclease activities of *Escherichia coli* DNA polymerase I (2, 14). In order to answer these questions we compared the requirements for RNase H to those of the avian polymerase. The requirements for optimal DNA synthesis by the reverse transcriptase were previously described (12, 17). Both AMV polymerase and RNase H activities of the same fraction have similar requirements for Mg^{++} ions (Mn^{++} for RNase activity), sulfhydryl reagents, and pH. Both activities are salt-sensitive, *i.e.* DNA synthesis is inhibited 50% by 0.08 M KCl, while RNase H activity is inhibited 50% by 0.13 M KCl. RNase H activity does not require dTTP for activity and is independent of DNA synthesis. A requirement for dTTP is observed when poly(dT) is replaced by oligo(dT)$_{12-14}$ in the reaction. With poly(A) (300 to 400 nucleotides in length) + poly(dT) the rate of nuclease action was linear until approximately 60% of the poly(A) was rendered acid-soluble, after which the rate decreased markedly. However, the reaction continued until more than 95% of the poly(A) was degraded. The polymerase and nuclease activities can be further distinguished by heat stability and sensitivity to N-ethylmaleimide (NEM). DNA polymerase is less heat-stable than RNase H. It has a

$t_{1/2}$ at 45°C of 2.5 min while the more stable RNase H activity has a $t_{1/2}$ at 45°C of 9.5 min. Treatment with NEM inactivates both activities, but DNA synthesis is ten times more sensitive to inhibition than RNase H activity (i.e. 10^{-3} M NEM inactivates DNA synthesis 50% while 1.1×10^{-2} M NEM inactivates RNase H activity 50%). The latter observations suggest that RNase H and DNA polymerase activities must have different active sites but whether they are part of one enzyme or two is not clear.

Purification and Properties of RNase H of *Escherichia coli*

RNase H activity has been purified 2000-fold from extracts of *E. coli* utilizing the same assay described for measuring RNase H activity of AMV preparations. The steps of purification involve polyethylene glycol-dextran partition, affinity chromatography on DNA cellulose, gel filtration on agarose, and chromatography on DEAE and phosphocellulose columns. Detectable RNase activity toward [^{3}H]poly(U) can be separated from RNase H by gel filtration on Sephadex G-100. The resulting RNase H preparations are free of measurable endo- or exonuclease activity toward DNA. The enzyme activity is dependent on Mg^{++} which cannot be replaced by Mn^{++}; the *E. coli* RNase H activity is NEM-sensitive and resistant to salt (up to 0.3 M NaCl).

Effect of Stimulatory Protein on RNase H

Since stimulatory protein increases DNA synthesis catalyzed by the AMV polymerase, we examined its effect on RNase H activity. As shown in Table 2, the rate of attack on synthetic hybrids by RNase H associated with the AMV polymerase was increased 4-fold in the presence of the stimulatory protein. The stimulatory protein, in the absence of AMV polymerase, was free of detectable RNas H activity. As observed for DNA synthesis, the stimulatory protein specifically affected the AMV RNase H activity and had no effect on the RNase H activity of *E. coli* under identical conditions.

Specificity of Action of AMV and *Escherichia coli* RNase H

A wide range of synthetic polymers are degraded in the presence of the corresponding complementary polydeoxynucleotide by either enzyme

Table 2. Influence of AMV stimulatory protein on degradation of poly(A) · poly(dT)

Additions	[³H]poly(A) rendered acid-soluble (pmoles/30 min)
AMV polymerase	56
AMV stimulatory protein	2
Polymerase + stimulatory protein	220
Polymerase + stimulatory protein*	56
E. coli RNase H	41
E. coli RNase H + AMV stimulatory protein	29

RNase H activity was measured as follows: reaction mixture (0.05 ml) containing 1 μmole Tris · HCl, pH 8.0, 0.5 μmole of $MgCl_2$, 0.3 μmole of DTE, 2 μg of bovine serum albumin, 0.52 nmole of poly(dT), 0.56 nmole of [³H]poly(A), polymerase (AMV, 0.22 unit or *Escherichia coli* RNase H 43 ng) and stimulatory protein (22.8 mμg), as indicated, were incubated for 30 min at 38°C. Reactions were stopped by addition of cold 0.1 ml of 0.1 sodium pyrophosphate, 0.4 mg of albumin, 52 nmoles of denatured salmon sperm DNA, and 0.5 ml of 5% trichloroacetic acid. The reaction mixtures were centrifuged at 2000 rpm in an international centrifuge and the supernatant collected and counted in 10 ml of Bray's scintillation fluid.
*Predigested with Pronase.

Table 3. Degradation of various homopolymers by RNase H from *Escherichia coli* and AMV

	Acid-soluble ³H produced by	
Substrate added	AMV RNase H (pmoles/30 min)	E. coli RNase H (pmoles/30 min)
[³H]poly(A) alone	3.1	6
[³H]poly(A) + poly(U)	<1	4.5
[³H]poly(A) + poly(dT)	265	144
[³H]poly(U) alone	<1	6
[³H]poly(U) + poly(A)	<1	<1
[³H]poly(U) + poly(dA)	<1	93
[³H]poly(C) alone	8	6
[³H]poly(C) + poly(I)	<1	<1
[³H]poly(C) + poly(dG)	110	254
[³H]poly(C) + poly(G)	<1	1.2

Purified AMV polymerase II (0.22 unit) and stimulatory protein (22.8 mμg) or *E. coli* RNase H (43 mμg) were incubated with various homopolymers as described in Table 2. With the AMV polymerase, the polymers were: [³H]poly(A) (32 cpm/pmole, 0.56 nmole), poly(U) (2.2 nmoles), and poly(dT) (0.52 nmole); [³H]poly(U) (5 cpm/pmole, 0.95 nmole), poly(A) (1 nmole), and poly(dA) (0.95 nmole); [³H]poly(C) (4.6 cpm/pmole, 0.84 nmole), poly(I) (0.9 nmole), poly(dG) (0.36 nmole), and poly(G) (0.59 nmole). Polymers added with *E. coli* RNase H were: [³H]poly(A) (1.19 nmoles), poly(dT) (1 nmole), poly(U) (0.94 nmole); [³H]poly(U) (0.95 nmole), poly(A) (1 nmole), poly(dA) (0.95 nmole); [³H]poly(C) (2.5 nmoles), poly(I) (3.3 nmoles), poly(dG) (2.5 nmoles), and poly(G) (2.85 nmoles).

as summarized in Table 3. Also shown in Table 3 is the specificity of both enzyme preparations. There is little hydrolysis of RNA in the absence of the complementary DNA strand or the presence of the complementary RNA strand. The only significant difference between the two enzymes appears to be the inability of the AMV RNase H to cleave [^{3}H]poly(U) · poly(dA). This specificity of the AMV enzyme is not understood but may have important consequences when the enzyme reaches regions rich in uridine. In addition to the substrates listed in Table 3, [^{3}H]poly(I) · poly(dC) and RNA polymerase products formed with single stranded circular fd DNA are attacked by both nucleases. In the experiments with RNA polymerase, RNA · DNA hybrids were isolated by isopycnic banding in Cs_2SO_4. The RNA in the hybrids was susceptible to digestion (85 to 90%) by either RNase H preparation. The RNase H-resistant RNA was susceptible to digestion by pancreatic RNase in the presence of 0.25 M NaCl. These experiments indicate that these enzymes can be used as reagents for quantitative estimation of RNA · DNA hybrids.

**Products
of the Reaction**

The products formed from poly(A) · poly(dT) after exhaustive digestion (95% of poly(A) was rendered acid-soluble) in the presence of excess AMV RNase H have been characterized by the following experiments. The acid-soluble products were chromatographed on DEAE-cellulose in 8 M urea. The eluted nucleotides appeared as a series of well defined peaks ranging from di- to octanucleotides in length. The radioactivity was distributed as follows: di-, 22%; tri-, 8.5%; tetra-, 12%; penta-, 15%; hexa-, 16%; hepta-, 17%; and octa-, 9%. AMP was not detected in the acid-soluble products. A similar analysis has been made by paper electrophoresis at pH 3.5. The hepta-oligonucleotide fraction was digested to AMP by the action of snake venom diesterase, suggesting the presence of free 3'-hydroxyl ends. Treatment of the hexa-oligonucleotide with bacterial alkaline phosphatase (BALP) decreased the negative charge on the oligonucleotide and the resulting oligoadenylate comigrated with unlabeled standard $(Ap)_5 A$ on electrophoresis. These results suggest that the oligoadenylate products possess free 3'-OH and 5'-phosphate termini. The acid-soluble products formed with *E. coli* RNase H have been identified by similar procedures and consist of oligonucleotides size 2 to 6 with only 4% AMP. It is not clear at present if the production of AMP is a direct consequence of RNase H action or a result of trace levels of another enzyme system.

Mode of Action of RNase H

In order to determine whether RNase H acts as an exo- or endonuclease, poly(A) was labeled at the 5' terminus with [γ-^{32}P]ATP and polynucleotide kinase and the 3' end was extended by treatment with [^{3}H]ADP and polyncudleotide phosphorylase. If either RNase H is an exonuclease then a preferential release of the label from one end or the other should occur. However, if label from both ends is released at the same rate then the enzyme could act either endonucleolytically or processively from an end. The results of experiments with doubly labeled poly(A) · poly(dT) are summarized in Fig. 6. As controls, the doubly labeled poly(A) · poly(dT) was treated with snake venom diesterase and DNA polymerase I. Venom diesterase, a 3' $\rightarrow$ 5' exonuclease, preferentially releases tritium from the 3' end. E. coli DNA polymerase I has a 5' $\rightarrow$ 3' RNase H activity associated with it. This activity was noted during the purification of RNase H from E. coli Pol A$^+$ strains. The activity is insensitive to NEM, coincides chromatographically with DNA polymerase activity measured with activated salmon sperm DNA, and is neutralized by antibodies to DNA polymerase I. With either E. coli or AMV RNase H preparations both ^{3}H and ^{32}P labels were released at the same rate. We conclude from these experiments that these RNase H activities might act endonucleolytically, exonucleoytically from both ends, or processively.

These mechanisms can be differentiated by using an RNA without ends. Poly(A) (40 ncucleotides in length) labeled at the 5' end with ^{32}P was treated with T4 RNA ligase, an enzyme which generates circular poly(A) products (21). After the reaction, unreacted poly(A) was removed by degradation with the 3' $\rightarrow$ 5' exonuclease, RNase II (20). The remaining poly(A) is circular in structure and contains ^{32}P in an internal phosphodiester bond insusceptible to attack by bacterial alkaline phosphatase or RNase II (Table 4). E. coli RNase H degrades the circular poly(A) to an acid-soluble form in the presence of poly(dT), while AMV RNase H has no effect (Table 4). Since circular poly(A) has no available ends we conclude that E. coli RNase H is an endonuclease and that the AMV enzyme is probably a processive nuclease. Consistent with this latter interpretation is the following experiment. [^{3}H]poly(A) · poly(dT) was treated with the AMV RNase H until 0, 15, and 60% of the poly(A) was rendered acid-soluble. The reaction mixtures were treated with 10% HCHO at 70°C for 10 min and then analyzed on HCHO-sucrose gradients. The labeled poly(A) sedimented either as untreated poly(A) or as oligonucleotides at the top of the gradient. No large oligonucleotide intermediates were observed in the gradients.

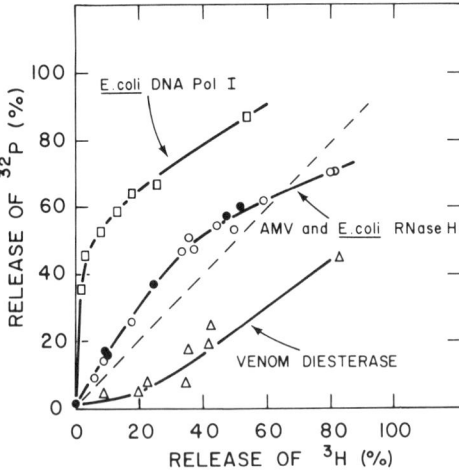

Fig. 6. [5'-^{32}P], [3'-^{3}H]poly(A) (1.1 mμmoles) and poly(dT) (1.04 mμmoles) were incubated under the conditions described in Table 2 for various lengths of time with each of the following enzymes: AMV polymerase (0.2 unit) and stimulatory protein (22.8 mμg), *Escherichia coli* RNase H (21 and 43 ng), snake venom phosphodiesterase (0.015 to 0.15 unit) treated by the method of Sulkowski and Laskowski (26) or *E. coli* DNA polymerase I (13) (0.12 to 1.2 units, Fraction VII) and the amount of acid-soluble ^{3}H and ^{32}P measured. The differentially labeled poly A (500 to 1000 nucleotides) in length was prepared as follows: 5'-hydroxyl-terminated poly(A) (0.18 μmole of nucleotide) prepared by treatment with BALP (2.8 units) was labeled with ^{32}P in a reaction mixture (1 ml) containing 50 μmoles of Tris · HCl, pH 8.0, 3 μmoles of sodium phosphate, 3 μmoles of DTE, 75 μg of albumin, 5 μmoles of MgCl$_2$, [γ-^{32}P] ATP (3240 cpm/pmole, 0.012 μmole), and polynucleotide kinase (13 units). The mixture was incubated at 38°C for 15 min after which time an additional 13 units of kinase, 0.012 μmole of [γ-^{32}P]ATP and 3 μmoles of DTE were added. Quantitative phosphorylation was achieved after a total of 30 min of incubation. The reaction was stopped by extraction with phenol (saturated with 1 M Tris · HCl, pH 7.5) and three times with ether. Unreacted ATP was removed by gel filtration on Sephadex G-50. [^{3}H]AMP was added to the 3' end of the [5'-^{32}P]poly (A) by treatment with polynucleotide phosphorylase as follows: the [5'-^{32}P]poly (A) dried *in vacuo* was dissolved in a reaction mixture (0.05 ml) containing Tris · HCl, pH 8.0, 5 mμmoles; MgCl$_2$, 0.5 μmole; albumin, 3.7 μg; [^{3}H]ADP (1.2 μmoles, 34 cpm/pmole); and polynucleotide phosphorylase (22) (*M. lysodeikticus*, 0.07 unit). The reaction was stopped after incubation for 20 min at 38°C by heating to 80°C for 4 min. The unreacted [^{3}H]ADP was removed by gel filtration on Sephadex G-50. *E. coli* DNA Pol I (□——□); AMV RNase H (○——○); *E. coli* RNase (H) (●——●); snake venom diesterase (△——△).

Table 4. Degradation of circular poly(A) by RNase H preparation.

Additions	Acid-soluble ^{32}P (%)
1. Circular poly(A)	<1
2. Circular poly(A) + AMV RNase H	<1
3. Circular poly(A) + AMV RNase H + poly(dT)	<1
4. Circular poly(A) + E. coli RNase H	5
5. Circular poly(A) + E. coli RNase H + poly(dT)	81
6. Circular poly(A) + RNase II (0.1 unit)	<1
7. Circular poly(A) + BALP (0.70 unit)	<1

Purified AMV polymerase (0.2 unit) and stimulatory protein (22.8 ng) or *Escherichia coli* RNase H (24 ng) was incubated with circular [^{32}P]poly(A) (2000 cpm, 0.35 mµmole) · poly(dT) (1 mµmole) as described in Table 2 except that acid-insoluble radioactivity was determined. Circular poly(A) · poly(dT) was shown to be in hybrid structure by banding density in Cs_2SO_4. Circular poly(A) was formed as follows: a reaction mixture (0.1 ml) containing 7 nmoles of [5'-^{32}P]poly(A) (0.2 nmole of 5'-^{32}P termini, 1420 cpm/pmole), 5 µmoles of Tris · HCl, pH 7.5, 1.0 µmole of $MgCl_2$, 10 nmoles of ATP, 5 µg of albumin, 5 µmoles of dithiothreitol, and RNA ligase (21) (0.04 unit) was incubated for 30 min at 38°C. The ligase reaction was terminated by heating at 100°C for 2 min after which 0.05-ml aliquots were added to a reaction mixture (0.1 ml final volume) containing 5 µmoles of Tris · HCl, pH 8.0, 10 µmoles of KCl, 7 µg of albumin, 0.2 µmoles of $MgCl_2$, and RNase II (0.1 unit). After incubation for 30 min at 30°C an additional amount of RNase II (0.1 unit) was added and the incubation was continued for another 10 min. The reaction was stopped by heating at 65°C for 4 min. Twenty per cent of the ^{32}P label remained acid-insoluble after treatment with RNase II.

Discussion

The present studies demonstrate an important role of an auxiliary protein in RNA and DNA transcription by the reverse transcriptase. The activity of this stimulatory protein was followed during purification by measuring its ability to increase the yield of DNA synthesis in reactions containing AMV RNA and highly purified polymerase preparations. The purified stimulatory protein fractions increase both the rate and yield of DNA synthesis by the polymerase. At the maximally stimulated rate, the yield of DNA synthesis relative to RNA added reached 40%.

In addition to increasing the rate of RNA directed reactions, the stimulatory protein increased the rate of DNA synthesis in reactions primed with d(AT) copolymer or λ DNA which had been treated extensively with exonuclease III. The latter observation is similar to those made with the *E. coli* unwinding protein in the presence of *E. coli* DNA polymerase II (Alberts and coworkers, personal communication) and the T4 gene 32 protein with T4 DNA polymerase (11). Like these bacterial proteins, the avian stimulatory protein appears to affect its homologous polymerase specifically. It should be pointed out that unlike the T4 gene 32 protein (11), the AMV stimulatory protein permits the AMV polymerase to utilize DNA containing single strand breaks.

A disturbing feature of the reverse transcriptase reaction has been the relatively small size of the DNA polymers produced (5-7 S). With highly purified enzyme, free of RNase and DNase, the products formed with RNA should depend upon the length of template regions transcribed. Previous studies on DNA priming indicated that the AMV polymerase was ineffective in repairing extensive single stranded regions and in this respect resembled DNA polymerase II and other DNA polymerases (11, 33). If this observation is applicable to RNA, the extensive single stranded regions of tumor viral RNAs should be copied to a limited extent. In the presence of the AMV stimulatory protein, however, the size of the DNA product should increase. A detailed study of the influence of the stimulatory protein on the size of the DNA product is now under investigation.

DNA products formed with viral RNA and purified AMV polymerase band at densities in Cs_2SO_4 (or CsCl) expected of RNA, indicating hybrid structures containing primarily RNA. When products were prepared in the presence of the stimulatory protein and analyzed by isopycnic banding, DNA appeared in both the RNA and DNA regions. The product which appeared in the DNA region after further analysis by isopycnic banding in CsCl was found to contain covalently attached RNA fragments. Thus the appearance of DNA covalently linked to small pieces of RNA and free of the 60 S RNA complex is mediated by the action of the stimulatory protein. We believe that these small RNA fragments are hydrogen-bonded to the 60 S AMV RNA. How these fragments arise is unclear, but we have demonstrated with our RNA preparations that DNA synthesis is initiated at 12 to 15 3'-OH ends per 60 S RNA as measured by nearest neighbor frequency experiments. Double stranded DNA has been found among the products of the reverse transcriptase isolated from virions (6, 8, 9, 18, 24, 25). We have observed the formation of some double stranded DNA with purified polymerase and AMV RNA as reported by Bishop and coworkers (personal communication). In addition, DNA prepared by alkaline digestion of RNA · DNA hybrids supports deoxynucleotide incorporation with purified AMV polymerase preparations. Most of the product of this reaction is resistant to digestion by nuclease S1 before and after heating. Since nuclease S1 specifically attacks single stranded DNA, these results suggest that double stranded and nondenaturable DNA products most likely arise through intramolecular hairpin structure as postulated in the DNA polymerase I system (15). The denaturable duplex DNA generated by permeabilized virions could arise through endonuclease action on nondenaturable DNA products.

Preparations of AMV polymerase purified in our laboratory by ammonium sulfate fractionation and chromatography on phosphocellulose have detectable RNase H activity as reported by Mölling et al. (19) and Baltimore (personal communication). Though RNase H is intimately associated with reverse transcriptase during purification, its activity does not coincide exactly with viral polymerase activity. This observation coupled with differences in the properties of the two enzyme activities suggests that they do not share a single common active site. Whether they are part of one enzyme molecule that is composed of different subunits or are present in physically separated proteins remains to be elucidated.

The avian RNase H activity rapidly hydrolyzes RNA in RNA · DNA hybrids formed with RNA polymerase and fd DNA as well as a wide variety of ribohomopolymers in the presence of their complementary deoxyribopolymers. A notable exception is the homopolymer pair [^{3}H]poly(U) · poly(dA) which is resistant to digestion by the avian nuclease, though susceptible to digestion by RNase H purified from E. coli. [^{3}H]Poly(A) in the duplex poly(A) · poly(dT) is quantitatively rendered acid-soluble by the action of the avian nuclease. The products of the reaction have been identified as a series of oligonucleotides containing between 2 and 8 adenylate residues with 3'-OH and 5'-phosphate termini. No AMP is detected among the reaction products. Poly(A), differentially labeled at the 3' and 5' ends, is degraded without the preferential release of one label over the other. Similar results have been obtained by Baltimore (personal communication). Although the degradation products are oligonucleotides, the nuclease appears to be processive in action. It shows an absolute requirement for ends of RNA chains for activity as shown by its inability to digest circular poly(A), prepared with RNA ligase. In addition, uniformly labeled poly(A) is degraded to an acid-soluble form without producing detectable large oligonucleotide fragments as intermediates. At this time, the polarity of exonucleolytic action by this enzyme, i.e. 5' → 3' or 3' → 5', is unknown.

Despite the fact that RNase H generates 3'-OH termini which are potential primer sites for the viral polymerase, the requirement for ends of RNA chains for activity suggests a limited role for this activity during the transcription of *intact* viral 60 S RNA. RNA in RNA · DNA hybrids present in internal positions will not be cleaved by the avian RNase H. However, the nuclease has the potential to provide a mechanism for release of DNA once transcription is completed and an RNA terminus becomes available for exonuclease action. Since DNA chains are synthesized in a 5' → 3' direction by the polymerase, if RNase H is to play any role (coupled only to the polymerase) we would predict that the RNase H must attack RNA in an RNA · DNA hybrid in a 5' → 3' direction. Experiments to define the direction of hydrolysis are in progress.

Acknowlegments

We gratefully acknowledge the folowing generous gifts: AMV, Dr. J. Beard, Duke University; [^{3}H]thymidine-labeled colicin El DNA, Dr. M. Wright; [^{3}H]thymidine-labeled λ DNA treated with exonuclease III, Dr. R. Wickner; HeLa cell DNA polymerase, Dr. A. Weissbach, Roche Institute of Molecular Biology; T4 DNA ligase and RNA ligase-treated poly(A), Doctors R. Silber and V. Malathi; antisera to DNA polymerase I, Dr. A. Kornberg, Stanford University; *Eschichia coli* RNase II and polynucleotide phosphorylase (*Micrococcus lysodeikticus*) Dr. M. Singer, National Institutes of Health. We are indebted to Dr. D. Baltimore for informing us of his studies on RNase H prior to publication and Dr. B. Alberts for providing us with the purification procedure of *E. coli* unwinding protein prior to publication. We wish to thank Mr. Robert Santini for his excellent technical assistance. This work was supported by grants from the Special Virus Cancer Program of the National Cancer Institute, National Institutes of Health (71-2251); National Institutes of Health (GM-13344); and the American Cancer Society (NP-89-1). J. P. L. is a fellow of the Damon Runyon Cancer Foundation (DRF 659). I. B. is a Medical Scientist trainee.

References

1. Baltimore, D. 1970. RNA-dependent DNA polymerase in virions of RNA tumour viruses. *Nature* 226: 1209.

2. Brutlag, D. L., Atkinson, M. R., Setlow, P., and Kornberg, A. 1969. An active fragment of DNA polymerase produced by proteolytic cleavage. *Biochem. Biophys. Res. Commun.* 37: 982.

3. Cozzarelli, N. R., Kelly, R. B., and Kornberg, A. 1968. A minute circular DNA from *Escherichia coli* 15. *Proc. Nat. Acad. Sci. U. S. A.* 60: 992.

4. Duesberg, P., Helm, K. V. D., and Canaani, E. 1971. Comparative properties of RNA and DNA templates for the DNA polymerase of Rous sarcoma virus. *Proc. Nat. Acad. Sci. U.S.A.* 68: 2505.

5. Duesberg, P., Helm, K. V. D., and Canaani, E. 1971. Properties of a soluble DNA polymerase isolated from Rous sarcoma virus. *Proc. Nat. Acad. Sci. U. S. A.* 68: 747.

6. Faras, A., Fanshier, L., Garapin, A., Levinson, W., and Bishop, J. M. 1971. Deoxyribonucleic acid polymerase of Rous sarcoma virus. Studies on the Mechanism of double-stranded deoxyribonucleic acid synthesis. *J. Virol.* 7: 539.

7. Fridlender, B., Fry, M., Bolden, A., and Weissbach, A. 1972. A new synthetic RNA-dependent DNA polymerase from human tissue culture cells. *Proc. Nat. Acad. Sci. U. S. A.* 69: 452.

8. Fujinaga, K., Parsons, J. T., Beard, J. W., Beard, D., and Green, M. 1970. Mechanism of carcinogenesis by RNA tumor viruses, III. Formation of RNA

· DNA complex and duplex DNA molecules by the DNA polymerase(s) of avian myeloblastosis virus. *Proc. Nat. Acad. Sci. U. S. A.* 67: 1432.

9. Garapin, A., Fanshier, L., Leong, J., Jackson, J., Levinson, W., and Bishop, J. M. 1971. Deoxyribonucleic acid polymerases of Rous sarcoma virus. Kinetics of deoxyribonucleic acid synthesis and specificity of the products. *J. Virol.* 7: 227.

10. Hausen, P., and Stein, H. 1970. Ribonuclease H. An enzyme degrading the RNA moiety of DNA-RNA hybrids. *Eur. J. Biochem.* 14: 278.

11. Huberman, J., Kornberg, A., and Alberts, B. 1971. Stimulation of T4 bacteriophage DNA polymerase by the protein product of T4 gene 32. *J. Mol. Biol.* 62: 39.

12. Hurwitz, J., and Leis, J. P. 1972. RNA-dependent DNA polymerase activity of RNA tumor viruses. I. Directing influence of DNA in the reaction. *J. Virol.* 9: 116.

13. Jovin, T. M., Englund, P. T., and Bertsch, L. L. 1969. Enzymatic synthesis of deoxyribonucleic acid. XXVI. Physical and chemical studies of a homogeneous deoxyrobonucleic acid polymerase. *J. Biol. Chem.* 244: 2996.

14. Klenow, H., and Overgaard-Hansen, K. 1970. Proteolytic cleavage of DNA polymerase from *Escherichia coli* B into an exonuclease unit and a polymerase unit. *FEBS Lett.* 6: 25.

15. Kornberg, A. 1969. Active center of DNA polymerase. *Science* 163: 1410.

16. Kornberg, T., and Gefter, M. 1971. Purification and DNA synthesis in cell-free extracts: Properties of DNA polymerase II. *Proc. Nat. Acad. Sci. U. S. A.* 68: 761.

17. Leis, J., and Hurwitz, J. 1972. RNA-dependent DNA polymerase activity of RNA tumor viruses. II. Directing influence of RNA in the reaction. *J. Virol.* 9: 130.

18. Manly, K., Smoler, D. F., Bromfeld, E., and Baltimore, D. 1971. Forms of deoxyribonucleic acid produced by virions of the ribonucleic acid tumor viruses. *J. Virol.* 7: 106.

19. Mölling, K., Bolognesi, D. P., Bauer, H., Büsen, W., Plassmann, H. W., and Hausen, P. 1971. Association of viral reverse transcriptase with an enzyme degrading the RNA moiety of RNA-DNA hybrids. *Nature New Biol.* 234: 240.

20. Nossal, N. G., and Singer, M. F. 1968. The processive degradation of individual polyribonucleotide chains. I. *Escherichia coli* ribonuclease II. *J. Biol. Chem.* 243: 913.

21. Silber, R., Malathi, V. G., and Hurwitz, J. 1972. Reaction at ends of RNA chains. I. Purification and properties of T4 induced RNA ligase. *Proc. Nat. Acad. Sci. U. S. A.* 69: 3009.

22. Singer, M. 1966. Polynucleotide phosphorylase. In Cantoni, G. L., and Davies, D. R. (eds.) *Procedures in nucleic acid research*, p. 245, Harper and Row, New York.

23. Smoler, D., Molineux, I., and Baltimore, D. 1971. Direction of polymerization by the avian myeloblastosis virus deoxyribonucleic acid polymerase. *J. Biol. Chem.* 246: 7697.

24. Spiegelman, S., Burny, A., Das, M. R., Keydar, J., Schlom, J., Travnicek, M., and Watson, K. 1970. Characterization of the products of RNA-directed DNA polymerase in oncogenic RNA viruses. *Nature* 227: 563.
25. Spiegelman, S., Burny, A., Das, M. R., Keydar, J., Schlom, J., Travnicek, M., and Watson, K. 1970. DNA-directed DNA polymerase activity in oncogenic RNA viruses. *Nature* 227: 1029.
26. Sulkowski, E., and Laskowski, M. 1971. Inactivation of 5'-nucleotidase in commercial preparations of venom exonuclease (phosphodieterase). *Biochim. Biophys. Acta* 240: 443.
27. Taylor, J. M., Faras, A. J., Varmus, H. E., Levinson, W. E., and Bishop, J. M. 1972. Ribonucleic acid directed deoxyribonucleic acid synthesis by the purified deoxyribonucleic acid polymerase of Rous sarcoma virus. Characterization of the enzymatic product. *Biochemistry* 11: 2343.
28. Temin, H. M., and Mizutani, S. 1970. RNA-dependent DNA polymerase in virons of Rous sarcoma virus. *Nature* 226: 1211.
29. Verma, I. M., Meuth, N. L., Bromfeld, E., Manly, K. F., and Baltimore, D. 1971. Covalently linked RNA-DNA molecule as initial product of RNA tumour virus DNA polymerase. *Nature New Biol.* 233: 131.
30. Weiss, B., and Richardson, C. C. 1967. Enzymatic breakage and joining of deoxyribonucleic acid. I. Repair of single-strand breaks in DNA by an enzyme system from *Escherichia coli* infected with T4 bacteriophage. *Proc. Nat. Acad. Sci. U. S. A.* 57: 1021.
31. Weissbach, A., Schlabach, A., Fridlender, B., and Bolden, A. 1971. DNA polymerases from human cells. *Nature New Biol.* 231: 167.
32. Wickner, R. B., Ginsberg, B., Berkower, I., and Hurwitz, J. 1972. Deoxyribonucleic acid polymerase II of *Escherichia coli*. I. The purification and characterization of the enzyme. *J. Biol. Chem.* 247: 489.
33. Wickner, R. B., Ginsberg, B., and Hurwitz, J. 1972. Deoxyribonucleic acid polymerase II of *Escherichia coli*. II. Studies of the template requirements and the structure of the deoxyribonucleic acid product. *J. Biol. Chem.* 247: 498.

Discussion

Faras. You didn't mention anything about the size of the DNA product with 70 S RNA as template.
Leis. In our hands the size of the DNA product with the purified enzyme in the absence of stimulatory protein is small, that is, possessing sedimentation coefficients between 6 and 7 S on alkaline sucrose gradients. The analysis of the size of the products in the presence of stimulatory protein is in progress.
Faras. What about the λ DNA?
Leis. The experiments with λ DNA indicate that in the presence of the stimulatory protein, longer regions of DNA are being transcribed.

RNA-DNA Bonds Formed by DNA Polymerases from Bacteria and RNA Tumor Viruses

R. M. Flügel, J. E. Larson, P. F. Schendel,
R. W. Sweet,* T. R. Tamblyn, and R. D. Wells

Department of Biochemistry
College of Agricultural and Life Sciences
University of Wisconsin
Madison, Wisconsin 53706

Comparative studies were performed on the enzymatic mechanisms of the highly purified DNA polymerases from *Escherichia coli* and *Micrococcus luteus* and of a partially purified DNA polymerase from virions of avian myeloblastosis virus (AMV). Some important features are: (*a*) with polymeric templates-primers, the three enzymes are approximately equally effective "reverse transcriptases"; (*b*) all three enzymes require both a primer and a template; and (*c*) all three enzymes readily form RNA-DNA bonds.

RNA-DNA bonds are formed *in situ* by the AMV DNA polymerase. The endogenous reaction was performed with ether-disrupted virions in order to preserve the *in vivo* template-primer relationship insofar as possible. Of the 16 possible dinucleotide sequences at the RNA-DNA junction, the predominant linkage is -rU-dC-. A small amount of -rA-dA- linkage is also observed. When NP-40-disrupted virions are used instead of ether-disrupted particles, only the -rA-dA- bond is formed.

RNase H and RNase activities are present in ether-disrupted virus. Studies indicate that RNase can influence the types of RNA-DNA bonds which are formed when steps are taken to express this activity. However, RNase H does not influence the types of RNA-DNA linkages which are formed.

Appreciable differences between ether-disrupted virions and NP-40-disrupted virions are documented.

*Present address, Columbia University, College of Physicians and Surgeons, Institute of Cancer Research, New York, N.Y. 10032.

Mechanism of Avian Myeloblastosis Virus DNA Polymerase

Polymeric Templates-Primers

Since DNA polymerase was first recognized to be associated with RNA tumor viruses 2 years ago, a number of studies have been reported on the purification and mechanisms of these enzymes (for a review see references 13, 33, 35). The initial emphasis of some of our work was to perform a comparative study on the highly purified DNA polymerases from *Escherichia coli* and *Microccocus luteus*, and on a partially purified DNA polymerase from avian myeloblastosis virus (AMV). We set out to ascertain if the "reverse transcriptase" had truly unique properties in terms of its enzymatic mechanism, or if it was similar to the well characterized bacterial DNA polymerases.

We have recently reported (35) the capacity of the three DNA polymerases to utilize 20 different high molecular weight templates (Table 1). The templates employed were double stranded DNAs, double stranded RNAs, double stranded DNA-RNA hybrids, and single stranded polynucleotides. The studies showed that the three polymerases have approximately the same capacity to utilize the templates-primers.

Our conclusions with the high molecular weight templates were the following: (*a*) double stranded DNAs are more active than single stranded DNAs with all three enzymes; (*b*) double stranded RNAs are weakly active and single stranded RNAs are ineffectual as templates; (*c*) double stranded DNAs are more active than homologous double stranded RNAs; (*d*) for the hybrid polymers, the poly(r-purine) · poly(d-pyrimidine) is more active than the poly(d-purine) · poly(r-pyrimidine) in virtually every case; (*e*) for the DNA-RNA hybrids, the DNA strand serves as a template at least as well, or better, than the RNA strand; (*f*) for the homopolymeric

Table 1. Polymers used as template-primers for the DNA polymerases of *Escherichia coli*, *Micrococcus luteus*, and avian myeloblastosis virus

dA · dT	dG · dC
rA · dT	rG · dC
dA · rU	dG · rC
rA · rU	rG · rC
(dA-dT) · (dA-dT)	(dG-dC) · (dG-dC)
(rA-rU) · (rA-rU)	(rG-rC) · (rG-rC)
dA	dG
dT	dC
rA	rG
rU	rC

DNA-DNA duplexes, the poly(d-pyrimidine) strand is generally synthesized to a lesser extent than the poly(d-purine) strand. Thus these studies clearly implied that RNA may serve as a primer, or a template, or both, for DNA synthesis.

Polymer-Oligomer Template-Primers

On the other hand, the AMV DNA polymerase shows a dramatic difference in template-primer specificity from the *M. luteus* DNA polymerase when certain polymer-oligomer mixtures are employed. We recently reported (35) that the *M. luteus* DNA polymerase could use either poly(dA) · oligo(dT) or poly(rA) · oligo(dT) as template-primer for dNTP incorporation (Fig. 1). On the other hand, the AMV DNA polymerase effectively uses only poly(rA) · oligo(dT) and not poly(dA) · oligo(dT) (Fig. 1). It should be noted that poly(dA) · oligo(dT) is a surprisingly ineffective template-primer for the AMV DNA polymerase. The difference in rates is not due to poly(rA) · oligo(dT) serving as an extraordinarily good template-primer.

Another difference between the two enzymes is that when dATP was supplied in addition to dTTP as substrate with the AMV polymerase,

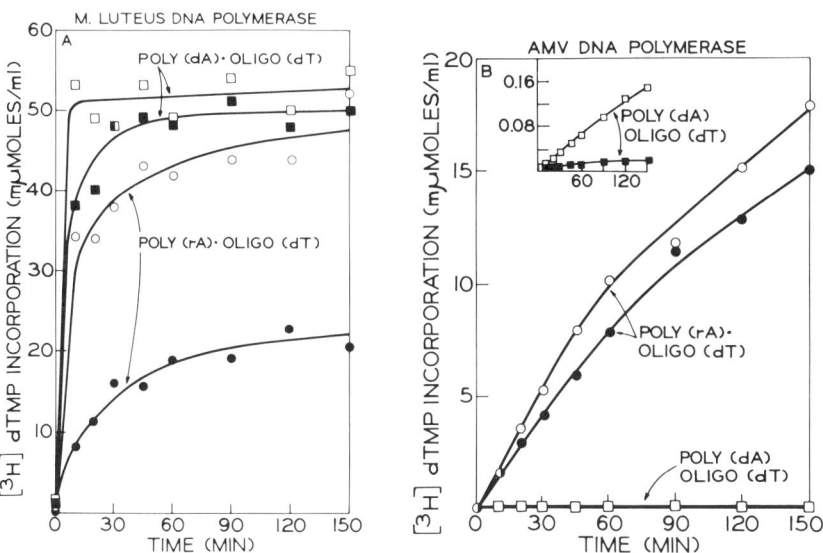

Fig. 1. Kinetics of $(dT)_n$ synthesis with oligo(dT) as a primer and either $(dA)_n$ or $(rA)_n$ as template. For both parts A and B, reactions were performed with $(dA)_n$, 43 μM (□, ■) and $(rA)_n$, 44 μM (○, ●) at two levels of $(pT)_{13}$ which were 32 μM (open symbols) and 1.1 μM (filled symbols). Note the expanded scale of the insert in B. Other reaction conditions were as described (35) and included buffer and salts, dTTP, nucleic acids (see above), and DNA polymerase.

no synthesis of poly(dA) was observed with either mixture of templates. However, when dATP was supplied in addition to dTTP with the *M. luteus* DNA polymerase, poly(dA) was formed with both mixtures of template-primers.

It is not clear, mechanistically, why the AMV DNA polymerase should use poly(dA) · oligo(dT) so poorly as a template-primer mixture. One possible explanation could be the presence of an associated deoxyribonuclease which is specific for double stranded DNA and not DNA-RNA hybrids. If this is the case, the DNase activity must be a very specific one. We had previously determined that our AMV DNA polymerase preparations had no detectable nuclease activity which could hydrolyze poly(dA-dT) · poly(dA-dT) (data not shown). Thus, if a DNase activity is responsible for the poor templating activity of poly(dA) · oligo(dT), it must act only on oligodeoxynucleotides to destroy the primer or on polypyrimidine sequences in the newly synthesized DNA strand. To test these possibilities, the fate of ^{3}H-labeled oligo(dT)$_8$ (6×10^4 cpm/nmole) primer was followed chromatographically during the poly(dA) · oligo(dT) or poly(rA) · oligo(dT) reactions. In both cases, the oligo(dT) primer was not degraded during the course of the experiment (1 hr). The fate of the newly synthesized poly(dT) strand was also followed chromatographically to determine whether labeled dTTP was incorporated into poly(dT) and immediately hyrolyzed again to give dTMP but no acid-precipitable radioactivity. This was also not the case. The poly(rA) · oligo(dT)-templated reaction showed transfer of counts from dTTP to high molecular weight poly(dT), but the poly(dA) · oligo(dT)-templated reaction gave only dTTP and no dTMP or polymer during the 1-hr reaction.

One unique feature of the AMV DNA polymerase is the absence of the $5' \rightarrow 3'$ exonuclease which was found in all bacterial DNA polymerases studied (15, 20). Thus we tested whether this activity could allow bacterial DNA polymerases to use the poly(dA) · oligo(dT) template more efficiently than the AMV polymerase. This alternative was ruled out by finding that both the large fragment of *E. coli* DNA polymerase I and *E. coli* DNA polymerase II, neither of which contain the $5' \rightarrow 3'$ exo activity, still use the poly(dA) · oligo(dT) template very well and in fact prefer it to poly(rA) · oligo(dT) (results not shown).

We still cannot rule out the possibility that the $3' \rightarrow 5'$ exonuclease activity associated with all bacterial DNA polymerases, but not detectable in the AMV polymerase, plays some role in the unique template specificity of tumor virus enzymes. We are currently preforming a systematic study on a variety of polymer-oligomer mixtures as template-primers for DNA polymerases in an attempt to clarify this problem. A recent paper (8) has reported studies similar to those presented above.

DNA Polymerase Mechanisms: Summary

In summary, we know of no fundamental mechanistic difference between the action of the two bacterial DNA polymerases and the AMV DNA polymerase (35). A similar conclusion has been reached by other workers (13). On the basis of our studies with high molecular weight synthetic templates-primers, we concluded that the bacterial DNA polymerases are at least as effective "reverse transcriptases" as the AMV DNA polymerase. This observation should not be construed as implying that the bacterial DNA polymerases necessarily carry out "reverse transcription" *in vivo*. However our results (35) and those of others (2, 13) make this a possibility. In addition, this statement should not be taken to imply that the DNA polymerases from RNA tumor viruses are unrelated to the process of cellular transformation. In fact, available evidence suggests the contrary (see review of Temin in this book). The biological process of reverse transcription may be controlled by a variety of regulatory factors. Our conclusion is based on enzymatic mechanisms alone.

Mechanism of Initiation

Background Studies

The studies described above strongly indicated that RNA can serve as a primer for *in vitro* DNA synthesis. The well characterized DNA polymerases from bacterial sources had previously been shown to initiate DNA synthesis by elongating a primer molecule. The *de novo* initiation of a new strand is either a very rare event or never occurs (1, 9, 10). Thus, we devised an assay to determine whether RNA-DNA covalent bonds could be formed *in vitro* (Fig. 2). This assay permits detection of a single internucleotidic bond at the RNA-RNA junction.

We recently reported (35) that both the *M. luteus* DNA polymerase and the AMV DNA polymerase readily form RNA-DNA covalent bonds. Hence, mechanistically it is possible for RNA to serve as a primer for DNA synthesis. Although these studies demonstrated that the AMV DNA polymerase could covalently join a newly synthesized strand onto a primer molecule, it was at least feasible that the enzyme could initiate *de novo* synthesis of a new strand. However, this possibility was ruled out with experiments with γ-^{32}P-dGTP in a poly(dG) · poly(dC)-templated reaction. No *de novo* initiation was detected in this reaction (35).

Hence it appears that the mechanism of initiation of *in vitro* DNA synthesis with the RNA tumor virus DNA polymerase is similar to that

```
     A      C      C      G      U      C      T      A      G
    ⊢OH   ⊢OH    ⊢OH    ⊢OH   ⊢OH    ⊢H     ⊢H    ⊢H     ⊢H
 ---\   \      \      \      \ 32P   \      \      \      \---
     P     P      P      P            P      P      P
```

     1. ALKALINE HYDROLYSIS
     2. CHROMATOGRAPHY

     U
    ⊢O⎫ ^{32}P + NONRADIOACTIVE 3' (2') RIBOMONOPHOSPHATES
    ⊢O⎭   + UNDEGRADED ^{32}P–DNA
   HO

Fig. 2. Outline of method used to elucidate the nucleotide sequence at the RNA-DNA covalent bond. The sequence is arbitrary. The template is not shown.

with the bacterial DNA polymerases. Later studies have verified our findings (14, 16). As indicated above, the tumor virus DNA polymerase is very similar to, if not identical mechanistically with, the well studied bacterial DNA polymerases. This conclusion can also be drawn from the work of other investigators (13, 16, 29).

Studies on the RNA-DNA Covalent Bond Formed by the Avian Myeloblastosis Virus DNA Polymerase

Since it was found that both bacterial DNA polymerases and the AMV DNA polymerase could form RNA-DNA bonds *in vitro*, it was obvious that we should assay for the presence of this linkage when the viral RNA functioned as a template (primer?). Other workers had shown (32) that viral RNA was the template from hybridization studies but it was not certain if some RNA species could serve as a primer for DNA synthesis. Also it was of interest, if an RNA-DNA bond was formed, to ascertain if a specific type of linkage was formed or if the junction was randomly formed.

We chose the endogenous system for these studies since it was obvious that every attempt should be made to preserve the *in vivo* primer-template relationship. If a highly purified system had been used, it is possible that this relationship would have been lost. In addition, if the study is performed with purified viral RNA, or purified DNA polymerase, or both, certain factors (either nucleic acids or proteins) could be lost during the purification steps. The endogenous DNA-synthesizing system has certain inherent disadvantages, namely the potential presence of deleterious factors such as nucleases. However it is apparent that a biased answer might be obtained if a purified system was used instead of the *in situ* system which should provide an answer which more closely mimics the *in vivo* situation.

The over-all scheme used for determining the RNA-DNA bonds is

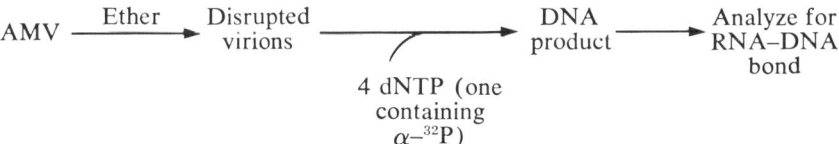

Fig. 3. Schematic outline of procedure for analyzing for RNA-DNA covalent bond (or bonds) formed by endogenous reaction with AMV.

shown in Fig. 3. Purified virus was disrupted with ether, and deoxyribonucleoside triphosphates were then added to permit DNA synthesis. After suitable work-up of the DNA product, analyses were performed to analyze for the RNA-DNA bonds. These analyses were similar to that shown in Fig. 2. The results (5) showed that the AMV DNA polymerase initiates *in vitro* synthesis by forming a specific covalent bond onto tumor virus RNAs.

Figure 4 shows the radiochromatogram observed when the endogenous reaction was performed with α-^{32}P-dCTP plus three other nonradioactive dNTPs. The only radioactive peak which was observed was at the position of 2'(3')-rUMP. To verify that the product was in fact rUMP, chromatography was performed in three other systems in which the relative mobility of the nucleotides was changed. In all cases the only product observed was rUMP.

When α-^{32}P-dATP is the radioactive triphosphate (the other three triphosphates are present but nonradioactive), after analysis the ^{32}P comigrates with the 2'-rAMP and 3'-rAMP markers (Fig. 5) which are well separated in this solvent system. Again the product after alkaline degra-

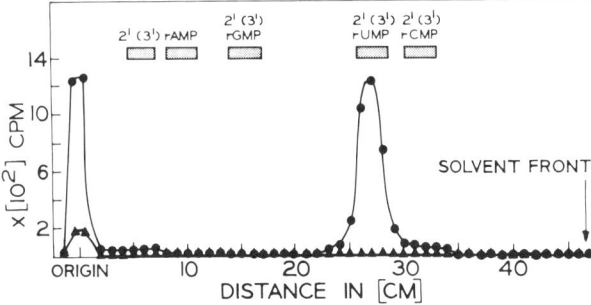

Fig. 4. Radiochromatogram of alkaline hydrolyzates of reaction products formed with α-^{32}P-dCTP substrate. Chromatography was performed in the system 0.1 M sodium phosphate (pH 6.8) (100 ml)-ammonium sulfate (60 g)-*n*-propanol (2 ml). The synthetic reaction, product purification, and analyses were described (5). ▲, Zero time sample; ●, 60-min sample.

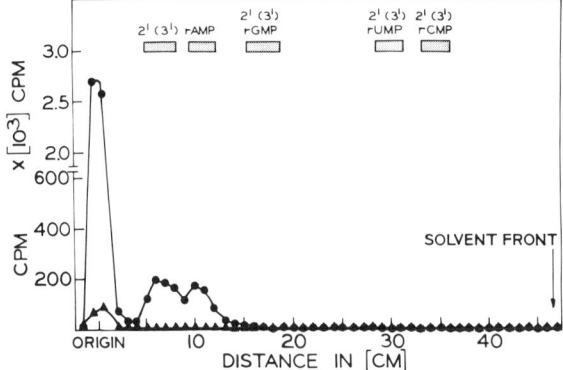

Fig. 5. Radiochromatogram of alkaline hydrolyzates of reaction products formed with α-^{32}P-dATP substrate. The synthetic reaction, product purification, and analyses were described (5). Chromatography was in the same system described in Fig. 4. ▲, Zero time sample; ●, 60-min sample.

dation was run in two other chromatography systems to demonstrate unequivocally that the products were rAMP.

Similar analyses were performed with α-^{32}P-dTTP and α-^{32}PdGTP in separate experiments. Whereas these substrates were incorporated into a high molecular weight polymer, there was no transfer of radioactivity to 2'(3')-ribomonophosphates after alkaline hydrolysis. Hence, it is apparent that dTMP and dGMP do not serve as the first deoxynucleotide in the newly formed DNA strand.

Comparative experiments were performed to determine the relative amount of the two RNA-DNA bonds. Initiation events of the type dCMP linked to rUMP occurred approximately 10 times more frequently than those of the type dAMP to rAMP under the conditions used. Also data of this type permit an estimation of a number of average molecular weights of the newly synthesized DNA. Our estimate (5) of 250 nucleotides per chain is consistent with sedimentation values which have been reported from other laboratories. In addition these estimates are consistent with virtually all of the initiation events being primed with RNA and not DNA since our analyses would not detect DNA-DNA linkages.

Thus the DNA polymerase of AMV initiates DNA synthesis *in vitro* by covalently joining the newly synthesized DNA strand to the 3' termini of viral RNAs in a specific manner. Out of 16 possible RNA-DNA bonds only two are observed. Figure 6 shows the rU-dC linkage. It may, or may not, be significant that a rU-dC bond is found in toluenized *E. coli* cells (see Okazaki in this volume). Since it is clear that both a primer and a template must be present for the AMV DNA polymerase to function (35 and previous papers cited therein), one can write the dinucleotide sequence on the template strand. In addition, the rU-dC

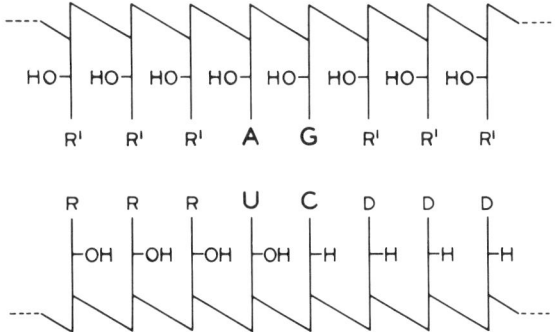

Fig. 6. Predominant dinucleotide sequence at RNA-DNA junction formed by endogenous reaction with ether-disrupted AMV.

bond is formed 10 times as often as the rA-dA linkage. While these studies were in progress, a report (34) appeared which indicated from density gradient studies that an RNA-DNA junction was formed in the tumor virus system.

Several different RNA species have been found in RNA tumor viruses (for a review see 32). It is not clear which of these RNAs serves as a primer for DNA synthesis. Three research groups (3, 17, 18) have recently reported that the 3' terminus of 35 S RNA in AMV is uridylic acid for a high percentage of all molecules. Hence we commented (5) that our experiments were consistent with the notion that 35 S RNA was the primer. However, a recent paper (30) provides evidence that the 3' end of 35 S AMV RNA is adenylic acid. Resolution of this dilemma will require further sequencing studies. Thus we can make no definite conclusion at the present time as to which of the tumor virus RNAs is serving as a primer.

After publication of our studies we learned that Verma and Baltimore had performed a similar experiment with AMV which was disrupted with NP-40. Personal communication with these authors indicated that they performed experiments in the detergent-disrupted endogenous system as well as using purified 70 S RNA and partially purified DNA polymerase from AMV. In both systems these authors indicated that they could observe only the dA to rA transfer. In order to resolve this discrepancy we obtained the details of their experiments from these workers. We performed experiments using α-^{32}P-dCTP as well as α-^{32}P-dATP as the labeled triphosphates in separate reactions. Using their reaction conditions and product work-up we have verified their finding that only a dAMP to rAMP transfer is observed. As indicated in a later section of this manuscript, the use of detergent instead of ether may give rise to several types of problems. In addition, the product work-up described by these workers (phenol extraction) provides for appreciable losses of high molecular weight product in our hands. Hence the over-all recovery of newly synthesized product is not as great as in the procedure which has been published (5). When $\alpha\text{-}^{32}\text{P-dCTP}$ was the radioactive substrate (in the presence of three nonradioactive dNTPs) a high molecular weight product was in fact formed. However, an appreciable amount of this material is extracted directly into the phenol layer; thus appreciable losses of certain types of transfers are apparently lost by this method.

Thus it is clear that the type of transfer which is observed varies with the details of the experimental procedure which is used. The use of NP-40 to disrupt the virus as well as phenol extraction during work-up shows only the dA-rA transfer, whereas ether extraction of the virus plus a somewhat more extensive work-up (which provides higher recovery) shows the dC to rU as well as the dA to rA transfer. Later in this paper we show that AMV RNase can influence transfer reactions. Also, we show that NP-40-disrupted virions contain a phosphatase, whereas little or none of this activity is present in ether-treated virus. If AMV RNase produces 3′-phosphoryl groups, these could not serve as primers unless the 3′-phosphate was removed. Due to the difference in phosphatase level between the two disruption steps, a variance in the types of transfers observed might be anticipated. It is also possible that the different reaction conditions provide for significant variances in the structures of the RNAs which permit only certain types of primers to be effective. That Verma and Baltimore find only the dA to rA transfer in the endogenous system as well as the purified system does not suggest that this is the true *in vivo* situation. As stated above, when studies are performed with purified RNA and partially purified polymerases, it is conceivable that certain types of primers, or other factirs, or both, may be lost. Clearly further studies are necessary. Since our laboratory and Baltimore's laboratory agreed to perform this experiment using the other laboratory's procedure, it will be of interest to learn the results of Verma and Baltimore when they perform the experiment under our published reaction conditions.

Studies on the Potential Influence of the Ribonucleases on the Observed RNA-DNA Linkage

Nonspecific RNase

In addition to DNA polymerase, a variety of other nucleic acid-metabolizing enzymes have been reported to be associated with RNA tumor viruses. These include DNA ligase (23), DNases (13, 21), nucleoside diphosphokinase (19), other nucleotide kinases and phosphotransferases (22, 28), protein kinase (12, 31), tRNA synthetases (4), RNA methylase

(7), and RNase H (24). Since ribonuclease H is present and since a variety of other nucleic acid-metabolizing enzymes are associated with these viruses, it is logical to imagine that a ribonuclease (13) may also be present.

Hence, it was of interest to determine whether the possible presence of a ribonuclease could alter the transfer results which are described above. Since we found a specific transfer (predominantly one transfer out of a possible total of 16) this seemed like an unlikely possibility. However, we preincubated the virus prior to addition of deoxyribonucleoside triphosphates to determine whether ribonuclease was a potential problem. The results of a transfer experiment performed on a product from a reaction after preincubation are shown in Table 2. The transfer from dCTP to rUMP is still present but in addition a transfer is also observed to rCMP. A similar preincubation experiment was performed using α-^{32}P-dATP as the radioactive substrate. In this case, transfers in addition to the rA-dA transfer were observed (data not shown). These results are consistent with the notion that a ribonuclease activity is present and can cause degradation of the template-primer, thus giving rise to additional isotope transfers. Hence steps should be taken to obviate potential degradation of the viral RNA when performing experiments of this type. Since we observed predominantly one nucleotide transfer in the absence of preincubation (see above), we conclude that the associated ribonuclease activity probably has not influenced our results as reported (5).

Potential Influence of Ribonuclease H

A previous report (24) indicated the presence of RNase H in AMV. Thus we tested to see if this activity could have an effect on the transfer reaction which is described above. Figure 7 shows a scheme for the potential influence of ribonuclease H on the DNA synthesis reaction. After DNA synthesis has occurred ribonuclease H could cleave the RNA strand complementary to the newly synthesized DNA. These nicks

Table 2. Distribution of radioactivity after alkaline hydrolysis of products of AMV DNA polymerase endogenous reaction after preincubation of ether-disrupted AMV

Substrate	2' (3')-Ribomonophosphates (cpm)			
	rGMP	rUMP	rCMP	rAMP
α-^{32}P-dCTP	0	1619	1969	0

Reaction conditions were identical with those previously described (5), with the exception that all components but dNTPs were preincubated at 41°C for 60 min. The deoxyribonucleoside triphosphates were then added to initiate the reaction. Product work-up and analysis was as described (5). Paper chromatography after alkaline hydrolysis was in isobutyric acid-concentrated NH_4OH-H_2O (66 : 1 : 33). Product work-up included a charcoal adsorption step.

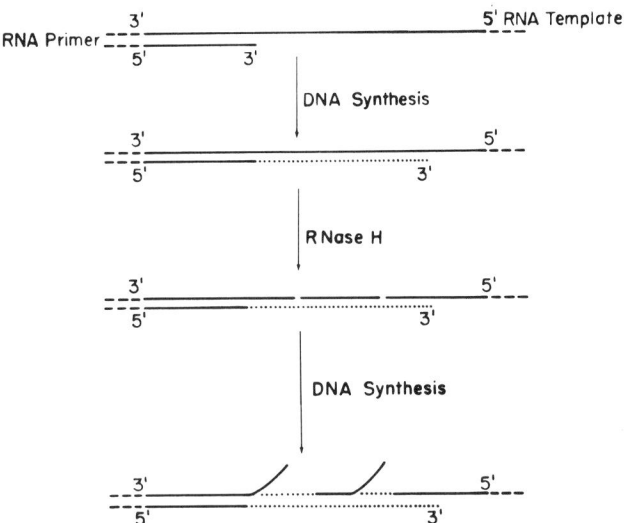

Fig. 7. Hypothetical scheme of potential involvement of RNase H in DNA synthesis. Solid lines represent AMV RNA; dotted lines represent newly synthesized DNA.

on the RNA strand could then provide new primer sites for DNA synthesis.

It should be noted that one would expect the template strand to be cleaved by ribonuclease H and not the primer strand. Thus when RNA serves as both a primer and a template (as for the RNA tumor virus system), an activity other than ribonuclease H is appropriate for cleaving the RNA primer from the newly synthesized DNA. The difference between this system and the bacterial system for which DNA is the template should be kept in mind (see paper by R. Okazaki in this book).

Previous workers (24) have shown the presence of a ribonuclease H activity associated with AMV when the virus was disrupted with detergent and have demonstrated the activity in a partially purified preparation of the enzyme. However, it was not certain if the activity was demonstrable in ether-treated virus. Figure 8 shows that RNase H activity is present in ether-disrupted AMV. Little or no degradation of poly(rA) · poly(rU) or poly(rA) is observed. However when the polyribopurine strand is complexed with either poly(dT) or $(dT)_{13}$, appreciable degradation of the RNA strand is observed. Other investigators have simultaneously investigated RNase H (see manuscripts in this book by Leis and Hurwitz, and by Baltimore).

A chromatographic analysis of the products of degradation of poly(rA) · poly(dT) indicates that the products are largely oligonucleotides which have a sharp distribution at approximately the octanucleotide

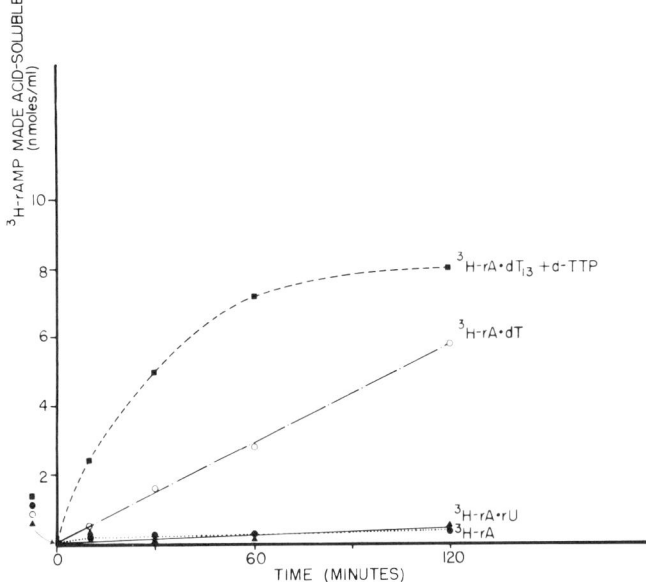

Fig. 8. Kinetics of degradation of hybrid and RNA polymers by ether-disrupted avian myeloblastosis virus. ^{3}H-Poly(rA), 14.6 μmoles, (specific activity 1.585×10^4 cpm/μμmole) was used in each experiment and was preannealed to the following amount of the complementary strand: 12.0 μmoles of poly (rU), 9.3 μmole of poly(dT), and 13.8 μmoles of $(dT)_{13}$. Otherwise, incubation conditions were identical with the endogenous reaction conditions (5) and conditions of nuclease assays were as described previously (20). α-^{32}P-dTTP (specific activity 3.51×10^5 cpm/μμmole) was present only in the one reaction as indicated. ^{32}P-dTMP, 520 μμmoles/ml, was incorporated into acid-insoluble polymer after 120 min of incubation.

stage. For the experiment shown in Fig. 9, 67% of the starting polymer was rendered acid-soluble. Approximately 36% of all the acid-soluble product migrated in the relatively sharp peak at the octanucleotide position. In addition to oligonucleotides, radioactive adenosine was observed which is no doubt derived from adenylic acid since a phosphatase activity is known to be associated with AMV (see below).

In order to determine whether the presence of ribonuclease H had an influence on the transfer reaction as described above, we wished to selectively inhibit this activity while retaining the DNA polymerase activity. Figure 10 demonstrates that a suitable inhibitor is sodium fluoride. In the presence of 7.5 mM sodium fluoride, the endogenous DNA polymerase activity is not measurably inhibited whereas the ribonuclease H activity is essentially completely inhibited. A variety of other compounds were examined as possible selective inhibitors including actinomycin D, netropsin, rifampicin, and 2′, 3′-cyclic CMP. The most suitable inhibitor found was sodium fluoride. (We wish to thank Dr.

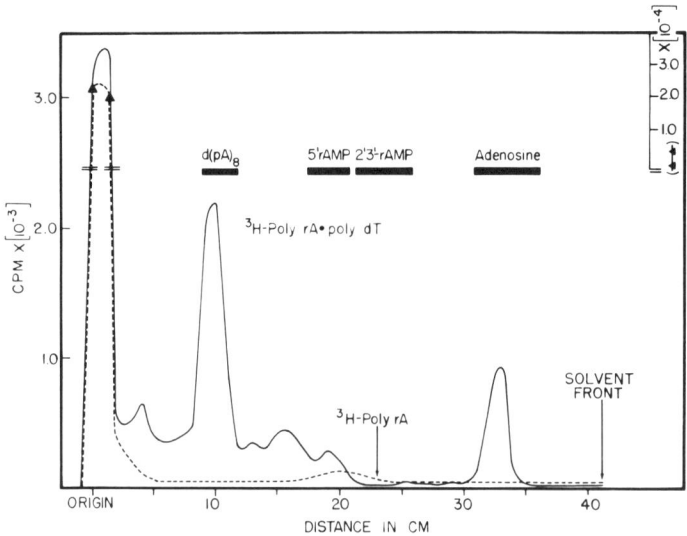

Fig. 9. Chromatographic analysis of products of degradation of ^{3}H-poly(rA) · poly(dT) by ether-disrupted AMV. A reaction mixture identical with that described in Fig. 8 was incubated at 37°C for 3 hr and was chromatographed on Whatman No. 1 paper in isobutyric acid-NH$_3$-H$_2$O (66:1:33). The chromatogram was dried, cut into 1.0-cm strips, and counted in a liquid scintillation counter. The radioactivity comigrating with an octamer represents 25% of the total activity. The black areas represent markers. ^{3}H-Poly(rA) was treated in an identical manner and was run as a control.

Robert C. Gallo, National Cancer Institute, for suggesting this compound as a phosphatase inhibitor.)

Since an inhibitor was found which would selectively eliminate the ribonuclease H reaction but would not inhibit the DNA polymerase reaction, we then proceeded to study the transfer reaction in the presence of sodium fluoride. Table 3 shows an experiment in which α-^{32}P-dCTP was used as a radioactive triphosphate for preparation of DNA polymer; the disrupted virions were preincubated in the presence of sodium fluoride in order to eliminate ribonuclease H activity. Table 3 shows that the rU-dC bond is still formed in the absence of RNase H. In addition the presence of an rC-dC linkage is also observed and is no doubt due to the necessary preincubation step in the presence of sodium fluoride. This rC-dC linkage is also observed when the disrupted virus is preincubated in the absence of sodium fluoride (Table 2). A similar experiment also has been performed with α-^{32}P-dATP as the radioactive substrate (results not shown). The transfer of radioactivity from dATP to rAMP was still observed in the presence of sodium fluoride but, in addition, other transfers were also seen. As indicated above this is no doubt due to the necessary preincubation step.

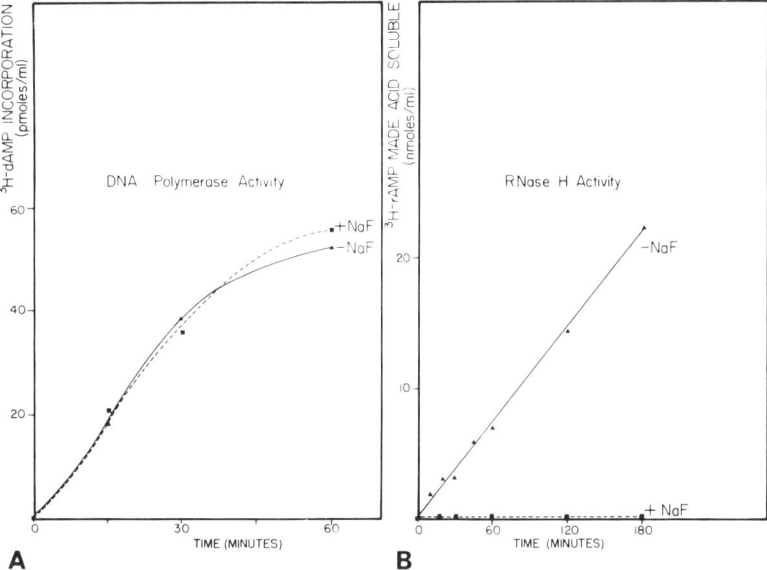

Fig. 10. A, Kinetics of DNA synthesis by ether-disrupted avian myeloblastosis virus. The dotted line shows the incorporation in the presence of 7.5 mM NaF. The disrupted virions were preincubated with NaF for 30 min at 0°C. Reaction conditions were as described previously (5), except that NaCl was omitted. The specific activity of ^{3}H-dATP was 1.51×10^6 cpm/mµmole. B, Kinetics of degradation of ^{3}H-poly(rA) · poly(dT) by ether-disrupted avian myeloblastosis virus. ^{3}H-Poly(rA) (specific activity 1.585×10^4 cpm/µµmole), 36.5 µmoles, was preannealed with 46.5 µmoles of poly(dT). The poly(dT) was depurinated as described (10). Reaction conditions were as described (5). The dotted line shows the release of radioactivity in the presence of 7.5 mM NaF. In this case, the disrupted virions were preincubated with NaF for 30 min at 0°C prior to addition of the reaction components. The nuclease assay was as described (20).

Table 3. Distribution of radioactivity after alkaline hydrolysis of products of AMV DNA polymerase endogenous reaction after preincubation of ether-disrupted AMV in the presence of NaF

Substrate	2'(3')-Ribomonophosphates (cpm)			
	rGMP	rUMP	rCMP	rAMP
α-^{32}P-dCTP	0	4200	6950	0

Reaction conditions were identical with those previously described (5), with the exception that all components but dNTPs were preincubated at 41°C for 60 min in the presence of 7.5 mM NaF. The triphosphates were then added to initiate the reaction. Product work-up and analysis were as described (5). Paper chromatography after alkaline hydrolysis was in 0.1 M sodium phosphate (pH 6.8) (100 ml)-ammonium sulfate (60 g)-n-propanol (2 ml). Product work-up included a charcoal adsorption step (5).

In conclusion, when the transfer experiment is performed in the presence of a ribonuclease H inhibitor, the rU-dC bond and the rA-dA bond are still observed, suggesting that the presence of RNase H activity does not influence our original observation on these transfer reactions (5).

Studies on Different Agents for Virus Disruption: Ether *versus* NP-40

A variety of laboratories have used the non-ionic detergent NP-40 as a means of disrupting RNA tumor viruses whereas our laboratory has primarily used ether disruption. Ether disruption was reported (11) to be effective for expressing DNA polymerase activity. We examined a variety of disrupting agents for their ability to express DNA polymerase activity. Nonidet P-40 and ether disruption were the most effective agents

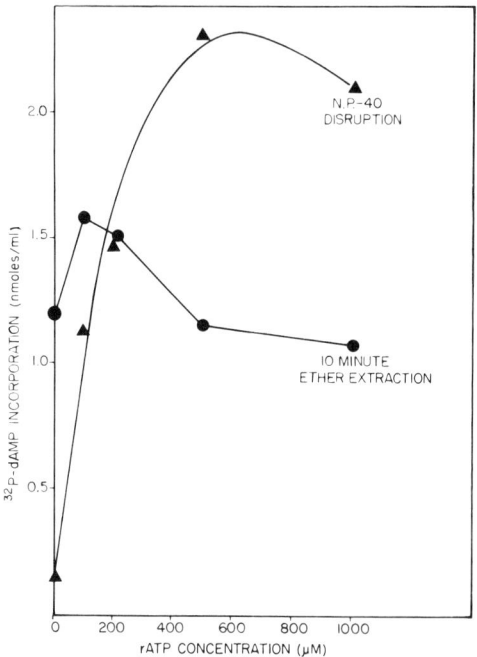

Fig. 11. Effect of rATP on reaction with Nonidet P-40 or ether-disrupted AMV. A sample of AMV was disrupted with 0.25% Nonidet P-40 in 0.05 M dithiothreitol at 0°C. A second sample was vortexed with ether for 10 min at room temperature in the presence of 0.05 M DTT. The final volume of aqueous DNA polymerase solution was the same in both cases. Samples of these enzyme preparations were assayed in a poly(dA-dT) · poly(dA-dT)-templated reaction at 37°C under normal assay conditions (35) in the presence of varying amounts of rATP. The data presented are from 60-min time points.

whereas Brij-58 was approximately half as effective. Tween 80, toluene, and dimethyl sulfoxide were not effective.

Also, we found that the presence of rATP caused a marked stimulation in the rate of the endogenous reaction when NP-40 was used as the disrupting agent. Figure 11 demonstrates that the rate of DNA synthesis is stimulated approximately 15-fold in the presence of an optimum amount of rATP (500 mM). On the other hand, addition of rATP to the ether-treated particles provided little or no stimulation in the rate of reaction. Thus it was clear from the outset that there is a marked difference in these two systems. When no rATP was added (Fig. 11), 6 times more DNA synthesis was observed for the ether-disrupted virus than for the NP-40-treated AMV. Initially we were greatly interested in this phenomenon since other workers had demonstrated in crude bacterial systems that rATP had a stimulatory role in DNA synthesis (6, 26, 27). However as demonstrated below, the rATP preserves the dNTPs by serving as an alternate substrate for the phosphatase which is present in the NP-40-disrupted particles. Conversely, the ether-treated virus has only a very low phosphatase activity which explains the fact that rATP provides no stimulation in the rate of DNA synthesis.

Figure 12 shows the ability of other phosphorylated compounds to

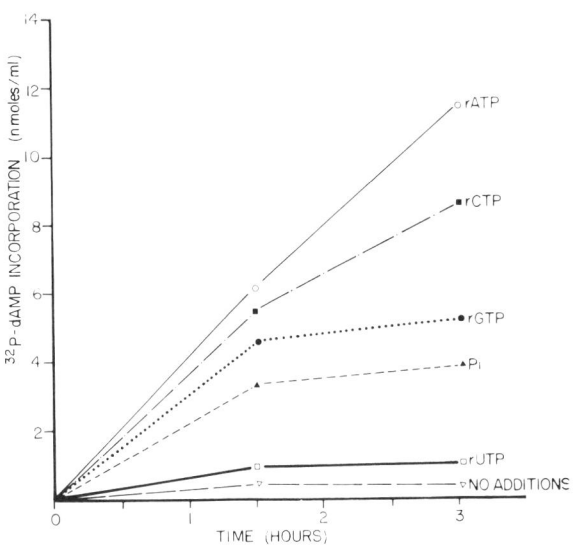

Fig. 12. Stimulatory effect of phosphorylated compounds on DNA synthesis by NP-40-disrupted AMV. Reaction mixtures (0.10 ml) were as previously described (35) with poly(dA-dT) · poly(dA-dT) template. AMV was disrupted with NP-40 as described in Fig. 11. Additions as indicated were at 0.5 mM, except for inorganic phosphate, which was at 1.5 mM. Samples (0.040 ml) were removed after 1.5 and 3 hr. Synthesis was compared with a similar reaction containing no rNTP or P_i.

substitute for rATP as a stimulator with NP-40-disrupted AMV. Other ribonucleoside triphosphates can at least partially replace rATP in this role. In addition, inorganic phosphate is partially effective in stimulating DNA synthesis. However, riboadenosine has no stimulatory effect.

To understand better the mechanism of this stimulatory effect of ribonucleoside triphosphates, we determined the fate of rATP after incubation in the system described in Fig. 12. The chromatographic system employed (isobutyric acid-ammonia-water; 66 : 1 : 33) was capable of separating rATP, rADP, rAMP, and adenosine. After 60 min of incubation in either the presence or absence of DNA template, over 90% of ^{14}C-rAMP was converted to riboadenosine. Also, other chromatographic determinations have proven that radioactive dATP and dCTP are dephosphorylated by NP-40-disrupted AMV (data not shown). Hence we conclude from these studies that a potent phosphatase is present in NP-40-disrupted virions.

We have performed a direct phosphatase assay on NP-40-disrupted AMV as well as ether-disrupted AMV to establish this point more conclu-

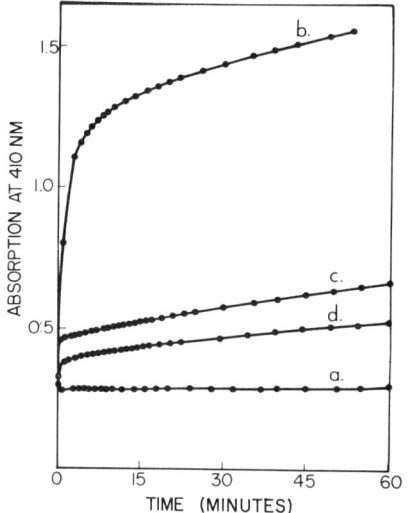

Fig. 13. Comparison of phosphatase activity for NP-40-disrupted *versus* ether-disrupted AMV. Reaction mixtures (0.5 ml final volume) contained 5.5 mM *p*-nitrophenylphosphate, 0.55 M Tris-HCl buffer (pH 8), and additions (see below). All reaction components except virus were added to stoppered cuvettes and kept at 37°C in a thermostatted Gilford spectrophotometer. Formation of *p*-nitrophenol was measured directly by adsorption at 410 nm. Readings began after the addition of virus. a, 0.1 ml of 0.05 M dithiothreitol (DTT)-0.25% NP-40; b, 0.05 ml of AMV (4 mg of protein/ml) plus 0.05 ml of 0.1 M DTT-0.5% NP-40; c, 0.050 ml of nondisrupted AMV; d, 0.050 ml of 1 min ether disrupted AMV (5).

sively (Fig. 13). As suspected, the NP-40-disrupted virus contains a phosphatase activity whereas little or no activity is found with ether-disrupted particles. It is well known that a phosphatase is associated with avian myeloblastosis virus (25) and the enzyme has been implicated as being associated with the envelope. It is apparent that ether disruption of AMV either inactivates the phosphatase activity or, if the activity is associated with the envelope, may be extracted into the ether layer. This is apparently not the case when NP-40 is used as the disrupting agent.

In summary, these results document one difference between NP-40-disrupted particles and ether-disrupted particles. Since a phosphatase activity clearly could have a deleterious effect on DNA polymerase reactions, results obtained with NP-40-disrupted virus must be interpreted with caution.

In addition, the difference in levels of phosphatase activity observed when the virus is disrupted by the two different procedures is but one example of a variety of differences that may exist between ether-disrupted and NP-40-disrupted virus. Differences may also exist in the structure, or integrity, or RNA molecules, or all three, as well as in the relative levels of various enzyme activities. In view of this finding it is not surprising that different transfer results are found when the different disruption techniques are used (see above).

Acknowledgments

This work was supported in part by funds from the National Science Foundation (GB-30528X) and the National Institutes of Health (71-2275).

We thank Dr. Joseph W, Beard, Duke University, for generous gifts of avian myeloblastosis virus.

Figure 1 is reproduced from reference 35, and Figs. 2, 4, 5, and 6 are reproduced from reference 5. We are grateful to the publishers of *Biochemistry* and *Virology* for permission to publish these figures.

References

1. Bollum, F. J. 1962. Oligodeoxyribonucleotide-primer reactions catalyzed by calf thymus polymerase. *J. Biol. Chem.* 237: 1945.

2. Cavalieri, L. F., and Carroll, E. 1970. RNA as a template with *E. coli* DNA polymerase. *Biochem. Biophys. Res. Commun.* 41: 1055.

3. Erickson, R. L., Erickson, E., and Walker, T. A. 1971. The identification of 3'-hydroxyl nucleoside terminus of avian myeloblastosis virus RNA. *Virology* 45: 527.

4. Erickson, E., and Erickson, R. L. 1972. Transfer ribonucleic acid synthetase activity associated with avian myeblastosis virus. *J. Virol.* 9: 231.

5. Flügel, R. M., and Wells, R. D. 1972. Nucleotides at the RNA-DNA covalent bonds formed in the endogenous reaction by the avian myeloblastosis virus DNA polymerase. *Virology* 48: 394.

6. Ganesan, A. T. 1971. Adenosine triphosphate-dependent synthesis of biologically active DNA by azide-poisoned bacteria. *Proc. Nat. Acad. Sci. U. S. A.* 68: 1296.

7. Gantt, R. R., Stromberg, K. J., and de Oca, F. M. 1971. Specific RNA methylase associated with avian myeloblastosis virus. *Nature* 234: 35.

8. Goodman, N. C., and Spiegelman, S. 1971. Distinguishing reverse transcriptase of an RNA tumor virus from other known DNA polymerases. *Proc. Nat. Acad. Sci. U. S. A.* 68: 2203.

9. Goulian, M. 1968. Initiation of the replication of single-stranded DNA by *Escherichia coli* DNA polymerase. *Cold Spring Harbor Symp. Quant. Biol.* 33: 11.

10. Harwood, S. J., and Wells, R. D. 1970. *Micrococcus luteus* deoxyribonucleic acid polymerase. Studies on the initiation of deoxyribonucleic acid synthesis in vitro. *J. Biol. Chem.* 245: 5625.

11. Hatanaka, M., Huebner, R. J., and Gilden, R. V. 1970. DNA polymerase activity associated with RNA tumor viruses. *Proc. Nat. Acad. Sci. U. S. A.* 67: 143.

12. Hatanaka, M., Twiddy, E., and Gilden, R. V. 1972. Protein kinase associated with RNA tumor viruses and other budding RNA viruses. *Virology* 47: 536.

13. Hurwitz, J., and Leis, J. P. 1972. RNA-dependent DNA polymerase activity of RNA tumor viruses. *J. Virol.* 9: 116.

14. Keller, W. 1972. RNA-primed DNA synthesis *in vitro*. *Proc. Nat. Acad. Sci. U. S. A.* 69: 1560.

15. Kornberg, A. 1969. Active center of DNA polymerase. *Science* 163: 1410.

16. Leis, J. P., and Hurwitz, J. 1972. RNA-dependent DNA polymerase activity of RNA tumor viruses: directing influence of RNA in the reaction. *J. Virol.* 9: 130.

17. Lewandowski, L. J., Content, J., and Leppla, S. H. 1971. The characterization of the subunit structure of the ribonucleic acid genome of influenza virus. *J. Virol.* 8: 701.

18. Maruyama, H. B., Hatanaka, M., and Gilden, R. V. 1971. The 3'-terminal nucleosides of the high molecular weight RNA of C-type viruses. *Proc. Nat. Acad. Sci. U. S. A.* 68: 1999.

19. Miller, L. K., and Wells, R. D. 1971. Nucleoside diphosphokinase activity associated with DNA polymerases. *Proc. Nat. Acad. Sci. U. S. A.* 68: 2298.

20. Miller, L. K., and Wells, R. D. 1972. Properties of the exonucleolytic activities of the *Micrococcus luteus* deoxyribonucleic acid polymerase. *J. Biol. Chem.* 247: 2667.

21. Mizutani, S., Boettiger, D., and Temin, H. M. 1970. A DNA-dependent DNA polymerase and a DNA endonuclease in virions of Rous sarcoma virus. *Nature* 228: 424.

22. Mizutani, S., and Temin, H. M. 1971. Enzymes and nucleotides in virions of Rous sarcoma virus. *J. Virol.* 8: 409.

23. Mizutani, S., Temin, H. M., Kodama, M., and Wells, R. D. 1971. DNA ligase and exonuclease activities in virions of Rous sarcoma virus. *Nature New Biol.* 230: 232.

24. Mölling, K., Bolognesi, D. P., Bauer, H., Büsen, W., Plassmann, H. W., and Hausen, P. 1971. Association of viral reverse transcriptase with an enzyme degrading the RNA moiety of RNA-DNA hybrids. *Nature New Biol.* 234: 240.

25. Mommaerts, A. B., Eckert, E. A., Beard, D., Sharp, D. G., and Beard, J. W. 1952. Dephosphorylation of adenosine triphosphate by concentrates of the virus of avian erythromyeloblastic leucosis. *Proc. Soc. Exp. Biol. Med.* 79: 450.

26. Mordoh, J., Hirota, Y., and Jacob, F. 1970. On the process of cellular division in *Escherichia coli*. V. Incorporation of deoxynucleotide triphosphates by DNA thermosensitive mutants of *Escherichia coli* also lacking DNA polymerase activity. *Proc. Nat. Acad. Sci. U. S. A.* 67: 773.

27. Moses, R. E., and Richardson, C. C. 1970. Replication and repair of DNA in cells of *Escherichia coli* treated with toluene. *Proc. Nat. Acad. Sci. U. S. A.* 67: 674.

28. Roy, P., and Bishop, D. H. L. 1971. Nucleoside triphosphate phosphotransferase: a new enzyme activity of oncogenic and non-oncogenic "budding" viruses. *Biochim. Biophys. Acta* 235: 191.

29. Smoler, D., Molineux, I., and Baltimore, D. 1971. Direction of polymerization by the avian myeloblastosis virus deoxyribonucleic acid polymerase. *J. Biol. Chem.* 246: 7697.

30. Stephenson, M. L., Wirthlin, L. S., Scott, J. F., and Zamecnik, P. C. 1972. The 3'-terminal nucleosides of the high molecular weight RNA of avian myeloblastosis virus. *Proc. Nat. Acad. Sci. U. S. A.* 69: 1176.

31. Strand, M., and August, J. T. 1971. Protein kinase and phosphate acceptor proteins in Rauscher murine leukemia virus. *Nature New Biol.* 233: 137.

32. Temin, H. M. 1971. Mechanism of cell transformation by RNA tumor viruses. *Annu. Rev. Microbiol.* 25: 609.

33. Temin, H. M., and Baltimore, D. 1972. RNA-directed DNA synthesis and RNA tumor viruses. *Advan. Virus Res.* 17: 129.

34. Verma, I. M., Meuth, M. L., Bromfeld, E., Manley, K. F., and Baltimore, D. 1971. Covalently linked RNA-DNA molecule as initial product of RNA tumor virus DNA polymerase. *Nature New Biol.* 233: 131.

35. Wells, R. D., Flügel, R. M., Larson, J. E., Schendel, P. F., and Sweet, R. W. 1972. Comparison of some reactions catalyzed by deoxyribonucleic acid polymerase from avian myeloblastosis virus, *Escherichia coli*, and *Micrococcus luteus*. *Biochemistry* 11: 621.

Discussion

Fidanian. I'm a little bit confused with the RNase activity that you just described. With our previous speaker, we understood that the enzyme was a processive enzyme and here you are telling us that it was cleaving one strand and not the other. Would you explain this please.

Flügel. I don't get the point.

Fidanian. The point is, apparently, that I don't understand why RNase H would create new ends to begin with. I thought that it would need a free end and then cleave in a processive way.

Flügel. I have not asked the question whether or not it is an exo- or endonuclease. I was simply interested in whether or not the RNase H influences the result of the transfer experiment. But it does not contradict Dr. Leis' results. We were not aware of Dr. Leis' results when we did the experiment.

Leis. I don't think the results are contradictory. You require an end for the ribonuclease H to work and if you have a situation where the RNA-DNA hybrid is internal and there is no end present, you won't see any degradation so it won't interfere with his reaction. He was able to knock out the ribonuclease H with the sodium fluoride and it still didn't affect the transfer. Can I ask Dr. Flügel a question about the oligonucleotides? What's the size of the oligonucleotides in your synthetic reactions? And what is the temperature of your incubation for the priming experiments?

Flügel. The size of the oligomers is 13 nucleotides long.

Leis. Was the incubation at 41°C?

Flügel. Yes.

Leis. I think the reason poly(dA) · oligo(dT) was not working was the you may not have a duplex structure at that temperature and I'd suggest carrying out your incubation at lower temperatures, such as 20 or 15°C. Poly(rA) works well with oligo(dT) with a chain length of about 12 to 14 when the reaction is carried out at temperatures of about 37°C, but if you use an oligo(dT) oligomer of chain length 3, it will not work. You can use the oligomer of 3 when you bring your temperature down to 15°C because you have to get the duplex structure to form.

Flügel. Do I understand you correctly? You tried to explain the inability of the AMV polymerase to use poly(dA) · oligo(dT) by a temperature effect?

Leis. That's right.

Flügel. I'm not sure if we did the experiment at 37°C but maybe Paul Schendel can answer that question.

Schendel. I don't think the argument holds in that the other polymerase reactions were done in exactly the same way and they used the poly(dA) · oligo(dT) very well. They were done under the same salt conditions and temperature.

Gallo. We have done it at 37°C. I think other laboratories probably also do with the same exact result, that the AMV or other viral DNA polymerases do not work at all or very minimally on that template. Maybe just one other point: in the summary slide, you mentioned that the properties of the purified *Escherichia coli* and *Micrococcus luteus* enzyme were like a reverse transcriptase in being able to utilize polymeric RNA molecules. And I think the conclusion would be again the same; perhaps it's worth qualifying it for the homopolymers.

Yang. I think Dr. Leis' comment is very interesting. We have compared AMV polymerase with Rauscher leukemia virus polymerase and we found actually that the AMV will use poly(rA) · oligo(dT)$_8$ at 37°C while Rauscher leukemia virus will use it only at 20°C. With poly(dA) · oligo(dT) as template and AMV DNA polymerase at 40°C, we find 8% of the template activity of poly(rA) · oligo(dT).

Flügel. Do you think you can explain the factor of a 100-fold more dT synthesis with poly(rA) · oligo(dT) by your data?

Yang. I say that possibly Dr. Leis' comment is not very likely in the situation. If he used a Rauscher leukemia virus polymerase, then it's likely that the temperature explained the effect.

Avian Myeloblastosis Virus DNA Polymerase

Initiation of DNA Synthesis and an Associated Ribonuclease

David Baltimore, Inder M. Verma,
Donna F. Smoler, and Nora L. Meuth

Department of Biology
Massachusetts Institute of Technology
Cambridge, Massachusetts 02139

Initiation of DNA synthesis by the avian myeloblastosis virus DNA polymerase requires a preformed primer. When 70 S viral RNA is used as template, a small RNA associated with the major species of viral RNA acts as primer. This has been proved both by analysis of the density of the initial product and by identification of the RNA-DNA joint in the product. This joint is ApdA.

Associated with the DNA polymerase is a ribonuclease which specifically degrades the RNA of a DNA-RNA hybrid. This enzyme has been inseparable from the DNA polymerase. It is readily assayed by the conversion of ^{3}H in [^{3}H]poly(A) · poly(dT) to an acid-soluble form. The cleavage reaction leaves 3'-OH and 5'-phosphoryl termini and the enzyme appears to be an endonuclease.

In the 2 years since the first identification of a DNA polymerase in virions of the RNA tumor viruses much progress has been made in the understanding of the enzyme (reviewed in reference 6). Initially an *endogenous* reaction was observed where the DNA polymerase was producing a copy of the 70 S RNA of the tumor virus particles. It was observed, however, that a much more extensive synthesis of DNA could be stimulated by added DNA and RNA templates. Using such templates for assay, the soluble DNA polymerase has been purified from virion preparations. The purified enzyme can copy both DNA and RNA and no separation of the ability to copy these two templates has been observed. The enzyme is primer-dependent, that is, unable to initiate DNA synthesis without a preformed piece of DNA (or RNA) to initiate the reaction. Short oligomers of DNA, complementary in base sequence to the template, are very convenient primers. Aside from its polymerizing activity the enzyme purified from avian myeloblastosis virus (AMV) also has an associated ribonuclease activity.

Two topics will be discussed in this paper: the initiation of DNA synthesis when 70 S viral RNA is the template and the nature of the ribonuclease activity associated with the DNA polymerase.

Initiation of DNA Synthesis

A number of different methods have shown that the initial product formed when 70 S RNA is used as a template for the DNA polymerase of AMV is a covalently bonded DNA-RNA molecule. The initial observations were made using cesium sulfate gradients where the heat-denatured product was shown to band as a covalently joined molecule of DNA and RNA (8). From the density and size of the product, it appeared

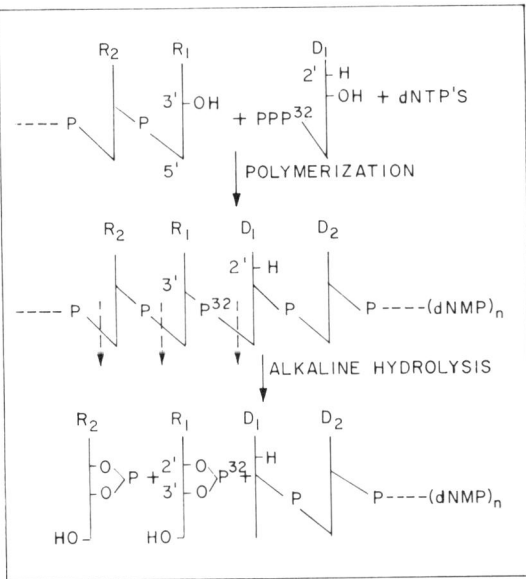

Fig. 1. Schematic representation of the transfer of a ^{32}P atom from the α-position of a deoxyribonucleotide to the 2′(3′)-position on a ribonucleotide. R_2 and R_1 represent penultimate and ultimate ribonucleotides of the presumed RNA primer. D_1 and D_2 represent the initial two deoxyribonucleotides incorporated.

that the initiator was a short RNA molecule (about 50 to 100 bases). More recently we have been studying the bond which joins DNA to the RNA primer using as an assay the transfer of ^{32}P from an α-^{32}P-deoxyribonucleoside triphosphate to a ribonucleotide after alkaline hydrolysis of the product (7). Figure 1 shows the rationale for this type of experiment.

The electrophoretograms of alkali-hydrolyzed material are shown in Fig. 2. It is evident that of the four deoxyribonucleotide precursors, only α-^{32}P-dATP is able to transfer a phosphorus to a ribonucleotide, and the only ribonucleotide which accepts the ^{32}P is 2' (3')-AMP. Table 1 quantitates the transfer data and shows that about 15% of the incorporated label can be recovered as ribonucleotide after a 60-min reaction. Identical results were obtained whether purified RNA and enzyme were used or if detergent-disrupted virions were assayed. These data indicate that the DNA product is very small (50 to 100 nucleotides) and is in agreement with analysis of the product by sucrose gradient centrifugation.

The fact that of the 16 possible DNA-RNA joints the only one found is ApdA indicates that the specificity of the initiation reaction is very

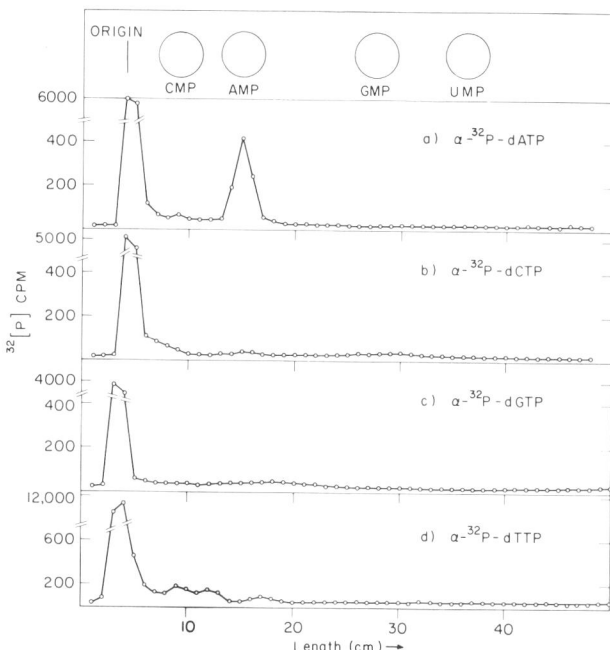

Fig. 2. α-^{32}P transfer with individual deoxyribonucleoside triphosphates in the reconstructed system.

Table 1. Transfer of ^{32}P to ribonucleotides

Nature of reaction	Substrate α-^{32}P-dNTP	Percentage recovered in			
		CMP	AMP	GMP	UMP
AMV virions	dCTP	<1	<1	<1	<1
(endogenous system)	dATP	<1	19-23	<1	<1
	dGTP	<1	<1	<1	<1
	dTTP	<1	<1	<1	<1
AMV polymerase plus	dCTP	<2	<1	<1	<1
60-70 S AMV RNA	dATP	<1	11-15	<1	<1
(reconstructed system)	dGTP	<1	<1	<1	<1
	dTTP	<1	<1	<1	<1

Data are taken from a series of experiments like that depicted in Fig. 2. All α-^{32}P-dNTPs were examined at least twice. The data for dCTP exclude two experiments where transfer to 2'(3')-CMP and 2'(3')-AMP was observed. This was not reproducible.

great. Where on the 70 S RNA molecule this initiation occurs, how many initiators there are per 70 S RNA, etc., are questions for the future.

The Ribonuclease Associated with the Avian Myeloblastosis Virus DNA Polymerase

A ribonuclease associated with the AMV DNA polymerase was first observed by Mölling et al. (5). They found that the ribonuclease had the properties of a ribonuclease H (3), that is, it would degrade the RNA moiety of a DNA-RNA hybrid but would not degrade either free RNA or double stranded RNA. We have confirmed their data with a somewhat different assay for the ribonuclease (2).

We initially observed the ribonuclease H when we were studying the fate of poly(A) after it was copied by the AMV DNA polymerase to form poly(A) · poly(dT). We found that when [^{3}H]poly(A) was used as template, the ^{3}H label was rapidly converted to an acid-soluble form.

Table 2. Degradation of poly(A) as a consequence of its use as a template for the AMV DNA polymerase

	Poly(A) degraded (%)	Poly(dT) synthesis (pmoles)
Complete	71	31
−oligo(dT)	<5	<2
−dTTP	<5	
−polymerase	<5	<2

Standard reaction conditions for AMV polymerase were used with (per 0.1 ml) 82 pmoles of [^{3}H]poly(A) (22 cpm/pmole), 12 pmoles of (dT)$_{14-18}$, 3.8 nmoles of [^{32}P]dTTP (18 cpm/pmole), and 15 units of AMV DNA polymerase. Incubation was for 60 min at 37°C.

Table 3. Catalysis of poly(A) degradation by poly(dT)

	Poly(A) degraded (%)
Complete	70
−magnesium acetate	<5
−poly(dT)	<5
−poly(dT); plus poly(U)	<5

Reaction conditions were (in 0.1 ml) 0.05 M Tris-HCl, pH 8.3; 20 mM magnesium acetate; 10 mM dithiothreitol; 340 pmoles of [^{3}H]poly(A) (10 cpm/pmole); 186 pmoles of poly(dT) and, where indicated, 168 pmoles of poly(U) with 20 units of AMV DNA polymerase. Incubation was for 60 min at 37°C.

Table 2 compares the synthesis of [^{32}P]poly(dT) and the degradation of [^{3}H]poly(A) during this reaction. Both synthesis and degradation are dependent on the presence of TTP in the reaction mixture.

This coupled synthesis of poly(dT) and degradation of poly(A) is difficult to use as an assay for ribonuclease H and we therefore turned to using preformed poly(A) · poly(dT) as a substrate. Table 3 shows that the degradation of poly(A) requires poly(dT) and magnesium and that the poly(dT) cannot be replaced by poly(U). This confirms the identification of the nuclease as a ribonuclease H. Using this assay we have followed the ribonuclease activity through our standard purification for the AMV DNA polymerase (2). We have found that polymerization and nuclease activity go hand-in-hand and that even glycerol gradient centrifugation of the purified material does not separate the two activities (Fig. 3).

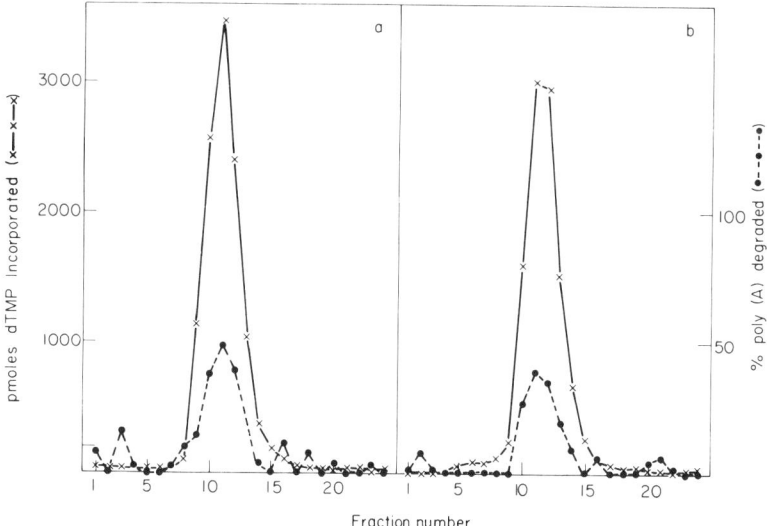

Fig. 3. Glycerol gradient centrifugation of AMV DNA polymerase and associated nuclease. a, 0.2 M KCl; b, 0.5 M KCl.

Studies on the nature of the polymerase-associated ribonuclease H indicate that it is an endonuclease which produces 3'-OH and 5'-phosphoryl termini at the cleavage site (2). Its ability to generate 3'-OH groups means that RNA which has been cleaved by ribonuclease H has primer sites for the DNA polymerase. This suggests that one of the functions of ribonuclease H might be to produce such primer sites.

The exact function of the ribonuclease H and the way that it participates in the DNA polymerase reaction are topics for future research. At present, we only know that it has not been separable from the polymerase, although future work might show that it is separable. The observation of two polypeptide chains in the purified AMV DNA polymerase (4) suggests that the nuclease and polymerase might be on different polynucleotide chains.

References

1. Baltimore, D., and Smoler, D. 1971. Primer requirement and template specificity of the RNA tumor virus DNA polymerase. *Proc. Nat. Acad. Sci. U. S. A.* 68: 1507.

2. Baltimore, D., and Smoler, D. A. Association of an endoribonuclease with the avian myeloblastosis virus DNA polymerase. *J. Biol. Chem.* In press.

3. Hausen, P., and Stein, H. Ribonuclease H. 1970. An enzyme degrading the RNA moiety of DNA-RNA hybrids. *Eur. J. Biochem.* 14: 278.

4. Kacian, D. L., Watson, K. F., Burny, A., and Spiegelman, S. 1971. Purification of the DNA polymerase of avian myeloblastosis virus. *Biochim. Biophys. Acta* 246: 365.

5. Mölling, I., Bolognesi, D. P., Bauer, H., Büsen, W., Plassmann, H. W., and Hausen, P. Association of the viral reverse transcriptase with an enzyme degrading the RNA moiety of RNA-DNA hybrids. *Nature New Biol.* 234: 240.

6. Temin, H., and Baltimore, D. RNA-directed DNA synthesis and RNA tumor viruses. *Advan. Virus Res.* In press.

7. Verma, I. M., Meuth, N. L. and Baltimore, D. The covalent linkage between RNA primer and DNA product of the avian myeloblastosis virus DNA polymerase. *J. Virol.* In press.

8. Verma, I. M., Meuth, N. L., Bromfeld, E., Manly, K. F., and Baltimore, D. 1971. A covalently linked RNA-DNA molecule as the initial product of the RNA tumor virus DNA polymerase. *Nature New Biol.* 233: 131.

Discussion

Sarin. I wanted to find out whether you have noticed any ribonuclease activity in any other tumor virus.

Baltimore. I think Dr. Leis can answer that.

Leis. I've tested the Rauscher leukemia virus preparations which we have in our hands and the amount of RNase H degradation is less than the amount of degradation that you would see with just the poly(A). So there is very little, if any, in these preparations.

Sarin. So apparently the ribonuclease H activity may not be common to all the RNA tumor viruses?

Leis. Well, I should put in a statement here that the Rausher polymerase which we get after the phosphocellulose step has lost its ability to use RNA as a template. It works very well with DNA, so I don't know.

Characteristics of the Transcription of RNA by the DNA Polymerase of Rous Sarcoma Virus

J. M. Bishop, A. J. Faras,
A. C. Garapin, H. M. Goodman,
W. E. Levinson, J. Stavnezer,
J. M. Taylor, and H. E. Varmus

Department of Microbiology
University of California
San Francisco, California 94122

The principal templates for DNA synthesis by virions of Rous sarcoma virus are the high molecular weight "subunits" of the viral 70 S RNA. Several low molecular weight RNAs associated with virions are also transcribed at a low frequency. The enzymatic product contains nucleotide sequences representing at least the bulk of the viral genome, but most of these sequences are present in very small proportions. By contrast, the single stranded DNA synthesized in the presence of actinomycin D contains a complete and relatively uniform representation of viral genome sequences. Three naturally occurring messenger RNAs, the genome RNA of poliovirus and the messenger RNAs for ovalbumin and mouse myeloma immunoglobulin κ chain, have been transcribed into DNA. In each instances, only a portion of the entire template is represented in the nucleotide sequences of double stranded enzymatic product. DNA synthesis with RSV 70 S RNA as template initiates by the covalent attachment of a dA residue to the 3′-adenosine of a 4 S RNA associated with the template. Initiation continues throughout the course of a 90-min reaction, and there is a perceptible lag between initiation and chain propagation.

The DNA polymerases associated with RNA tumor viruses are capable of transcribing DNA from a variety of RNA templates (27). We are studying the characteristics and mechanisms of this transcription, using both detergent-activated virions (holoenzyme) of the Schmidt-Ruppin strain of Rous sarcoma virus (RSV) and enzyme purified from RSV (9). This communication summarizes our present knowledge of the templates used by holoenzyme, the extent to which these templates are transcribed, and the mechanism by which DNA synthesis is initiated when the 70 S RNA of RSV is used as template. In addition, we report evidence that the purified enzyme transcribes 70 S RNA in the same manner and to approximately the same extent as does holoenzyme. Finally, we confirm previous findings (15, 20, 33) that the RNA-directed DNA polymerases of oncornaviruses can be used to transcribe part or all of a variety of naturally occurring RNAs if a suitable site can be provided for the initiation of DNA synthesis, and we extend the successfully transcribed RNAs to include two more messenger RNAs of eukaryotic origin.

The Templates for Holoenzyme

The principal template for DNA synthesis by detergent-activated virions of RSV has been identified as the 70 S RNA of the virus (10, 22). However, 70 S RNA is a noncovalent complex of high and low molecular weight RNAs, the latter of which (4 and 5 S RNAs) share nucleotide sequences with similar RNAs found in the host cell (7; Faras et al., manuscript in preparation). In addition, purified preparations of virions contain reproducible amounts of several low molecular weight RNAs (4, 5, and 7 S) free of the 70 S complex (3; Faras et al., manuscript in preparation). We have determined the extent to which each of these various RNA populations are transcribed by holoenzyme. The principal templates for DNA synthesis are the high molecular weight subunits of 70 S RNA, as judged by their ability to hybridize with virtually all of the single stranded enzymatic product (Fig. 1b) and approximately 50% of the double stranded product (Fig. 1, c and d). The DNAs in question are therefore homologous to the viral genome and constitute valid reagents for the detection of virus-specific DNA (12, 30) and RNA (13, 18) in normal and infected cells.

Hybridization of RNA with DNA in vast excess permits detection of very minor portions of the DNA population which share nucleotide sequences with the test RNA. Using this procedure, we find that all three forms of low molecular weight viral RNA (4, 5, and 7 S) are represented in the product of holoenzyme (Table 1). However, we have yet to achieve DNA : RNA ratios capable of driving these hybridizations to completion, and are therefore uncertain as to whether any of the low molecular weight RNAs are completely transcribed.

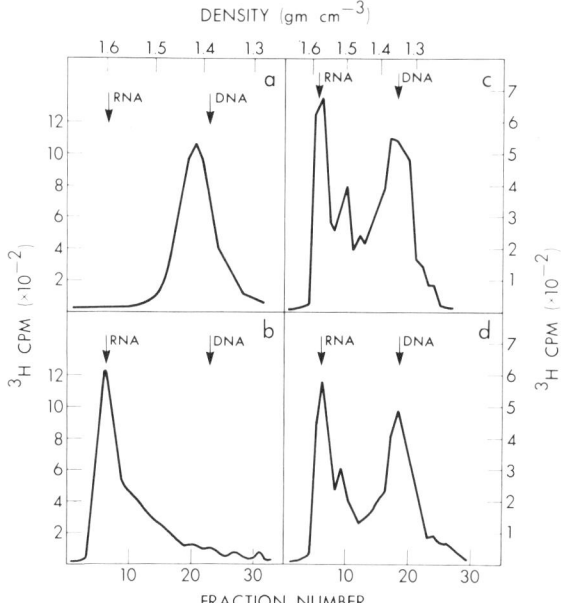

Fig. 1. Hybridization of enzymatic products to high molecular weight subunits of viral RNA. ³H-Labeled DNA was synthesized with detergent-activated RSV and fractionated into single and double stranded forms (8, 29). The double stranded material was in turn divided into rapidly and slowly reassociating fractions as described elsewhere (29). Purified 70 S RNA of RSV was denatured with $(CH_3)_2SO$ (26) and the high molecular weight subunits isolated by rate-zonal centrifugation (26). The purified DNAs were treated with NaOH, hybridized with large excesses of the subunit RNAs, and centrifuged to equilibrium in Cs_2SO_4. Details of these experiments will be published elsewhere (11). Each centrifugation included ³²P-labeled density references for RNA and DNA. a, Control: single stranded enzymatic product mixed with RNA and centrifuged without prior hybridization; b, hybridized single stranded enzymatic product; c, hybridized double stranded enzymatic product, rapidly reassociating fraction; d, hybridized double stranded enzymatic product, slowly reassociating fraction.

Table 1. Transcription of low molecular weight viral RNAs

	Fraction of RNA resistant to hydrolysis by RNase	
RNA	Control (%)	Hybridized (%)
4 S (70 S-associated)	4	80
5 S (70 S-associated)	3	82
7 S	9	65
70 S	3	100
HeLa ribosomal (18 and 28 S)	3	4

Single stranded DNA synthesized by detergent-activated RSV in the presence of actinomycin D (100 µg/ml) was hybridized with each of the indicated RNAs, all labeled to specific activities of 250,000 to 500,000 cpm/µg with ³²P. Hybridization was carried out in 0.4 M $NaPO_4$ at 68°C for 68 hr, followed by RNase treatment (50 µg/ml, 37°C, 30 min) in 0.3 M NaCl and acid precipitation. Controls were carried through the entire procedure in the absence of DNA. The ratio of DNA : RNA was 100 for the low molecular weight RNAs, 20 for 70 S and ribosomal RNAs.

The Extent of Transcription from 70 S RNA by Holoenzyme

The published data of Duesberg and Canaani (6) indicate that detergent-activated virions of RSV transcribe at least the bulk (65 to 85%) of the viral genome into DNA. These results were obtained by hybridizing viral RNA with vast excesses of DNA, and do not discriminate between single and double stranded product. Similar data have also been reported for murine leukemia and sarcoma viruses (23). We have repeated these experiments, but using purified fractions of single and double stranded enzymatic product. Large excesses of DNA are required to achieve an appreciable amount of hybridization (Fig. 2), and neither single nor double stranded enzymatic product can hybridize with all of the viral

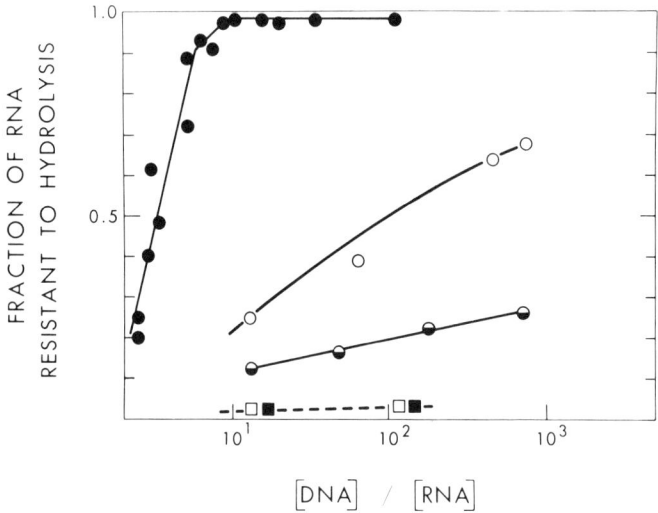

Fig. 2. Hybridization of RNA with increasing amounts of enzymatic product. DNA was synthesized with detergent-activated RSV in either the absence or presence of actinomycin D (100 μg/ml) and fractionated into single and double stranded forms (8, 29). Double stranded DNA represented approximately 85% of the enzymatic product prepared in the absence of actinomycin D, 20% of that synthesized in the presence of the antibiotic. A constant quantity (approximately 10 ng) of ^{32}P-labeled RNA was hybridized with different amounts of the purified, denatured (NaOH) DNAs and then treated with RNase. Details are given with Table 1. The amount of acid-insoluble radioactivity (^{3}H-DNA and ^{32}P-RNA) remained constant throughout the course of hybridization. ○, 70 S RNA hybridized with double stranded enzymatic product from a standard reaction; ◐, 70 S RNA hybridized with single stranded DNA from a standard reaction; ●, 70 S RNA hybridized with single-stranded DNA synthesized in the presence of actinomycin D; □, 70 S RNA hybridized with denatured DNA of phage λ; ■, ribosomal RNA of HeLa cells hybridized with single stranded DNA synthesized in the presence of actinomycin D.

RNA at the highest DNA : RNA ratio attained. The amounts of hybridization obtained with single and double stranded DNA are not additive. There is no detectable hybridization between either enzymatic product and unrelated RNA (*e.g.* HeLa ribosomal RNA) or 70 S RSV RNA and heterologous DNA (phage λ). We conclude that the DNA (and in particular, the double stranded fraction thereof) synthesized by holoenzyme contains nucleotide sequences representing at least the bulk of the viral genome, but most of those sequences are present in very small proportions. By contrast, the DNA synthesized by virions of RSV in the presence of actinomycin D will completely hybridize 70 S viral RNA at relatively low (5 to 10) DNA : RNA ratios (Fig. 2), and must therefore contain a far more complete and uniform representation of viral genome sequences than does the product of a standard enzymatic reaction. The mechanism by which this occurs is presently unknown, but the DNA in question is a valuable reagent in two types of analyses: (*a*) determination of the precise portion of the 70 S viral genome represented in various fractions of cellular RNA and DNA, and (*b*) measurement of the extent of homology between the nucleotide sequences of various oncornavirus genomes.

The preceding data provide no information regarding the frequency at which various portions of the viral genome are represented in enzymatic product. We therefore determined the extent of transcription into double stranded DNA by measuring the reassociation kinetics of this DNA in the manner first described by Britten and Kohne (4). The results indicate that limited portions of the viral genome are preferentially transcribed into double stranded DNA (Fig. 3 and Table 2). Thus, the double stranded DNA synthesized by virions of RSV consists principally of two populations (29). One comprises approximately 85% of the double stranded product, and contains nucleotide sequences representing perhaps 5% of the total viral genome (referred to as "rapidly reassociating" DNA). The other comprises 10 to 15% of the double stranded product and contains neucleotide sequences representing approximately 30% of the viral genome (referred to as "slowly reassociating" DNA). A third, minor fraction (approximately 5%) of the double stranded product is too complex to be reassociated at available concentrations ("nonreassociating" DNA). This DNA will not hybridize with 70 S viral RNA, but can interact extensively with DNA from normal avian cells (unpublished observations). We conclude that the template for synthesis of nonreassociating DNA is the small amount of cellular DNA found in purified preparations of RSV (19).

Preferential transcription has also been observed with murine leukemia virus (MLV) (12) and with mouse mammary tumor virus (Table 2 and reference 28). We have observed one interesting exception: RAV-0, the avian leukosis virus produced spontaneously by certain avian

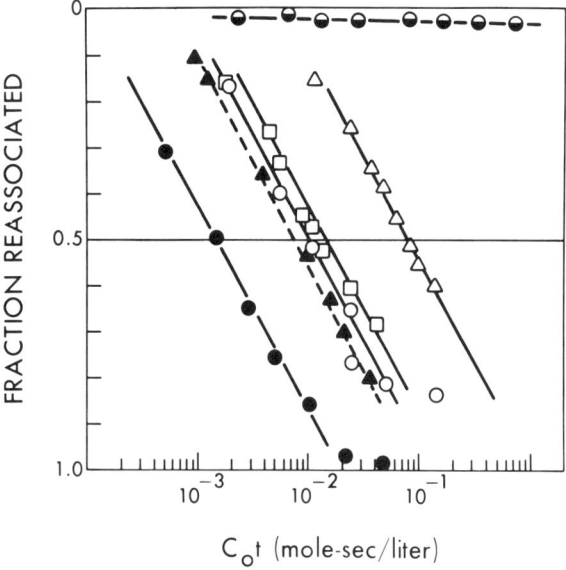

Fig. 3. Reassociation kinetics of double stranded DNAs. DNA was synthesized with detergent-disrupted virions of RSV and RAV-O, and fractionated into single and double stranded forms (8, 29). The double stranded DNA from the RSV reaction was then divided into rapidly and slowly reassociating fractions (rrDNA and srDNA, respectively), and a small residue of nonreassociating DNA (nrDNA), as described elsewhere (29). The replicative form DNA of phage fd and the DNA of phage λ were provided by H. Boyer. Rates of reassociation for each of the denatured DNAs were measured as described (29, 30) and are expressed here as a function of the convention C_0t (concentration of DNA times time) as suggested by Britten and Kohne (4). ●, rrDNA of RSV; □, srDNA of RSV; ○, DNA of RAV-O; ▲, fdRF; △, λ DNA; ⊖, nrDNA of RSV.

Table 2. Complexity of double stranded DNAs synthesized by holoenzyme

DNA	Half-C_0t	Complexity (nucleotide pairs)	Proportion of viral genome (%)	Proportion of double stranded product (%)
RSV(rr)	$1-2.5 \times 10^{-3}$	1500	5	85
RSV(sr)	$1-2 \times 10^{-2}$	9000	30	10
RSV(nr)	?	$>1 \times 10^6$		5
RAV-0	1×10^{-2}	9000	30	85-100
MMTV	$1-2 \times 10^{-3}$	1500	5	85-100

DNA was synthesized with detergent-activated virions of RSV, RAV-0, and mouse mammary tumor virus (MMTV), the double stranded products isolated, and rates of reassociation determined as illustrated in Fig. 3. The point of half-reassociation (Half-C_0t) is taken as a standard measure of the complexity of unique nucleotide sequences present in the DNA (4, 12, 29). Computations of the portion of the viral genome represented in these sequences are based on an estimated 30,000 nucleotides in 70 S viral RNA. Abbreviations denote rapidly reassociating (rr), slowly reassociating (sr), and nonreassociating (nr) fractions of RSV double stranded DNA (29). The RAV-0 was kindly supplied by P. Vogt and R. Friis, the MMTV by N. Sarkar and R. Nowinski.

fibroblasts (34). The double stranded DNA synthesized by the holoenzyme of RAV-O is composed almost entirely of nucleotide sequences representing approximately 30% of the viral genome (Fig. 3; Table 2). We have purified the DNA polymerase of RAV-O, have transcribed various RNAs (e.g. RAV-O, RSV, MLV) into double stranded DNA, and are measuring reassociation kinetics for these various products. When complete, the results should indicate whether the enzyme, the template, or some other element in the virion is responsible for preferential transcription.

Transcription of RNA by Purified DNA Polymerase of RSV

DNA polymerase purified from detergent-disrupted RSV readily transcribes 70 S RNA of RSV and other oncornaviruses into single and double stranded DNA (26). The nucleotide sequences in the double helical product represent a very limited portion of the template (Table 3), and do not include sequences other than those transcribed by holoenzyme (Table 4). Other naturally occurring RNAs are poor templates for the polymerase (Table 5), but those containing regions of polyadenylic acid can be activated by the addition of $(dT)_{12-18}$ (Table 5). This oligomer

Table 3. Extent of transcription of RNA templates into double stranded DNA

RNA template (2 µg/ml)	$(dT)_{12-18}$ (0.1 µg/ml)	Complexity of dsDNA product (nucleotide pairs)	Predicted complexity
RSV	–	820	30,000
RSV	+	130	30,000
RSV (denatured)	+	130	30,000
Poliovirus	–	1480	7500
Poliovirus	+	490	7500
Ovalbumin (avian)	+	300	1800
Myeloma κ chain (mouse)	+	800	1200

DNA polymerase was purified from detergent-disrupted virions of RSV (9) and used to transcribe the indicated RNAs into DNA. Where indicated the exogenous primer $(dT)_{12-18}$ was used to facilitate initiation of DNA synthesis (see text and Table 5). Poliovirus RNA was extracted from purified virus and further purified by rate-zonal centrifugation. The mRNA for avian ovalbumin was purified by D. Sullivan and R. Palacios in the laboratory of R. Schimke. The mRNA for mouse myeloma κ chain was prepared by J. Stavnezer in the laboratory of R. C. Huang (24). Double stranded enzymatic products were purified, the rates of their reassociation determined, and their sequence complexity computed, all as described previously (see Table 2 and Fig. 3). The complexities of ovalbumin and mouse κ chain are predicted on the basis of their total chain length (18 S and about 16 S, respectively) rather than the estimated complexity of their genetic message (1200 and 650 nucleotides, respectively).

Table 4. Homologous sequences among double stranded DNA products

Double stranded DNAs in reassociation mixture	% reassociation of ^{3}H-DNA	^{32}P-DNA
Experiment 1		
a. ^{3}H-DNA (synthesized by purified polymerase using 70 S RSV RNA template)	18(15)	
b. As above + ^{32}P-DNA (synthesized by endogenous reaction of holoenzyme)	80(82)	77(75)
Experiment 2		
a. ^{3}H-DNA (synthesized by purified polymerase using denatured RSV RNA plus (dT)$_{12-18}$)	13	
b. As above + ^{32}P-DNA (synthesized by endogenous reaction of holoenzyme)	76	83

Labeled DNA was synthesized in standard reactions of 18 hr duration, with conditions of enzyme, template, and primer as described in the table. Double stranded DNA was isolated using hydroxyapatite and subsequently passed through a column of Sephadex G-50. The ^{3}H-DNA in the presence or absence of a 400-fold excess of ^{32}P-DNA was denatured and then allowed to reassociate in 0.6 M NaCl to a ^{32}P-DNA C_0t of 0.16 and 0.012 mole-sec/liter for experiments 1 and 2, respectively. The extent of reassociation of the double stranded DNA was then assessed by use of a single strand-specific nuclease from *Aspergillus oryzae* (18). For experiment 1, reassociation was also measured at the increased ^{32}P-DNA C_0t of 0.4 mole-sec/liter and the results are shown in parentheses. The reassociation of ^{3}H-DNA in the presence of ^{32}P-DNA had reached a plateau at the lower C_0t value. The ability of the ^{3}H-DNAs to interact with the ^{32}P-DNAs, as indicated by the similar extent of reassociation attained by both forms when mixed together, is taken to indicate extensive sequence homology between the two DNAs (see reference 26).

Table 5. Stimulation of DNA synthesis by (dT)$_{12-18}$

RNA template	Relative incorporation	
	$-$ (dT)$_{12-18}$	$+$ (dT)$_{12-18}$
RSV	1.00	4.2
RSV (denatured)	0.04	2.8
Poliovirus	0.09	10.1
Ovalbumin mRNA	0.05	8.0
Myeloma κ chain	0.06	2.1
R-17	0.02	0.03
Ribosomal RNA	0.04	0.03

Purified DNA polymerase of RSV (9, 26) was used to transcribe the indicated RNAs in either the absence or presence of (dT)$_{12-18}$. Templates were used at 2 µg/ml, (dT)$_{12-18}$ at 0.1 µg/ml (25). R-17 RNA was extracted from purified virus, ribosomal RNA (18 and 28 S) from HeLa cells. The source of additional materials was described for Table 3. Results are expressed as the amount of DNA synthesis relative to that obtained with native 70 S RNA (RSV) in the absence of (dT)$_{12-18}$. RSV RNA was denatured with heat (80°C, 0.01 M EDTA-0.02 M Tris-HCl, pH 7.4, 2 min).

apparently binds to the polyadenylate region of the template and provides sites (*i.e.* 3'-hydroxyl groups) for the initiation of DNA synthesis (21). The complexities of nucleotide sequences contained in double stranded DNA synthesized from several RNA templates are summarized in Table 3. In every instance the principal double stranded product appears to represent less than the entire template.

The 70 S RNA of RSV contains polyadenylic acid (16), and $(dT)_{12-18}$ therefore stimulates transcription from both native and denatured viral RNA (Table 5). The double stranded DNA made under these circumstances is less complex than that synthesized in the absence of $(dT)_{12-18}$ (Table 3), but the nucleotide sequences used as template in both instances are either homologous or identical with those transcribed by holoenzyme (Table 4). The double stranded DNA transcribed from poliovirus RNA in the presence of $(dT)_{12-18}$ is also less complex than that transcribed in the absence of the oligomer. We cannot presently explain the reductions in extent of transcription which accompany the use of $(dT)_{12-18}$.

Initiation of DNA Synthesis

DNA synthesis by oncornavirus polymerase initiates on the 3'-hydroxyl terminus of a *primer* moleucle irrespective of whether the *template* is RNA or DNA (1, 14, 17). In the case of tumor virus 70 S RNA template, the primer molcule is apparently a low molecular weight polyribonucleotide to which nascent DNA is covalently bound (2, 5, 32). We have purified the primer molecule (2), identified its 3'-terminal nucleoside as adenosine, and examined the variety of deoxynucleotide sequences with which DNA synthesis initiates. The basic protocol, designed to select for nascent DNA chains with minimum loss, is illustrated in Fig. 4. The primer was released from nascent DNA with DNase and analyzed by electrophoresis in gels of polyacrylamide (2). RNAs obtained from two separate sets of DNA-RNA complexes have identical mobilities, migrating only slightly slower than 4 S RNA (Fig. 5). These observations conform to previous suggestions that DNA synthesis with 70 S RNA as template initiates on a low molecular weight RNA (5, 32). However, both the nucleotide composition (guanosine + cytosine = 64%) of purified primer and the preliminary results of partial nucleotide sequence determinations indicate that the primer RNAs are a special set of molecules rather than the complete population of 70 S-associated 4 S RNAs (guanosine + cytosine content = 58%).

Enzymatic products from separate reactions using each of the α-^{32}P-deoxynucleoside triphosphates were hydrolyzed with alkali. This degrades all RNA and leaves ^{32}P on the 3'-terminal ribonucleoside of the primer molecule. We find radioactivity in Ap when α-^{32}P-dATP is

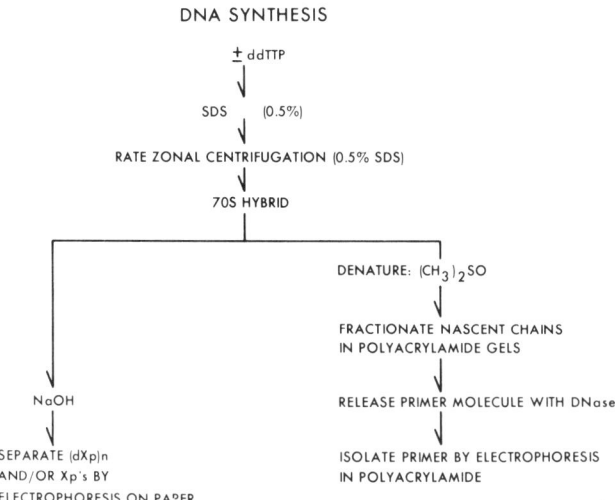

Fig. 4. Isolation and characterization of the covalent complex between nascent DNA and primer RNA. Enzymatic reactions were carried out with either detergent-disrupted (0.01% NP-40) RSV or purified RSV polymerase and 70 S RNA as template. Details for all the analytical procedures and their results are given elsewhere (25, 26). Where used dideoxythymidine triphosphate (ddTTP) replaced TTP at a concentration of 1×10^{-6} M.

used as the labeled precursor with both holoenzyme (Table 6) and purified enzyme transcribing 70 S RNA (Table 7). No ribonucleotides are labeled when any of the other α-^{32}P-deoxynucleoside triphosphates are used (Tables 6 and 7). Similar results have been obtained for avian myeloblastosis virus by Verma et al. (31). The addition of $(dT)_{12-18}$ as an exogenous primer does not change the extent of terminal ribonucleotide labeling if the data are properly normalized (Table 7). Thus, $(dT)_{12-18}$ provides more sites of initiation, but does not interfere with initiation on the intrinsic primer (or primers) of RSV RNA. No labeling of terminal ribonucleotides is found when denatured 70 S RNA is used as template (Table 7), although extensive DNA synthesis occurs with this template in the presence of $(dT)_{12-18}$ (Table 5). We conclude that denaturation removes the intrinsic primer sites and DNA synthesis is then initiated only on the $(dT)_{12-18}$.

Dideoxythymidine triphosphate (ddTTP) inhibits DNA synthesis by RNA-directed DNA polymerase (21). Incorporation of the nucleoside analogue into the propagating DNA chain synchronously arrests polymerization at the first thymidine residue (Fig. 6). We have used this phenomenon to label the RNA primer molecule with oligodeoxynucleotides, thereby confirming the size of the primer and obtaining a set of oligodeoxynucleotides which represent the sequences with which

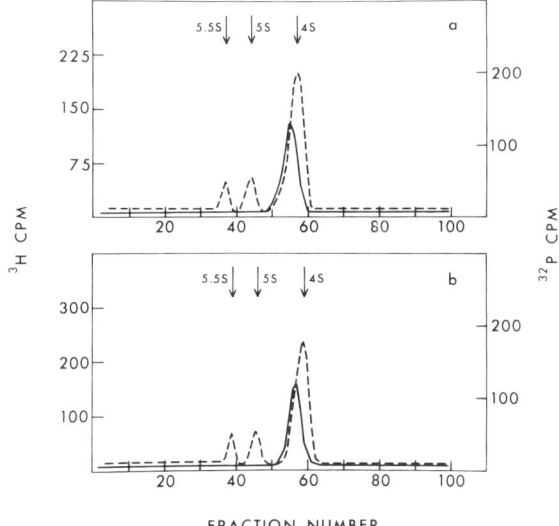

Fig. 5. Electrophoresis of purified primer RNA. Covalent complexes of nascent DNA-primer RNA were prepared as outlined in Fig. 4, using purified DNA polymerase of RSV and ^{32}P-labeled 70 S RSV RNA (100 μg, 250,000 cpm/μg) as template. The isolated complexes were treated with DNase rigorously freed of RNase. Details are given elsewhere (2). The remaining RNA was isolated by electrophoresis in 10% polyacrylamide, eluted from the gel, and reanalyzed in the presence of ^{3}H-labeled marker RNAs (4 and 5.5 S). The two panels illustrate the primers obtained from two sets of DNA-RNA complexes, one of which migrated at 5 to 6 S in 10% polyacrylamide, the other at 7 to 9 S. ———, ^{32}P cpm; - - -, ^{3}H cpm.

Table 6. Initiation of DNA synthesis by holoenzyme (30-min reaction)

Labeled precursor	% Transfer of α-^{32}P to			
	Cp	Ap	Gp	Up
dATP	<0.07	9.93	<0.09	<0.07
dCTP	<0.03	<0.02	<0.02	<0.01
dGTP	<0.08	<0.03	<0.03	<0.02
dTTP	<0.09	<0.04	<0.03	<0.02

Nascent DNA was prepared with detergent-disrupted virions as outlined in Fig. 4, hydrolyzed with alkali (0.3 N NaOH, 18 hr, 37°C), subjected to high voltage electrophoresis, and the separated mononucleotides were identified and measured, all as described previously (25, 26). The percentage of transfer of α-^{32}P to Ap is calculated on the basis of the total radioactivity recovered from the electrophoretogram, all of which label (with the exception of Ap) remains at the origin in the form of oligodeoxynucleotides and DNA.

Table 7. Initiation of DNA synthesis by purified polymerase

RNA template (2 μg/ml)	$(dT)_{12-18}$ (μg/ml)	% Transfer of α-^{32}P from dATP to			
		Cp	Ap	Gp	Up
RSV	0	<0.03	3.68	<0.02	<0.02
RSV	0.1	<0.5	3.69	<0.5	<0.5
RSV (denatured)	0.1	<0.6	<0.6	<0.6	<0.6

DNA was synthesized with purified polymerase of RSV and 70 S RSV RNA as template in a standard reaction of 1 hr. The product was treated with sodium dodecyl sulfate, passed through a column of Sephadex G-50, treated with NaOH, and analyzed by electrophoresis as described for Table 6. Transfer of α-^{32}P to ribonucleotide was observed only with α-^{32}P-dATP, and then only to Ap as illustrated. Results with the other α-^{32}P-deoxynucleoside triphosphates are therefore not shown. The results obtained in the presence of $(dT)_{12-18}$ have been normalized to correct for the stimulation of DNA synthesis by the exogenous primer (see Table 5).

DNA synthesis is initiated. DNA synthesized with holoenzyme in the presence of ddTTP migrates in the vicinity of 4 S RNA in gels of polyacrylamide irrespective of either the radioactive precursor or the duration of the reaction (Fig. 7). The accumulation of radioactivity in this DNA between 10 and 30 min (totals with α-^{32}P-dATP given in Fig. 7) suggests that initiation of DNA synthesis has continued throughout this time. Verma et al. have reached the same conclusion on entirely different grounds (31).

Alkaline hydrolysis of the enzymatic products prepared in the presence of ddTTP releases a different (but overlapping) set of oligodeoxynucleotides for each labeled deoxynucleoside triphosphate (Fig. 8). A minimum of six unique sequences are represented in these sets, and DNA synthesis may therefore initiate at a minimum of six unique sites

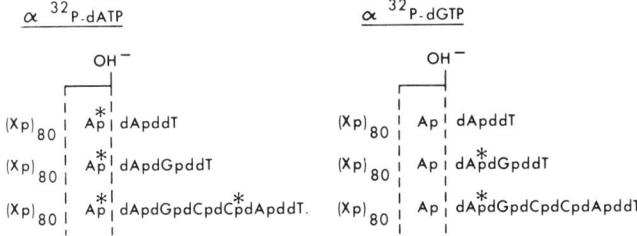

Fig. 6. Identification of initiation sequences with ddTTP. The diagram illustrates the consequences of synchronously arresting the polymerization of DNA at the 1st dT residue by using dideoxythymidine triphosphate (ddTTP) (generously supplied by D. Brutlag). Synthesis was carried out using dATP, dGTP, and dCTP (one of which was labeled as α-^{32}P) at 2×10^{-6} M and ddTTP at 1×10^{-6} M. The diagram is based on the assumption that more than one nucleotide sequence occurs at initiation sites, and also illustrates the pertinent consequences of alkaline hydrolysis. $(Xp)_{80}$ represents the RNA primer molecule, terminating with 3'-pA. ddT represents a residue of dideoxythymidine. Asterisks denote the position of radioactive phosphorus in the polynucleotides.

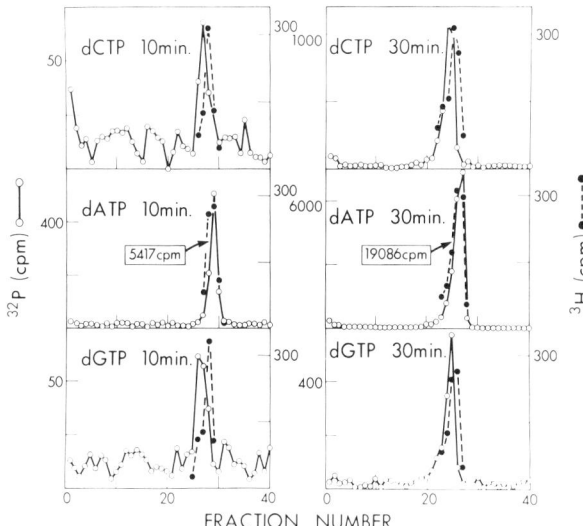

Fig. 7. Electrophoresis of covalent complexes between nascent DNA and primer RNA formed in the presence of ddTTP. Reactions were carried out with detergent-activated RSV in the presence of ddTTP as described for Fig. 6. The covalent complexes of nascent DNA and primer RNA were isolated and analyzed by electrophoresis in 10% polyacrylamide (see Fig. 4). The α-^{32}P-nucleoside triphosphate precursors and reaction times are indicated in the figure. ^{3}H-Labeled 4 S RNA was used as a marker. Total amounts of radioactivity recovered in nascent DNA labeled with α-^{32}P-dATP were computed and normalized for reaction volumes, and are given in the figure.

on 70 S RNA. However, these results are not definitive. At least one of the oligodeoxynucleotides contains a residue of pT (unpublished observation), indicating that complete arrest of polymerization by ddTTP has not been achieved. The pT is presumably derived from trace amounts of TTP present in the other nucleoside triphosphates.

The incorporation of pdA is at least 10 times more frequent than that of pdG and pdC in the presence of ddTTP (Fig. 7), and Ap is the principal radioactive product obtained by alkaline hydrolysis of the DNA-RNA complex labeled with α-^{32}P-dATP (Fig. 8). Only small quantities of oligomers containing dA in internal positions are seen. We conclude as before that all DNA chains initiate with dA, and that the bulk (approximately 80 to 90%) of these are either terminated in the second position by ddT or are not extended beyond the first nucleotide residue under present conditions. The oligomers containing labeled dG and dG will therefore be longer than the majority of those labeled with dA. The small but reproducible differences of R_F observed between the ddT-arrested DNA-RNA complexes labeled with pdA and those labeled with pdG and pdC conform to this view (Fig. 7). In every instance, the electrophoretic mobility of the RNA-DNA is determined principally

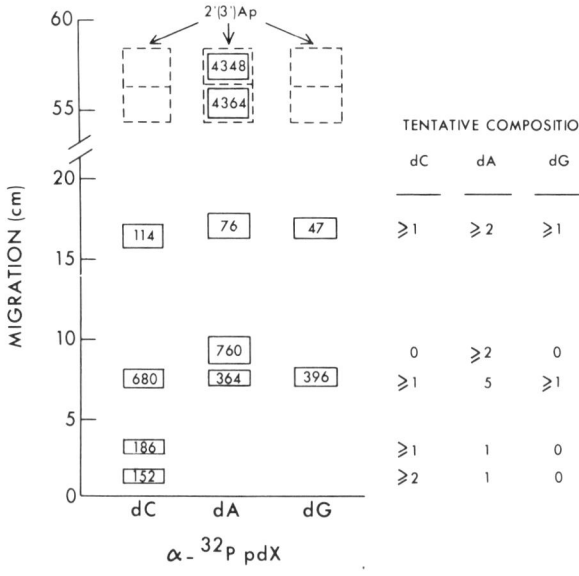

Fig. 8. The oligodeoxynucleotides synthesized in the presence of ddTTP. DNA was synthesized with detergent-disrupted RSV in the presence of ddTTP, using the α-^{32}P form of each of the other deoxynucleoside triphosphates in separate reactions. The covalent complexes of nascent DNA and primer RNA were isolated as desribed for Fig. 7, eluted from the polyacrylamide gels, hydrolyzed with NaOH (0.3 N, 37°C 18 hr), and subjected to high voltage electrophoresis on DEAE-paper in 7% (v/v) formic acid. ^{3}H-Labeled 2' (3')-Ap was included as an internal marker. The radioactive (^{32}P) products were located by autoradiography, cut out, and counted in a scintillation spectrometer. The positions of the markers, the amounts of ^{32}P recovered in the various monomer and oligomer residues, and tentative nucleotide compositions of the oligomers are given in the figure. The last are based on unpublished results of further analyses performed in this laboratory. The major, rapidly moving material obtained with α-^{32}P-dATP has been designated 2' (3')-Ap on the basis of both the migration of the internal marker (dashed boxes) and the predicted consequences of this procedure (Fig. 6).

by the RNA constituent (we estimate that none of the oligodeoxynucleotides visualized in Fig. 8 contains more than 12 residues). These data therefore confirm our previous conclusion that all DNA synthesis with 70 S RNA as template is initiated on the 3' terminus of 4 S RNA.

Identification of an Initiation Complex

We have found that initiation of DNA synthesis can also be visualized without the use of the analogue ddTTP. The product of a brief enzymatic reaction (10 min) using holoenzyme and limiting concentrations of precursor contains principally dA and migrates with 4 S RNA in polyacrylamide (Fig. 9). Analysis of such material following alkaline hydrolysis indicates

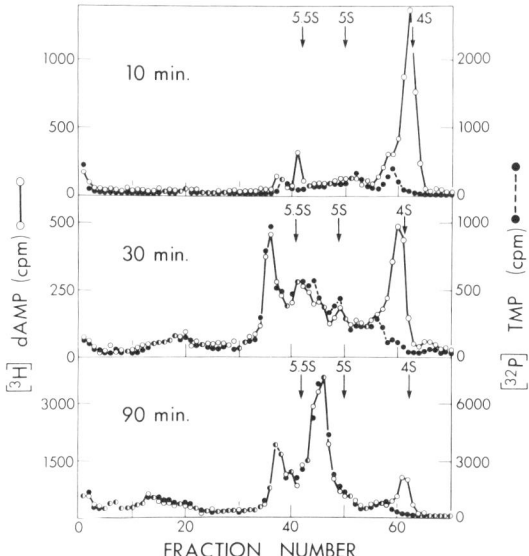

Fig. 9. Time course of DNA synthesis using two radioactive nucleoside triphosphates. DNA was synthesized with detergent-disrupted RSV using α-^{32}P-TTP and ^{3}H-dATP (and unlabeled dGTP, dCTP). At the indicated times, nascent DNA was prepared as described in Fig. 4 and analyzed in gels of 10% polyacrylamide following denaturation with $(CH_3)_2SO$. The position of marker RNAs in a separate gel is given in each panel. Similar experiments performed with α-^{32}P-dGTP and α-^{32}P-dCTP gave identical results: the 4 S material is appreciably labeled only with pdA. In an identical enzymatic reaction, but using α-^{32}P-dATP, the DNA in the 4 S region at 10 min was eluted from the gel, hydrolyzed with NaOH, and subjected to electrophoresis as for Table 6. Over 90% of the ^{32}P was recovered in Ap.

that at least 90% of the pdA residues are 5'-terminal in the DNA and attached to the 3'-terminal adenosine of the primer molecule (see legend, Fig. 9). As the enzymatic reaction progresses, longer DNA chains accumulate (Fig. 9), but the DNA at 4 S never contains appreciable amounts of deoxynucleoside other than dA.

The persistence of 4 S material labeled with dA as late as 90 min again suggests that initiation of DNA synthesis must continue throughout the reaction. We have confirmed this conclusion by performing a pulse-chase experiment (Fig. 10). Labeled dA is removed from the 4 S region by a chase with unlabeled dATP, indicating that chain propagation occurs on essentially all of the molecules, and that the persistence of label in the 4 S population during a continuous label is attributable to ongoing initiations (Fig. 9). The removal of labeled pdA from the 4 S population to the longer chain DNAs has also been demonstrated by alkaline hydrolysis of various fractions isolated from the gels illustrated in Fig. 10. The amount of label found in Ap isolated from the 5 to 10 S regions

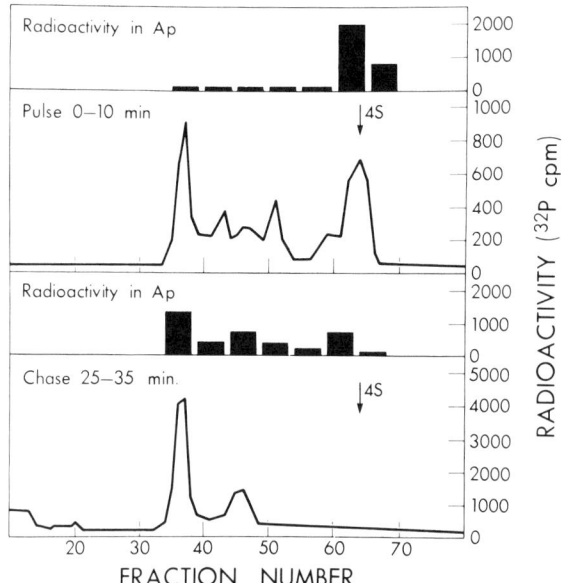

Fig. 10. Pulse-chase labeling with $\alpha\text{-}^{32}\text{P-dATP}$. Reactions, product isolation, and analysis were carried out as for Fig. 9. $\alpha\text{-}^{32}\text{P-dATP}$ was the sole radioactive precursor, and this was chased from 25 to 35 min after initiation of the reaction by the addition of a 550-fold excess of unlabeled dATP. Slices from the indicated regions of the gels were pooled, the nucleic acids eluted, hydrolyzed with NaOH, and analyzed by paper electrophoresis as for Table 6. The amount of ^{32}P recovered in Ap from each pool is plotted in the histograms above the gels. Equal reaction volumes were used for each gel. Similar results have been obtained with a pulse of radioactivity from 0 to 10 min, followed by a chase from 10 to 30 min.

of the gels increases many-fold following the chase (see histograms in Fig. 10). We conclude that DNA synthesis with 70 S RNA as template initiates by the covalent attachment of a dA residue to the 3'-adenosine of a 4 S RNA, that the formation and accumulation of this "initiation complex" continues throughout the course of a 90-min reaction, and that essentially all of the complexes eventually participate in chain elongation.

Acknowledgments

We thank L. Levintow for advice and support; D. Brutlag for a generous gift of ddTTP; Merck, Sharpe and Dohme, Inc., for actinomycin D; N. Quintrell, J. Jackson, L. Fanshier, and B. Evans for technical assistance. We also acknowledge the collaboration of D. Sullivan, R. Palacios, and R. Schimke in portions of the work reported here. This work was supported by United States Public Health Service Grants AI 08864, CA 12380, CA 12705, AI 06862, AI 00299,

American Cancer Society Grant VC-70, and Contract 71-2147 within the Special Virus-Cancer Program of the National Cancer Institute, National Institutes of Health, Public Health Service. J. M. T. and H. E. V. acknowledge the support by Senior Dernham Fellowships (D-201 and D-164, respectively) of the American Cancer Society, California Division.

References

1. Baltimore, D., and Smoler, D. 1971. Primer requirement and template specificity of the RNA tumor virus polymerase. *Proc. Nat. Acad. Sci. U. S. A.* 68: 1507.

2. Bishop, J. M., Faras, A. J., Garapin, A. C., Hansen, C., Jackson, N., Levinson, W., Taylor, J. M., and Varmus, H. E. 1972. RNA-directed DNA polymerase and the replication of Rous sarcoma virus. *The Molecular Basis of Neoplasia*, M. D. Anderson Symposium. In press.

3. Bishop, J. M., Levinson, W., Sullivan, D., Fanshier, L., Quintrell, N., and Jackson, J. 1970. The low molecular weight RNAs of Rous sarcoma virus. II. The 7 S RNA. *Virology* 42: 927.

4. Britten, R. J., and Kohne, D. E. 1968. Repeated sequences in DNA. *Science* 161: 520.

5. Canaani, E., and Duesberg, P. 1972. Role of subunits of 60–70 S avian tumor virus RNA in its template activity for the viral DNA polymerase. *J. Virol.* 10: 23.

6. Duesberg, P. H., and Canaani, E. 1970. Complementarity between Rous sarcoma virus RNA and the *in vitro* synthesized DNA of the virus-associated DNA polymerase. *Virology* 42: 783.

7. Erikson, E., and Erikson, R. L. 1972. Association of 4S ribonucleic acid with oncornavirus ribonucleic acid. *J. Virol.* 8: 254.

8. Fanshier, L., Garapin, A. C., McDonnell, J. P., Levinson, W. E., and Bishop, J. M. 1971. DNA polymerase associated with avian tumor viruses. Secondary structure of the DNA product. *J. Virol.* 7: 76.

9. Faras, A. J., Taylor, J. M., McDonnell, J. P., Levinson, W. E., and Bishop, J. M. 1972. Purification and characterization of the deoxyribonucleic acid polymerase associated with Rous sarcoma virus. *Biochemistry* 11: 2334.

10. Garapin, A. C., McDonnell, J. P., Levinson, W. E., Quintrell, N., Fanshier, L., and Bishop, J. M. 1970. Deoxyribonucleic acid polymerase associated with Rous sarcoma virus and avian myeloblastosis virus. Properties of the enzyme and its product. *J. Virol.* 6: 589.

11. Garapin, A. C., Varmus, H. E., Faras, A. J., Levinson, W. E., and Bishop, J. M. 1972. RNA-directed DNA synthesis by virions of Rous sarcoma virus. Further characterization of the templates and the extent of their transcription. *J. Virol.* Submitted.

12. Gelb, L., Aaronson, S., and Martin, M. 1971. Heterogeneity of murine leukemia virus *in vitro* DNA. Detection of viral DNA in mammalian cells. *Science* 172: 1353.

13. Green, M., Rokutanda, H., and Rokutanda, M. 1971. Virus-specific RNA in cells transformed by RNA tumor viruses. *Nature New Biol.* 230: 229.
14. Hurwitz, J., and Leis, J. P. 1972. RNA-dependent DNA polymerase activity of RNA tumor viruses. I. Directing influence of DNA in the reaction. *J. Virol.* 9: 116.
15. Kacian, D. L., Spiegelman, S., Bank, A., Terada, M., Metafora, S., Dow, L., and Marks, P. A. 1972. *In vitro* synthesis of DNA components of human genes for globins. *Nature New Biol.* 235: 167.
16. Lai, M. M. C., and Duesberg, P. H. 1972. Adenylic acid-rich sequence in RNAs of Rous sarcoma virus and Rauscher mouse leukemia virus. *Nature* 235: 383.
17. Leis, J. P., and Hurwitz, J. 1972. RNA-dependent DNA polymerase activity of RNA tumor viruses. II. Directing influence of RNA in the reaction. *J. Virol.* 9: 130.
18. Leong, J., Garapin, A. C., Jackson, N., Fanshier, L., Levinson, W. E., and Bishop, J. M. 1972. Virus-specific ribonucleic acid (RNA) in cells producing Rous sarcoma virus. Detection and characterization. *J. Virol.* 9: 891.
19. Levinson, W., Varmus, H. E., Garapin, A. C. and Bishop, J. M. 1972. DNA of Rous sarcoma virus. Its nature and significance. *Science* 175: 76.
20. Ross, J., Aviv, H., Scolnick, E., and Leder, P. 1972. *In vitro* synthesis of DNA complementary to purified rabbit globin mRNA. *Proc. Nat. Acad. Sci. U. S. A.* 69: 264.
21. Smoler, D., Molineux, I., and Baltimore, D. 1971. Direction of polymerization by the avian myeloblastosis virus deoxyribonucleic acid polymerase. *J. Biol. Chem.* 246: 7697.
22. Spiegelman, S., Burny, A., Das, M. B., Keydar, J., Schlom, J., Travnicek, M., and Watson, K. 1970. Characterization of the products of RNA-directed DNA polymerases in oncogenic RNA viruses. *Nature (London)* 227: 563.
23. Stephenson, J. R., and Aaronson, S. 1971. Murine sarcoma and leukemia viruses. Genetic differences determined by RNA-DNA hybridization. *Virology* 46: 480.
24. Stavnezer, J., and Huang, R. C. 1971. Synthesis of a mouse immunoglobulin light chain in a rabbit reticulocyte cell-free system. *Nature New Biol.* 230: 172.
25. Taylor, J. M., Faras, A. J., Varmus, H. E., Goodman, H. M., Goodman, W. E. Levinson, W. E., and Bishop, J. M. 1973. Transcription of RNA by the RNA-directed DNA polymerase of Rous sarcoma virus and DNA polymerase I of *E. coli. Biochemistry* 12: 466.
26. Taylor, J. M., Faras, A. J., Varmus, H. E., Levinson, W. E., and Bishop, J. M. 1972. Ribonucleic acid-directed deoxyribonucleic acid synthesis by the purified deoxyribonucleic acid polymerase of Rous sarcoma virus. Characterization of the enzymatic product. *Biochemistry* 11: 2343.
27. Temin, H. M., and Baltimore, D. 1972. RNA-directed DNA synthesis and RNA tumor viruses. *Advan. Virus Res.* In press.
28. Varmus, H. E., Bishop, J. M., Nowinski, R., and Sarkar, N. 1972. Detection of mammary tumor virus-specific nucleotide sequences in the DNA of high and low incidence mouse strains. *Nature New Biol.* 238: 189.

29. Varmus, H. E., Levinson, W. E. and Bishop, J. M. 1971. Extent of transcription by the RNA-dependent DNA polymerase of Rous sarcoma virus. *Nature New Biol.* 233: 19.

30. Varmus, H. E., Weiss, R. A., Friis, R. R., Levinson, W., and Bishop, J. M. 1972. Detection of avian tumor virus-specific nucleotide sequences in avian cell DNA. *Proc. Nat. Acad. Sci. U. S. A.* 69: 20.

31. Verma, I. M., Meuth, N. L., and Baltimore, D. 1972. The covalent linkage between RNA primer and DNA product of the avian myeloblastosis virus DNA polymerase. *J. Virol.* 10: 662.

32. Verma, I. M., Meuth, N. L., Bromfeld, E., Manly, K. F., and Baltimore, D. 1971. Covalently linked RNA-DNA molecules as initial product of RNA tumor virus DNA polymerase. *Nature New Biol.* 233: 131.

33. Verma, I. M. Temple, G. F., Fan, H., and Baltimore, D. 1972. In vitro synthesis of DNA complementary to rabbit reticulocyte 10S RNA. *Nature New Biol.* 235: 163.

34. Vogt, P. K., and Friis, R. R. 1971. An avian leukosis virus related to RSV(o). Properties and evidence for helper activity. *Virology* 43: 223.

DNA Polymerase in Association with Intracisternal A-type Particles

S. H. Wilson, E. L. Kuff,
E. W. Bohn, K. K. Lueders, and A. Matsukage

Laboratory of Biochemistry
National Cancer Institute
National Institutes of Health
Bethesda, Maryland 20014

Intracisternal A-type particles (A-particles) are abundant intracellular virus-like structures occurring in a variety of mouse tumors (2). They are distinguished from C-type and B-type RNA tumor viruses both morphologically and because they occur in association with the endoplasmic reticulum rather than the plasma membrane (3). There are no reports of their occurrence extracellularly. Despite their obvious virus-like appearance the isolated particles that have an average density of 1.22 g/cc have not been shown to possess characteristics of viruses such as infectivity or other evidence of biological activity (5). However, the main structural protein of A-particles isolated from several different mouse tumors contained a common antigen (6).

We have recently shown (9) that preparations of A-particles from murine myeloma MOPC-104E contained DNA polymerase activities. The preparations exhibited very low but detectable levels of activity in reactions containing no added template-primer. Among the added template-primers and conditions tested, a high level of DNA polymerase activity was found only for poly(rA)-directed poly(dT) synthesis. A DNA primer such as $(dT)_{14}$ was required for activity. Rabbit globin messenger RNA and DNA template-primers such as activated DNA, poly[d(A-T)], poly(dA) · $(dT)_{14}$, poly(dA), and poly(dG) · (dC) all were ineffective in promoting DNA synthesis. The poly(rA)-directed poly(dT) synthesis was optimal between 100 to 250 mM potassium chloride and 10 to 15 mM magnesium acetate, was sensitive to sulfhydryl inhibitors and was not blocked by antiserum against murine leukemia virus DNA polymerase.

After observing DNA polymerase activity in A-particle preparations we first sought to determine its subcellular distribution in order to gain insight into whether the enzyme responsible was specifically associated with the A-particle or was perhaps a cellular constituent contaminating the particle preparations. In order to determine the subcellular distribution of the poly(rA)-dependent activity, an assay was developed that permitted its specific, quantitative measurement in crude tissue fractions. The specificity of the assay, which was based on the unusual properties of the activity in A-particle preparations, was investigated in the experiment shown in Table 1. Unfractionated homogenates and Triton X-100-resistant particulate fractions from cells that do not contain A-particles, JLS V9 cells, 3T3 cells, HTC cells, adult mouse liver and spleen, possessed no detectable poly(rA) · $(dT)_{14}$-dependent activity under the conditions used to measure activity of A-particle preparations (referred to as A conditions). No substantial inhibition was detected when a previously characterized sample of A-particles was mixed with these preparations and then assayed again. In contrast, similarly prepared homogenates from A-particle-containing cells such as myeloma MOPC-104E, neuroblastoma N4, and a neuroblastoma-L cell hybrid possessed activity. Myeloma MOPC 321, which contained very low numbers of A-particles, possessed a low almost undetectable level of activity. Three cellular DNA polymerases isolated from myeloma MOPC-104E were not active under the A reaction conditions. These included preparations of two partially purified DNA-dependent DNA polymerases corresponding to the high and low molecular weight DNA polymerases in rat liver (1), and a preparation of highly purified poly(rA)-directed DNA polymerase (7) with activity similar to the enzyme reported in HeLa cells by Fridlender et al. (4).

The subcellular distribution of DNA polymerase activity measured under the A reaction conditions described in Table 1 was determined by assay of the fractions produced during the isolation of A-particles from myeloma MOPC-104E. The procedure (5) was based upon purification of material that remained particulate after treatment of a cytoplasmic membrane fraction with 1.7% Triton X-100. The results are shown in Table 2. The purified particles contained 0.5% of the original homogenate protein and 30% of the DNA polymerase, resulting in a 62-fold purification of the activity. Very little activity was detected in the two supernatant fractions derived from the postnuclear supernatant.

The purification of both DNA polymerase activity and A-particles during the isolation procedure suggested that a portion of the total cellular enzyme activity was associated with the particle. However, substantial loss of activity occurred when A-particles were purified from the Triton

Table 1. Specificity of DNA polymerase assay

Enzyme preparation	Amount of protein per reaction (μg)	Modification (pmoles [^{3}H]-TMP incorporated per reaction)		
		(−) poly(rA)·(dT)$_{14}$	(+) poly(rA)·(dT)$_{14}$	(+) poly(rA)·(dT)$_{14}$ and 1 μg A-particles
A-particles	1.0	0.5	15.7	
None	0	0.6	0.5	19.0
A. Unfractionated homogenates:				
1. Containing A-particles				
Myeloma MOPC-104E	1	0.5	1.2	24.2
Neuroblastoma N-4	2.3		1.7	
Neuroblastoma-L cell	1.7		2.5	
hybrid NL1	1.0	0.5	0.7	17.4
2. A-particle negative:				
JLSV9 cells	2.6	0.5	0.5	16.6
HTC cells	1.0	0.5	0.4	20.9
3T3 cell	1.2	0.5	0.5	21.1
Adult mouse spleen	0.3	0.5	0.4	21.1
Adult mouse liver	0.4	0.5	0.5	21.1
B. Triton X-100-resistant particulate fractions:				
JLSV9 cells	2.3	0.6	0.5	
Adult mouse spleen	3.3	0.6	0.5	
Adult mouse liver	2.6	0.6	0.5	
C. Myeloma MOPC-104E DNA polymerases:				
DEAE peak I (low mol wt polymerase)	0.4		0.5	
DEAE peak III (high mol wt polymerase)	4.5		0.5	
(rA)$_n$·(dT)$_{14}$-dependent polymerase	1.5		0.5	
D. Isolated organelles:				
Myeloma MOPC-104E nuclei	1.0		0.5	
Adult mouse liver mitochondria	3.1		0.6	

Intracisternal A-type particles were isolated according to the method of Kuff et al. (5). Homogenates of confluent cultures of the indicated cell lines were prepared according to the method of Wilson et al. (10). Myeloma MOPC-104E DNA-dependent DNA polymerases were purified through the phosphocellulose step according to the procedure of Baril et al. (1). Reactions contained in a final volume of 25 μl: 50 mM Tris · HCl buffer (pH 8.3 at 37°C), 18.5% (v/v) glycerol, 350 μg/ml of bovine serum albumin, 250 mM KCl, 12.5 mM magnesium acetate, 1 mM dithiothreitol, 1% Tween-80, 0.5 mM ATP, 10 μg/ml of nucleoside diphosphokinase (EC 2.7.4.6), 460 μg/ml of poly(rA) · (dT)$_{14}$(1 : 1) where indicated, and 0.5 mM[^{3}H]-thymidine 5′-triphosphate (280 cpm/pmole). Reactions containing all components except nucleotides and nucleoside diphosphokinase were incubated at 37°C for 20 min before incorporations were begun by addition of the remaining components. Incorporations were at 37°C for 75 min in silicon-treated soft glass tubes, 10 × 75 mm. Acid-precipitable radioactivity was determined by previously described (9).

Table 2. Activity of poly(rA) · $(dT)_{14}$-dependent DNA polymerase in subcellular fractions of myeloma MOPC-104E*

Step and tissue fraction	Amount of protein recovered (mg)	(%)	Amount of DNA polymerase recovered (units)†	(%)	(units/mg protein)
I. Whole homogenate	439	100	4210	100	9.6
II. Nuclear pellet	59.6	13.6	416	9.9	7.0
III. Membrane supernatant	164	37.4	196	4.6	1.2
Membrane pellet (10,000 × g pellet)	74	16.9	2740	65	37
IV. Supernatant after treatment with Triton X-100-citrate	64.4	14.7	2	0.1	0.05
Pellet after treatment with Triton X-100-citrate (78,000 × g pellet)	23.8	5.4	2780	66	117
V. 48% sucrose pellet	5.3	1.2			
VI. A-type particles after equilibrium gradient centrifugation	2.1	0.5	1250	30	595

*Data taken from three experiments in which 3.2 g of myeloma MOPC-104E were used. Steps I to IV were with passage 127-0 and Steps V and VI with passage 120-1.

†One unit of DNA polymerase activity = 1 pmole of [^{3}H]-dTMP incorporated per min at 37°C.

X-100-potassium citrate-resistant particulate fraction. In order to investigate the possibility that this loss was due to association of enzyme with material other than A-particles, sedimentation properties of the enzyme activity and A-particles were compared using each of the three particulate fractions produced during similar isolation procedures, Steps III, IV, and V. The results of this experiment are shown in Fig. 1. In Fig. 1A, a membrane pellet (Step III) was subjected to isopycnic centrifugation using a 9 to 45% linear sucrose gradient formed over a cushion of 68% sucrose. In this system A-particles banded at or near the interphase between the 45 and 68% sucrose solutions, fractions 19 to 23. It may be seen that essentially all of the enzyme activity recovered from the gradient was found in these fractions. Mixing of A-particles with other fractions from the gradient revealed no inhibitor of activity.

Isopycnic centrifugation in 33 to 68% linear sucrose gradients was performed in the experiments shown in Fig. 1, B and C, using, respectively, a Triton X-100-citrate-resistant 78,000 × g pellet (Step IV) and a pellet (105,000 × g pellet) recovered after sedimentation through 48% sucrose (Step V). A-particles were located in the gradients by microtiter complement fixation assay of the A-particle common antigen (6). In both cases all of the starting enzyme activity was recovered, and over

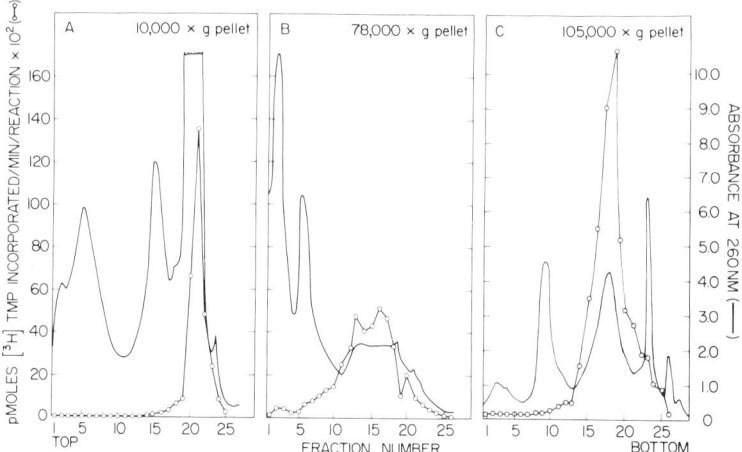

Fig. 1. Density equilibrium centrifugation of the pellet fractions from Steps III, IV, and V of the A-particle isolation procedure. In panel A, a portion of a 10,000 × g membrane pellet (from myeloma MOPC-104E passage 128-2) containing 6.87 mg of protein and 237 units of DNA polymerase activity was adjusted to 1.7% Triton X-100, sheared by expression through a 23-gauge needle, then adjusted to 135 mM potassium citrate and layered over a 3.8-ml 9 to 45% linear sucrose gradient that had been formed over a layer of 0.8-ml 68% sucrose. Centrifugation was at 300,000 × g av for 3 hr at 4°C. A total of 125 units of DNA polymerase was recovered in gradient fractions. In panel B a portion of a Triton X-100-citrate-resistant 78,000 × g pellet (from myeloma MOPC-104E passage 126-1) containing 3.05 mg of protein and 96 units of DNA polymerase activity was layered over a 5.0-ml 33 to 68% linear sucrose gradient and centrifuged for 3 hr at 300,000 × g at 4°C. A total of 179 units of DNA polymerase was recovered in gradient fractions. Sucrose solutions used in both gradients contained 50 mM Tris · HCl, pH 7.4, at 25°C and 100 mM potassium chloride. In panel C a portion of a 105,000 × g pellet (from myeloma MOPC-104E passage 127-2) containing 1.99 mg of protein and 295 units of DNA polymerase was layered over a 5.0-ml 33 to 68% linear sucrose gradient containing 50 mM potassium citrate. Centrifugation was for 4 hr at 234,000 × g. A total of 313 units of DNA polymerase was recovered in gradient fractions. DNA polymerase assays were performed with 1 to 5 µl of each 200-µl gradient fraction. One unit of DNA polymerase activity was equal to 1 pmole of [^{3}H]-TMP incorporated per min at 37°C. In the experiment shown in panel B, A-particle antigen was determined using 25µl of each gradient fraction according to a modification (6) of the microtiter assay of Sever (8).

90% of the activity cosedimented with the A-particle antigenic marker (data not shown).

In summary the results indicate that DNA polymerase activity of the type found in highly purified preparations of intracisternal A-particles from murine myeloma fractionated as a cytoplasmic constituent that was resistant to treatment with 1.7% Triton X-100. Only low levels of activity were found in nonparticulate fractions. Essentially all of the particulate activity cosedimented with A-particles upon isopycnic centrifugation. Cells that do not contain A-particles possessed no detectable DNA polymerase activity under the assay conditions used. Thus, the poly(rA) · (dT)$_{14}$-dependent DNA polymerase activity in A-particle preparations appears to be a specific particle constituent.

Acknowledgments

We are grateful to Mr. Larry Gottlieb for technical assistance and Doctors E. B. Thompson and J. Minna for providing some of the cell lines used.

References

1. Baril, E. F., Brown, O. E., Jenkins, M. D., and Laszlo, J. 1971. Deoxyribonucleic acid polymerase with rat liver ribosomes and smooth membranes. Purification and properties of the enzymes. *Biochemistry* 10: 1981.
2. Dalton, A. J., Potter, M., and Merwin, R. M. 1961. Some ultrastructural characteristics of a series of primary and transplanted plasma-cell tumors of the mouse. *J. Nat. Cancer Inst.* 26: 1221.
3. Dalton, A. J. 1962. Micromorphology of murine tumor viruses and of affected cells. *Fed. Proc.* 21: 936.
4. Fridlender, B., Fry, M., Bolden, A., and Weissbach, A. 1972. A new synthetic RNA-dependent DNA polymerase from human tissue culture cells. *Proc. Nat. Acad. Sci. U. S. A.* 69: 452.
5. Kuff, E. L., Wivel, N. A., and Lueders, K. K. 1968. The extraction of intracisternal A-particles from a mouse plasma-cell tumor. *Cancer Res.* 28: 2137.
6. Kuff, E. L., Lueders, K. K., Ozer, H. L., and Wivel, N. A. 1972. Some structural and antigenic properties of intracisternal A particles occurring in mouse tumors. *Proc. Nat. Acad. Sci. U. S. A.* 69: 218.
7. Matsukage, A., Bohn, E. W., and Wilson, S. H. 1973. *Fed. Proc.* 32: 451.
8. Sever, J. L. 1962. Application of a microtechnique to viral serological investigations. *J. Immunol.* 88: 320.
9. Wilson, S. H., and Kuff, E. L. 1972. A novel DNA polymerase activity found in association with intracisternal A-type particles. *Proc. Nat. Acad. Sci. U. S. A.* 69: 1531.
10. Wilson, S. H., Schrier, B. K., Farber, J. L., Thompson, E. J., Rosenberg, R. N., Blume, A. J., and Nirenberg, M. W. 1972. Markers for gene expression in cultured cells from the nervous system. *J. Biol. Chem.* 247: 3159.

Discussion

Yang. We have made similar studies with A-particles from other kinds of lines instead of the 104E you have used. We used MOPC 31C and haven't found any activity like yours. When your paper appeared in PNAS, I went back and used your conditions, except I omitted the deoxynucleotide kinase, but I used high salt and couldn't find any activity. Ine one of my control experiments, I used a line which had been passed in tissue culture and put back into the animal; in the crude extract of the tumor you can find something like RNA-dependent DNA polymerase which is very similar to Rauscher tumor virus polymerase.

If you transplant the tumor for longer times, say 9 days, then you see more activity. So there is, therefore, a danger in using transplantable tumors to look for the reverse transcriptase.

Faras. Under your conditions of assay, that is, high salt, KCl, and high pH, does the enzyme prefer an rA · dT or ribopolymer? Have you tried the same assay under low salt and low pH conditions and does it now, in fact, prefer a deoxytemplate to a normal template? This is in relation to Chargaff's *Escherichia coli* polymerase I system; under certain conditions he can push everything to prefer and rA · dT.

Wilson. We've tried studies like that and we have always found that the rA · dT_{14} is the most effective template-primer. This was true under several types of magnesium and manganese conditions. We haven't checked at different pH.

Faras. How about salt?

Wilson. Yes, with low salt, high pH.

Inhibition of Leukemia Virus Replication by Vinyl Analogs of Polynucleotides

P. M. Pitha, N. M. Teich, D. R. Lowy, and J. Pitha

The Johns Hopkins University School of Medicine
Baltimore, Maryland 21205
and the National Institutes of Health
Bethesda, Maryland 20014

Of the compounds known to influence the virus-specific enzymes, the polynucleotides and their analogs are perhaps those most suitable for a rational drug design aimed at improving their pharmaceutical applicability. In this presentation we want to report that even some very distant analogs of polynucleotides are effective *in vitro* and *in vivo* in interfering with virus replication. The active compounds we studied are vinyl analogs of polynucleotides, poly(1-vinyluracil) (poly VU) and poly(9-vinyladenine) (poly VA); the formulas are in the left portion of Fig. 1. These compounds are electrically neutral, stable to chemical and enzymatic hydrolysis, and interact specifically with polynucleotides with the expected complementarity (3). The geometrical dissimilarity between these analogs and nucleic acids (as visualized on the right-hand portion of Fig. 1) and the absence of charged phosphate groups are the probable reasons that the vinyl polymers have failed to show any template activity in every biological reaction we have tested up to now, for example cell-free protein synthesis (4). That does not imply that vinyl polymers are without any influence on syntheses involving templates: when complexed with polynucleotide they effectively block the template activity of the latter.

Studies of the effects of vinyl polymers on eukaryotic cells in tissue culture revealed that these analogs, in contrast to the corresponding polynucleotides, are not adsorbed extensively on the cell membrane; nevertheless, they enter the cell, presumably by transport through the pinocytotic vesicles. That some fraction of the analogs eventually finds its way into the cytoplasm of cells is documented by the selective toxici-

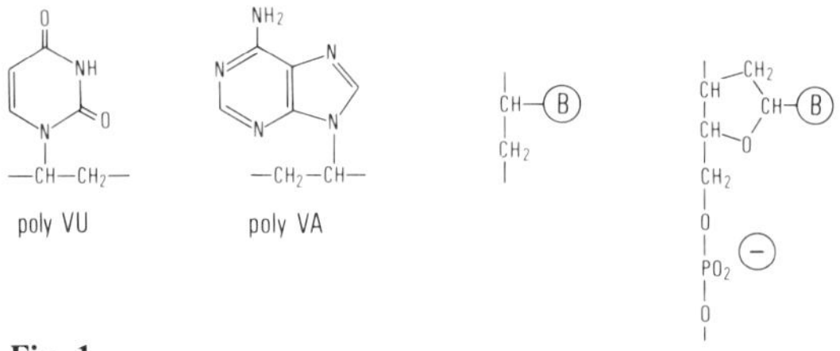

Fig. 1

ties of the compounds. While poly VA is nontoxic up to concentrations of 10^{-3} M (in monomer units) poly VU at the same concentration shows some toxicity. This effect is apparent only on the rapidly growing cells, as no serious toxicity was observed on the cell population whose growth is contact-inhibited; the degree of toxicity thus seems to depend on the metabolic activity of the cells. As we know from extensive interaction studies that poly VU reacts mainly with complementary polynucleotides, it seems reasonable to assume that its toxicity may be somehow related to interference with the adenylic acid-rich stretches which are known to be present in messenger RNA and heterogenous nuclear RNA of eukaryotic cells.

The possibility that poly VU can interfere with the adenylic acid-rich stretches of RNA is an intriguing one as some tumor viruses contain a considerable amount of such stretches: up to 50 times more than the nononcogenic viruses and up to 9 times more than cellular RNA (2). Consequently we tested the effect of poly VU on the murine leukemia virus Moloney strain and found that its replication is slowed down; in contrast to that observation the effects of poly VU on replication of vesicular stomatis virus, which does not have adenylic acid-rich stretches in its genome, are considerably smaller. Surprisingly enough we found that poly VA also is effective in the suppression of replication of murine leukemia virus. The concentration of analog that substantially slows down the virus replication does not have any effects on the host cell multiplication as measured by cell counts and by incorporation of thymidine into DNA of cells. The inhibitory effects of poly VA are of the same order as those of the polyadenylic acid which have been disclosed recently (5).

The effects of poly VA are very difficult to rationalize. It is extremely unlikely that vinyl polymer can serve in cells as a template competing with viral RNA in the same way as polyadenylic acid. The possibility that the analog blocks an enzyme necessary for replication by simply binding to the active site is also not too probable, as the analog lacks the charged phosphate groups which were found necessary for many polynucleotide-enzyme interactions that have been studied thus far. Consequently, to elucidate somewhat the mode of poly VA action *in vivo*, we studied the *in vitro* effects of both analogs on the enzymatic systems which are contained directly in the virions of murine leukemia virus. If virus particles are only partially disrupted by low concentration of neutral detergent, the natural endogenous template system yields measurable incorporation of deoxyribonucleotide triphosphates into the macromolecular fraction; under these conditions the analogs do not show any considerable effect on the incorporation. When a higher concentration of detergent is used (compare reference 1), then in order to achieve effective incorporation an exogenous template with the proper primer must be used. The results then indicate that poly VU effectively blocks thymidine triphosphate incorporation when polyadenylic acid is used as a template and oligothymidylic acid as a primer. Also poly VA is effective in this *in vitro* system and in quite a surprising way it increases the incorporation of thymidine triphosphate. As the poly VA alone or with primer does not function as a template, the reaction must involve all the components: polynucleotide primer and template, analog, and enzyme. We have been unable to explain the mechanism of the observed stimulation as yet. The very fact that poly VA has an *in vitro* effect on the enzymatic system that is present in the virion is in correspondence with the observed interference which this analog has on virus replication *in vivo*.

In conclusion, let us point out the two differences which may favor vinyl polymers when they are compared with polynucleotide inhibitors. Our analogs are extremely stable compounds and will thus persist in cells for a considerable time. The analogs do not interfere in a direct way with cell metabolism; nor do they serve by themselves as a template for nucleic acid replication and transcription and they do not bind to ribosomes. Last, but not least, the analogs are electrically neutral, and that may be helpful in the increased, and hopefully selective, uptake by eukaroytic cells.

References

1. Baltimore, D., and Smoler, D. 1971. Primer requirement and template specificity of the DNA polymerase of RNA tumor viruses. *Proc. Nat. Acad. Sci. U. S. A.* 68: 1507.

2. Gillespie, D., Marshall, S., and Gallo, R. C. 1972. RNA of RNA tumor viruses contains poly A. *Nature New Biol.* 236: 227.

3. Pitha, J., Pitha, P. M., and Stuart, E. 1971. Vinyl analogs of polynucleotides. *Biochemistry* 10: 4595.

4. Reynolds, F., Grunberger, D., Pitha, J., and Pitha, P. M. 1972. *Biochemistry* 11: 3261.

5. Tennant, R. W., Kenney, F. T., and Tuominen, F. W. 1972. Inhibition of leukemia virus replication by polyadenylic acid. *Nature New Biol.* 238: 51.

Discussion

Heidelberger. What was the molecular weight of the polymer?

Pitha. Polymers were obtained by free radical polymerization and the average molecular weight was over 100,000. The sample of poly VU was excluded from Sephadex G-200. Poly VA is chemically adsorbed to Sephadex, but its sedimentation properties are comparable to that of poly VU and thus the adenine polymer should also have a rather high molecular weight.

Baltimore. Did you test whether poly VA acts as a template for the reverse transcriptase?

Pitha. Yes, we tested this polymer and found it inactive even in the presence of oligo-thymidylic acid.

Baltimore. I would like to caution against a simple interpretation of poly VA effects based on the presence or absence of polyadenylic acid-rich stretches in viruses. Dr. A. Huang found recently that replicative RNA of VSV contains some poly A stretches.

Pitha. The differences may be more quantitative than qualitative. At the end the host cells survived the treatment with vinyl polymers and also had some poly A stretches.

In Vivo/in Vitro
DNA Synthesizing Systems

Complementation Analysis of Mutations at the *dnaB*, *dnaC*, and *dnaD* Loci

James A. Wechsler

Department of Biological Sciences
Columbia University
New York, New York 10027

Seventeen *dnaB* alleles have been tested for the ability to complement each other using recombination-deficient F' merodiploids. None of the *dnaB* alleles complements any other *dnaB* allele. The *dnaB* locus is, therefore, concluded to be a single cistron. Two *dnaC* alleles and the only *dnaD* mutation extant have been similarly analyzed for complementation. Evidence is given that the previously designated loci, *dnaC* and *dnaD*, are likely to be only one cistron. It is suggested that the notation *dnaD* be deleted.

The *dna* mutations in a large number of *Escherichia coli* mutants temperature-sensitive for DNA replication have been placed in seven loci in six regions on the *E. coli* chromosome (8, 15). The number of genes at each location has, however, not been determined. To determine the number of cistrons at each of the various loci, complementation tests have been conducted. This report details the results of complementation tests at the *dnaB* locus and at the *dnaC*, *dnaD* region.

Some of the bacterial strains employed in these experiments are listed in Table 1.

The *dnaB* Locus

It was previously shown that F'118 carries the *dnaB* gene (15). An F'118 episome containing an *ampA* mutation was constructed by transduction. The *ampA1* mutation, which confers resistance to low levels of ampicillin, is dominant (Wechsler, unpublished; 3). Strain NY100

Table 1. *Escherichia coli* strains

Strain	Mating character	Genotype*
NY100	F'118	*ampA1 pyrB⁺ / ampA⁺ pyrB thr leu thi his lac recA56 val-r strA*
NY101	F'118	*ampA1 pyrB⁺ dnaB391 / ampA⁺ dnaB391 thr leu thi thy lac tonA strA*
NY102	F'118	*ampA1 pyrB⁺ dnaB391 / ampA⁺ pyrB thr leu thi his lac recA56 val-r strA*
NY114	F'118	*ampA1 pyrB⁺ dnaB391 / ampA⁺ dnaB391 thr leu thi lac tonA recA56 strA*
F'101/AB2463	F'101	*thr⁺ leu⁺ / thr leu thi his argE proA ara lac gal xyl mtl recA13 strA T6-r*
F' dnaC/NY105	F'101	*thr⁺ leu⁺ dnaC / leu argG met lacY gal recA56 strA*
F' dnaD/NY105	F'101	*thr⁺ leu⁺ dnaD / leu argG met lacY gal recA56 strA*
E391	F⁻	*thr leu thi thy dnaB391 lac tonA str*
NY105	F⁻	*leu argG met lacY gal recA56 strA*
PC2	F⁻	*leu thy dnaC2 strA*
PC7	F⁻	*leu thy dna-7 strA*

*The genetic symbols are those of Taylor (14). Strain F'101/AB2463 was the gift of Dr. K. B. Low. PC2 and PC7 were obtained from Dr. P. Carl.

is a Rec⁻ strain containing F'118 *ampA1*. A schematic drawing of this F' is shown in Fig. 1.

A merodiploid strain heterozygous for *ampA* and *dnaB* was isolated from a cross between NY100 and E391. The presumptive homozygous merodiploid strain, NY101, was isolated from cultures of the heterozygote by screening for colonies which were ampicillin-resistant and temperature-sensitive. Strain NY101 was used as the donor to construct NY102. To prove that the F' in NY101 and NY102 actually carries the *dnaB391* allele as a result of a recombination event rather than a deletion which includes the locale of *dnaB391*, the presumptive homoallelic merodiploid NY114, F'118 *ampA1 dnaB391/ampA⁺ dnaB391 recA56*, was constructed. Two of three spontaneous temperature-resis-

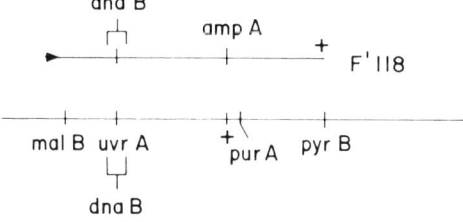

Fig. 1. The genetic complement of F'118 *ampA1* showing the relative position of relevant genetic markers. The origin and direction of transfer of the F' are denoted by an arrow.

Table 2. Temperature sensitivity of heteroallelic *dnaB* merodiploids

Relevant merodiploid genotype	Growth characteristics	
	43°C	30°C
dnaB391 ampA1 / dnaB391 recA56	−	+
dnaB391 ampA1 / dnaB173 recA56	−	+
dnaB391 ampA1 / dnaB42 recA56	−	+
dnaB391 ampA1 / dnaB43 recA56	−	+
dnaB391 ampA1 / dnaB313 recA56	−	+
dnaB391 ampA1 / dnaB454 recA56	−	+
dnaB391 ampA1 / dnaB500 recA56	−	+
dnaB391 ampA1 / dnaB6 recA56	−*	+
dnaB391 ampA1 / dnaB27 recA56	−	+
dnaB391 ampA1 / dnaB107 recA56	−	+
dnaB391 ampA1 / dnaB125 recA56	−	+
dnaB391 ampA1 / dnaB279 recA56	−	+
dnaB391 ampA1 / dnaB8 recA56	−*	+
dnaB391 ampA1 / dnaB194 recA56	−	+
dnaB391 ampA1 / dnaB368 recA56	−	+
dnaB391 ampA1 / dnaB70 recA56	−	+
dnaB391 ampA1 / dnaB59 recA56	−	+

Five merodiploids of each type were isolated and purified. A culture of each was streaked on nutrient media at 43 and 30°C. +: growth; −: absence of growth.
*These strains were tested for temperature sensitivity at 40°C rather than 30°C because the original wild type parent strain does not grow at 43°C (for source references for *dnaB* alleles see review by Gross (8)).

tant revertants of NY114 simultaneously become temperature-sensitive, ampicillin-sensitive, and unable to transfer $pyrB^1$ or $purA^+$ to recipient strains after curing of the F′ by treatment with either acridine orange or ethidium bromide (2, 9). This result proves that the *dnaB* lesion on the episome in NY114 and, therefore, also in NY102, is not a deletion.

Strain NY102 was mated with a series of *dnaB, recA56* recipients at 30°C with selection for ampicillin resistance. Five heteroallelic merodiploid strains of the genotype *ampA1 dnaB391/ampA$^+$ dnaBx recA56* were isolated and purified from each mating. The results listed in Table 2 show that each of these 16 heteroallelic merodiploid strains is temperature-sensitive. Since all of these *dnaB* alleles are complemented by F′118 *ampA1 dnaB$^+$* and none is complemented by S′118 *ampA1 dnaB391*, they define a single complementation group and the *dnaB* locus represents a single cistron.

The *dnaC*, *dnaD* Region

Three *dna* mutations cotransducible with *dra* were differentially assigned to the *dnaC* and *dnaD* loci based on the inability of one mutant, *dna-7*, to support the growth of λ bacteriophage (4, 15). The fact that some, but not all, *dnaB* mutants are *groP*⁻, unable to support the growth of λ at the permissive temperature, cast doubt on the suitability of the rationale for separating linked mutations into two genes based on their ability to propagate λ (7). To determine whether the mutations designated *dnaC* and *dnaD* represent one or more cistrons, complementation tests were conducted.

F'101 is known to cover the *dnaC* and *dnaD* mutations (Fig. 2) (15). In fact, *dnaC* and *dnaD* mutations are complemented by wild type genes on F'101 (Wechsler, unpublished). Merodiploid strains, homozygotic for either the *dnaC2* or *dnaD7* mutations, were isolated using the same method employed at the *dnaB* locus except that in this case the episome carried *leu*⁺ instead of *ampA* and the chromosome contained a *leu* mutation (Fig. 2). F'101 *dnaC2* and F'101 *dnaD7* were transferred into the intermediate *recA* strain, NY105, and subsequently transferred to strains carrying one of the *dnaC* or *dnaD* mutations, *recA56* and *leu* with selection for Leu⁺.

It has not been possible, as yet, to demonstrate unequivocally that either the *dnaC* or *dnaD* apparent homozygotes are actually homozygous for their respective *dna* alleles, as treatment of temperature-resistant revertants of these strains with acridine orange or ethidium bromide gives anomalous results. Growth is strongly inhibited in the presence of curing agents and though most of the colonies obtained after treatment show no apparent loss of the F', some exhibit a randomization of markers that is most easily explained as loss of the F' following or during integration and excision from the chromosome. Integration of an F' into the chromosome of a *recA* strain can be shown to occur (W. Maas, personal communication). The reason for this apparent selection process in acridine orange or ethidium bromide has not been investigated.

The results in Table 3 show that merodiploid strains heteroallelic for the *dnaC*, *dnaD* mutations are unable to grow at the restrictive temperature. These mutations, therefore, define a single complementation group and, unless both apparent homozygotes are due to deletions on the episome, a single cistron. It is suggested that the *dnaD* notation be deleted and the *dna-7* allele be designated *dnaC7*.

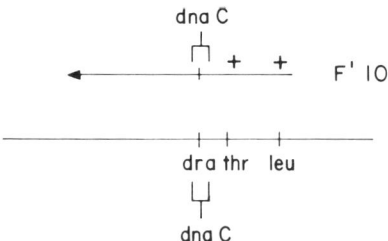

Fig. 2. The genetic complement of F'101 showing the relative position of relevant genetic markers. The origin and direction of transfer of the F' are denoted by an arrow.

A map reflecting the abolition of the *dnaD* gene and incorporating some other new information is shown in Fig. 3.

Complementation studies at the *dnaA*, *dnaE*, and *dnaG* loci are not yet completed but preliminary results suggest that the *dnaG* mutations also lie within one cistron.

Table 3. Temperature sensitivity of heteroallelic *dnaC*, *dnaD* merodiploids

Relevant merodiploid genotype	Growth characteristics	
	40°C	30°C
dnaD leu$^+$ / dnaD7 leu recA56	−	+
dnaD leu$^+$ / dnaC1 leu recA56	−	+
dnaD leu$^+$ / dnaC2 leu recA56	−	+
dnaC leu$^+$ / dnaC2 leu recA56	−	+
dnaC leu$^+$ / dnaC1 leu recA56	−	+
dnaC leu$^+$ / dnaD7 leu recA56	−	+

Growth of the merodiploid strains was tested in the same manner specified for Table 2.

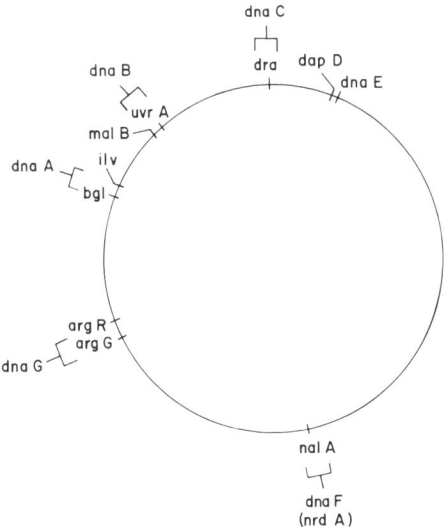

Fig. 3. Map locations of the *dna* mutations on the *Escherichia coli* chromosome. Standard chromosome map after Taylor (14). The *dnaE* locus has been placed clockwise from *dapD* according to Richardson *et al.* (12). The *dnaF* gene product has recently been shown to be the B1 subunit of ribonucleoside diphosphate reductase (6). These authors have suggested that the notation *nrdA* should replace dnaF.

Discussion

Seventeen *dnaB* mutations have been shown to constitute a single complementation group, and three mutations previously divided between the *dnaC* and *dnaD* loci have been similarly shown to define one complementation group.

Since the majority of temperature-sensitive *dna* mutations in *E. coli* have been mapped at the *dnaB* locus, it is somewhat surprising that this locus contains only one gene. Some possible explanations for the preponderance of *dnaB* mutations are: (*a*) the gene is very large relative to the other *dna* genes and, therefore, is more likely to be altered by mutation, (*b*) the gene contains one or more "hot spots" which result in an abnormally high number of mutations at specific points within the gene (1), and (*c*) the protein specified by the *dnaB* gene is unusually subject to temperature sensitivity, rather than to inactivity, as a consequence of mutation.

The first two of these explanations for the bias toward the isolation of *dnaB* mutants are essentially trivial. In regard to the third explanation, a protein which must be complexed, or must interact, with one or more other proteins in order to be active might be expected to be more subject to the effects of temperature than are most independently acting proteins. There is some evidence that the *dnaB* gene product may interact with bacteriophage proteins.

Some *dnaB* mutants are Gro$^-$ for λ (specifically, some *dnaB* mutations are groP$^-$ and all groP$^-$ mutations are *dnaB*) (7). A groP$^-$ mutant will not propagate λ wild type but will propagate one or two classes of λP$^-$ mutants; λP$^-$ mutants are unable to grow in wild type hosts (7). Also some mutants of P1 suppress all *dnaB* mutations tested (A. Jaffe-Brachet, D. Schwartz, and M. Yarmolinsky, personal communication). Though neither of these results specifically demonstrates the existence of a protein-protein interaction between the *dnaB* product and a phage protein, both results do show that alteration of a phage protein can moderate the phenotypic consequences of a bacterial mutation, and, conversely, in the case of λ, that an altered *dnaB* protein can affect the action of a phage protein. (The P gene product of λ is required for DNA replication (5, 10).) It would not be unreasonable to expect that the *dnaB* protein can also affect the action of other host proteins involved in DNA synthesis, and, consequently, many *dnaB* mutations might cause temperature sensitivity of the DNA replication machinery.

Two *dnaC* mutations have been shown to be defective in initiation (4, 13, and personal communication). A mutant carrying the *dnaC7* allele does not allow λ growth at the restrictive temperature (4). An inability to grow λ has been used as a criterion for DNA chain elongation mutants, whereas initiation mutants are expected to permit λ growth (8). As *dnaC7* is a leaky mutation, DNA synthesis kinetics experiments have not been able to show whether this mutation actually affects chain elongation or initiation. Since two *dnaC* alleles, however, have been shown to affect initiation, it is expected that the *dnaC* product is directly involved in initiation. Also, *dnaC2*, *dnaC7*, and *dnaC28*, isolated by Davern and co-workers, but not mutants carrying lesions at other *dna* loci, are all unable to effect the sustained initiation described by Kogoma and Lark (11, K. G. Lark, personal communication).

If one assumes that *dnaC7* alters an initiation protein, its inability to grow λ suggests that the *dnaC* protein, unlike the *dnaA* product, is perhaps generally used for initiation of both bacterial and phage DNA replication. Alternatively, if the inability of *dnaC7* strains to grow λ is due to a defect in chain elongation, presumably a defective *dnaC* protein can alter either initiation or chain elongation processes. This would imply that the *dnaC* protein may interact with other DNA replicating proteins.

Acknowledgments

I gratefully acknowledge the excellent technical assistance of Mr. G. Grandusky. This work was supported by Public Health Service Research Grant CA12590 from the National Cancer Institute and by Grant GB32417 from the National Science Foundation.

References

1. Benzer, S. 1961. On the topography of the genetic fine structure. *Proc. Nat. Acad. Sci. U. S. A.* 47: 403.

2. Bouanchaud, D. H., Scavizzi, M. R., and Chabbert, Y. A. 1969. Elimination by ethidium bromide of antibiotic resistance in Enterobacteria and Staphylococci. *J. Gen. Microbiol.* 54: 417.

3. Burman, L. G., Nordström, K., and Boman, H. G. 1968. Resistance of *Escherichia coli* to penicillins. V. Physiological comparison of two isogenic strains, one with chromosomally and one with episomally mediated ampicillin resistance. *J. Bacteriol.* 96: 438.

4. Carl, P. 1970. *Escherichia coli* mutants with temperature-sensitive synthesis of DNA. *Mol. Gen. Genet.* 109: 107.

5. Eisen, H. A., Fuerst, C. R., Siminovitch, L., Thomas, R., Lambert, L., Pereira da Silva, L., and Jacob, F. 1966. Genetics and physiology of defective lysogeny in K12 (λ): studies of early mutants. *Virology* 30: 224.

6. Fuchs, J. A., Karlström, H. O., Warner, H. R., and Reichard, P. 1972. Identification of the gene product defective in a *dnaF* mutant of *Escherichia coli*. *Nature New Biol.* 238: 69.

7. Georgopoulos, C. P., and Herskowtiz, I. 1971. *Escherichia coli* mutants blocked in lambda DNA synthesis. *In* Hershey, A. D. (ed.), *The bacteriophage lambda*, pp. 553-564, The Cold Spring Harbor Laboratory, Cold Spring Harbor, New York.

8. Gross, J. 1972. DNA replication in bacteria. *Curr. Top. Microbiol. Immunol.* 57: 39.

9. Hirota, Y. 1960. The effect of acridine dyes on mating type factors in *Escherichia coli*. *Proc. Nat. Acad. Sci. U. S. A.* 46: 57.

10. Joyner, A., Isaacs, L. N., Echols, H., and Sly, W. S. 1966. DNA replication and messenger RNA production after induction of wild-type λ bacteriophage and λ mutants. *J. Mol. Biol.* 19: 174.

11. Kogoma, T., and Lark, K. G. 1970. DNA replication in *Escherichia coli*. Replication in absence of protein synthesis after replication inhibition. *J. Mol. Biol.* 52: 143.

12. Richardson, C. C., Campbell, J. L., Chase, J. W., Hinkle, D. C., Livingston, D. M., Mulcahy, H. L., and Shizuya, H. 1972. DNA polymerases of *E. coli*. This symposium.

13. Schubach, W. M., Whitmer, J. D., and Davern, C. I. 1973. Genetic control of DNA initiation in *Escherichia coli*. In press.

14. Taylor, A. L. 1970. Current linkage map of *Escherichia coli*. *Bacteriol. Rev.* 34: 155.

15. Wechsler, J. A., and Gross, J. D. 1971. *Escherichia coli* mutants temperature-sensitive for DNA synthesis. *Mol. Gen. Genet.* 113: 273.

Discussion

Oeschger. From the work of Speyer and Drake and more recently biochemically by Bessman, there is definite evidence that DNA polymerases are involved in mutation frequency, and there are a number of mutator genes in *Escherichia coli*. How did the mutator genes correlate with these DNA temperature-sensitive mutants?

Wechsler. The two known mutators, *mutT* and *mutS*, which have been mapped well, map some minutes away from any known DNA mutation. I understand that at least some of the *dnaE* mutants, the polymerase III mutants which we isolated, show a slight mutator effect at intermediate temperatures, but it is not dramatic like the ones that Speyer looked at before.

Reznikoff. It seems to me another explanation for the preponderance of *dnaB* temperature-sensitive mutants is that perhaps B function is only required at high temperatures, so that any mutation in B creates a temperature-sensitive phenotype.

Wechsler. Yes, that would be a reasonable alternative except that the Gro⁻ effect with λ is an effect at the permissive temperature, in fact. The *dnaB*s which show a Gro⁻ phenotype, that is, will not grow λ wild type, show that phenotype at 30°C even though the host seems to be perfectly healthy at that temperature.

Reznikoff. It would fit in also because perhaps λ needs that function at all temperatures.

Wechsler. Well, I suppose that is possible.

In Vivo and *in Vitro* Chromosome Replication in *Bacillus subtilis*

Noboru Sueoka,* Tatsuo Matsushita,
Seigou Ohi, Aideen O'Sullivan, and Kalpana White

Departments of Biochemical Sciences and Biology
Princeton University
Princeton, New Jersey 08540

Studies on *in vitro* DNA replication are often liable to artifacts. For a meaningful interpretation of the results, therefore, several criteria should be met. Using the genetic transformation of *Bacillus subtilis* and comparing it with *in vivo*, the *in vitro* criteria for normal DNA replication and initiation are formulated, and simple experimental tests are devised. For example, toluenized cells satisfy the criteria for elongation but not for initiation. The penetration of pancreatic ribonuclease into toluenized cells is established while that of DNA polymerase I is not clear. Some characterization of *B. subtilis* DNA polymerases and ATP-dependent nucleases has been made as a preparation toward *in vitro* reconstitution of chromosome replication.

The chromosome replication of bacteria is an organized result of the interplay of various enzymes and structural components of the cell and is necessarily quite complex. The results of *in vitro* studies, therefore, are liable to be subjected to various artifacts leading to misinterpretations.

The availability of genetic transformation in *Bacillus subtilis* provides a unique opportunity for *in vitro* study by allowing us to analyze the product directly in terms of its chromosomal location as well as its biological (transforming) activity, and to compare it with that of *in vivo*

---

*Present address, Department of Molecular, Cellular and Developmental Biology, University of Colorado, Boulder, Colorado 80302.

synthesis. First we will summarize salient facts currently available on *in vivo* chromosome replication in *B. subtilis*.

In Vivo Replication

Replication of the Bacillus subtilis Chromosome Is Sequential

It is a well established fact that there is a definite temporal sequence of replication. This has been demonstrated by both marker frequency analysis and synchro-transfer experiments. The temporal replication order is shown in Fig. 1. There is little doubt that the replication origin occupies a unique position on the chromosome close to a marker, *ade*16 (or *pur*A16), isolated and mapped by O'Sullivan and Sueoka (19). It is likely that the terminus of replication is also fixed, since *met*5-*ileu*5 always ends up as the last to replicate among the markers shown in Fig. 1. However, less synchrony of replication in this region makes it impossible to decide whether the terminus occupies a unique locus or an area of the chromosome.

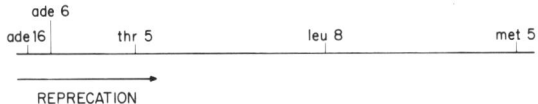

Fig. 1. A genetic map of *Bacillus subtilis* based on the temporal order of gene replication. The map positions of *ade*6, *thr*5, *leu*8, and *met*5 are based on marker frequency analysis (31) and the replication order has been confirmed by synchro-transfer experiments (19, 32). The *ade*16 position is assigned by the synchro-transfer experiment of O'Sullivan and Sueoka (19).

Chromosome Initiation Is Symmetrical

This point is firmly established in the case of normal multifork replication in *B. subtilis* (22), and also in *Escherichia coli* (6).

Replication Is at Least Partially Bidirectional

The temporal replication sequence of genes on the *B. subtilis* chromosome is fixed. This time sequence of marker replication can be consistent with unidirectional, bidirectional, or any other arrangement of replication segments of the chromosome, as long as the temporal sequence of gene replication is unique (31). Recent radioautographic studies of Wake (27) and Gyurasits and Wake (9) on the *B. subtilis* chromosome leave little

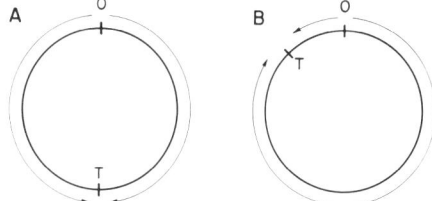

Fig. 2. Bidirectional replication of a circular chromosome. A: complete bidirectional replication; B: partial bidirectional replication; O: origin; T: terminus.

doubt that on each side of the origin up to 20% of the replication proceeds both ways from the origin. The simplest model, then, is the one shown in Fig. 2A of complete bidirectional replication. However, a dilemma exists. The largest linkage group constructed by transduction consists of 70 to 75% of the whole linkage map, and the replication time order in this group is unidirectional (see 4, 33). This leads us to the picture presented in Fig. 2B of partial bidirectional replication. If, on the other hand, the present linkage map has sizeable silent regions in which few markers have been isolated, we lose the logical ground for partial bidirectionality. Settlement of this point should come from further studies of autoradiography covering longer portions of the chromosome and gene analysis on the temporal and physical relationships of markers on both sides of the origin.

One interesting clue for bidirectionality, at least around the origin, is the sucrose marker. The gene for levan-sucrose (*suc*A) replicates after *ade*16 by synchro-transfer experiments (11, 19). Lepesant *et al.* (14) report that the map position analyzed by genetic linkage is to the left of *ade*16. This discrepancy is resolved if replication is bidirectional, if the origin lies between the two markers, and if *ade*16 is closer to the origin than *suc*A.

The Origin and Terminus as well as
the Replication Point May Be Attached on the Membrane

There is evidence indicating that the origin and terminus are permanently attached on the membrane in both *B. subtilis* and *E. coli* (5, 10, 23, 25). Further studies on this point have given consistent results supporting the conclusion (20, 24). A recent study on the dichotomous configuration of the chromosome indicates that all four origins are associated with the membrane fraction (20). The membrane attachment of the replication fork in *B. subtilis* (7, 25, 30) seems to be of a different nature, and there is no solid evidence against attachment.

Effort toward in Vitro Replication

Criteria of Real Replication

For elongation, two criteria should be fulfilled. (a) There should be semiconservative replication for reasonably long stretches of DNA (at least, for example, 10^7 molecular weight); (b) there should be normal sequential replication of markers.

For the *in vitro* study of initiation, it is necessary to obtain the induction of initiation substantially over that of *in vivo* at the time of cell sampling. This point is important but somewhat elusive, and will be discussed later.

Toluenized Cells

As in *E. coli* (16), toluenized cells of *B. subtilis* cells, with the presence of ATP and Mg^{++}, have been shown to satisfy the above two criteria for elongation, and the amount of DNA synthesized comes routinely to about 10% of the chromosome length per replication fork, although the rate of replication is approximately one-tenth the *in vivo* rate (15). The system, then, is worth further analysis. Our current recipe for toluenization of *B. subtilis* is shown in Table 1. With the presence of ATP

Table 1. DNA Synthesis in toluenized cells of *Bacillus subtilis*

Reaction mixture	Activity (%)
Complete	100
−ATP	10
−MG^{++}	2
+N-Ethylmaleimide (NEM) (2 mM)	15
+p-Chloromercurial benzoate (PCMB) (0.05 mM)	17
+Chloramphenicol (CM) (150 µg/ml)	95
+Dithiothreitol (2 mM)	120
+CTP, GTP, UTP (1 mM each) + ATP (2mM)	
15 min	85
30 min	124
45 min	129

A typical procedure. *B. subtilis* strain 168TT (*thy try*) was grown at 30°C to a concentration of 7×10^7 cells/ml in 50-ml medium C$^+$ containing 10 µg/ml thymine and 50 µg/ml tryptophan. Cells were collected by filtration, washed in 0.1 M potassium phosphate buffer (pH 7.4), centrifuged at room temperature and resuspended in phosphate buffer at a concentration of 4×10^9 cells/ml. The cells were then agitated 10 min at 25°C with 1% toluene and incubated for 5 min at 4°C. The complete reaction mixture (0.5 ml) contained 70 mM potassium phosphate buffer (pH 7.4), 13 mM MgSO$_4$, 1.3 mM ATP, 33 µM dGTP, dCTP, dATP, TTP, 1 µM ^{3}H-dATP, and 4×10^8 toluene-treated cells. Reactions were carried out at 37°C with ATP and without ATP for 60 min. Dithiothreitol was used in most cases except in the NEM and PCMB experiments.

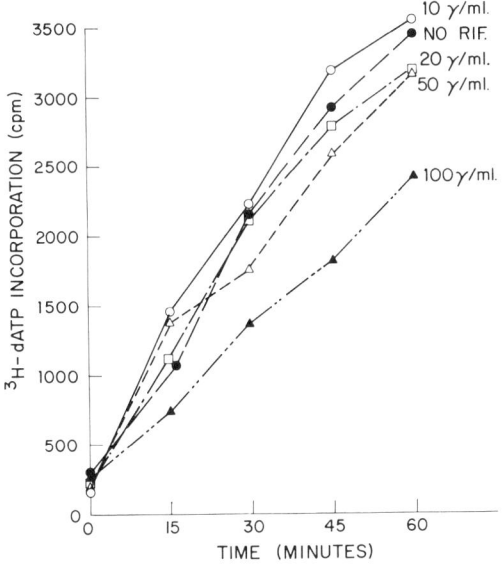

Fig. 3. Effect of rifampycin on ATP-stimulated DNA replication in toluenized *Bacillus subtilis* (Matsushita and Sueoka, in preparation). Strain 168TT (*thy try*) was grown at 30°C to a concentration of 7×10^7 cells/ml in medium C^+. Cells were toluenized as described in Table 1. The treated cells were incubated at 37°C in the complete reaction mixture (described in Table 1) with 2 mM dithiothreitol and various concentrations of rifampycin.

and Mg^{++}, toluene-treated cells undergo protein synthesis at a rate of about 10% of intact cells. The protein synthesis is completely chloramphenicol-sensitive. The effects of N-ethylmaleimide (NEM), *p*-chloromercurialbenzoate (PCMB), and chloramphenicol (CM) on the DNA synthesis are shown in Table 1. The effect of rifampycin in various concentrations is shown in Fig. 3. Also examined was the addition of ribonucleoside triphosphates (CTP, GTP, UTP) (Table 1). These results indicate either that RNA synthesis is not necessary for elongation, or that RNA synthesis is necessary but that the synthesis is rifampycin-resistant and sufficient ribonucleoside triphosphates are generated in the toluene system without being supplied from the outside. According to the work presented at this symposium by Okazaki and his collaborators, the latter seems to be the case.

No Initiation Occurs in Toluenized Cells. When half the exponentially growing cells (168 thy^-, trp^-) are subjected to the toluenization system with dBUTP (*in vitro*) and the other half are transferred to the BU medium (*in vivo*), the chromosome replication in the two systems is different in one important aspect, as revealed in the manner of the density transfer of markers (Fig. 4A). The *in vivo* transfer shows that

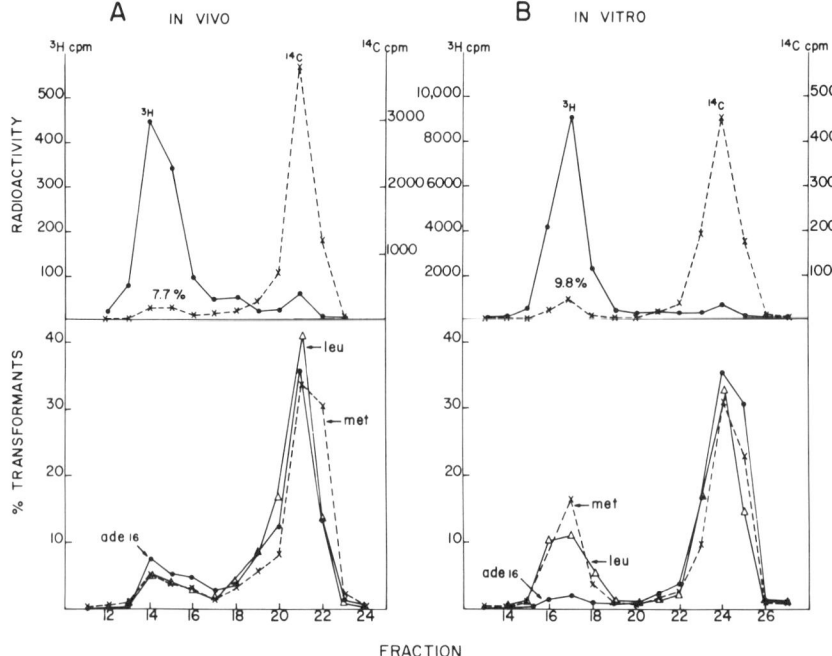

Fig. 4. Evidence for the absence of initiation in toleunized cells (White, Matsushita, and Sueoka, in preparation). *B. subtilis* strain 168TT (*thy try*) was grown at 30°C to a concentration of 7×10^7 cells/ml in 50 ml of medium C$^+$, containing 12.5 μCi of ^{14}C thymine, 3 μg/ml cold thymine, and 50 μg/ml tryptophan. Cells, 25 ml, were collected and toluenized as described in Table 1. This *in vitro* sample (B) was treated as in Table 1, except that dBUTP was substituted for TTP and 4 mM phosphoenolpyruvate and 13 I.U./ml pyruvate kinase were added to 1 ml of the reaction mixture. The remaining 25 ml of cells for the *in vivo* control (A) were collected by filtration and resuspended in 25 ml of C$^+$ media containing 50 μg/ml tryptophan and 10 μg/ml bromouracil. This sample was grown for 20 min at 37°C. Cells were collected by filtration and resuspended in 5 ml of 0.1 M KPO$_4$ (pH 7.4). Lysates of 1 ml of this *in vivo* suspension and 1 ml of the *in vitro* reaction mixture were prepared by separately adding 0.6 ml of 0.3 M NaCl, 0.2 M EDTA (pH 8.2), and 0.15 ml of 5 mg/ml lysozyme, and incubated for 30 min at 37°C. Sodium dodecyl sulfate (0.042 ml of 20%) was added and incubated for 15 additional min. Added to 1.74 ml of the lysate and shaken with 4.6 g of CsCl were 1.95 ml of 0.01 M Tris, 0.001 M EDTA (pH 8.4). After 72 hr centrifugation at 35,000 rpm (25°C) in a Spinco SW 50.1 rotor, three-drop fractions were collected, precipitated, and counted. Transformations were performed by adding 0.1-ml aliquots to 1 ml of competent cells, shaking at 37°C for 40 min, and plating on selective plates. Recipient 168 *leu*8-*met*5-*ade*16 was made competent by the method of Bott and Wilson (2). A: fractions 12 to 17 represent hybrid DNA and fractions 18 to 23, parental DNA. B: fractions 15 to 19 represent hybrid DNA and fractions 22 to 26, parental DNA. In both cases, no DNA peaks were found in the double replication region (not shown).

all markers are transferred to hybrid molecules more or less equally. This means that the steady state replication and initiation are kept normal before and after the transfer. Theoretically, any markers, regardless of their positions on the chromosome, should be transferred at the same rate (26, 28). On the other hand, in the toluenized system there is much less of the *ade*16 marker transferred to the hybrid peak than the other markers, as is evident from the diagram shown in Fig. 4B. This is what is expected if there is no new initiation during replication in the toluenized system. Theoretically, a marker should be transferred to the hybrid position at the same rate as other markers until the period of synthesis without initiation has reached the time equivalent to the travel time of the replication fork from the origin to the marker, and then no further transfer of the marker should occur (28). For *ade*16 this time equivalent would be about 2% and for larger amounts of synthesis the *ade* transfer would be less than other later markers. The result is consistent with this expectation and unequivocally shows that there is no new initiation in the toluenized system. This fact provides a condition for an *in vitro* test of initiation as discussed later.

DNA Synthesis in the Toluenized Cells Is Localized in the Cell. Cells made permeable to otherwise impermeable compounds are liable to release macromolecules including DNA. A single localization test is described here for toluenized cells as well as for toluene-Brij-58-treated cells. The test consists of diluting each cell sample with buffer, agitating the suspension with a Vortex mixer, precipitating the cells, and analyzing the synthesized DNA in the precipitated cells and in the supernatant. As is evident from Fig. 5, A1 and A2, most of the newly synthesized DNA in toluenized cells is confined in the cell, although in later samples some is found outside the cell. On the other hand, in the toluene-Brij-58-treated cells, practically all the synthesized DNA is found outside the cell (Fig. 5, B1 and B2).

This Vortex centrifugation experiment (localization test) is a simple way of locating DNA synthesis in toluenized cells, or more generally in the chromosomal complex (for lysate systems without agitating with a Vortex mixer).

Toluenized Cells Are Permeable to Some Proteins. In *E. coli*, the originators of the toluenized system, Moses and Richardson (16), showed results which indicated the penetration of DNase I (pancreatic) and less convincingly, the antipolymerase I antibody. Since the permeability of macromolecules is critical to the later development of this system, we have examined this point further.

For the first experiment, a *pol* I$^+$ strain (168 *thy$^-$ trp$^-$*) was used and incorporation of ^{3}H-deoxyadenylate was examined on toluenized cells in the presence of pancreatic DNase I (0.4 µg/ml) and PCMB

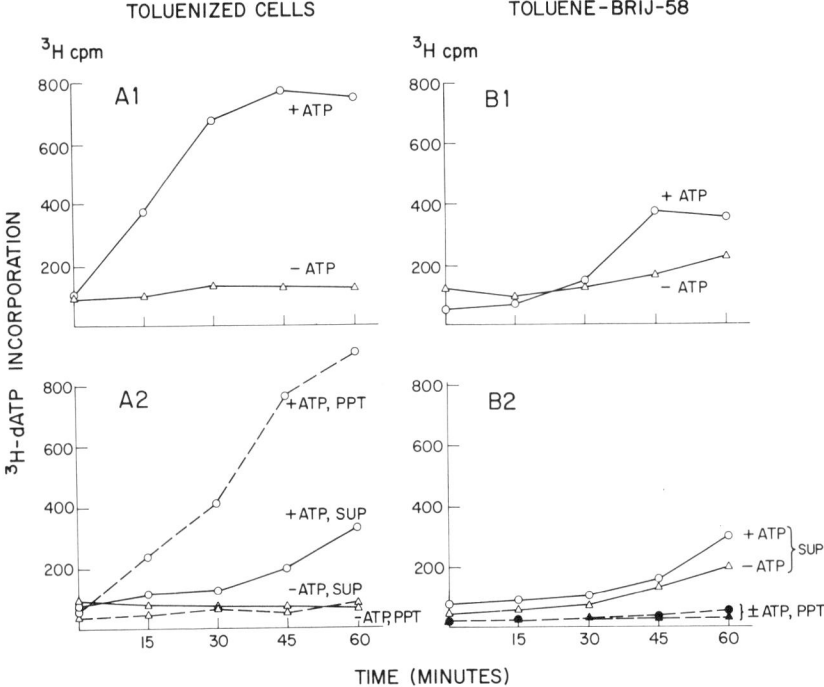

Fig. 5. Localization test of DNA synthesis in toluenized cells and in toluene-Brij-58-treated cells (Matsushita and Sueoka, in preparation). A1: exponentially growing cells of strain NB 841 (*pol* I$^-$, *thy*) in 50 ml of culture were concentrated 50-fold by centrifugation and toluenized and incubated with the deoxynucleoside triphosphates and ATP, as described in Table 1. For each time sample, 50 µl of the reaction mixture were added to 2 ml of 10% trichloroacetic and counted in a liquid scintillation counter. This represents the total trichloroacetic acid-precipitable counts. A2 (localization test): another 50-µl sample of the above reaction mixture was added to 2 ml of cold potassium phosphate buffer (0.1 M, pH 7.4), agitated for 15 sec with a Vortex mixer at highest speed, and centrifuged 5 min at 10,000 rpm at 4°C. The supernatant was poured into 4 ml of 10% trichloroacetic acid. The remaining precipitated cells were resuspended in 2 ml of the buffer and poured into 4 ml of 10% trichloroacetic acid. Both samples were counted for radioactivity in a liquid scintillation counter as described previously (19). Sup: Supernatant fraction; PPT: cell fraction. B1: cells from a similar culture of NB 841 as that above were treated in the same way except that the reaction mixture contained 1% Brij-58. The total trichloroacetic acid-precipitable radioactivity of each sample was measured as described in A1. B2 (localization test): each sample was treated as described in A2.

(0.05 mM). Because of the use of PCMB, dithiothreitol was omitted from the reaction mixture shown in Table 1. This condition is called the "repair" condition. ^{3}H-Deoxyadenylate incorporation occurs with and without the presence of ATP, while in a control without DNase and PCMB there is no incorporation without ATP (Fig. 6). Since PCMB alone inhibits toluenized synthesis (Table 1), the result shows that DNase (mol wt 31,000) has penetrated into the toluenized cell. Furthermore,

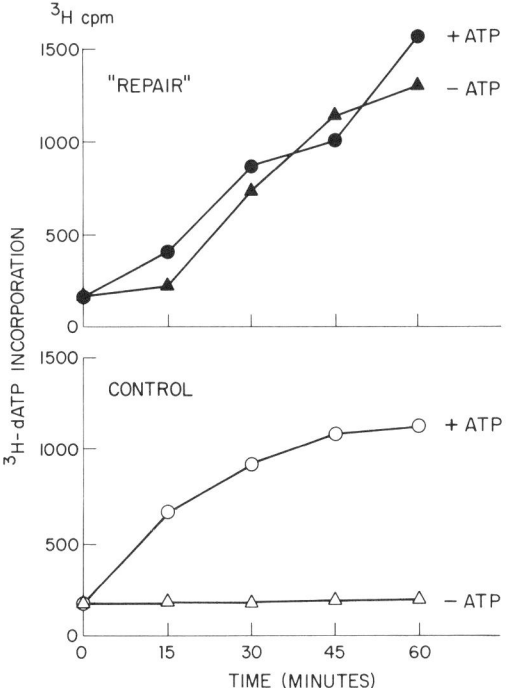

Fig. 6. "Repair" synthesis induced by DNase I in toluenized cells (Matsushita and Sueoka, in preparation). Toluenized cells of 168 *thy try* (*pol* I$^+$) were prepared as described in Table 1. No dithiothreitol was added. In the "repair" experiment, 0.05 mM PCMB and 0.4 μg/ml pancreatic DNase I were added to the complete reaction mixture of Table 1. The nonlinear incorporation seen in the control experiment is due to the lack of dithiothreitol.

for this incorporation DNA polymerase I is apparently necessary. Thus, neither the *pol* I$^-$ mutants from N. Brown (NB 841) nor from B. Strauss (HA 101(59)F; (8)) respond to the "repair" condition (Fig. 7). The result also indicates that other polymerases of *B. subtilis* cannot replace polymerase I or are not available for this "repair" synthesis. The penetration of DNase I is further supported by a *localization test* experiment, as shown in Fig. 8. Samples of the "repair" synthesis of 168 *thy*$^-$ *trp*$^-$ similar to Fig. 6 were suspended in cold buffer, agitated by a Vortex mixer, and centrifuged briefly. The supernatant and precipitated material were separately treated with 10% trichloroacetic acid and radioactivity was measured. Prior to toluenization, cells were labeled with ^{14}C-thymine for several generations to label DNA uniformly. The result shows that the incorporation is confined to the precipitated toluenized cells, with only a little found in the supernatant. Moreover, almost all the ^{14}C-DNA remained in the cell fraction, indicating that the degradation of DNA

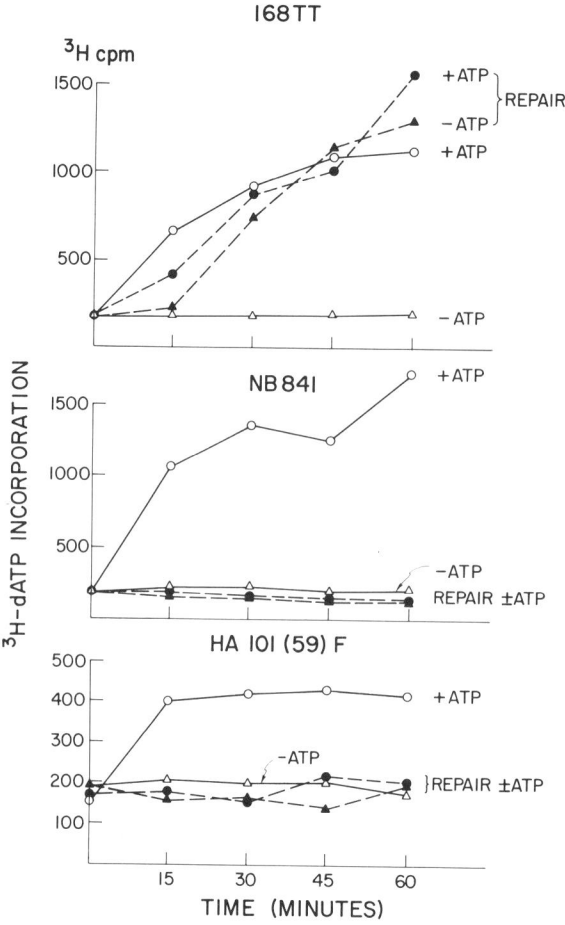

Fig. 7. "Repair" synthesis in pol I⁻ strains. An experiment similar to the one described in Fig. 6 was applied to two pol I⁻ strains, NB 841 and HA 101(59)F. As a comparison, the result of pol I⁺ (168 TT) is also shown.

in this system is not substantial. These results fit the "repair" synthesis which must have been stimulated by the penetration of DNase I.

The penetration of macromolecules to toluenized cells was further tested. An extract of pol I⁺ cells (168 thy trp), from which nucleic acids were removed by DEAE-cellulose chromatography (fraction III), was added to toluenized pol I⁻ cells (NB841) under "repair" conditions. As shown in Fig. 9A, fraction III clearly stimulates "repair" DNA synthesis and, furthermore, a large fraction of it is stimulated by ATP. It is clear that some macromolecular component penetrated and stimulated the synthesis. The localization test in this case, however, shows

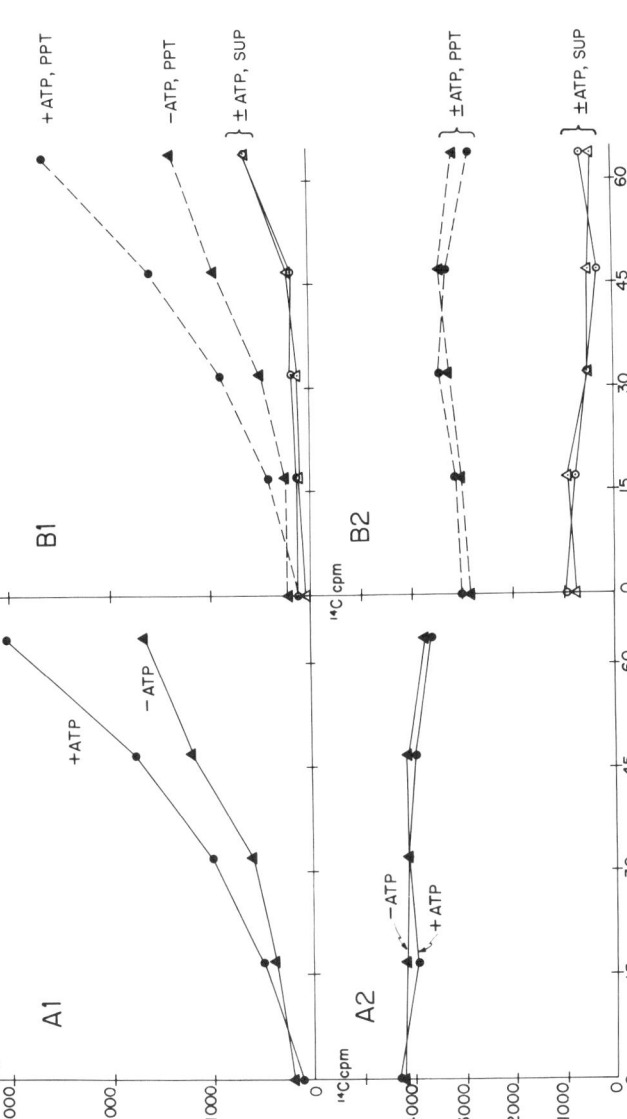

Fig. 8. Localization test on parental DNA and the "repair" DNA synthesis of 168 *thy try* (*pol* I$^+$) strain (Matsushita and Sueoka, in preparation). The 168 *thy try* strain was grown in ^{14}C-thymine until the cell density reached 40 Klett-Summerson colorimeter units for uniform labeling. Time samples of "repair" synthesis as described in Fig. 5 were treated as follows. For each time, 50 μl of the reaction mixture were added to 2 ml of 10% trichloroacetic acid. Simultaneously, another 50 μl were added to 2 ml of cold 0.1 M KPO$_4$ (pH 7.4), vortexed vigorously for 15 sec, and centrifuged 5 min at 10,000 rpm at 4°C. The supernatant was poured into 4 ml of 10% trichloroacetic acid. The remaining precipitated cells were resuspended in 2 ml of KPO$_4$ buffer and poured into 4 ml of 10% trichloroacetic acid. The three trichloroacetic acid precipitates from each sample were then counted, as described previously (19). A1: total ^{3}H-deoxyriboadenylate incorporation with and without ATP; A2: total parental DNA measured by ^{14}C-thymine; B1: ^{3}H-deoxyriboadenylate incorporated into the cell fraction (PPT) and the supernatant fraction (SUP); B2: parental DNA (^{14}C-thymine) in the cell fraction (PPT) and the supernatant fraction (sup).

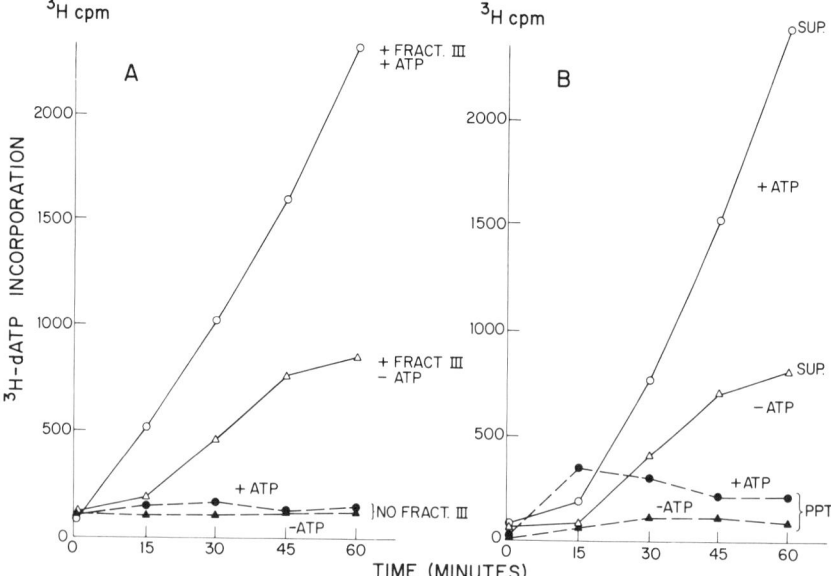

Fig. 9. Stimulation of "repair" synthesis of a toluenized $pol\ I^-$ strain (NB 841) by the external addition of Fraction III of $pol\ I^+$ (168 thy try) cell extract (Matsushita and Sueoka, in preparation). A: exponentially growing cells of NB 841 were toluenized and incubated in 0.5 ml of the "repair" reaction mixture as in Fig. 6, with and without Fraction III (0.1 ml). B: the localization test was performed, as described in Fig. 6. Fraction III was prepared according to Kornberg and Gefter (13). Cells were harvested from 8-liter culture of 168 thy try in Penassay medium at 80 Klett-Summerson colorimeter units and suspended in cold 60 ml of 0.02 M Tris acetate buffer (pH 8.2) with 5 mM mercaptoethanol, 0.01 M magnesium acetate, and 5 mM EDTA. The cell suspension was passed through a French press at 9000 p.s.i. and centrifuged at 10,000 rpm for 5 min at 4°C. The supernatant was centrifuged at 15,000 rpm for 30 min at 4°C. The third centrifugation was applied to the supernatant at 100,000 $\times$ g for 90 min at 4°C. The final supernatant was dialyzed overnight at 4°C against 0.01 M KPO_4 buffer (pH 7.5) plus 20% glycerol, 5 mM mercaptoethanol and 0.5 mM EDTA. Fifty per cent ammonium sulfate solution (2.4 M) was added dropwise to the final concentration of 0.2 M. To remove nucleic acids, the mixture was then applied to a DEAE-cellulose column which had been equilibrated with the dialysis solution. Fractions with substantial absorption at 260 and 280 nm were combined and dialyzed against the dialysis solution overnight at 4°C. This is Fraction III.

that practically all newly synthesized DNA, both ATP-stimulated and nonstimulated, is found in the supernatant (Fig. 9B). This test then made the penetration of polymerase I ambiguous. The possibility exists that a nuclease or nucleases, some stimulated by ATP, break down DNA whose fragments leak out of the cell, and that the DNA synthesis by polymerase I may have occurred outside the cell. This point will be further examined by adding a purified fraction of polymerase I in place of fraction III.

An in Vitro Elongation Test Is Devised. Our previous elongation or sequential replication test (15) involved toluenization of synchronized cells by using a temperature-sensitive initiation mutant, *dna*-1 (White and Sueoka, in press). Another method, simpler and more sensitive, has been devised. The design of the experiment is shown in Fig. 10A. DNA synthesis of an exponentially growing culture is halted temporarily. When DNA synthesis is resumed, initiation occurs in high frequency. This premature initiation has been known in *E. coli* (21, 29). An excess initiation is clearly shown in *B. subtilis* by the density transfer experiment (Fig. 11). When the cells are exposed to toluenization immediately after starvation with thymine, an apparent reinitiation similar to that *in vivo* is observed, as seen by preferential transfer of the *ade*16 marker into the hybrid peak.

The initiation in this case, however, is different from that in the experiment described in Fig. 4. The experiment in Fig. 4 tested the complete processes of initiation, which did not occur in the toluenized system. In the present experiment, all the processes of initiation prior to the actual start of DNA replication (preinitiation) are completed, possibly even including polymerization of the first few nucleotides. The real start is blocked due to the absence of thymine. Preferential replication of the marker, *ade*16, therefore, originates from the synchronous start of elongation at the origin. This condition seems to be ideal for *in vitro* elongation study. Cells should be harvested at the end of thymine starvation, or at the end of other methods of halting DNA synthesis, and prepared for the *in vitro* system. The replication of *ade*16 more than other markers into hybrid molecules should be an indication of meaningful *in vitro* replication. Since the marker *ade*16 locates within approxi-

Fig. 10. Experimental design of tests of *in vitro* elongation (A) and *in vitro* initiation (B). In both cases parental DNA is uniformly labeled with ^{14}C-thymine; newly replicated DNA after density transfer is labeled with ^{3}H-ether of thymine (*in vivo*) or of dATP (*in vitro*). The density transfer was made by using BU (*in vivo*) or dBUTP (*in vitro*). The comparison of *in vivo* and *in vitro* is an essential part of the tests. The success of failure of both *in vitro* elongation (A) and *in vitro* initiation (B) can be judged by transfer of the *ade*16 marker into hybrid DNA, as in both control samples.

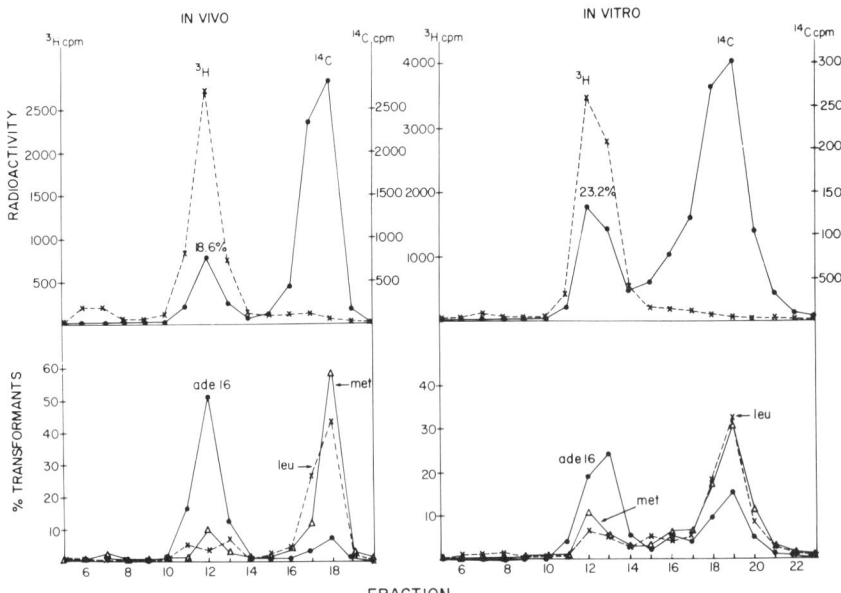

Fig. 11. Excess initiation after thymine starvation (elongation test). A culture of 50 ml of 168TT (*thy try*) was uniformly labeled at 30°C with ^{14}C-thymine in a C$^+$ medium plus L-tryptophan (50 μg/ml) and thymine (5 μg/ml) and quick-filtered. Cells were washed and shifted to 37°C in C$^+$ medium without thymine. At the end of 30 min the two 25-ml samples were withdrawn. One sample (for *in vivo*) was suspended in BU-C$^+$ medium (C$^+$ medium plus 30 μg/ml BU, 3 μg/ml thymine, 2μCi/ml ^{3}H-thymine, and 50 μg/ml L-tryptophan) and incubated for 20 min at 37°C, before the lysate was prepared. The other sample (for *in vitro*) was toluenized and incubated for 45 min at 37°C in the reaction mixture described in Table 1, except that the TTP was replaced by dBUTP before the lysate was prepared. Both lysates were centrifuged in 55% CsCl of 68 hr at 35,000 rpm at 25°C. Fractions were collected from the bottom of the tube and analyzed for radioactivity and transforming activities for *ade*16, *leu*8, and *met*5 markers. Fractions 5 to 9 represent double density labelled DNA, fractions 10 to 15 hybrid DNA, and fractions 16 to 22, parental DNA.

mately 2% from the origin (19), less than 5% synthesis should be sufficient for the test. The apparent absence of lag *in vivo* and also *in toluo* in the replication start after the halting of DNA synthesis also gives a favorable condition for the test.

An in Vitro Initiation Test Is Devised. The fact that in an exponentially growing culture only a small fraction of chromosomes has replication forks before the marker *ade*16 (Fig. 4B) provides ideal testing for *in vitro* initiation. As shown in Fig. 10B, exponentially growing cells should be harvested and prepared for *in vitro* initiation tests with simultaneous density transfer using, for example, dBUTP. A significant transfer of *ade*16 is the sign of *in vitro* initiation. A control for this test is the straight toluenization system as shown in Fig. 4B, in which *ade*16 is the least marker to be transferred to hybrid.

DNA Polymerases and ATP-stimulated DNases in *B. subtilis*

As explored in *E. coli* (12, 13, 17), *B. subtilis* also has DNA polymerase or polymerases in addition to polymerase I which are being observed in various laboratories (N. Brown, N. Cozzarelli, and A. T. Ganesan, personal communication). Figure 12 shows our DEAE-cellulose column

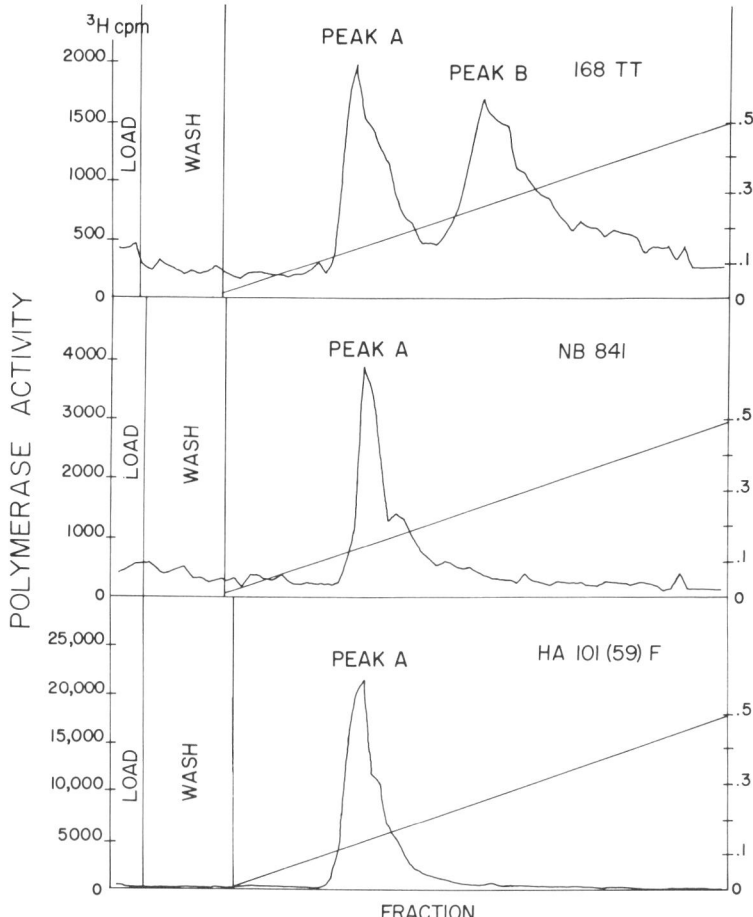

Fig. 12. Fractionation of *Bacillus subtilis* DNA polymerases on a DEAE-cellulose column (Matsushita and Sueoka, in preparation). Fraction III's (see the legend of Fig. 9) of 168 TT (*thy try pol* I$^+$), HA 101(59)F (*pol* I$^-$), and NB 841 (*thy$^-$pol* I$^-$) were fractionated on the DEAE-cellulose column with a gradient of 0.01–0.5 M potassium phosphate buffer (pH 6.5 constant). Peak B activity corresponds to DNA polymerase I. DNA polymerase activity was assayed according to Kornberg and Gefter (13) with calf thymus DNA.

Table 2. Inhibition of polymerase activities of Figure 12 by
p-chloromercuribenzoate (PCMB)

Strain	Potassium phosphate elution (M)	Original activity (%)*
168TT	0.17	23
	0.19	24
	0.31	128
	0.32	166
NB 841	0.14	26
	0.18	42
	0.19	28

*DNA polymerase activities of various fractions shown in Fig. 12 were tested with and without 0.05 mM PCMB.

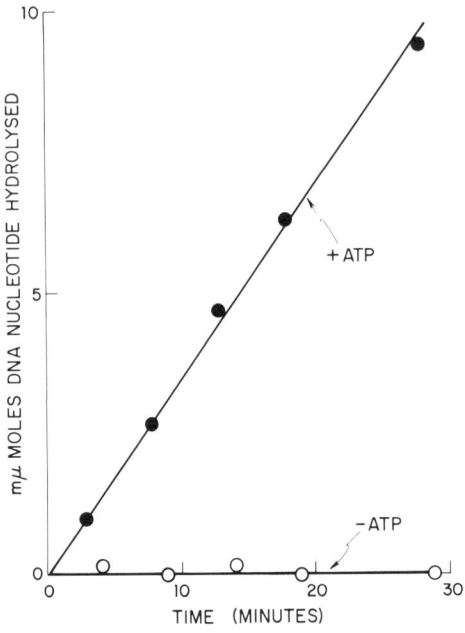

Fig. 13. ATP-stimulated DNase activity in *Bacillus subtilis* (Ohi and Sueoka, in preparation). ATP-stimulated DNase activity was purified up to 580-fold from cells of W168 by use of streptomycin sulfate precipitation, ammonium sulfate fractionation, and agarose, DEAE-cellulose column chromatographic fractionation. The reaction mixture contained 25 mM Tris-maleate buffer, pH 7.5; 7 mM MgCl$_2$; 6 mM 2.0 mercaptoethanol; 0.5 mM Na$_2$ATP; 30 μM nucleotide equivalent ^{3}H-thymine-labeled DNA and 1.0 to 5.0 units/ml reaction mixture of purified enzyme. One unit of enzyme activity hydrolyzes 0.1 nmole of DNA substrate during 30-min incubation at 32°C. At the time indicated, each 0.2-ml aliquot is withdrawn, 0.1 ml of a carrier (crude salmon sperm DNA and bovine serum albumin) is added and coprecipitated by 0.3 ml of 3.5% perchloric acid (PCA). After standing for 10 min at 0°C the sample is filtered on Whatman GF/A filter paper and washed with 0.2 ml of 3.5% PCA twice. The collected filtrate is neutralized with one drop of 5N-KOH. Ten milliliters of Bray's solution are added and the activity measured by Packard Liquid Scintillation Counter.

profiles of polymerase activities in 168 *thy trp* (*pol* I$^+$), NB 841 (*pol* I$^-$), and HA 101(59)F (*pol* I$^-$). Some properties of the two activity peaks are summarized in Table 2. At the moment we cannot say whether the residual activity in the *pol* I$^-$ mutants corresponds to polymerase II or polymerase III, or both, of *E. coli*. ATP-stimulated DNase activity similar to *recB* exonuclease of *E. coli* (1, 3, 18) exists also in *B. subtilis* (Fig. 13; Ohi and Sueoka, in preparation). The existence of this enzyme activity and DNA polymerase or polymerases other than polymerase I makes *in vitro* replication liable to various kinds of artifacts. Therefore, it will be useful to have rigid criteria for real replication such as those raised in this article (*localization test, elongation test,* and *initiation test*).

Conclusion

Currently, the *in vivo* replication of *B. subtilis* has the following features. (*a*) The temporal order of gene replication is fixed, suggesting that the origin of the chromosome, and most likely the terminus as well, occupies specific sites on the chromosome. (*b*) The initiation under normal growth conditions is symmetrical, leading to a symmetrical configuration of the chromosome. (*c*) The replication is bidirectional. The temporal order of gene replication and the linkage data are consistent with a partial bidirectionality as shown in Fig. 2.

The present status of the *B. subtilis in vitro* studies in our laboratory can be summarized as follows. (*a*) ATP-stimulated DNA synthesis in toluenized cells satisfies two criteria of normal DNA replication: semiconservative replication and sequential elongation. (*b*) So far, the maximum replication in toluenized cells is about 10% of the chromosome length per replication fork in 60 min at 37°C. (*c*) There is no new initiation in toluenized cells. (*d*) Effective and rigorous tests for elongation and initiation have been established. (*e*) A simple test for locating the site of synthesis has been formulated. (*f*) At least one PCMB-sensitive DNA polymerase has been found. (*g*) ATP-stimulated DNase has been characterized.

Acknowledgments

This work was supported by Grant GB19560 from the National Science Foundation and Grant GM10923 from the United States Public Health Service. One of us (T. M.) is the recipient of a United States Public Health Service National Cancer Institute Fellowship, and one of us (K. W.) is a United States Public Health Service Predoctoral Trainee, GM962. We are grateful for the gifts of mutant NB 841 from Dr. N. Brown, and mutant HA 101(59) from Dr. B. Strauss.

References

1. Barbour, S. D., and Clark, A. J. 1970. Biochemical and genetic studies of recombination proficiency in *E. coli*. I. Enzymatic activity associated with recB$^+$ and recC$^+$ genes. *Proc. Nat. Acad. Sci. U. S. A.* 65: 955.
2. Bott, K. F., and Wilson, G. A. 1967. Development of competence in the *Bacillus subtilis* transformation system. *J. Bacteriol.* 94: 562.
3. Buttin, G., and Wright, M. 1968. Enzymatic DNA degradation in *E. coli*; its relationship to synthetic processes at the chromosomal level. *Cold Spring Harbor Symp. Quant. Biol.* 33: 259.
4. Dubnau, D. 1970. Linkage map of *Bacillus subtilis*. In Sober, H. A. and Harte, R. A. (eds.), *Handbook of Biochemistry*, 2nd ed., pp. I-39–I-45. Chemical Rubber Co., Ohio.
5. Fielding, P., and Fox, C. F. 1970. Evidence for stable attachment of DNA to membrane at the replication origin of *Escherichia coli*. *Biochem. Biophys. Res. Commun.* 41: 157.
6. Fritsch, R., and Worcel, A. 1971. Symmetric multifork chromosome replication in fast growing *Escherichia coli*. *J. Mol. Biol.* 59: 207.
7. Ganesan, A. T., and Lederberg, J. 1965. A cell-membrane bound fraction of DNA. *Biochem. Biophys. Res. Commun.* 18: 824.
8. Gass, K. B., Hill, T. C., Goulian, M., Strauss, B. S., and Cozzarelli, N. R. 1971. *J. Bacteriol.* 108: 364.
9. Gyurasits, E. B., and Wake, R. G. 1973. Bidirectional chromosome replication in *Bacillus subtilis*. *J. Mol. Biol.* 73: 55.
10. Ivarie, R. D., and Pene, J. J. 1970. Association of the *Bacillus subtilis* chromosome with the cell membrane: Resolution of free and bound deoxyribonucleic acid on renografin gradients. *J. Bacteriol.* 104: 839.
11. Kennett, R. H., and Sueoka, N. 1971. Gene expression during outgrowth of *Bacillus subtilis* spores. The relationship between gene order on the chromosome and temporal sequences of enzyme synthesis. *J. Mol. Biol.* 60: 31.
12. Knippers, R. 1970. DNA polymerase II. *Nature* 228: 1050.
13. Kornberg, T., and Gefter, M. 1971. DNA synthesis in cell-free extracts: purification and properties of DNA polymerase II. *Proc. Nat. Acad. Sci. U. S. A.* 68: 761.
14. Lepesant, J. A., Kunst, F., Carayon, A., and Dedonder, R. 1969. Localisation génétique de mutants du système métabolique du saccharose chez *Bacillus subtilis*. Localisation par transduction à l'aide du phage PBS1. *C. R. Hebd. Seances Acad. Sci. Paris* 269: 1712.
15. Matsushita, T., White, K. P., and Sueoka, N. 1971. Chromosome replication in toluenized *Bacillus subtilis* cells. *Nature New Biol.* 232: 111.
16. Moses, R. E., and Richardson, C. C. 1970. Replication and repair of DNA in cells of *Escherichia coli* treated with toluene. *Proc. Nat. Acad. Sci. U. S. A.* 67: 674.

17. Moses, R. E., and Richardson, C. C. 1970. A new DNA polymerase activity of *Escherichia coli*. II. Properties of the enzyme purified from wild-type *E. coli* and DNA_{ts} mutants. *Biochem. Biophys. Res. Commun.* 41: 1565.

18. Oishi, M. 1969. An ATP-dependent DNase from *E. coli* with a possible role in genetic recombination. *Proc. Nat. Acad. Sci. U. S. A.* 64: 1292.

19. O'Sullivan, A., and Sueoka, N. 1967. Sequential replication of the *Bacillus subtilis* chromosome. IV. Genetic mapping by density transfer experiment. *J. Mol. Biol.* 27: 349.

20. O'Sullivan, A., and Sueoka, N. 1972. Membrane attachment of the replication origins of a multifork (dichotomous) chromosome in *Bacillus subtilis*. *J. Mol. Biol.* 69: 237.

21. Pritchard, R. H., and Lark, K. G. 1964. Induction of replication by thymine starvation at the chromosome origin in *Escherichia coli*. *J. Mol. Biol.* 9: 288.

22. Quinn, W. G., and Sueoka, N. 1970. Symmetric replication of the *Bacillus subtilis* chromosome. *Proc. Nat. Acad. Sci. U. S. A.* 67: 717.

23. Snyder, R. W., and Young, F. E. 1969. Association between the chromosome and the cytoplasmic membrane in *Bacillus subtilis*. *Biochem. Biophys. Res. Commun.* 35: 354.

24. Sueoka, N., Bishop, R. J., Harford, N., Kennett, R., Quinn, W. G., and O'Sullivan, A. 1973. Chromosome replication and cell metabolism in *Bacillus subtilis*. In Vaněk, Z., Hošťálek, Z., and Cudlín, J. (eds.), *Genetics of Industrial Microorganisms*, pp. 73-87. Academia, Prague.

25. Sueoka, N., and Quinn, W. G. 1968. Membrane attachment of the chromosome replication origin in *Bacillus subtilis*. *Cold Spring Harbor Symp. Quant. Biol.* 33: 695.

26. Sueoka, N., and Yoshikawa, H. 1965. The chromosome of *Bacillus subtilis*. I. Theory of marker frequency analysis. *Genetics* 52: 747.

27. Wake, R. G. 1972. Visualization of reinitiated chromosomes in *Bacillus subtilis*. *J. Mol. Biol.* 68: 501.

28. White, K., and Sueoka, N. 1973. Temperature sensitive DNA synthesis mutants of *Bacillus subtilis*. Appendix. Theory of density transfer for symmetric chromosome replication. *Genetics* In press.

29. Worcel, A. 1970. Induction of chromosome re-initiations in a thermosensitive DNA mutant of *Escherichia coli*. *J. Mol. Biol.* 52: 371.

30. Yamaguchi, K., Murakami, S., and Yoshikawa, H. Studies on the *Bacillus subtilis* chromosome-membrane association. I. Release of newly synthesized DNA from membrane fraction in preference to the chromosomal origin. *Biochem. Biophys. Res. Commun.* In press.

31. Yoshikawa, H., and Sueoka, N. 1963a. Sequential replication of *Bacillus subtilis* chromosome. I. Comparison of marker frequencies in exponential and stationary growth phases. *Proc. Nat. Acad. Sci. U. S. A.* 49: 559.

32. Yoshikawa, H., and Sueoka, N. 1963b. Sequential replication of *Bacillus subtilis* chromosome. II. Isotopic transfer experiments. *Proc. Nat. Acad. Sci. U. S. A.* 49: 806.

33. Young, F. and Wilson, G. A. 1972. Genetics of *Bacillus subtilis* and other gram-positive sporulating bacilli. *In* Halvorson, H. O., Hanson, R., and Campbell, L. L. (eds.), *Spores V*, pp. 77–106. American Society for Microbiology.

In Vitro DNA Synthesis and Function of DNA Polymerases in *Bacillus subtilis**

A. T. Ganesan, P. J. Laipis,† and C. O. Yehle

Department of Genetics
Lt. Joseph P. Kennedy, Jr. Laboratories for Molecular
  Medicine
Stanford, California 94305

An *in vitro* DNA-synthesizing system in *Bacillus subtilis* replicates biologically active DNA semiconservatively using externally added substrates. The occurrence of repair synthesis in this system is minimized by the use of a well characterized DNA polymerase I-deficient mutant. This strain enabled the characterization of two other polymerizing activities, DNA polymerases II and III, which are masked in the wild type cells by the abundance of DNA polymerase I. DNA polymerase I is involved in the repair of X-ray-irradiated DNA as shown by *in vitro* complementation studies. A mode of repair accompanies normal replication of DNA, as suggested by *in vitro* studies using a temperature-sensitive DNA synthesis mutant. Phage SPO1 induces a new DNA polymerase upon infection of *B. subtilis,* whose properties and role in *in vitro* phage DNA synthesis are presented.

DNA replication in bacteria is semiconservative and proceeds with the simultaneous replication of both strands in parallel starting from one end. Although sequential replication of the chromosome in *Bacillus subtilis* has been demonstrated by gene frequency analysis (54), recent evidence in *Escherichia coli* and phage λ suggests that the replication might be bidirectional (3, 30, 44). The actual mechanism of replication still remains obscure. It was earlier suggested (13) that the association

---
*Dedicated to Professor C. B. van Niel.
†Dept. of Biochemical Sciences, Princeton University, Princeton, New Jersey 08540.

of DNA with organized structures in the cell such as membrane, polymerizing enzymes, and other cofactors may provide the essential basis for continued orderly replication. This view has become more significant in recent years, especially with the isolation of several conditional lethal mutants in DNA synthesis, which suggest involvement in DNA replication of at least six clusters of genes at distinct locations on the *E. coli* chromosome (49). The study of conversion of single strand DNA templates to double stranded replicative forms in simple phage systems like M-13 and φX-174 of *E. coli* suggests that there are several host proteins that are involved in this mode of conversion (16, 43, 50). Furthermore, new DNA polymerases have been isolated from a mutant of *E. coli* that virtually lacks the DNA polymerase I (9), the classical enzyme once thought to be the only catalyst in DNA replication (22, 24, 25, 36, 37). Our own studies were mainly directed toward obtaining a system that would satisfy two criteria: (*a*) a semiconservative replication and (*b*) genetic integrity of the synthesized product free of template atoms in an *in vitro* system. *B. subtilis*, being a transformable organism, was chosen as an ideal system for this type of study. Once a satisfactory but relatively pure *in vitro* system of DNA replication is achieved in *B. subtilis*, it should be possible to study the components of such a system and approach the isolation and reconstitution of the DNA-replicating complex (11).

Previous studies in our laboratory (11) have established the following. An enzyme complex has been isolated from the fast sedimenting fraction of *B. subtilis* lysates. This has been purified 200-fold and was shown to contain separate nuclease activities, in addition to polymerase activity, by polyacrylamide gel electrophoresis. DNA synthesis was promoted by this complex that utilized preferentially bihelical DNA. Using density, radioactivity, and genetic activity as labels to distinguish the template DNA from the product, we have shown that replication is semiconservative and the product was biologically active. However, the synthesized molecules were small, the majority of them carrying only activity for single genes. In this reaction an excess of polynucleotide ligase was shown to block DNA synthesis, suggesting that nicks on the template molecules might serve to unwind the DNA during replication. This type of system is useful, but lacks several features that are characteristic of *in vivo* DNA synthesis. We have recently resorted to newer methods that will fulfill several criteria needed for a perfect *in vitro* DNA-synthesizing system. These inferred requirements are: (*a*) The chromosome should be intact. (*b*) The cells should be nonviable and permeable to externally added substrates and large proteins, if complementation studies were to be made using different mutants that are blocked in DNA synthesis. (*c*) Cellular nucleases which degrade DNA at random and interfere in the reaction by promoting an excess of repair synthesis should be eliminated. (*d*) The cell membrane damage should be minimal if such sites are involved in DNA replication (46).

Role of Nonionic
Detergents as a Reagent for the Study

The initial discovery of a DNA polymerase I-deficient mutant of *E. coli* (polA1) was achieved by the use of the nonionic detergent Brij-58 (Polyoxyethylene (20) Cetyl Ether), by De Lucia and Cairns (9). They made the observation that DNA polymerase I was released into the supernatant of detergent-treated cells while the DNA remained intact with the cell. We have used a similar system (12) to see whether enzymes were still present with the chromosome that can promote orderly replication of DNA using externally supplied substrates. It was also observed that all nonionic detergents like Tween-80, Triton-X, Lubrol, and Brij-58 were found to exercise a similar effect, releasing cellular proteins into the supernatant. Brij-58 was used in the following work as a reagent to obtain an *in vitro* DNA-synthesizing system.

In Vitro DNA Synthesis
Using Nonviable Permeable Cells

Growing cells of *B. subtilis* were shown to stop DNA synthesis on poisoning with sodium azide, an inhibitor of oxidative phosphorylation (13). The poisoned cells lose their viability within 20 min. More than 99.99% of the treated cells were nonviable. When these dead cells were treated with the detergent at a concentration of 2% final volume at 0°C, 60% of the cellular proteins leached out into the supernatant starting at 30 min. Ninety-five per cent of the DNA polymerase I was also released into the supernatant along with nuclease activities. During this treatment very little of the DNA was released from the cells in the initial 2 to 3 hr (Fig. 1). The pattern of release of proteins was also monitored by polyacrylamide gel electrophoresis and is given in Fig. 2. Along with polymerase, more than 80% of the nuclease activity was also released into the supernatant during a period of 1.5 hr. Azide-poisoned cells do not incorporate deoxynucleosides and hence we decided to add ATP and the four deoxyribonucleosides (dN) to the Brij-treated cells at different times. It is clear from Fig. 3 that as soon as the cells became permeable to ATP, DNA synthesis in the dead cells resumed. On prolonged incubation at 0°C with the detergent, the ability of the system to use dN with ATP was lost and could be replaced only by the four deoxyribonucleoside triphosphates (dNTP), presumably due to the loss of kinases from the cells. The detergent does not seem to affect

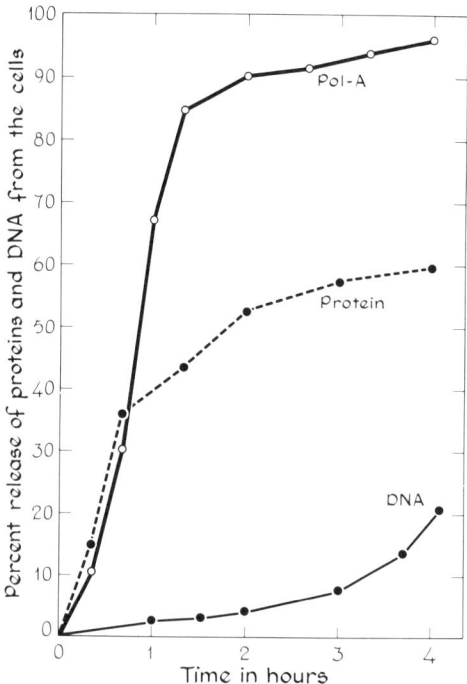

Fig. 1. Release of proteins and DNA during detergent treatment. Five milliliters of azide-poisoned cells (3×10^9 cells/ml) were incubated at 0°C with Brij-58 (12) and aliquots of 0.5 ml were withdrawn at indicated times. The cells were centrifuged and the supernatant was assayed for protein, DNA and poly [d(A-T)]-primed synthesis to detect DNA polymerase I (polA) activity (12). Nuclease assays were also performed with these fractions using a transformation system (11).

the enzyme activities. It soublizes parts of cell membrane that accumulate on the cell surface as droplets (Fig. 4). The fragile permeable cells incorporate dNTP, which could be enhanced several-fold by the addition of ATP as a cofactor. The requirement for ATP was originally observed by Smith et al. (45) in carefully prepared lysates of E. coli polA1 cells embedded in an agar base prior to lysis. These lysates were shown to synthesize DNA semiconservatively using externally added dNTP. Knippers and Stratling (23) have studied φX-174 replication in the same mutant, in which the replicating enzyme was shown to be bound to the cell membrane. This activity, unlike the DNA polymerase I, was inhibited by sulfhydryl-blocking agents. Thus, the ATP stimulation and inhibition by agents like p-hydroxymercuribenzoate (pHMB) became new probes in the study of normal DNA replication and its required enzymes.

Toluene-treated cells of E. coli and B. subtilis were shown to incorporate dNTP which could be stimulated by ATP (31, 35). Moses and Ri-

Release of Proteins from the cells

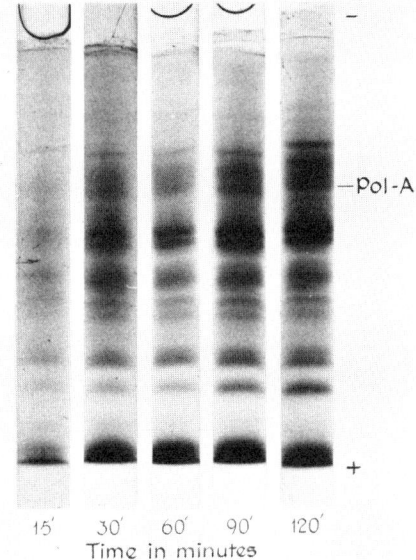

Fig. 2. Polyacrylamide gel electrophoresis of proteins released from cells. Protein, 100 μg, from the supernatant (Fig. 1) of the Brij-58-treated cells was subjected to polyacrylamide gel electrophoresis as in reference 11. One set of the gels was stained while the other set was sectioned and assayed for DNA polymerase I activity (11).

chardson (35) observed that repair replication proceeded in the absence of ATP, whereas semiconservative replication required the cofactor. Toluene-treated cells of thermosensitive DNA synthesis (DNA_{ts}) mutants preserved their *in vivo* phenotype, and failed to synthesize DNA at the restrictive temperature. The non-ATP-dependent synthesis, presumably repair, continued at the restrictive temperature (31, 34). The toluenized bacteria are not permeable to externally added large molecules like proteins. The role of the cofactor still remains obscure. We have used ATP labeled at the γ position with ^{32}P (8.5 × 10^3 cpm/μμM) and failed to see any exchange occur during DNA synthesis and, thus, no radioactivity was detected in the product or in the template DNA in our system. Its function might be involved in unwinding of DNA coupled with a sequential reaction that involves pathways of vectorial metabolism (33). The permeable cell system of *B. subtilis* requires all four dNTP, ATP, and is sensitive to Pronase and deoxyribonuclease-1. These enzymes are able to enter and exit the cells easily. All ribonucleoside triphosphates were found to stimulate synthesis, although ATP was the best cofactor in this reaction. Sulfhydryl antagonists like N-ethylmaleimide (NEM) and *p*HMB inhibited the ATP-stimulated synthesis, but

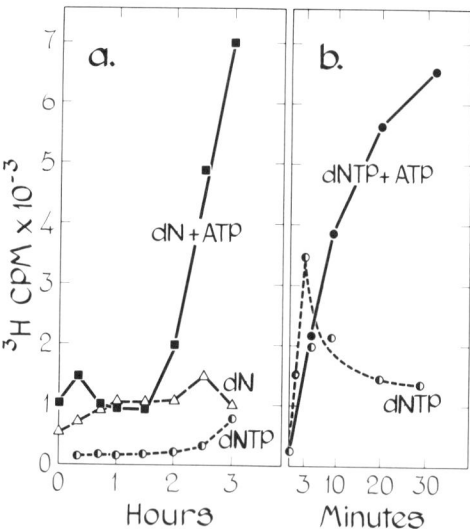

Fig. 3. Effect of ATP on incorporation of dN and dNTP. a, 3×10^9 azide-poisoned cells in 1 ml were incubated at 0°C with Brij-58 and aliquots of 0.1 ml were taken at the indicated time and centrifuged. The sedimented cells were assayed at 37°C for 30 min with dN, dN + ATP, and dNTP (12). After the reaction, the mixture was acid-precipitated and counted for radioactivity. b, Permeable cells obtained as above after 3 hr of incubation at 0°C were assayed for time course incorporation in this assay. One-tenth milliliter of cells was incubated at 37°C for different lengths of time with dNTP or dNTP + ATP, as above.

did not have any effect on the repair synthesis, *i.e.* the ability to use the substrate in the absence of ATP (11).

In Vitro DNA Synthesis in a Permeable Cell System

In order to verify the nature of the synthesized product, *B. subtilis* cells were grown in a media containing heavy isotopes ^{15}N and ^{2}H and the DNA was labeled with ^{14}C-thymidine, supplied as a precursor. The density of the DNA was 1.756 g/ml in cesium chloride solution. In an analysis of the nature of the observed incorporation, permeable cells were made from the above culture. Synthesis was allowed in the presence of the four dNTP and the four dNTP plus ATP. In this case, density-labeled cells were allowed to grow in a light medium (^{14}N, ^{1}H) for 15 min so that about 10% of the template DNA became hybrid. The logic behind this protocol was, if semiconservative replication occurred, more hybrid material would accumulate, as evidenced by the density shift accompanying the incorporation of the externally supplied radioactive light substrates. On the other hand, repair synthesis would

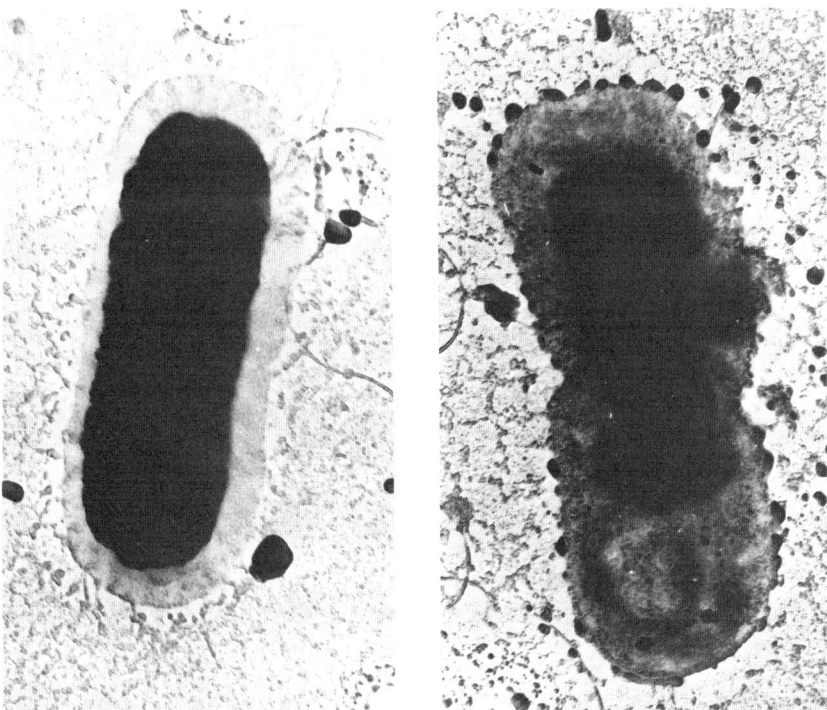

Fig. 4. Electron micrograph of cells treated with Brij-58. The method of preparation of the samples was the same as in reference 11. The top photograph shows a cell at the beginning of the treatment reaction with detergent at 0°C and the bottom one shows a cell at 2 hr. The large heavily shadowed area in the bottom picture represents the nuclear area. Magnification × 46,000.

be reflected by poorer incorporation, with little change in density profiles, except for the broadening of the DNA band width. The results were clear that in the presence of ATP there was a greater than 4-fold stimulation of DNA synthesis compared to the one without the cofactor, and that the DNA profile moved to hybrid (Fig. 5). In the absence of the cofactor there was poor synthesis and the incorporation reflects a pattern of repair synthesis, which in alkaline CsCl gradients (Fig. 6) showed the nonseparation of template and product molecules. In the case of ATP-stimulated synthesis one can observe significant amounts of product light strands free of template atoms. However, even here, there was an excess of repair-type synthesis that could be detected in the position of the template. Unlike the replication pattern observed *in vivo* with *E. coli* (32), we failed to see the gradual shift of the template DNA to hybrid and light density strata. It appeared that there was a gradual, total migration of all of the template DNA molecules to lighter

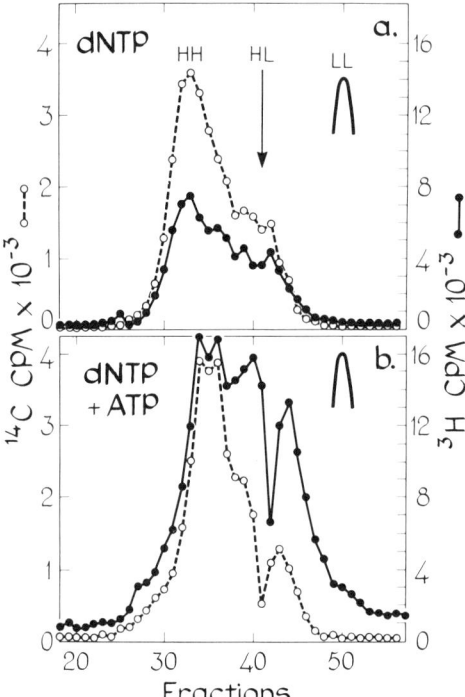

Fig. 5. Pycnography of DNA synthesized in the presence of dNTP (a), and dNTP + ATP(b). The assay conditions are described in reference 12. The heavy cellular DNA was labeled with ^{14}C-thymidine. The dNTP contained ^{3}H-labeled dTTP. The density positions of heavy, hybrid, and light DNA molecules correspond to densities 1.756, 1.729, and 1.703 g/ml in CsCl and are designated as HH, HL, and LL. The cellular DNA from heavy SB 168 (trp$_2$) contained 10% hybrid DNA as described in the text. The control DNA gave the same ^{14}C distribution as observed in Fig. 5a. One microgram of light *Bacillus subtilis* DNA was added to each gradient as a standard. Its position was determined by a transformation assay of light fractions involving a marker activity that did not interfere with the rest of the marker assays.

positions in the gradient. This could be due to an excess of repair occurring in the semiconservatively replicating DNA, to an extent that 70% of the synthesis observed could be attributed to repair synthesis. We have analyzed the biological activity of the product using the transformation system. The recipient bacteria used for the assay of products and template activity contained two mutations, ade$_{16}$ and met$_5$, which were shown to be located at the origin and terminus of the *B. subtilis* chromosome, respectively (54). In Fig. 7, the control, completely heavy DNA showed a unimodal distribution for the two gene activities. With the four dNTPs alone (not shown), there was no change in the profile except that it was broader and the specific biological activity was lower due

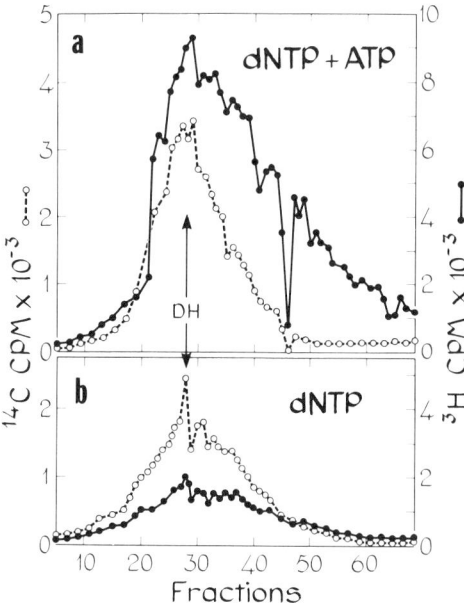

Fig. 6. Pycnography of denatured DNA after synthesis with dNTP in the presence (a) and absence (b) of ATP. Permeable cells made from completely heavy SB 168 cells were used. The conditions for synthesis were described before (12). The DNA was denatured at pH 12.5, centrifuged in alkaline CsCl gradients at the same pH, and analyzed for ^{3}H and ^{14}C counts after acid precipitation of the fractions. DH refers to the denatured heavy DNA position which is 1.810 g/ml in CsCl.

to nuclease degradation. The profile of activity with dNTP plus ATP was different. Both of the markers have moved to a hybrid density position. The distribution of both gene activities compared to the control was drastically changed, suggesting not only that one may observe the finishing of replication of certain chromosomes, but that there were others that have initiated new rounds of DNA synthesis as observed by the migration of ade$_{16}^+$ to hybrid and lighter strata. It is possible that some of the hybrid, and even the completely light activity profile one observed, could be due to continued synthesis from forks which were ahead of the gene ade$_{16}^+$, at the time permeable cells were prepared. This type of continuation of synthesis to finish the replication cycle that was currently in progress has been observed in toluenized bacteria (7, 31). A kinetic analysis during the *in vitro* replication and analysis of the product suggested that at least certain chromosomes were replicating sequentially and normally (11). In all these experiments there was a considerable amount of repair that occurred even in the presence of ATP. As mentioned earlier, as much as 70% of the incorpo-

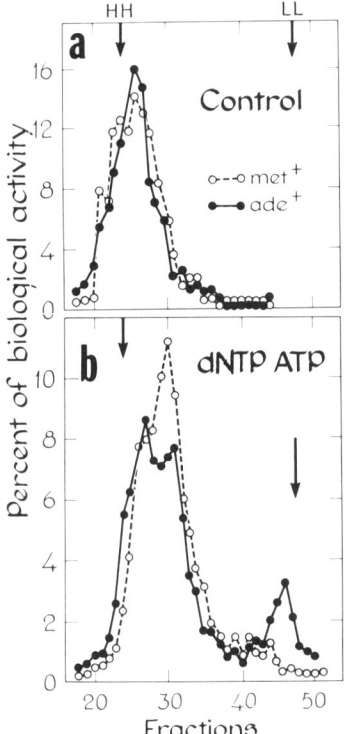

Fig. 7. Pycnographic analysis of DNA after synthesis with dNTP + ATP for biological activity. Heavy SB 168 (trp$_2$) permeable cells were allowed to synthesize DNA with ATP (12) for 40 min at 37°C. DNA was purified and centrifuged in CsCl. Each fraction, 0.01 ml, was assayed for biological activity using competent cells that contained mutations for ade$_{16}$ and met$_5$ (12). One hundred percent of ade$_{16}$+ and met$_5$+ represents 2×10^3 and 7×10^2 colonies for the control (a) and 1.5×10^4 and 6.3×10^3 colonies for (b).

ration could be ascribed to repair, judging from the alkaline CaCl gradients. Bazill et al. (1) had shown with E. coli lysates that ATP-stimulated synthesis was due to the participation of recB nuclease. One paradox presented by our system is that very little (5 to 10%) of the template DNA was degraded during the reaction, whereas synthesis was far more than the observed amount of degradation (11). It is possible that some branched molecules might be formed in the absence of template degradation.

Isolation, Characterization of DNA Polymerase I-deficient Mutant in *Bacillus subtilis*

As discussed above, much of the incorporation observed in this *in vitro* system seems to be of the repair type. We decided to isolate a DNA polymerase I-negative mutant that would permit a clear study of semiconservative replication, if this enzyme was partly responsible for the observed excessive repair. A methylmethane sulfonate (MMS)-sensitive

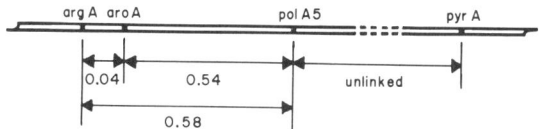

Fig. 8. Genetic map of pyrA and argA region of *Bacillus subtilis* chromosome. Three factor and two factor crosses were used to derive the map (26). Numbers represent recombination frequency (1, frequency of cotransfer).

mutant (26) was isolated after treatment of the cells with the mutagen nitrosoguanidine. This mutant (polA5) appeared to lack DNA polymerase I activity and revealed the presence of another polymerizing activity in the crude lysate of the cells. It was found to be more sensitive to UV and X irradiation and was inefficient in reactivation of UV-irradiated SPO1 bacteriophage.

The polA5 mutation seemed to be a single step mutation as evidenced by the reversion frequency of 8×10^{-8} for loss of MMS sensitivity. Transduction mapping with phage PBS1 revealed linkage of this gene to a previously mapped arginine locus (arg_A) on the *B. subtilis* chromosome (Fig. 8). This mutation did not affect the recombinational process or cotransfer of linked genes during transformation. Hence, DNA polymerase I may not play a major role in either recombination or DNA replication, a result very similar to that observed in *E. coli* polA1 mutant (19).

DNA Polymerase Activities in the Mutant polA5

Table 1 presents the observed levels of polymerase activity in the wild type *B. subtilis* and the polA5 mutant using various DNA templates in the reaction. Mixing experiments using lysates of the mutant and its wild type ruled out the possibility of an increased nuclease level or a soluble inhibitor as the cause of low polymerase activity. Lysates of the mutant still showed significant polymerase activity using cellular DNA templates and externally added calf thymus DNA or activated salmon sperm DNA. These results and the inhibition of the enzyme activity in the presence of pHMB are shown in Table 2. All activity was suppressed in the mutant in the presence of the inhibitor, suggesting that the catalytic activity depends on sulfhydryl groups of the enzyme. N-Ethylmaleimide (NEM) gave similar results although at a much higher concentration. The DNA polymerase I activity in the wild type was not inhibited by pHMB, so that using poly[d(A-T)] as a template, no inhibition was obtained with pHMB, whereas only 40% of the double stranded DNA-priming activity was inhibited. To rule out the possibility

Table 1. DNA polymerase activities

Template	Bacillus subtilis SB1058	B. subtilis SB1060	Escherichia coli W3110T	E. coli W3110T⁻ polA1
	(units/mg protein)			
Poly[d(A-T)]	2.22	0.068	0.52	0.015
	S.D. 0.55	S.D. 0.022	(66%)	(45%)
	(100%)	(100%)		
Endogenous DNA	0.149	0.056	0.142	0.004
	S.D. 0.064	S.D. 0.036		
Calf thymus DNA	0.435	0.084		
Activated salmon sperm DNA	0.292	0.162		
Denatured salmon sperm DNA	0.334	0.086		

Preparation of lysates and the DNA polymerase assay were described in references 26 and 39. Figures in parentheses under poly[d(A-T)] assays refer to the percentage of ^{3}H-labeled poly[d(A-T)] remaining acid-precipitable after incubation under the standard assay conditions with 30 μg of protein, an amount within limits for linear assay response. For comparison, *E. coli* wild type and polA1 mutant are included.

that the residual polymerizing activity is due to a leaky mutation, we have purified the enzyme after freeing it from DNA template in a step gradient of CsCl and then subjecting the resulting protein from the wild type and polA5 mutant to zone sedimentation in sucrose gradients. Separation of two distinctive polymerizing activities, one inhibited by *p*HMB, is possible in the wild type. The mutant has only the fastest sedimenting *p*HMB-inhibitable enzyme activity (Fig. 9). The activity represents 10 to 15% of the DNA polymerase activity in the cells and is termed DNA polymerase II, as in *E. coli* (36, 37). Like *E. coli*, the fast sedimenting fraction in the sucrose gradient (DNA polymerase II) might contain another activity very similar to the DNA polymerase

Table 2. DNA polymerase activities in the presence of *p*-hydroxymercuri-benzoate

Assay substrate	Wild type (SB1058)	polA5 mutant (SB1060)
	(units/mg protein)	
Poly[d(A-T)]	2.23	0.049
Plus *p*HMB	2.19	0
Endogenous DNA	0.190	0.089
Plus *p*HMB	0.118	0

Preparation of crude lysates and assays were previously described (26) except that β-mercaptoethanol was omitted. *p*HMB, dissolved in the dark in 10 mM glycylglycine, pH 8.5, was added to 0.1 mM. Endogenous DNA refers to the cellular DNA present in the lysate.

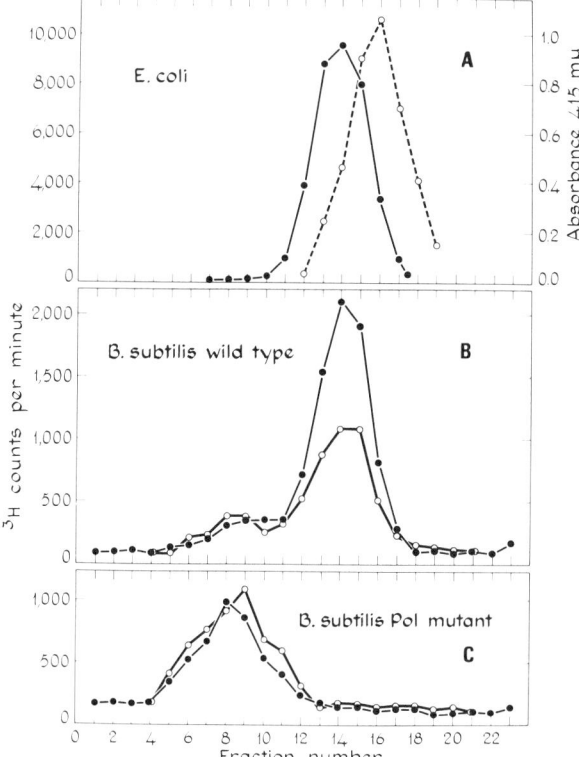

Fig. 9. Sucrose gradients of template-free lysates of wild type and polA5 mutant cells of *Bacillus subtilis*. Of the lysates, 0.08 ml was centrifuged for 30 hr at 35,000 rpm at 4°C in 5 ml of 5 to 20% linear sucrose gradients. A separate control tube contained 70 units of purified *Escherichia coli* DNA polymerase I and 300 μg of human hemoglobin included as a standard. Aliquots of gradient fractions were assayed for DNA polymerase using various templates as substrates (26). A, *E. coli* DNA polymerase I; B, *B. subtilis* wild type; and C, *B. subtilis* polA5 mutant. ○- - -○, human hemoglobin; ●———●, poly[d(A-T)] template; ○———○, activated salmon sperm DNA template.

III. This seemed to be very likely because the DNA polymerase II profile was broader in sucrose gradients and suggested a slight bimodality in the activity pattern (Fig. 9). To explore the possibility, lysates were made of wild type cells and the DNA was removed by phase partition in a polyethylene glycol-dextran system (39). The resulting protein fraction was fractionated on a DEAE-cellulose column and eluted with a linear KCl gradient (0.1 to 0.9 M). Here an apparent unimodal distribution of polymase activity was observed using activated salmon sperm DNA as a template. The peak activity was eluted at a concentration of 0.3 to 0.35 M KCl. When the same assay was performed in the presence of pHMB, both ends of the above activity distribution, *i.e.*

activity eluting at a salt concentration of 0.2 to 0.3 M and 0.35 to 0.43 M KCl, were inhibited by the reagent. The major pHMB noninhibitable fraction was DNA polymerase I. The fractions from both sides of the distribution were pooled and subjected to refractionation on a second DEAE-cellulose column. The proteins reproduced the same elution pattern as before. If these activities that elute at low and high salt concentrations were reflections of different polymerase activities in the cell that were masked by the excess of DNA polymerase I, then the polA5 mutant should clarify the situation. The protein from the mutant was subjected to the same regimen of purification. We clearly observed (14) two distinguisable activities that eluted at exactly the same salt concentrations as in the wild type cell extract. The proportion of these two activities present in the wild type cells relative to DNA polymerase I was 10 to 12% for DNA polymerase II and 1 to 2% for DNA polymerase III, very similar to that in *E. coli* (25). These three polymerases were subjected to zone sedimentation in sucrose gradients. Sedimentation profiles of the three enzymes and the properties of polymerases II and III are given in Table 3 and Fig. 10. Like *E. coli*, both the enzymes prefer exonuclease III-treated DNA templates, but did not show differential inactivation by heat or inhibition by salts. DNA polymerase II has a molecular weight of 160,000 to 180,000 daltons, whereas polymerase III has a molecular weight of 140,000 daltons. DNA polymerase I was reported to have a molecular weight of 46,000 daltons (10). The enzyme in crude lysates, as well as from DEAE-cellulose columns, when sedimented in sucrose gradients, was found to be very similar to *E. coli* DNA polymerase I, with a molecular weight of 110,000 daltons (Figs. 9 and 10). It is likely that fragmentation of this enzyme could occur due to proteases, when purified by standard procedures, as had been demonstrated for the enzyme from *E. coli* (5, 21).

Table 3. Properties of DNA polymerases II and III of *Bacillus subtilis*

Condition	Relative activity	
	DNA polymerase II (%)	DNA polymerase III (%)
Control	100	100
+ 0.1 mM pHMB	4	10
+ 0.05 M $(NH_4)_2SO_4$	113	140
+ 0.15 M KCl	106	125
+ Enzymes heated at 45°C for 10 min	35	62
+ Template treated with exonuclease III	153	120

DNA polymerases II and III were purified 140- and 200-fold, respectively. Activities are expressed relative to those observed with activated salmon sperm DNA template as 100% (0.04 and 0.025 nmole of acid-precipitable ^{3}H-dTMP for polymerase II and III, respectively). *Escherichia coli* exonuclease III treatment was sufficient to degrade 5 to 10% of the DNA template.

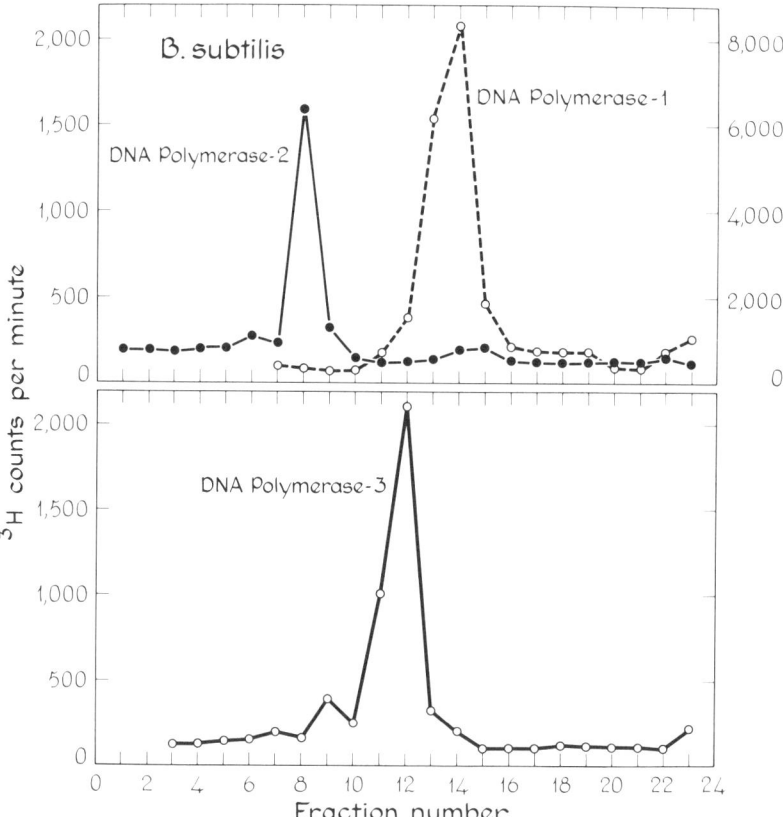

Fig. 10. Sedimentation analysis of DNA polymerases from *Bacillus subtilis*. Four units of DNA polymerases I and II and three units of DNA polymerase III from DEAE-cellulose column fractions in 0.05 ml of Tris-HCl buffer (0.01 M) with 0.01 M β-mercaptoethanol were layered on top of a sucrose gradient (see Fig. 9 legend). After centrifugation, 23 fractions were collected and 0.025 ml of each fraction was assayed with activated salmon sperm DNA as template for DNA polymerase activity (14).

In Vitro Synthesis of DNA Using the polA5 Mutant

It was pointed out above that as much as 70% of the *in vitro* DNA synthesis observed in the presence of dNTP and ATP could be attributed to a mode of repair. If DNA polymerase I is responsible for this repair, the mutant should show less repair synthesis in this *in vitro* system. This should allow a clear detection of semiconservative replication. ^{15}N, ^{2}H, and ^{14}C-thymidine-labeled cells of the polA5 mutant were prepared and made permeable with the detergent. Synthesis was allowed to proceed for 30 min at 37°C. The DNA was purified and sedimented in a CaCl gradient with appropriate density makers. There was a 4-fold stimulation in synthesis when ATP was added as a cofactor in this experiment. The profile exhibited by the reaction product synthesized in the absence of ATP was similar to the one shown in Fig. 5a, except that the extent of incorporation was less than 20% of the incorporation observed using the wild type cells. Hence we can interpret that at least 80% of the incorporation that occurred in the absence of ATP was due to DNA polymerase I bound to cellular DNA that failed to leach out of the cells. The alkali-denatured sample of the above preparation showed a pattern (not shown) similar to the one in Fig. 6b, again suggesting that the incorporation reflects covalent linkage of repaired segments with the template heavy DNA strands.

In the presence of ATP and dNTP, a semiconservative replication pattern was very clear (Fig. 11B). In comparison, the distribution of

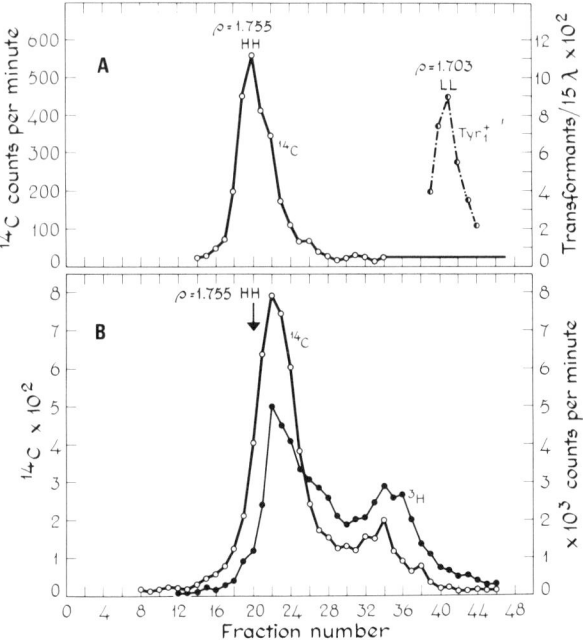

Fig. 11. Pycnography of DNA synthesized by permeable cells of polA5 mutant. Cells, 3×10^9, of polA5 mutant of *Bacillus subtilis* grown in ^{15}N, ^{2}H medium containing ^{14}C-thymidine were converted to permeable cells (12). The specific activity of the cellular DNA was 1.2×10^3 cpm/nmole. Synthesis of DNA by these cells was followed with dNTP + ATP, in which the TTP was labeled with ^{3}H (2×10^4 cpm/nmole). Each gradient contained 30 nmoles of DNA and 0.15-ml fractions were collected from an 8.0-ml gradient. Of each fraction, 0.03 ml was acid-precipitated and counted for ^{3}H and ^{14}C. One microgram of light DNA from *B. subtilis* (ade$_{16}$, leu$_2$, and met$_5$) was added to each gradient. The light DNA position was determined by transformation for tyr$_1^+$ activity using tyr$_1$-competent cells. A, control heavy DNA from the permeable cells with the standard DNA; B, profile of the template and the product after synthesis with dNTP + ATP.

^{3}H label in the wild type cell system (Fig. 5B) occurred throughout the template DNA position and all the DNA molecules moved to lighter densities. The mutant showed a clear stepwise transition, in which 15% of the template DNA moved to hybrid density by the synthesis of light complementary strands. The rest of the label remained with the main heavy DNA peak, slightly shifted toward a lighter stratum. We feel that there was still some incorporation that reflected a mode of repair in the presence of ATP. The presence of DNA in the hybrid position suggested that at least this synthesis was semiconservative. In order to see whether these DNA strands were separable, we pooled the fractions from the hybrid position, denatured them in alkali, and sedimented the resulting DNA in a CsCl gradient following neutralization. The clear separation of heavy, template (^{14}C) and product, light (^{3}H) strands of DNA (Fig. 12B) suggested that repair was negligible in these fractions.

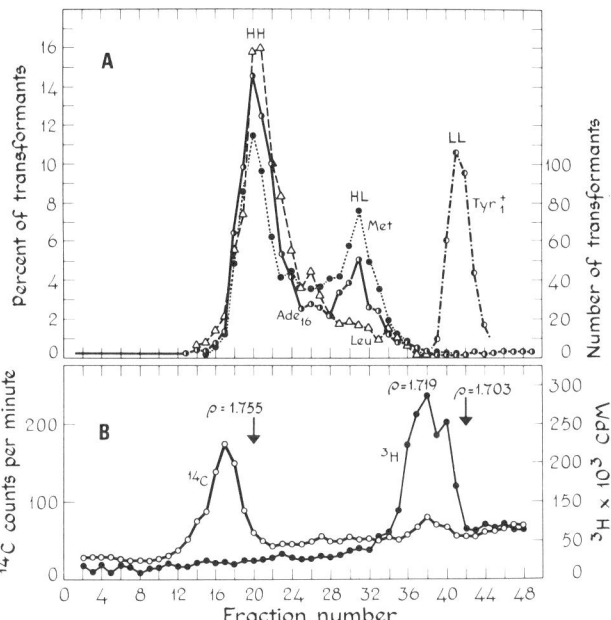

Fig. 12. Biological activity profile of the gradient shown in Fig. 11B. The peak of activity in the HH position is the fraction 22 in Fig. 11B. Of each fraction, 0.015 ml was incubated with 0.5 ml of competent *Bacillus subtilis* cells carrying ade$_{16}$, leu$_2$, and met$_5$ (1.5 × 10^8 cells/ml). The three gene activities were scored by plating the transformed bacteria on appropriately supplemented plates (12). One hundred percent activity represents 3.5 × 10^3 colonies for ade$_{16}$$^+$, 1.2 × 10^3 colonies for leu$_2$$^+$, and 2.4 × 10^3 colonies for met$_5$$^+$ (A). The remainder of each fraction from 30 to 37, corresponding to the hybrid position, was pooled and denatured in alkali at pH 12.5 (12). The denatured DNA sample was centrifuged in a neutral CsCl gradient at pH 8.0 after neutralization. Collected fractions were acid-precipitated and counted for ^{3}H and ^{14}C. The denatured heavy DNA sediments were at 1.769 g/ml density position (B).

However, the main heavy DNA, when denatured in alkali and analyzed as above, showed light (^{3}H) segments linked with template, heavy (^{14}C) DNA strands (data not shown). This might reflect in some part continuation of chromosome replication that had a growing point with fully heavy atoms. We feel that part of the synthesis reflected a repair type of incorporation similar to the one reported earlier (1).

As shown in Fig. 7, the control, biological activity of the DNA showed a unimodal distribution for three genes located at the origin, middle and the terminus of the chromosome (ade$_{16}$, leu$_2$, met$_5$). It was of interest to see whether the synthesized hybrid DNA had any biological activity. If so, what is the profile of the above three gene activities? We found (Fig. 12A) that 35% of the met$_5{}^+$ showed a clear transition to the hybrid position, whereas only 24% of ade$_{16}{}^+$ was in this position. We can interpret this as a continuation of synthesis to finish the cycle of DNA replication in a majority of the chromosomes. Since the major activity (74%) of ade$_{16}{}^+$ still remained with the heavy position, the remaining activity in the hybrid strata for this gene is interpreted as either new initiations or a continuation of the growing point that existed ahead of this gene at the time of preparation of permeable cells. These two gene activities in concert with the pattern of activity for leu$_2{}^+$ which is located in the middle region of the chromosome are clear enough to conclude that at least 15% of the template DNA in this system replicates semiconservatively and that polA5 mutation significantly reduces the repair observed with the wild type cells.

Function of DNA Polymerase I

From the UV and X-ray sensitivity data it was clear that *B. subtilis* DNA polymerases II and III may be present at higher levels or perhaps be more efficient than the corresponding *E. coli* enzymes. The presence of these additional polymerases might account for the lessened sensitivity of the *B. subtilis* polA5 mutant to UV and X irradiation, assuming that DNA polymerase I is involved in a repair process. Apparently *E. coli* normally has less polymerizing activity than *B. subtilis*, but the template in poly[d(A-T)]-primed synthesis was rapidly degraded in *E. coli* whereas, under similar conditions, no loss of template was observed in *B. subtilis* (39). Our results are consistent with those of Boyle *et al.* (4) that one effect of polA1 mutation in *E. coli* is to allow extensive exonucleolytic degradation of the cellular DNA. The results are consistent with the hypothesis that the decreased sensitivity of the *B. subtilis* polA5 mutant to irradiation and UV-stimulated degradation might be related to the decreased degradation of DNA observed, compared to *E. coli*.

Instead of elaborating on repair pathways, we would like to present data on the *in vivo* demonstration of repair of X-ray-damaged regions, since it pertains to one of the functions of DNA polymerase I. Town et al. (48) demonstrated that polA1 cells of *E. coli* are extremely sensitive to X irradiation and appeared to lack the fast DNA repair synthesis that will restore the damaged regions. We have earlier shown (28) that single strand breaks or gaps in a bihelical DNA could be repaired by treating with purified DNA polymerase I from *E. coli* and polynucleotide ligase. The transforming activities of these molecules were restored and were correlated with an increase in the single strand molecular weight. DNA from X-irradiated polA5 cells has more single strand breaks and poorer transforming activity, both for single and multiply linked markers, compared to irradiated wild type cells (29). However, these breaks could be repaired by the addition of a lysate of wild type cells together with purified polynucleotide ligase or by the addition of purified DNA polymerase I and polynucleotide ligase. Polynucleotide ligase was added to the reaction mixture since we have earlier shown that in *B. subtilis* this enzyme is extremely unstable in crude lysates. Table 4 and Figure 13 show that there is almost full recovery of transforming activity with a parallel increase in molecular weights of DNA single strands of irradiated DNA. These results directly demonstrate one of the major roles of DNA polymerase I. A similar repair synthesis could occur by the mediation of polymerases II and III, which were shown to function on DNA with gaps in the molecule (14, 24, 36, 37). Since there are fewer molecules of these enzymes per cell (Table 3) compared to polymerase I, their role in repair may not be very efficient.

Table 4. *In vitro* repair of X-irradiated DNA with lysates

Treatment of DNA	Biological activity for single and linked markers		
	No addition	SB202 lysate	SB1060 lysate
25 krads	52 (38)	54 (38)	49 (44)
+ Ligase	49	86 (59)	47 (45)
+ Polymerase	53	58 (37)	62 (43)
+ Polymerase + ligase	78 (52)	81 (58)	86 (78)
Control	100 (100)		
+ Polymerase + ligase	103 (106)		

The 100% value for control unirradiated DNA represents 1.5×10^3 trp_2^+ transformants whereas 166 colonies exhibited linkage (in parentheses). SB202 carries the four genetically linked mutations and hence the lysate did not contribute any biological activity for the markers assayed. Maximum error is $\pm 13\%$.

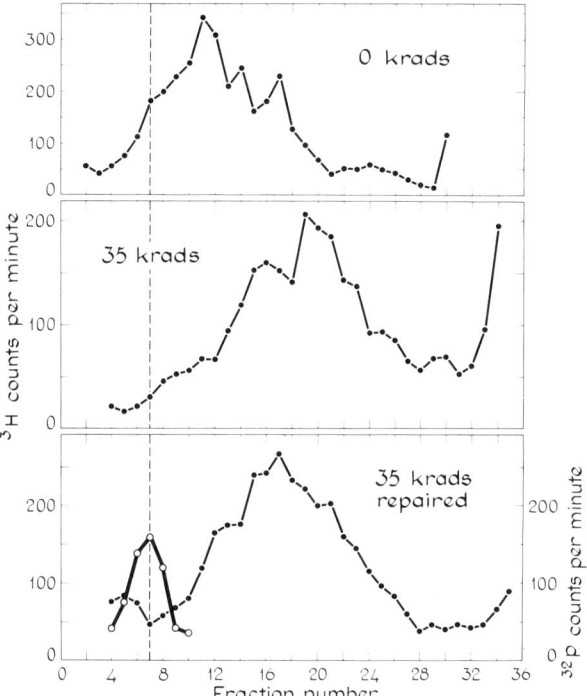

Fig. 13. Alkaline sucrose gradients of irradiated and repaired transforming DNA. SB 1060 (polA5) cells were grown in minimal medium with required supplements (27) and ^{3}H-thymidine (6 Ci/nmole) at 1 μCi/ml for 3 hr. These cells were irradiated with X-rays and the DNA was extracted along with a control unirradiated sample. The reaction mixture (27) contained 200 μM in each dNTP; 20 μM DPN; 9 units of purified *Escherichia coli* polynucleotide ligase and 25 units of DNA polymerase I. Fifteen nanomoles of irradiated DNA were added to the reaction mixture. After 15 min at 30°C, the reaction was stopped by the addition of NaOH (27). The 35-krad irradiated DNA, repaired DNA, and the control (0 krad) denatured DNA was sedimented in alkaline sucrose gradients (27) with ^{32}P-labeled T_4 phage DNA as a standard. The weight-average molecular weight of denatured T_4 DNA was 5×10^6 daltons. ●——●, (^{3}H) *Bacillus subtilis* DNA; ○——○, (^{32}P) T_4 DNA. The molecular weight for unirradiated, irradiated, and irradiated-repaired DNA samples were 33.7, 16.4, and 22.4×10^6 daltons, respectively.

Function of DNA Polymerases II and III

Bonhoeffer and colleagues have developed an elegant system (42) for the study of *in vitro* DNA synthesis in *E. coli*, in which complementation studies could be made using different thermosensitive mutants in DNA synthesis (38). The *in vitro* system could be divided into a soluble (A) and insoluble (B) fraction. Both were required for *in vitro* synthesis. The fraction B consisted of DNA and membraneous structures with their bound proteins. Fraction A contained components which are needed at *in vivo* concentrations to promote semiconservative replication. Thermosensitive DNA mutants were found to have the thermolabile components in either A or B fractions by *in vitro* complementation studies. A group of mutants, dnaE, clustered in one region of the *E. coli* map were unable to complement each other. The labile protein has a molecular weight of 150,000 and has been shown to be polymerase III (15, 38). The enzyme is present in wild type cells at about 1% polymerase I activity, and seems to be required at its *in vivo* concentration for DNA replication. So far, this is the first demonstration of an enzyme that is directly involved in DNA synthesis. A similar activity has also been purified from *B. subtilis* cells (14).

In all the dna mutants of *E. coli* tested (15) polymerase II activity appeared normal. The mutant of polymerase II grows normally and there was no detectable defect in its efficiency to perform various functions (see the articles by Richardson *et al.* and Gefter *et al.* in this symposium). The antileukemic agent 1-β-D-arabinofuranosyl cytosine inhibits DNA synthesis in prokaryotic and eukaryotic cells (8). The ATP-stimulated DNA synthesis in toluene-treated polA1 cells of *E. coli* was inhibited by the triphosphate of the above analog, ara-CTP (41). Purified polymerases I and III did not show marked inhibition by ara-CTP in *in vitro* assays, while polymerase II was inhibited (41). Polymerase II specifically interacts with *E. coli* "unwinding" protein in an *in vitro* system (data of Gefter *et al.* in this symposium). It is difficult to reconcile these observations and the normal behavior of the polB mutant of *E. coli*. Our studies indicate that the loss of polymerases II and III from permeable cells into the supernatant results in the loss of DNA synthesis in the presence or absence of ATP. Since the *E. coli* mutant of polymerase II does not show any defect, unlike polA1, it is likely that the mutation resulted in an unstable enzyme which might function normally *in vivo* but undergoes conformational change *in vitro*. Neither of these enzymes requires ATP, which seems to be required for normal DNA replication, in *in vitro* experiments. At present it is not clear as to the exact role of these enzymes in DNA synthesis. It is likely that detailed complementation analysis of the dna mutants (38, 51) should clairfy the roles of these enzymes and other cofactors.

DNA Synthesis in a Thermosensitive Mutant of *Bacillus subtilis*

A thermosensitive DNA mutant of *B. subtilis* continues to synthesize DNA at 30°C, but stops at elevated temperatures. This is also a polA mutant and was kindly given to us by Dr. Bazill. This mutant has no detectable defect in precursor synthesis or other observable defects in

RNA or protein synthesis at nonpermissive temperatures. It does not degrade its DNA at elevated temperatures. It behaves as though a factor which is continuously needed for DNA replication becomes inactivated at an increasing rate at temperatures above 40°C (2). When permeable cells were made from cells grown at 30°C there was incorporation of externally added dNTP, with and without the addition of ATP. The cofactor stimulation was 2-fold compared to the control. Incorporation at the permissive temperature is inhibited by nalidixic acid (18) and pHMB. These permeable cells were then heated for 15 min at different temperatures up to 48°C. The interesting observation was that the incorporation of dNTP in the presence or absence of ATP stops at elevated temperatures (Table 5). The inactivation was gradual and was complete within 15 min at 48°C. These 48°C-inactivated cells do not show any incorporation when assayed either at 30 or 48°C. However, if these cells were either lysed or freeze-thawed several times, they did show incorporation that was stimulated by ATP at both 30 and 48°C. The phenotype which was observed in the dead, permeable cells was lost and could not be restored (Table 5). The synthesis observed in lysates or freeze-thawed permeable cells was not inhibited by nalidixic acid.

The above outlined experiment suggests the following.

1. In the first case, both types of DNA synthesis stop at the elevated temperature as in the *in vivo* pattern of DNA replication.

Table 5. *In vitro* DNA synthesis in a DNA_{ts}, polA mutant of *Bacillus subtilis*

| | Amount (cpm) incorporated | | | |
| | Assayed at 30°C | | Assayed at 48°C | |
Temperature	−ATP	+ATP	−ATP	+ATP
None	2803	5117	320	480
40°C	1414	2443		
45°C	977	1405		
48°C	199	264		
None plus pHMB	108	220		
None plus nalidixic acid	180	260		
48°C lysate	1400	3806	1600	4400
48°C freeze-thaw	1900	4120	2080	5086
48°C lysate + nalidixic acid	1660	4220		
40°C lysate + pHMB	300	260		

B. subtilis 2340 DNA_{ts}, polA mutant cells grown at 30°C were made permeable with Brij-58 at 0°C. Aliquots of the suspension were heated at the indicated temperatures for 15 min. The cell suspension, 0.02 ml, containing 3×10^8 cells (6.5 nmoles of DNA) was used in each assay. The reaction mixture (12) contained dNTP in which dTTP was labeled (4×10^4 cpm/nmole). ATP concentration was 2 mM. The reaction time was 30 min at 30°C and in some cases at 45 or 48°C for the same period. pHMB and nalidixic acid were 0.3 mM and 5 μg, respectively. Volume of the assay mixture was 0.1 ml.

2. Lysis, or freeze-thawing, removes the inhibition, so that one finds synthesis which occurs to the same extent at both temperatures. It is likely that some components of the cell membrane, or inhibitors that were normally present in the replication complex, were destroyed, or that the dissociation of cofactors in the replicating complex had occurred. This mode of synthesis is not inhibited by nalidixic acid and reflects mainly extensive degradation and repair of cellular DNA. The inhibition observed by pHMB suggests that polymerases II and III might be involved in this mode of synthesis. The stimulation observed by ATP here is mainly due to the mediation of nucleases that require ATP for their function (1, 17). Looking more closely at the nalidixic acid-inhibitable, thermosensitive DNA synthesis, again there are two types of synthesis, *i.e.* with and without the addition of ATP. Both these processes have been studied to some extent and permit us to postulate that there is a mode of repair synthesis that is also thermosensitive, which accompanies normal, ATP-requiring semiconservative replication. The recent discovery (47) by Okazaki and colleagues that RNA fragments are associated with nascent DNA chains in *E. coli* led to the extension of their model of discontinuous DNA replication reported earlier (40). According to this model, DNA synthesis occurs using RNA sequences as initiators, as was observed in the case of M-13 and ϕX-174 phages of *E. coli* (6, 20). After DNA synthesis, these sequences of RNA are degraded and the generated gaps are then filled by a DNA polymerase and joined by polynucleotide ligase to yield continuous DNA strands. Our data suggest that it may be this mode of repair that accompanies replication and is also thermosensitive in this mutant. In other words, it is difficult, if not impossible, to separate these two processes.

Complementation of Functions in the DNA_{ts} Mutant

The permeable cells made from the mutant stop DNA synthesis irreversibly on exposure to 48°C for 15 min. If DNA-free extracts of the wild type cells and those of the mutant grown at 48°C for 40 min are added to these cells, synthesis is stimulated several-fold by the wild type cell extract. This incorporation continues for only 40 min. The ability of the partially purified extract from the wild type cells and not the 48°C-grown mutant cells to stimulate DNA synthesis (Table 6) suggests that the extract contains the factor that can stimulate DNA synthesis. The partially purified factor seems to be a protein, is required in catalytic amounts, and is currently under investigation.

Table 6. Complementation studies with a DNA_{ts}, polA mutant of *Bacillus subtilis*

Treatment	Amount (cpm) incorporated	
	−ATP	+ATP
48°C control	200	280
48°C plus wild type extract (48°C)	2300	3680
48°C plus mutant extract (48°C)	120	260
48°C plus mutant extract (30°C)	368	416
30°C control	2872	5500

Permeable cell preparation, heat inactivation, and assays were done as described in the legend of Table 5. Extracts were the supernatants obtained by sedimenting sonicated lysates at 15,000 × g for 15 min at 4°C. Wild type refers to the polA mutant whereas mutant refers to polA, DNA_{ts} strains. Ten micrograms of protein used per assay.

DNA Polymerase from Phage SPO1-infected Cells of *Bacillus subtilis*

Another DNA polymerase has been purified and characterized from cells which have been infected by the virulent phage SPO1. Host DNA polymerase activity was greatly reduced by 8 min after infection. Subse-

Table 7. Properties of phage SP01 DNA polymerase

Requirements	Relative activity (%)
Reaction	
Complete system	100
− dATP	2
− Mg^{++}	1
− Mg^{++}, plus Mn^{++}	2
− β-Mercaptoethanol	60
− β-Mercaptoethanol, plus *p*HMB	4
− β-Mercaptoethanol, plus NEM	13
− DNA	1
Template	
Native SPO1 DNA	5
Denatured SP01 DNA	100
Native *Bacillus subtilis* DNA	1
Denatured *B. subtilis* DNA	9
Activated salmon sperm DNA	2
Denatured salmon sperm DNA	92
Poly[d(A-T)]	18

The complete DNA polymerase reaction mixture (11) (0.1 ml) contained 10 nmoles of denatured salmon sperm DNA as template and 0.04 unit of enzyme. Assays were performed at pH 7.5, the optimum for this enzyme. Where indicated, the concentration of Mn^{++} was 15 m*M*, *p*HMB was 1 m*M*, and NEM was 4 m*M*. Template preference was measured using 10 nmoles of the listed DNA preparations. No increase in activity was observed upon activation of native DNA by pancreatic deoxyribonuclease or by partial digestion with *Escherichia coli* exonuclease III.

quently, a new phage-induced enzyme appears. This protein has been purified at least 1000-fold and has been characterized and compared with the host enzymes and other phage-induced DNA polymerases (52, 53). The enzyme has a molecular weight of 125,000 daltons. Its properties are presented in Table 7.

The DNA of SPO1 contains hydroxymethyluracil in place of thymine which results in a natural buoyant density shift to 1.740 g/ml compared to that of thymine-containing DNA with a similar base composition (1.703 g/ml). However, lysates prepared from cells after phage infection were capable of synthesizing phage DNA when supplied with the usual dNTP. Since the *in vitro* system utilizes dTTP in the place of dHMUTP, the product molecules became lighter than the natural phage DNA template in CsCl. We have taken advantage of this observation to fractionate the template and product DNA without a complicated labeling regimen. As shown in Fig. 14, the product could be separated and assayed for

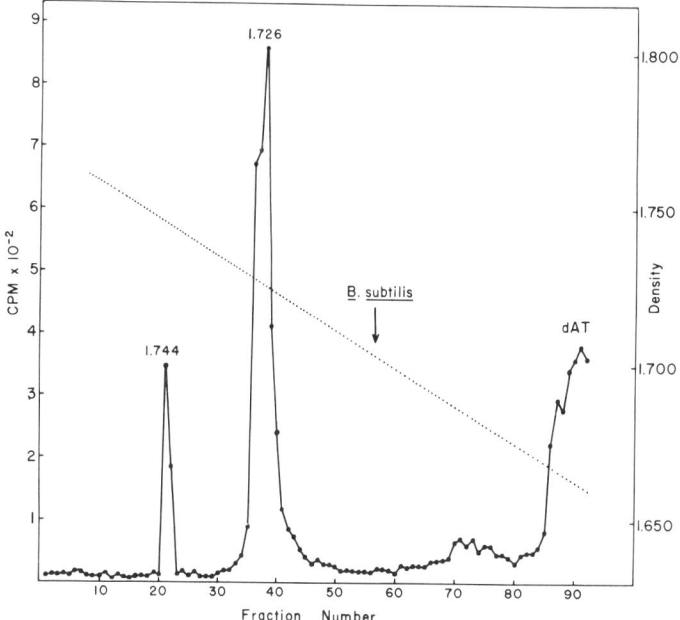

Fig. 14. Pycnographic fractionation of SP01 DNA template and product synthesized by the phage-induced DNA polymerase (52). An extract, 0.1 ml, of a *Bacillus subtilis* polA mutant (1420) prepared 20 min after infection with SP01 was added to 0.7 ml of the standard polymerase reaction mixture (11) containing ^{3}H-labeled TTP (1×10^4 cpm/nmole) and incubated at 37°C. ^{3}H-labeled poly[d(A-T)] was added to the mixture after the 30-min reaction was terminated as a standard for the density gradient. Solid CsCl was added to the mixture to bring the density to 1.710 g/ml. Nine milliliters of the solution in 0.01 M Tris, pH 8.0, were centrifuged (52) and 0.1-ml fractions were collected. These were then analyzed for refractive index and acid-precipitable counts of ^{3}H.

biological activity. When this was assayed in a transfection experiment, no infectious centers were recovered. However, when the transfection assay was coupled with a marker rescue system using suppressor-sensitive phage mutants, the product DNA was shown to contain wild type phage markers (53). The product is currently being analyzed in more detail.

Conclusion

There are several *in vitro* systems available for studying DNA synthesis. The permeable cell system discussed above, although it is somewhat limited, was shown to be useful for studying DNA synthesis. *In vitro* complementation by proteins from DNA mutants and wild type cells could be studied, since it was shown that large molecules easily enter the cells. The system is also useful for the study of repair replication and other aspects of DNA metabolism. It is clear that this system promotes semiconservative replication and that *in vitro* products are biologically active as observed by genetic transformation. These studies give strong support to our earlier results and conclusions that DNA replication is promoted by a large complex of enzymes and other factors *in vivo*.

Acknowledgments

This investigation was supported by Grants GM-14108 and 2 TO1-GM 295 from the National Institute of General Medical Sciences and by Grant GB 8739 from National Science Foundation. A. T. G. is a recipient of Public Health Service Research Career Program Award, GM-50199. C. O Y. is a postdoctoral fellow of the American Cancer Society (PF-578). We thank Doctors R. L. Baldwin, C. Yanofsky, I. R. Lehman, A. K. Ganesan, and Joshua Lederberg for their helpful suggestions and encouragement. We acknowledge the expert assistance of Mrs. R. E. Syverson and Caroline Yu. In addition, we thank the editors of the *Proceedings of the National Academy of Sciences, U.S.A.*, the *Journal of Virology*, the *Journal of Biological Chemistry*, and *Biochemical and Biophysical Research Communications* for their permission to reproduce some of the figures and tables presented in this article.

References

1. Bazill, G. W., Hall, R., and Gross, J. D. 1971. DNA synthesis in lysates of Rec B$^-$ and Rec$^+$ *E. coli* cells. *Nature New Biol.* 233: 281.

2. Bazill, G. W., and Retief, Y. 1969. Temperature-sensitive DNA synthesis in a mutant of *Bacillus subtilis*. *J. Gen. Microbiol.* 56: 87.

3. Bird, R. E., Lowarn, J., Martuscelli, J., and Caro, L. G. 1972. The origin and sequence of chromosome replication *Escherichia coli. J. Mol. Biol.* 70: 549.

4. Boyle, J. B., Paterson, M. C., and Setlow, R. B. 1970. Excision-repair properties of an *Escherichia coli* mutant deficient in DNA polymerase. *Nature (London)* 226: 708.

5. Brutlag, D., Atkinson, M. R., Setlow, P., and Kornberg, A. 1969. An active fragment of DNA polymerase produced by proteolytic cleavage. *Biochem. Biophys. Res. Commun.* 37: 982.

6. Brutlag, D., Schekman, R., and Kornberg, A. 1971. A possible role for RNA polymerase in the initiation of M13 DNA synthesis. *Proc. Nat. Acad. Sci. U. S. A.* 68: 2826.

7. Burger, R. M. 1971. Toluene treated *Escherichia coli* replicate only that DNA which was about to be replicated *in vivo. Proc. Nat. Acad. Sci. U. S. A.* 68: 2124.

8. Cohen, S. S. 1966. Introduction to the biochemistry of D-arabinosyl nucleosides. *Prog. Nuc. Acid Res. Mol. Biol.* 1: 88.

9. DeLucia, P., and Cairns, J. 1969. Isolation of an *E. coli* strain with a mutation affecting DNA polymerase. *Nature (London)* 224: 1164.

10. Falaschi, A., and Kornberg, A. 1966. Biochemical studies on bacterial sporulation. II. Deoxyribonucleic acid polymerase in spores of *Bacillus subtilis. J. Biol. Chem.* 241: 1478.

11. Ganesan, A. T. 1968. Studies on *in vitro* replication of *Bacillus subtilis* DNA. *Cold Spring Harbor Symp. Quant. Biol.* 33: 45.

12. Ganesan, A. T. 1971. Adenosine triphosphate-dependent synthesis of biologically active DNA by azide-poisoned bacteria. *Proc. Nat. Acad. Sci. U. S. A.* 68: 1296.

13. Ganesan, A. T., and Lederberg, J. 1965. A cell membrane bound fraction of bacterial DNA. *Biochem. Biophys. Res. Commun.* 18: 824.

14. Ganesan, A. T., Yehle, C. O., and Yu, C. C. 1973. Semiconservative DNA replication in a polA mutant and the identification of DNA polymerases II and III in *Bacillus subtilis. Biochem. Biophys. Res. Commun.* 50: 155.

15. Gefter, M. L., Hirota, Y., Kornberg, T., Wechsler, J. A., and Barnoux, C. 1971. Analysis of DNA polymerases II and III in mutants of *Escherichia coli* thermosensitive for DNA synthesis. *Proc. Nat. Acad. Sci. U. S. A.* 68: 3150.

16. Geider, K., Lechner, H., and Hoffmann-Berling, H. 1972. Structure of newly synthesized $\phi\chi$-174 replicative form DNA. *J. Mol. Biol.* In press.

17. Goldmark, P. J., and Linn, S. 1972. Purification and properties of the recBC DNase of *Escherichia coli* K-12. *J. Biol. Chem.* 247: 1849.

18. Goss, W. A., Deitz, W. H., and Cook, T. M. 1965. Mechanism of action of nalidixic acid on *Escherichia coli. J. Bacteriol.* 89: 1068.

19. Gross, J., and Gross, M. 1969. Genetic analysis of an *E. coli* strain with a mutation affecting DNA polymerase. *Nature (London)* 224: 1166.

20. Keller, W. 1972. RNA-primed DNA synthesis *in vitro*. *Proc. Nat. Acad. Sci. U. S. A.* 69: 1560.

21. Klenow, H., and Overgarrd-Hansen, K. 1970. Proteolytic cleavage of DNA polymerase from *Escherichia coli* B into an exonuclease unit and a polymerase unit. *FEBS Lett.* 6: 25.

22. Knippers, R. 1970. DNA polymerase II. *Nature (London)* 228: 1050.

23. Knippers, R., and Stratling, W. 1970. The DNA replicating capacity of isolated *E. coli* cell wall-membrane complexes. *Nature (London)* 226: 713.

24. Kornberg, T., and Gefter, M. L. 1970. DNA synthesis in cell-free extracts of a DNA polymerase-defective mutant. *Biochem. Biophys. Res. Commun.* 41: 1557.

25. Kornberg, T., and Gefter, M. L. 1971. Purification and DNA synthesis in cell free extracts: properties of DNA polymerase II. *Proc. Nat. Acad. Sci. U. S. A.* 68: 761.

26. Laipis, P. J., and Ganesan, A. T. 1972. A deoxyribonucleic acid polymerase I-deficient mutant of *Bacillus subtilis*. *J. Biol. Chem.* 247: 5867.

27. Laipis, P. J., and Ganesan, A. T. 1972. *In vitro* repair of X-irradiated DNA extracted from a polymerase I deficient *Bacillus subtilis*. *Proc. Nat. Acad. Sci. U. S. A.* 69: 3211.

28. Laipis, P. J., Olivera, B. M., and Ganesan, A. T. 1969. Enzymatic cleavage and repair of transforming DNA. *Proc. Nat. Acad. Sci. U. S. A.* 62: 289.

29. Laipis, P. J. 1972. Genetic and biochemical studies on repair and recombination in *Bacillus subtilis*. Ph.D. dissertation, Stanford University, Palo Alto, Calif.

30. Masters, M., and Broda, P. 1971. Evidence for the bidirectional replication of the *Escherichia coli* chromosome. *Nature New Biol.* 232: 137.

31. Matsushita, T., White, K. P., and Sueoka, N. 1972. Chromosome replication in toluenized *Bacillus subtilis* cells. *Nature New Biol.* 232: 111.

32. Meselson, M., and Stahl, F. 1958. The replication of DNA in *Escherichia coli*. *Proc. Nat. Acad. Sci. U. S. A.* 44: 671.

33. Mitchell, P. 1966. Chemiosmotic coupling in oxidative and photosynthetic phosphorylation. *Biol. Rev. (Cambridge)* 41: 445.

34. Mordoh, J., Hirota, Y., and Jacob, F. 1970. On the process of cellular division in *Escherichia coli*. V. Incorporation of deoxynucleoside triphosphates by DNA thermosensitive mutants of *Escherichia coli* also lacking DNA polymerase activity. *Proc. Nat. Acad. Sci. U. S. A.* 67: 773

35. Moses, R. E., and Richardson, C. C. 1970. Replication and repair of DNA in cells of *Escherichia coli* treated with toluene. *Proc. Nat. Acad. Sci. U. S. A.* 67: 674.

36. Moses, R. E., and Richardson, C. C. 1970. A new DNA polymerase activity of *Escherichia coli*. I. Purification and properties of the activity present in *E. coli* polA1. *Biochem. Biophys. Res. Commun.* 41: 1557.

37. Moses, R. E., and Richardson, C. C. 1970. A new DNA polymerase activity of *Escherichia coli*. II. Properties of the enzyme purified from wild type *E. coli* and DNA_{ts} mutants. *Biochem. Biophys. Res. Commun.* 41: 1565.

38. Nüsslein, V., Otto, B., Bonhoeffer, F., and Schaller, H. 1971. Function of DNA polymerase III in DNA replication. *Nature New Biol.* 234: 285.

39. Okazaki, T., and Kornberg, A. 1964. Enzymatic synthesis of deoxyribonucleic acid. XV. Purification and properties of a polymerase from *Bacillus subtilis*. *J. Biol. Chem.* 239: 259.

40. Okazaki, R., Okazaki, T., Sakabe, K., Sugimoto, K., Kainuma, R., Sugino, A., and Iwatasuki, N. 1968. In vivo mechanism of DNA chain growth. *Cold Spring Harbor Symp. Quant. Biol.* 33: 129.

41. Rama Reddy, G. V., Goulian, M., and Hendler, S. S. 1971. Inhibition of *E. coli* DNA polymerase II by ara-CTP. *Nature New Biol.* 234: 286.

42. Schaller, H., Otto, B., Nüsslein, V., Huf, J., Herrmann, R., and Bonhoeffer, F. 1972. Deoxyribonucleic acid replication *in vitro*. *J. Mol. Biol.* 63: 183.

43. Schekman, R., Wickner, W., Westergaard, O., Brutlag, D., Geider, K., Bertsch, L. L., and Kornberg, A. 1972. Initiation of DNA synthesis. III. Synthesis of $\phi\chi$-174 replicative form requires RNA synthesis resistant to rifampicin. *Proc. Nat. Acad. Sci. U. S. A.* 69: 2691.

44. Schnos, M., and Inman, R. B. 1970. Position of branch points in replicating λ DNA. *J. Mol. Biol.* 51: 61.

45. Smith, D. W., Schaller, H. E., and Bonhoeffer, F. J. 1970. DNA synthesis *in vitro*. *Nature (London)* 226: 711.

46. Sueoka, N., and Quinn, W. G. 1968. Membrane attachment of the chromosome replication origin in *Bacillus subtilis*. *Cold Spring Harbor Symp. Quant. Biol.* 33: 695.

47. Sugino, A., Hirose, S., and Okazaki, R. 1972. RNA-linked nascent DNA fragments in *Escherichia coli*. *Proc. Nat. Acad. Sci. U. S. A.* 69: 1863.

48. Town, C. D., Smith, K. C., and Kaplan, H. S. 1971. DNA polymerase required for rapid repair of X-ray-induced DNA strand breaks *in vivo*. *Science* 172: 851.

49. Wechsler, J. A., and Gross, J. D. 1971. *Escherichia coli* mutants temperature-sensitive for DNA synthesis. *Mol. Gen. Genet.* 113: 273.

50. Wickner, W., Brutlag, D., Schekman, R., and Kornberg, A. 1972. RNA synthesis initiates *in vitro* conversion of M13 DNA to its replicative form. *Proc. Nat. Acad. Sci. U. S. A.* 69: 965.

51. Wickner, R. B., and Hurwitz, J. 1972. DNA replication in *Escherichia coli* made permeable by treatment with high sucrose. *Biochem. Biophys. Res. Commun.* 47: 202.

52. Yehle, C. O., and Ganesan, A. T. 1972. Deoxyribonucleic acid synthesis in bacteriophage SPO1-infected *Bacillus subtilis*. 1. Bacteriophage deoxyribonucleic acid synthesis and fate of host deoxyribonucleic acid in normal and polymerase-deficient strains. *J. Virol.* 9: 263.

53. Yehle, C. O., and Ganesan, A. T. 1973. DNA synthesis in bacteriophage SPO1-infected *Bactillus subtilis*. II Purification and characterization of a DNA polymerase induced after infection. *J. Biol. Chem.* In press.

54. Yoshikawa, H., and Sueoka, N. 1963. Sequential replication of *Bacillus subtilis* chromosome. I. Comparison of marker frequencies in experimental and stationary growth phases. *Proc. Nat. Acad. Sci. U. S. A.* 49: 559.

Discussion

Brown. You certainly have done a lot in *subtilis* replication, but I couldn't follow you quite fast enough on your purification scheme for your polymerases I, II, and III. What were those peaks you showed us again? Was that a chromatogram or a gradient or what?

Ganesan. That was a sucrose gradient sedimentation profile of enzymes purified from a DEAE-cellulose column.

Brown. Have you figured out the proportion of molecules per cell based on these purified enzymes?

Ganesan. No, this is a crude measure of relative activities that are present in the cell lysate. It is based on the relative proportion of activities found for the three polymerases on fractionation of DNA-free lysate on a DEAE-cellulose column. We know the total polymerase activity and the amount of protein loaded on to a column. Keeping the total recovered polymerase I activity as 100%, we find about 10 to 15% equivalent of pol-I activity correspond to pol-II whereas only 2 to 3% correspond to pol-III.

Brown. You mentioned something about the salt sensitivity of what you call pol-III. Are you distinguishing any of these minor enzymes in your pol-I mutant on the basis of salt sensitivity?

Ganesan. No. The salt sensitivity was used simply to distinguish the *Bacillus subtilis* enzymes from the known *Escherichia coli* enzymes. It was at that concentration (0.15 M KCl) at which you don't find any preferential inhibition of pol-III compared to pol-II. This is Gefter's condition; but our preparations are still far from pure.

Brown. Well, so is pol-II at that concentration. That's around 200 mM.

Ganesan. I mean that under assay conditions of *E. coli* enzymes we don't find any differential inactivation between these two enzymes.

Brown. Right, and their temperature sensitivity to say 45°C is no different. This is kind of unusual because the enzymes I'm looking at do have Gefter's pol-III properties.

Kushner. I didn't follow what you were proposing for the recB enzyme.

Ganesan. I am not proposing anything. I am just saying that what you find as the ATP-stimulated synthesis in lysates could be interpreted as normal replication but it is not. You can also see that in the earlier gradients I showed there was a lot of ATP-stimulated repair going on. If you mix, let's say, carefully prepared permeable cells with no lysate and then mix with a lysate, you get a lot of ATP-stimulated repair synthesis in the lysate and not in the permeable cells.

Kushner. Oh, are you then proposing a polymerase function for this?

Ganesan. No, all I am saying is that DNA is degraded by the ATP-dependent nuclease and then maybe pol-II or pol-III might promote the observed repair synthesis, since all these enzymes use DNA with gaps.

Brown. I became interested in bacterial DNA replication somewhat indirectly while studying the mechanism of action of a novel arylazopyrimidine which selectively inhibits the growth of Gram-positive bacteria. This pyrimidine, the formula of which is shown below,

is called 6-(p-hydroxyphenylazo)-uracil, or in abbreviated form, HPUra. HPUra is a remarkably selective, reversible inhibitor of replicative DNA synthesis in Gram-positive organisms. The properties of HPUra, as elucidated in whole cell systems, suggest that it may be a tool useful for identifying and characterizing components of the apparatus responsible for bacterial DNA replication.

During the past year we have made considerable progress in elucidating the mechanism of HPUra action *in vitro*. We began our study by examining, in collaboration with Tatsuo Matsushita, the effects of HPUra on DNA synthesis catalyzed by toluene-treated *B. subtilis*. The results of this work, which have been published, indicated that HPUra inhibits ATP-dependent synthesis in a manner essentially indistinguishable from that observed in intact cells. Exploitation of the toluene-treated *B. subtilis* system provided two important facts concerning the mechanism of HPUra action. First, it became clear that HPUra is not inhibitory *per se*, but must be activated by metabolic reduction. We have now isolated this reduction product and have identified it as the hydrazo derivative, *i.e.* p-(hydroxyphenyl*hydrazo*)-uracil. We have also succeeded in preparing the hydrazo derivative directly by reduction with dithiothreitol or sodium dithionite.

The second important finding with the toluene-treated *B. subtilis* system was the observation that the drug-induced inhibition of DNA synthesis could be antagonized competitively by 2'-deoxyguanosine 5'-triphosphate (dGTP). The latter finding strongly suggested that HPUra inhibits DNA replication by interfering with the interaction of a component of the replication machinery (probably a DNA polymerase) and its dGTP substrate.

I would add that I have examined in the toluene-treated *B. subtilis* system the mechanism of 6-(p-hydroxyphenylazo)-isocytosine, a 2-amino analogue of HPUra, which has identical inhibitory effects on the replication of DNA of Gram-positive organisms. Interestingly, the inhibitory activity of reduced HPIsocytosine was not antagonized by dGTP, but instead by dATP. Thus, HPIsocytosine appears to be a specific antagonist of dATP function in replication.

After determining how to prepare the active, hydrazo form of HPUra and HPIsocytosine by direct reduction, Dr. Marilyn Neville, a postdoctoral fellow in my laboratory, initiated studies on cell-free extracts of *B. subtilis* in an effort to find a drug-sensitive polymerase. Dr. Neville began by examining the effect of the hydrazopyrimidines on the activity of DNA polymerase I, purified partially from crude extracts of Pol$^+$ *B. subtilis* by the method of Okazaki and Kornberg.

She found no inhibitory effect of the drug on this enzyme. Dr. Neville then examined the effect of the reduced drugs on the DNA polymerase activity of cell-free extracts of a DNA polymerase I-deficient mutant (strain NB841, derived from *B. subtilis* 168 $thy^- \, tryp^-$), which we had isolated. She found, in a high speed supernatant (S-100) of the mutant extract, DNA polymerase activity which was clearly drug-sensitive. The inhibitory effect of reduced HPUra, like that observed in toluene-treated cells, was specifically antagonized by high concentrations of dGTP; similarly the effect of reduced HPIsocytosine was reversed specifically by dATP.

Examining the drug effect more closely, Dr. Neville found that concentrations of reduced azopyrimidines which completely inhibited ATP-dependent replication in the toluene-treated system inhibited the crude polymerase activity by only approximately 50%. This finding suggested that this crude S-100 contained a mixture of drug-sensitive and drug-insensitive polymerase activity. Therefore, she attempted to fractionate the extract, beginning with ammonium sulfate treatment. By adjusting the extract to 50% saturation she obtained two fractions containing DNA polymerase activity. The precipitate (Fraction II) contained approximately 30 to 40% of the activity of the crude extract, and the supernatant (Fraction III) contained the remaining 50 to 60% of the starting activity. Fraction II appeared to be nearly completely drug *resistant*, whereas Fraction III was essentially completely drug-sensitive. We have examined, in addition to drug sensitivity, other properties of the polymerase activities of these crude fractions, and they appear to differ in several respects. The differences resemble, superficially, those reported for the purified polymerases II and III of *E. coli*. For example, the activity of Fraction III is clearly more sensitive to inhibition by salts such as KCl and NaCl. Further, the activity of Fraction III is much less stable than that of Fraction II, and requires high concentrations of glycerol for optimal activity. Fraction III activity is also more susceptible to destruction by incubation at moderately high (40–45°C) temperatures.

We have selected a number of mutants resistant to HPUra; through the use of toluene treatment, we have screened from these 3 mutants which apparently have a drug-resistant site in replication. In one of these mutants, we have found that the polymerase activity of ammonium sulfate fraction III is unusually resistant to reduced HPUra. At present, these rather preliminary results suggest that Fraction III contains a polymerase III-like enzyme which has a role in DNA replication *in vivo*. We are currently testing this possibility by purifying this activity from temperature-sensitive DNA replication mutants. We are also investigating with purer enzyme preparations the molecular mechanism of hydrazo pyrimidine action.

Sueoka. Does the reduced form of HPUra inhibit *E. coli* DNA replication of *E. coli* DNA polymerase III?

Brown. No. I haven't been able to find inhibition. I have examined intact *E. coli*, and toluene-treated or disc-immobilized preparations (Bonhoeffer method) of a polA mutant; in these three systems, drug concentrations which clearly inhibit comparable *B. subtilis* preparations have no significant effect.

D-Loops in Intracellular λ DNA

Ross B. Inman and Maria Schnös

Biophysics Laboratory and Biochemistry Department
University of Wisconsin
Madison, Wisconsin 53706

When λ bacteriophage, containing a suppressable mutation in either genes O or P (or O and P), infects W3350($sm^R su^-$), the intracellular phage DNA is found to be circular and a significant fraction of the circles have one or more D-loops. The single stranded sections of D-loops have lengths that extend from 0.06 μ up to about 0.5 μ and the loops are found at many different positions on the λ DNA molecule; loops are not found to be preferentially situated at the origin of replication.

These results are discussed from the point of view of a growing point model.

Earlier experiments from this laboratory showed that, during the first round of replication, λ phage DNA replicates bi-directionally from a unique origin (4). Electron microscopic examination of the fine structure at branch points revealed that growing points often had single stranded regions connecting daughter duplexes to the parental DNA (single stranded connections are shown at A in Fig. 1). Single stranded connections were often accompanied by single stranded segments (B in Fig. 1) and were separated from each other by short sections of double stranded DNA (2). Single stranded connections and single stranded segments were almost always (99 and 100%, respectively) associated with the daughter helical sections of replicating molecules. Single stranded segments did not always accompany single stranded connections. Due to the bi-directionality of the first round of λ replication there can be two growing points and either one, or both, can possess single stranded connections and segments. When both growing points are associated with single stranded connections and segments they are deployed as indicated in

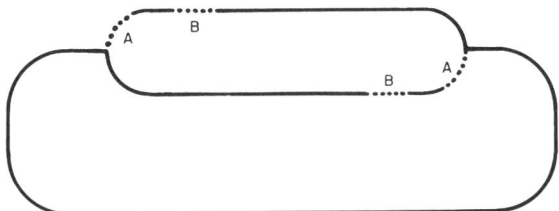

Fig. 1. Structure of λ DNA during the first round of replication. The molecule replicates bi-directionally and in the vicinity of the two growing points are single stranded connections (A) and single stranded segments (B). Single stranded regions are always associated with daughter helical strands.

Fig. 1 (one on each daughter segment and situated at opposite ends of daughter segments).

On the basis of the above results, and the work from other laboratories, a growing point model has been proposed (2) which is shown in Fig. 2. The DNA molecule is imagined to be divided into a number of replication units; for successful replication of any one unit, two modes of DNA synthesis are required. First there is a forward synthetic event (whose template is double stranded) which generates a single stranded connection (Fig. 2 (b)) and this continues until the growing point reaches the end of a replication unit. Second, there is a backward synthetic event (whose template is single stranded DNA; the single strand corre-

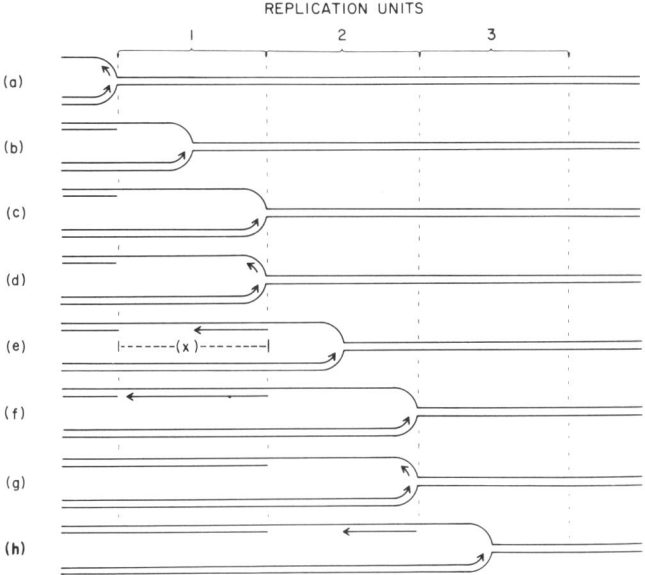

Fig. 2. Growing point model for DNA replication.

sponds to the single stranded connection generated by the forward synthetic event). Both backward and forward events must operate (as shown in Fig. 2(e)) to produce the fine structure observed in electron micrographs of λ growing points (Fig. 1).

The exact nature of the events that occurs at the terminus of each replication unit is not understood. We will consider two possibilities. First, the terminus might have the property of switching the growing point to the other parental strand. If this happened then a backward synthetic event could take place and a new forward synthetic event could then be initiated by nicking the newly synthesized DNA at the position of the switch (this mechanism will be called mode A). Alternatively, the same final configuration shown in Fig. 2(e) could be achieved if the forward moving growing point advanced continuously along its template and, as each terminus of a replication unit was made single stranded, a new backward synthetic event could be initiated (this mechanism will be called mode B).

The present report tests certain predictions that can be made from either mode A or B of the growing point model. The test involved interfering with the postulated forward or backward synthetic events.

Rationale for Experiments

We have tried to test the model (Fig. 2) by interfering with the postulated forward and backward synthetic events. Presumably certain mutations, in either λ or the host genome, that effect λ replication will influence these two types of DNA synthesis. The results to be discussed below appear to involve interference with the forward synthetic event.

Interference with Backward Synthetic Events

If we could prohibit backward synthetic events but allow forward synthesis then a number of possible structures could be anticipated. First, we can expect that the infecting λ DNA will circularize and that a forward synthetic event will be initiated at the origin of replication in exactly the same way as in a normal infection. Previous experiments have established that the origin of replication in λ DNA is situated $18.3 \pm 2.9\%$ from the right end of the mature molecule (4). In the present case, however, when the forward growing point reaches the terminus of the first replication unit two new possibilities exist. First, the growing point will simply stop; this can be expected from mode A (see the introduction) in the absence of backward synthetic events. The resulting structure would be a λ circle containing a short single stranded loop, one end situated at the origin of replication and the other at the terminus of the first replication unit. According to mode A, therefore, there should be one single stranded loop per molecule situated at a unique position and it should be of a unique size (assuming the terminus of the first replication unit is situated at an unique position). This type of structure is shown in diagrammatic form in Fig. 3(a). The second general possibility is covered by mode B (see the introduction). In this case the forward synthetic event can be expected to continue past the first and subsequent replication units and will be stopped at whatever position it had reached at the time the intracellular DNA was isolated. Structures similar to those in Fig. 3(a) could therefore be anticipated except that now the growing point could be at any position along the molecule. Again one end of the single stranded loop would be found at the origin of replication but the length of the loop would be variable from molecule to molecule and could be expected, in some cases, to extend over many replication units.

The structures so far discussed represent a simplified picture of what would actually be expected. It has been assumed that the original initiation of a forward synthetic event leads to synthesis in a rightward direction (Fig. 3(a)). Replication is bi-directional during at least the first round of replication in λ (4) and this will therefore lead to other possible predicted structures. Various initiation mechanisms can be considered

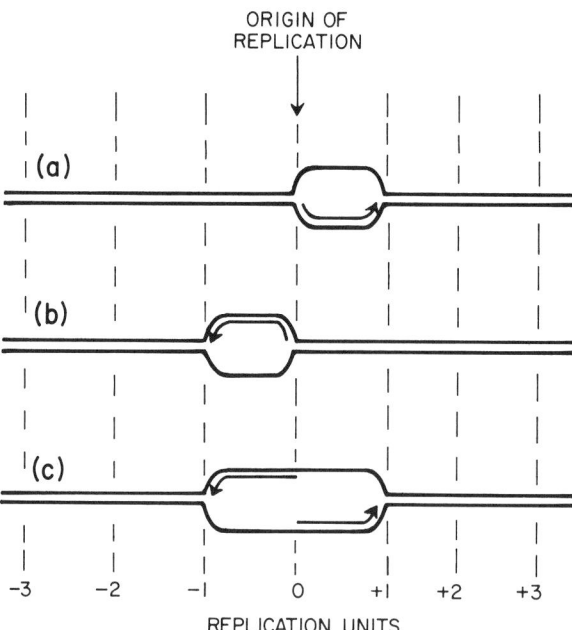

Fig. 3. Possible structures that might be expected if forward synthetic events were allowed and backward synthesis was blocked.

that would lead to bi-directionality. Synthesis to the right and to the left could be initiated by separate events that could occur at the same or different points in time, or a single original initiation could finally result in bi-directional replication. Fig. 3(b) and (c), show alternative structures that could be anticipated because of the bi-directional nature of λ replication. The structures given in Fig. 3 are those predicted from mode A (see the introduction); mode B would lead to similar structures except that the growing points would be situated at any position on the circular molecule rather than at the terminus of the first replication unit.

Interference with Forward Events

According to the simplest interpretation of the model shown in Fig. 2 no synthesis could occur if we could, in some way, block forward events because backward synthetic events (which are now allowed) are triggered or initiated by forward synthetic events. Under these circumstances, simple circular structures would therefore be expected. We assume here that the act of primary initiation itself does not cause any change that can be observed by electron microscopy.

Size of λ Replication Units

If, in fact, replication units do exist, then their size can be approximately measured from existing data. First, as can be seen from Fig. 2, during normal replication, single stranded connections are never longer than the corresponding replication unit and thus the maximum length observed in a large set of single stranded connections should yield a very approximate estimate of the average length of all replication units (we assume that replication units N_i and N_{i+1} can be of different size within 1 molecule but that replication unit N_i will be the same in all molecules). The maximum length of single stranded connections in replicating λ DNA has been found to be 0.40 μ for a total λ DNA length of 17.5 μ (2).

A second and more precise estimate can be obtained from the sum of the length of a single stranded segment and the length of the short double stranded region that exists between single stranded segments and single stranded connections (shown by length (x) in Fig. 2(e)). Because growing points can occur at different points on each molecule the second estimate will also yield an average of all λ DNA replication units. Measurements were made on 21 λ DNA growing points which exhibited single stranded segments and connections (similar to that shown in Fig. 2(e)) and the average length was 0.35 μ with a range extending from 0.15 to 0.59 μ.

The more precise estimate therefore leads to 50 replication units in λ DNA, while the first and very approximate calculation yields 44 replication units.

Results and Discussion

We are, at present, testing the ideas presented in the introduction and "Rationale for Experiments" by investigating the intracellular structures that result from λ infection when either the host or the phage carry mutations which prevent (or reduce to very low levels) the replication of λ DNA. The underlying idea is that some of the mutations might interfere with the backward or forward synthetic events.

In the present report we will discuss results in which phage replication was blocked by mutations in the λ genes O and P.

When λcI857 sus Oam29, λcI857 sus Pam3, or λcI857 sus Oam29 Pam80 infect *Escherichia coli* (su$^-$), growing in a heavy medium, there is very little replication as judged by the resulting intracellular buoyant density gradient pattern. Density gradient fractions corresponding to light ^{3}H-labeled parental λ DNA and also fractions from the heavy side of the light ^{3}H-labeled DNA were then examined by electron microscopy. In each of the experiments noted above (mutations in either O or P or both) a large proportion of the λ length molecules was circular and a significant number of these exhibited single stranded looped structures similar to that shown in Fig. 3, (a) and (b). Examples of single stranded loops are given in Fig. 4. The structures of the single stranded regions were similar to the D-loops already observed in mitochondrial DNA (3) and we will therefore refer to such structures as D-loops. We did not observe D-loops in mature λ DNA with any significant frequency.

Length Distribution of D-Loops

The length of the single stranded region was approximately the same as that of the corresponding double strand segment. Figure 5 shows the distribution of length found for the single stranded regions in a number of D-loops. The distribution is broader than can be accounted for by experimental error and does not correspond to any of the anticipated structures expected if backward synthetic events were blocked. For instance, according to mode A of the model the D-loops should all be of the same length and correspond to the length of the first replication unit. The data in Fig. 5 are not in accord with this expectation. According to mode B of the model the distribution of single strand length should be broad and extend up to lengths of many microns; again the data

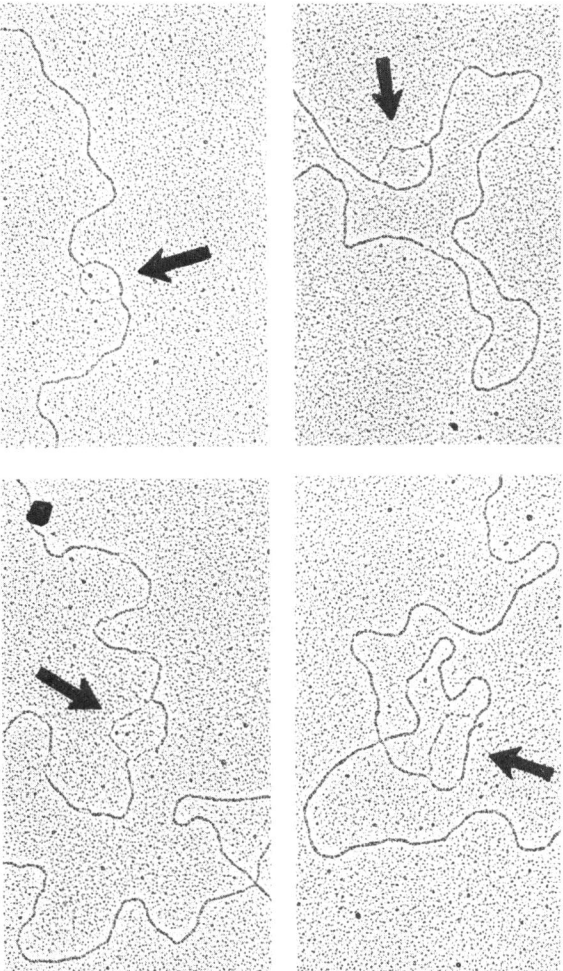

Fig. 4. Examples of λ D-loops.

in Fig. 5 are not consistent with this view because although the distribution is broad the single strands are always shorter than 0.5 to 0.6 μ.

Number of D-Loops per Molecule

A significant proportion of λ circles had more than one D-loop (Table 1) and there was no very large effect of multiplicity of infection or mutant type on this result. Again this observation was not anticipated on the basis of the ideas presented in the introduction or "Rationale for Experiments."

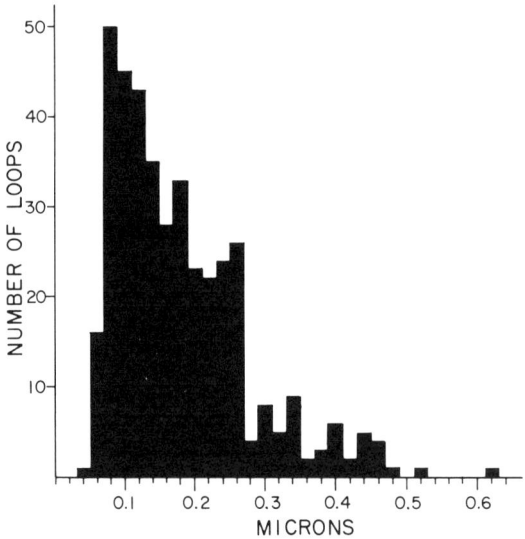

Fig. 5. Distribution of D-loop lengths observed after infection with λcI857 sus Oam29.

Table 1.

Experiment	m.o.i.*	% circles with (N) D-loops/molecule					
		N=0	N=1	N=2	N=3	N=4	N=5
λcI857 sus Oam29	1.6	76	12	10	1	1	
λcI857 sus Oam29	5	71	16	9	3	1	<1
λcI857 sus Oam29	16	69	23	6	2		
λcI857 sus Pam3	5	75	16	9			
λcI857 sus Oam29 sus Pam80	5	80	11	7	2	<1	

*Multiplicity of infection.

Position of D-Loops on λ DNA

If a mutation in O or P blocked backward synthetic events then, as has already been emphasized, D-loops should always be situated at the origin of replication. We have already presented data that show that mutations in either O or P can result in intracellular circles containing more than one loop per molecule and thus we cannot expect all D-loops to be situated at the origin of replication. Nevertheless we have determined the position of a large number of D-loops in an effort to understand the mechanism leading to their formation. The position of D-loops was found by the partial denaturation mapping technique but because structures of the type shown in Fig. 3, (a) and (b), would be destroyed on partial denaturation, the short piece of newly synthesized DNA had first to be cross-linked to the parental strand. The intracellular circles were therefore incubated in the presence of 5% HCHO for 2 days at room temperature prior to denaturation mapping. Such a procedure will, of course, produce cross-links at many positions in addition to the site of the D-loop. Cross-linking was presumed to be effective because the denaturation maps of cross-linked molecules showed many denatured sites which were smaller than usual; the over-all denaturation pattern, however, remained unchanged. The denaturation maps of the cross-linked D-loop molecules were then used to determine the position corresponding to the mature ends of each molecule and the D-loop position was then related to this point of reference. Figure 6(a) shows the deduced position of D-loops on the mature physical λ map. D-loops can be situated at a large number of positions along the λ molecule but there is an experimentally significant absence of D-loops between the 8.2 and 10.4 μ positions which corresponds closely to the position of the b_2 deletion which extends from 7.9 to 10.1 μ, taking 17.5 μ as the total length of λ DNA (1). The absence of D-loops at this position may be an artifact; this region corresponds to the AT richest section in the λ DNA molecule (the denaturation map is shown in Fig. 6c))

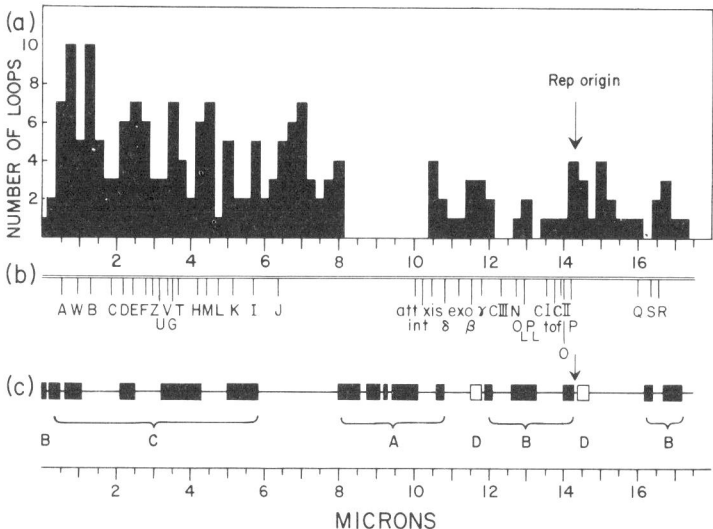

Fig. 6. Position of D-loops on λ DNA. (a) Position of D-loops with respect to the mature λ DNA map. The left mature end of λ is situated at 0.0 μ and the right end at 17.5 λ. (b) The physical position of various λ genes (1). (c) Grand average denaturation map of λ. The A + T content decreases in segments A, B, and C, respectively. The positions marked D represent the G + C richest regions in the right half of the molecule. The origin of replication is shown by the arrow.

and is therefore exactly the position from which D-loops would disappear on partial denaturation unless many cross-links were present on each molecule. No attempt was made to determine the frequency of cross-links and thus we cannot say at this time whether or not the absence of D-loops at the center of the molecule is significant.

The main conclusion is that D-loops are not found at high frequency near the origin of replication (shown by the arrow in Fig. 6) but rather at a large number of places along the molecule and perhaps are situated at random. The physical position of various λ genes is shown in Fig. 6(b) and is taken from the data of Davidson and Szybalski (1).

Conclusion

Assuming the model shown in Fig. 2 is correct then the results presented in Figs. 5 and 6 and Table 1 are not consistent with the assumption that mutations in *O* and *P* block backward but allow forward synthetic events. However, if the opposite were true (forward events blocked and backward allowed) then according to the model no structural change or DNA synthesis should be observed.

Evidently, therefore, the model must be modified to allow for the present results. If we preserve the notion of forward and backward synthetic events then there appear to be two possible ways to reconcile the results with the model. We could assume, for instance, that mutations in *O* and *P* prevent normal replication and that what we are seeing is simply a number of spurious initiations of forward events that start at random positions. A more interesting possibility is that mutations in *O* and *P* have, in fact, blocked forward synthetic events (either by stopping forward DNA synthesis or by preventing the primary initiation) and that in some way backward events can be initiated from the various termini of replication units. According to the model, backward events are triggered or initiated by the forward synthetic event; in the absence of a forward event we would have to make a further assumption about the initiation of backward events. It has been proposed that during normal replication the backward event is initiated when its parental template is made single stranded by forward synthesis (Fig. 2 and introduction). If this is the case then in the absence of forward synthesis perhaps other mechanisms can lead to a single stranded template and thus allow initiation of backward synthetic events. Transcription is one process that could lead to single stranded regions in the DNA molecule and this possibility is being investigated at present. If the D-loops do arise from a series of abortive backward synthetic events then each could be expected to have originated at the terminus of a replication unit. Purification of the small pieces of newly synthesized DNA in the D-loops would then provide a way of isolating the individual unique DNA segments corresponding to the 40–50 presumed replication units in λ.

Two very important pieces of information are lacking at the present time. It has been assumed that the observed D-loops are a result of DNA synthesis; we cannot offer strong evidence for this. The short section of nucleic acid that produces the D-loops might possibly therefore contain RNA. Similarly the possibility exists that the short piece of DNA arose from a parental fragment rather than from DNA synthesis.

Materials and Methods

Preparation of Bacteriophage

Bacteriophages λcI857 sus Oam29 and λcI857 sus Pam3 were obtained from Doctors Dove and Kaiser. Lysogens were isolated after infection of CR34(su$^+$) and then ^{3}H-labeled phages prepared by temperature induction of the lysogen in the presence of ^{3}H-thymidine. C-600 (λcI857 sus Oam29 Pam80)/λ was a gift from Dr. Dove and ^{3}H-labeled phage was prepared from this lysogen by temperature induction in the presence of adenosine (250 µg/ml) and ^{3}H-thymidine.

Preparation and Isolation of Intracellular λ DNA

W3350 (sm^R su$^-$) was grown in 40 ml of D-medium containing D and ^{15}N (4) to $A_{590} = 0.8$. Cells were centrifuged and washed with 10 ml of 0.01 M Tris, 0.01 M MgSO$_4$ (pH 7.2) in D$_2$O and, following a further centrifugation, suspended in 2 ml of 0.01 M Tris, 0.01 M MgSO$_4$ in D$_2$O. After starving cells for 30 min at 37°C, ^{3}H-labeled phages (in D$_2$O) were added, followed by incubation for 10 min at 37°C. Cells were then diluted with 20 ml of prewarmed D-medium and incubated at 37°C for 10 min. Twenty minutes after infection the cells were poured into an equal volume of killing solution (0.015 M KCl, 0.015 M Na$_3$N, 0.015 M EDTA in SSC (0.15 M sodium choride–0.015 M (sodium citrate) at pH 8.0) and then immediately centrifuged and resuspended in a small volume of killing solution. Lysozyme (500 µg/ml) was added and the mixture frozen in liquid nitrogen and then thawed in cold water and left at 0°C for 20 min. Cold Sarcosyl was added to a final concentration of 0.5% and the mixture was again left at 0°C for 15 min and then heated for 5 min at 50°C. The mixture was incubated for 1 hr at 37°C with nuclease-free Pronase (previously self-digested at 37°C for 4 hr). Finally the lysate was fractionated by two successive CsCl density gradient centrifugations. Fractions, corresponding to light λ DNA (and fractions a little heavier than light), were then examined in the electron microscope.

Electron Microscopy

Preparation of samples for electron microscopy has been described previously (4) The intracellular DNA was either spread directly from the CsCl gradient or was first dialyzed into 0.02 M NaCl, 0.005 M EDTA (pH 7.5). Partial denaturation was carried out at high pH as previously described (4) except that the DNA was cross-linked, prior to partial denaturation, by a 4-hr incubation at 23°C in the presence of 5% HCHO.

Acknowledgments

We thank Dr. W. F. Dove for his thoughtful criticism. This research was supported by grants from the United State Public Health Service, National Institutes of Health, the American Cancer Society, and the Graduate School, University of Wisconsin.

References

1. Davidson, N., and Szybalski, W. 1971. Physical and chemical characteristic of lambda DNA. Chap. 3. *In* Hershey, A. D. (ed.), *Bacteriophage lambda*, Cold Spring Harbor Laboratory.
2. Inman, R. B., and Schnös, M. 1971. Structure of branch points in replicating DNA; presence of single-stranded connections in λ DNA branch points. *J. Mol. Biol.* 56: 319.
3. Kasamatsu, H., Robberson, D. L., and Vinograd, J. 1971. A novel closed-circular mitochondiral DNA with properties of a replicating intermediate. *Proc. Nat. Acad. Sci. U. S. A.* 68: 2252.
4. Schnös, M., and Inman, R. B. 1970. Position of branch points in replicating λ DNA. *J. Mol. Biol.* 51: 61.

Discussion

Wolfson. David Dressler and I have used the electron microscope to study the structure of intracellular T7 DNA after infection with various viral mutants defective in DNA synthesis. Specifically, we have found that infection with T7 phage carrying an amber mutation in gene 4 (mechanism of action unknown) generates intracellular T7 DNA forms containing a high frequency (up to 50%) of randomly located D-loop-type structures, and also duplex molecules with single stranded tails.

The origin of these structures is not known. They could result from *de novo* synthetic events (for instance, an attempt at replication on one side of the parental DNA strands). Or they might result because of the invasion of a pre-existing nucleotide strand (an intermediate in recombination). And, of course, it is also possible that these structures are not related to any normal *in vivo* process.

In further studying the meaning of these structures, we have been concentrating on the mechanism of action of T7 gene 4. We isolated a gene 4 temperature-sensitive mutant and examined the effect *in vivo* of raising the temperature midway through the first T7 DNA replication cycle. During the initial part of the infection, when conditions were permissive, the T7 DNA was seen to be replicating in a normal fashion, with greater than 85% of the growing points containing a single stranded region on one of the daughter arms. However, when the infected cells were transferred to nonpermissive conditions for a brief time period, the single stranded regions in the growing points were then seen to be 2 to 3 times longer. This change was not observed in infections with wild type phage. Although there are several interpretations for the temperature shift result, the simplest possibility (and our working hypothesis) is that the gene 4 product is directly involved in the initiation of fragments on one side of the T7 growing point.

Sueoka. On the average, how many of these loops do you have in a molecule?

Inman. At multiplicities of infection of 1.6, 5, and 16 we found that 24, 29, and 31% of the circular molecules had D-loops. But some of these had more than one D-loop per molecule. For instance, at a multiplicity of infection of 5, there were 16, 9, 3, 1, and <1% of circles with 1, 2, 3, 4, and 5 D-loops, respectively.

Some Biochemical Elements in Bacteriophage T7 DNA Replication *in Vitro*

Rolf Knippers, Wolf Strätling, and Elke Krause

Friedrich-Miescher-Laboratorium der
　Max-Planck-Gesellschaft
Tübingen, Germany

Intact T7 DNA is used (in the presence of ATP) as a template for DNA synthesis by a protein extract from bacteriophage T7-infected cells but not by an extract from uninfected cells. The functions involved in viral DNA synthesis *in vitro* include: (*a*) a complex of the viral DNA polymerase with an unidentified macromolecule; (*b*) a viral RNA polymerase.

The replication of bacteriophage T7 DNA is controlled by at least five known viral gene products: (*a*) a DNA polymerase (product of gene 5; (5, 14)); (*b*) an endonuclease (product of gene 3; (1, 2); (*c*) an exonuclease (gene 6 product; (9)); (*d*) and (*e*) two unknown functions (coded for by genes 2 and 4) (review (18)). In addition to these functions, a polynucleotide ligase seems to be required for viral DNA replication. The viral ATP-dependent ligase can be replaced by the corresponding NAD-dependent host cell enzyme (12). It seems that bacteriophage T7 DNA replication does not depend on any of the products coded for by the bacterial genes dna A through dna G which are required for host cell DNA replication (review (6)). The burst size of infected cultures of representative temperature-sensitive *Escherichia coli* strains (6) is not significantly different when the infection is carried out at 30°C or at 42°C (Table 1). It is conceivable, of course, that other host cell functions for which mutations have not been detected participate in T7 DNA replication (*cf.* 7).

Table 1. Multiplication of T7 in temperature-sensitive *Escherichia coli* cells

Host strains	Gene locus	Average number of phages/cell produced at	
		30°C	42°C
F 117	A	79	49
E 279, 266	B	23	44
PC 1	C	31	36
PC 7	D	137	171
E 486	E	51	62
E 101	F	49	80
CR 34	G	140	116

The *E. coli* strains were grown at 30°C in 20 ml of tryptone-broth (plus 20 μg/ml of thymine). At a cell density of 2×10^8/ml, 10-ml samples were removed from the cultures and placed in a 42°C water bath. After 15 min with aeration, all cultures were infected with T7 wild type in a multiplicity of 0.2 infective particle/cell. After lysis had occurred, a few drops of chloroform were added to each culture. Progeny phages were plated with *E. coli* B as indicator strain. The *E. coli* strains used in this study were kindly provided by Dr. M. Kohiyama.

The Effects of Amber Mutations on T7 DNA Replication *in Vivo*

T7 wild type as well as amber mutant-infected cells were pulse-labeled with ^{3}H-thymidine late in infection (11). The radioactive viral DNA was extracted and analyzed in neutral sucrose gradients (8, 17). In wild type-infected cells ^{3}H-thymidine label was recovered (*a*) in DNA with the sedimentation rate of a ^{32}P-T7 DNA marker (32 S), (*b*) in a DNA species sedimenting slightly faster than the marker DNA, and (*c*) in a fast sedimenting DNA form (40 to 80 S). The fast sedimenting DNA forms are linear structures of 4 to 10 times the length of mature viral DNA. The physiological significance of these DNA structures is unknown. They seem to be obligatory intermediates in T7 DNA replication since they accumulate in gene 3 and gene 6 mutant-infected cells. The nucleases coded for by genes 3 and 6 are apparently necessary to cut these long linear DNA forms into unit-length viral DNA (17). The DNA form which sediments slightly faster than the viral DNA marker probably corresponds to the linear replicating T7 DNA molecules described by Dressler *et al.* (4). The DNA species described above were also found in gene 2 mutant-infected cells, although in much lower amounts. No DNA synthesis was observed in gene 4 and gene 5 mutant-infected cells.

T7 DNA Synthesis *in Vitro*

The following *in vitro* DNA-replicating systems were prepared from wild type- or amber mutant-infected cells and incubated with tritium-labeled deoxynucleoside triphosphates and ATP: toluene-treated cells (13); gently lysed cell extracts concentrated on cellophane discs (15); and DNA-membrane complexes (16). The sedimentation patterns of the T7 DNA labeled in freshly prepared *in vitro* systems were similar to those of *in vivo* pulse-labeled T7 DNA (see above). There was one exception. In toluenized gene 4 mutant-infected cells or in lysates from gene 4 mutant-infected cells ^{3}H label was incorporated into DNA fragments with sedimentation coefficients of 4–8 S. DNA fragments of this size were not observed when gene 4 mutant-infected cells were labeled with ^{3}H-thymidine *in vivo*. No DNA synthesis was found in any *in vitro* system prepared from gene 5 mutant-infected cells.

Complementation of Extracts from Gene 5 Mutant-infected Cells

To obtain some information about the biochemical role of the viral gene products in phage DNA replication we tried to complement *in vitro* systems from amber mutant-infected cells with purified or partially purified protein fractions from wild type-infected cells. For the complementation assays reported below we used cell lysates concentrated on cellophane discs (15) because this experimental system requires very little material and sustains DNA replication longer than the other known "open" cell preparations. An obvious first experiment to try was a complementation test between extracts from gene 5-infected cells with highly purified T7 DNA polymerase. Therefore, we added 0.01 unit of T7 DNA polymerase (14) in a 1-μl sample to 4 μl of a lysate from mutant-infected cells and concentrated the mixture in a stream of cold air on a cellophane disc (15). The disc was then incubated at 37°C on a drop of incubation buffer containing all four deoxynucleoside triphosphates including ^{3}H-TTP and ATP. The complemented lysate was compared to a control sample without T7 DNA polymerase. To our surprise we could not detect any stimulation of DNA synthesis in the complemented lysates. *Escherichia coli* polymerases I and II were also unable to complement lysates from gene 5 mutant-infected cells. We concluded that T7 DNA polymerase has to be in a special form or configuration in order to perform its function in viral DNA replication.

This assumption was corroborated by the results obtained from a DEAE-cellulose chromatogram of an extract from T7-infected H560 cells (endoI$^-$ polA$^-$) (20) (Fig. 1). The bacterial DNA polymerases can be identified in this chromatogram by their preference for double stranded DNA as templates. T7 DNA polymerase prefers single stranded DNA or poly dAT as template (14). Using these criteria, one polymerizing activity (peak B of Fig. 1) could be identified as *E. coli* DNA polymerase and the other two (peaks A and C) as viral DNA polymerases. The activities of peaks A and C were not found in uninfected cells. They are also absent in gene 5 mutant-infected cells.

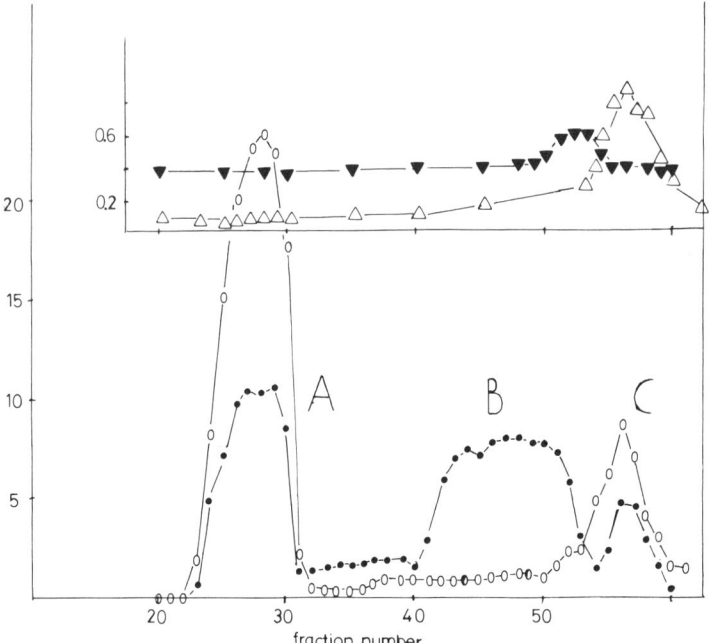

Fig. 1. DNA-polymerizing activities in T7-infected polA$^-$ cells. About 15 g of wild type T7 infected H560 cell (polA$^-$ endo I$^-$; (20)) were lysed with 10 mg of lysozyme in 30 ml of standard buffer (0.05 M tris-HCl, pH 7.6; 0.001 M EDTA; 0.005 M β-mercaptoethanol) and treated with the Branson Sonifier. Cellular debris was removed with low speed (10 min at 10,000 × g) centrifugation. The supernatant was applied to a DEAE-cellulose column (1.5 × 24 cm; equilibrated with standard buffer). The column was washed with 100 ml of standard buffer before a linear gradient of 150 ml of standard buffer and 150 ml of 0.5 M KCl in standard buffer was applied. DNA-polymerizing activity was tested in each fraction under standard conditions with poly dAT (○) as template for T7 DNA polymerase (14) and with sonicated calf thymus DNA (●) as template for *Escherichia coli* DNA polymerase (10). Complementation assays are shown in the upper panel of the graph. Equivalents of 10^8 gene 5 mutant-infected H560 cells (△) and of gene 4 mutant-infected cells (▼) were concentrated on cellophane discs according to the method of Schaller *et al.* (15). Samples of 2 μl from each fraction of the DEAE-cellulose chromatogram were added to these discs. Incubation was carried out according to Schaller *et al* (15). Ordinates: 10^{-12} moles of ^{3}H-thymine incorporated into acid-precipitable material.

The largest peak, peak A, elutes from the DEAE-cellulose column at about 0.15 M KCl, while the smallest peak, peak C, elutes at about 0.25 M KCl. Only the DNA polymerase of peak C successfully complemented lysates from gene 5 mutant-infected cells (upper panel, Fig. 1). The two T7 DNA polymerases were further purified by sucrose gradient sedimentation. The activity in peak A had a sedimentation coefficient of about 5.5 S, as has been described before for the purified T7 DNA polymerase (14). The DNA-synthesizing activity in peak C sediments with 6.7 S, suggesting that another macromolecular structure, possibly a protein, forms a complex with the DNA polymerases. We have not been able to produce a complementing form of T7 DNA polymerase by mixing highly purified DNA polymerase with protein fractions from T7-infected cells. It is not known whether the associated macromolecule is specified by the host cell or by the virus. The complex is most probably not a combination of the gene 4 product and the gene 5 product. An activity which complemented extracts of gene 4-infected cells eluted from the DEAE-column several fractions ahead of the fractions which complemented gene 5-infected cells (Fig. 1).

An enzymatic activity for the gene 4 product is not yet known. Complementation assays with lysates from gene 4-infected cells are difficult because of the rather high background caused by the synthesis of DNA fragments (see above).

Initiation of T7 DNA Replication *in Vitro*

The number of proteins involved in T7 DNA replication is small enough (see 18) to encourage experiments to reconstitute the viral replication apparatus *in vitro*.

In a series of experiments, DNA-free extracts were prepared from uninfected as well as from T7-infected H560 cells (Table 2). Native T7 DNA when added to the extract from infected cells served as a template for DNA synthesis. DNA synthesis was not observed when native T7 DNA was added to an extract from uninfected cells. This difference is not due to a smaller level of general DNA-polymerizing activity in uninfected cells since an addition of highly purified T7 DNA polymerase to the uninfected cell extract does not stimulate the T7 DNA-directed polymerization of deoxynucleoside triphosphates. It is also shown in Table 2 that ATP is essential for the DNA synthesis observed with the protein extract from infected cells. Since purified T7 DNA polymerase does not accept native T7 DNA as template (14) it is obvious that other functions besides T7 DNA polymerase are required for the DNA synthesis shown in Table 2. This conclusion is supported by the fact that an extract from uninfected cells did not sustain

Table 2. DNA synthesis in crude extracts from T7-infected H560 cells

	Extract from	
	T7-infected cells	Uninfected cells
	($\mu\mu$moles ^{3}H-TMP incorporated)	
Complete system	3.0	0.25
−ATP	0.18	0.15
−ATP; + rNTP (equimolar; 2 mM)	3.2	0.21
−T7 DNA	0.12	0.15
+ 0.1 unit T7 DNA polymerase	3.1	0.24
+ 0.5 unit T7 DNA polymerase	3.05	0.26

H560 cells (4 × 10^8 cells/ml in 1-liter tryptone broth) were infected with T7 wild type at a multiplicity of 10 infective phage particles/cell. About 8 min after infection at 37°C, the cells were rapidly cooled and pelleted. Uninfected H560 cells from a 1-liter culture served as control. The pelleted cells were resuspended in 10 ml of 1 M KCl in standard buffer (see legend to Fig. 1). Lysozyme (0.5 mg) was added to each cell suspension. The suspension was then frozen at −70°C and thawed at 30°C when needed. The cell lysate was centrifuged for 2 hr at 100,000 × g. More than 90% of the DNA was removed by this step. Two volumes of saturated ammonium sulfate were added to the supernatant. The proteins were collected by centrifugation (10 min at 10,000 × g) and resuspended in 0.5 to 1.0 ml of a buffer containing 100 mM KCl, 50 mM tris-HCl, pH 7.6; 1 mM EDTA; 1 mM Spermidine; 0.1 mM dithioerythritol; 20% glycerol. The protein preparation was then dialyzed for 3 to 4 hr against two changes of 200 ml of the buffer. Between 20 and 40% (different experiments) of the total proteins in the original cell suspension was finally recovered. Assay: 50 µl of dialyzed protein preparation; 5 µl of 50 mM MgCl$_2$ (dissolved in buffer); 1 µl of T7 DNA (0.02 µg); 1 µl of 50 mM ATP; 1 µl of a mixture of deoxynucleoside triphosphates (0.5 mM each of dATP, dGTP, and dCTP; 0.01 mM ^{3}H-dTTP; specific activity 60 000 cpm/µl). Incubation time was 15 min at 37°C. One-fourth milliliter 0.1 M ice-cold EDTA was added to stop the reaction. Incorporated radioactivity was determined by trichloroacetic acid precipitation on nitrocellulose filters.

DNA synthesis even in the presence of large quantities of highly purified viral DNA polymerase. One of the additional functions essential for T7 DNA-primed DNA synthesis seems to utilize ATP.

To obtain some information about the functions which are necessary for T7 DNA-directed synthesis, an extract from T7 wild type-infected H560 cells was fractionated on a DEAE-cellulose column. For this purpose 10 g of infected cells were mixed with about 0.4 g of T7 wild type-infected cells which had been labeled from 4 to 13 min after infection at 30°C with a ^{14}C-amino acid mixture. The DEAE-cellulose chromatogram is shown in Fig. 2. The chromatogram was, more or less arbitrarily, divided into 13 sections. Samples from each of these sections were combined and concentrated together with native intact T7 DNA on cellophane discs and incubated under standard conditions with 1 mM ATP and all four deoxynucleoside triphosphates including ^{3}H-TTP. An incorporation of radioactivity into acid-precipitable DNA was observed. No DNA synthesis was detected when purified T7 DNA polymerase or *E. coli* DNA polymerase II or III was mixed with any single protein

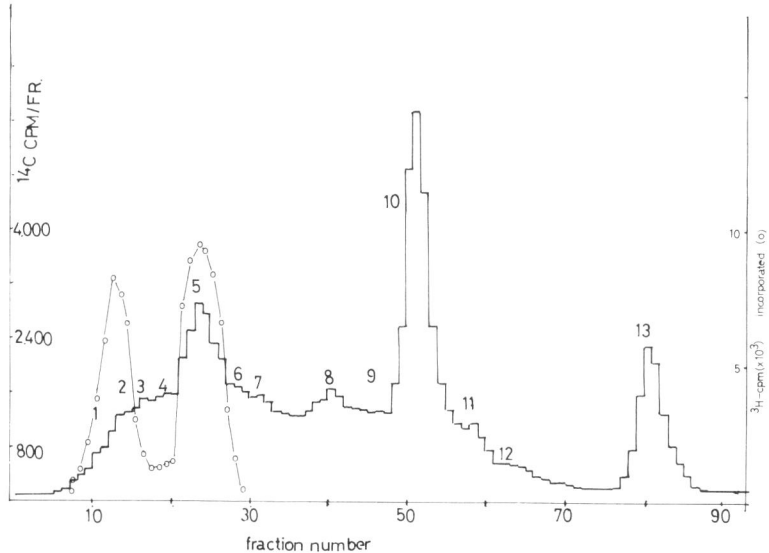

Fig. 2. Fractionation of ^{14}C-labeled T7 proteins on a DEAE-cellulose column. An UV-irradiated H560 culture (in 400 ml of minimal medium) was infected with T7 wild type. From the 4th to 13th min after infection at 30°C a mixture of ^{14}C-labeled amino acids (0.01 μCi/ml) was added to label the "early" T7 proteins (18). The pelleted cells (about 0.4 g) were mixed with 10 g of T7-infected H560 cells. All further preparation steps have been described in the legend to Fig. 1. ^{14}C radioactivity (solid line) was determined in 0.2-ml samples taken from each fraction. T7 RNA polymerase activity (O) was tested according to Chamberlin et al. (3).

fraction. Mixtures of protein fractions were then prepared in which one or more of the 13 fractions were omitted systematically. This search for the minimal set of proteins required for DNA synthesis with native T7 DNA as template showed that a combination of fractions 1 and 9 (Fig. 2) was sufficient to give the same total DNA synthesis as a combination of all 13 fractions. It was found after further purification by sucrose gradient sedimentation that the active component of fraction 9 was the T7 DNA polymerase complex described earlier (see Fig. 1). The active form of fraction 1 cosedimented in sucrose gradients with a T7 RNA polymerase activity.

As indicated in Fig. 2, we find two peaks of rifampicin-resistant RNA polymerase activity in T7-infected cells. Both activities have similar sedimentation coefficients (6-7 S) and both accept only T7 DNA as template. We do not yet know what the biochemical difference between the two forms of T7 RNA polymerase is (aside from the fact that they are eluted with different salt concentrations from the DEAE-cellulose column). It is, however, obvious that only one of the two forms stimulates T7 DNA synthesis.

As mentioned above, our assay was performed with dried samples on cellophane discs. The high concentration of components achieved with this procedure seems to be essential since we could barely detect a stimulation of T7 DNA synthesis by viral RNA polymerase when a liquid assay was employed. It should also be pointed out that ATP was required in our assay but no other ribonucleoside triphosphates. It is possible therefore that no RNA synthesis is necessary for the initiation of T7 DNA synthesis. It is not known whether the ATP is required for the attachment of the RNA polymerase to its promotor site on the T7 DNA (from where replication seems to start *in vivo*) (4), or for some biochemical process during chain elongation. One of the striking features of this initiation process *in vitro* is its specificity: it can be observed only with intact T7 DNA. Heavily nicked T7 DNA, DNA from calf thymus, *E. coli*, bacteriophage λ, and T4 are not accepted as a template. This result should be compared with the pronounced selectivity of the viral RNA polymerase (3) and suggests that the specificity of the reaction is introduced by the T7 RNA polymerase. The viral RNA polymerase cannot be replaced by the bacterial RNA polymerase (with or without σ factor).

In order to determine the size of the *in vitro* synthesized DNA and to see whether it was covalently connected to the primer DNA, ^{32}P-labeled T7 primer DNA was used in the usual incorporation assay. The ^{3}H-labeled products were analyzed by alkaline sucrose gradient centrifugation. It was found (Fig. 3) that most of the "parental" ^{32}P-DNA sedimented with the sedimentation rate of a viral DNA marker (centri-

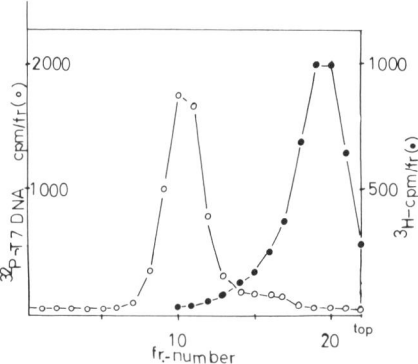

Fig. 3. Sedimentation analysis of *in vitro* synthesized DNA. ^{32}P-T7 DNA was used in an incorporation assay with T7 RNA polymerase (section 1 of Fig. 2) and a T7 DNA polymerase complex (section 9 of Fig. 2). The mixture was concentrated on cellophane discs and incubated according to Schaller *et al.* (15). The DNA was removed from the disc with 0.1 *M* NaOH-0.005 *M* EDTA and centrifuged through an alkaline sucrose gradient (20 to 5% sucrose in 0.2 *M* NaOH-0.005 *M* EDTA; SW 27 rotor of the Spinco centrifuge; 8 hr at 25,000 rpm and 2°C).

fuged in a parallel tube under identical conditions). The *in vitro* synthesized ^{3}H-DNA sedimented as fragments with 4–6 S. These fragments are probably not covalently bound to the "parental" DNA. Viral polynucleotide ligase when added to the reaction mixture did not join these fragments to significantly longer DNA strands. It appears as if the *in vitro* initiated DNA synthesis stops after completion of the first small section of the daughter DNA strands. Conditions have not yet been found to propagate *in vitro* DNA synthesis beyond this stage. It is interesting to note that DNA fragments of similar size are also produced in toluenized gene 4 mutant-infected cells (see above). It is conceivable therefore that the gene 4 product functions in an elongation mechanism.

Conclusions

The experiments reported in this paper show that the DNA-polymerizing function required for T7 DNA replication is a complex of the T7 DNA polymerase plus (an) associated macromolecule (macromolecules) of unknown nature. This complex, but not purified T7 DNA polymerase, complemented cell extracts from gene 5 mutant-infected cells, suggesting that complex formation does not easily occur *in vitro*. The T7 DNA polymerase complex could also be used to initiate DNA synthesis on native intact T7 DNA provided a viral-specific RNA polymerase was present.

It is not yet clear how the RNA polymerase is involved in initiating DNA synthesis. Since only ATP and no other ribonucleoside triphosphate is required for initiation, one may conclude that no (or only very limited) RNA synthesis is necessary for this reaction. A possible mechanism might be a local denaturation of the double stranded DNA template caused by the attachment of the RNA polymerase to its promoter region (regions). According to this proposition, *in vitro* DNA synthesis begins somewhere in these denatured regions and probably does not continue beyond these areas. ATP may assist in stabilizing this RNA polymerase DNA interaction. It is also conceivable that the ATP is held by the viral RNA polymerase in a position which may be functionally comparable to a 3'-OH primer end of a polynucleotide. These questions are currently under investigation.

Acknowledgment

We thank Doctors F. W. Studier and R. Hausmann for generously providing us with amber mutants of bacteriophage T7 (19).

References

1. Center, M. S. 1972. Bacteriophage T7-induced endonuclease. II. Purification and properties of the enzyme. *J. Biol. Chem.* 247: 146.
2. Center, M. S., Studier, F. W., and Richardson, C. C. 1970. The structural gene for a T7 endonuclease essential for phage DNA synthesis. *Proc. Nat. Acad. Sci. U. S. A.* 65: 242.
3. Chamberlin, M., McGrath, J., and Waskell, L. 1970. New RNA polymerase from *Escherichia coli* infected with bacteriophage T7. *Nature* 228: 227.
4. Dressler, D., Wolfson, J., and Magazin, M. 1972. Initiation and reinitiation of DNA synthesis during replication of bacteriophage T7. *Proc. Nat. Acad. Sci. U. S. A.* 69: 998.
5. Grippo, P., and Richardson, C. C. 1971. Deoxyribonucleic acid polymerase of bacteriophage T7. *J. Biol. Chem.* 246: 6867.
6. Gross, J. 1972. DNA replication in bacteria. *Curr. Top. Microbiol. Immunol.* 57: 39.
7. Hausmann, R., and Härle, E. 1971. Expression of the genomes of the related bacteriophages T3 and T7. Abstract, *First European Biophysics Congress*, p. 467.
8. Hausmann, R., and La Rue, K. 1969. Variations in sedimentation patterns among DNA synthesized after infection of *Escherichia coli* by different amber mutants of bacteriophage T7. *J. Virol.* 3: 278.
9. Kerr, C., and Sadowski, P. D. 1972. Gene 6 exonuclease of bacteriophage T7. *J. Biol. Chem.* 247: 305.
10. Knippers, R. 1970. DNA polymerase II. *Nature* 228: 1050.
11. Lindqvist, B. H., and Sinsheimer, R. L. 1967. Process of infection with bacteriophage $\phi\chi 174$. XIV. Studies on macromolecular synthesis during infection with a lysis-defective mutant. *J. Mol. Biol.* 28: 87.
12. Masamune, Y., Frenkel, G. D., and Richardson, C. C. 1971. A mutant of bacteriophage T7 deficient in polynucleotide ligase. *J. Biol. Chem.* 246: 6874.
13. Moses, R. E., and Richardson, C. C. 1970. Replication and repair of DNA in cells of *Escherichia coli* treated with toluene. *Proc. Nat. Acad. Sci. U. S. A.* 67: 674.
14. Oey, J. L., Strätling, W., and Knippers, R. 1971. A DNA polymerase induced by bacteriophage T7. *Eur. J. Biochem.* 23: 497.
15. Schaller, H., Otto, B., Nüsslein, V., Huf, J., Herrmann, R., and Bonhoeffer, F. 1972. Deoxyribonucleic acid replication *in vitro*. *J. Mol. Biol.* 63: 183.
16. Strätling, W., and Knippers, R. 1971. Properties of the DNA synthesizing activity in DNA-membrane complexes from bacterial cell extracts. *Eur. J. Biochem.* 20: 330.
17. Strätling, W., Krause, E., and Knippers, R. 1972. Fast sedimenting DNA in bacteriophage T7 infected cells. Submitted for publication.

18. Studier, F. W. 1972. Bacteriophage T7. *Science* 176: 367.

19. Studier, F. W., and Hausmann, R. 1969. Integration of two sets of T7 mutants. *Virology* 39: 587.

20. Vosberg, H.-P., and Hoffmann-Berling, H. 1971. DNA synthesis in nucleotide-permeable *Escherichia coli* cells. I. Preparation and properties of ether-treated cells. *J. Mol. Biol.* 58: 739.

Discussion

Lehman. I wondered whether your system was rifampicin- or streptolydigin-sensitive.

Knippers. No, you can't expect that because T7 RNA polymerase is not rifampicin-sensitive.

Lehman. When you used the lysate without fractionation, what was the size of the DNA that was synthesized?

Knippers. All small. I think that while preparing the extract (it takes a few hours) the "elongation factor" is decaying. Other explanations are also possible.

Autoradiographic Demonstration of Bidirectional Replication in *Escherichia coli*

P. L. Kuempel, D. M. Prescott, and P. Maglothin

Department of Molecular, Cellular and Developmental Biology
University of Colorado
Boulder, Colorado 80302

Chromosome replication in *Escherichia coli* is initiated in a bidirectional manner in cells phased by amino acid deprivation and in exponentially growing cells. Bidirectional replication continues until at least 40% (480 μ or 9.5×10^8 daltons) of the chromosome has been replicated, with the two forks traveling at approximately the same rate. A few autoradiographic configurations of the chromosomes from log cultures appear to reflect the approach of two forks at the terminus of replication.

The idea of unidirectional replication dominated early studies of DNA synthesis in bacterial chromosomes. The experimental evidence was insufficient to prove unidirectional replication, but most of the data were consistent with this mode of replication. In earlier studies the idea of bidirectional replication was usually not considered. Bidirectional replication in bacterial chromosomes became a more plausible hypothesis with the experiments of Huberman and Riggs indicating bidirectional replication in mammalian chromosomes (6) and the demonstration of bidirectional replication in bacteriophage λ by Schnös and Inman (12). Caro and Berg (4), Nishioka and Eisenstark (10), and Yahara (14) were among the first to consider the possibility that replication in bacterial chromosomes might be bidirectional.

Masters and Broda (9), Yahara (15), and Bird *et al.* (2) have recently provided strong genetic evidence that replication in *Escherichia coli* is bidirectional. Wake (13) and Gyurasits and Wake (5) have shown by autoradiography that replication is also bidirectional in *Bacillus subtilis*, at least during the early stages of the replication cycle. We have recently demonstrated by autoradiography that replication in *E. coli* is bidirectional for cells initiating DNA synthesis after amino acid deprivation (11). In this report we describe extensions of these previous experiments.

Materials and Methods

E. coli strain 15 TAU-bar was used in all experiments. The procedures used for growth of the cells (7) and for the preparation of the autoradiographs (11) have been previously described. ^{3}H-Thymine and ^{3}H-thymidine were purchased from New England Nuclear Corp.

Results and Discussion

Cells initiating chromosome replication were obtained by incubating *E. coli* strain 15 TAU-bar for 150 min in the absence of required amino acids, and then resupplying the amino acids. A number of experiments have demonstrated that initiation occurs shortly after readdition of the amino acids (7, 8). To obtain labeled DNA for autoradiography, the cells were labeled with ^{3}H-thymine (10 Ci/mM) for different intervals following amino acid readdition and then labeled with both ^{3}H-thymine and ^{3}H-thymidine (52 Ci/mM) for 2.5 min. This labels only the DNA between the replication origin and the replication fork (or forks), and the replication forks have a higher grain density due to the higher specific activity of the ^{3}H-thymidine.

We have concluded that replication is bidirectional (11), since the autoradiographic patterns consist almost entirely of grain tracks with a low density of grains in the middle (produced by ^{3}H-thymine), with a higher grain density on both ends (produced by ^{3}H-thymidine). Examples of the grain tracks obtained in this type of experiment are shown in Fig. 1. These grain tracks were obtained from cells labeled with ^{3}H-thymine for 16 min and then labeled with both ^{3}H-thymine and ^{3}H-thymidine for 2.5 min. The lengths of the grain tracks in the samples were heterogenous, since initiations occurred at different times after amino acid readdition. The average length of 100 of these grain tracks was 193 μm. When the cells were labeled with the ^{3}H-thymidine at 13 and 19 min after readdition of the amino acids, the average length of the grain tracks was 152 and 236 μm, respectively (11). The average length of the grain tracks increased, as would be expected for growing chromosomes. The longest grain track observed was 480 μm. We have not labeled the cells with ^{3}H-thymidine later than 19 min after readdition of the amino acids, and since the first initiations occurred approximately 6 min after amino acid readdition (7, 11), the oldest replication cycles

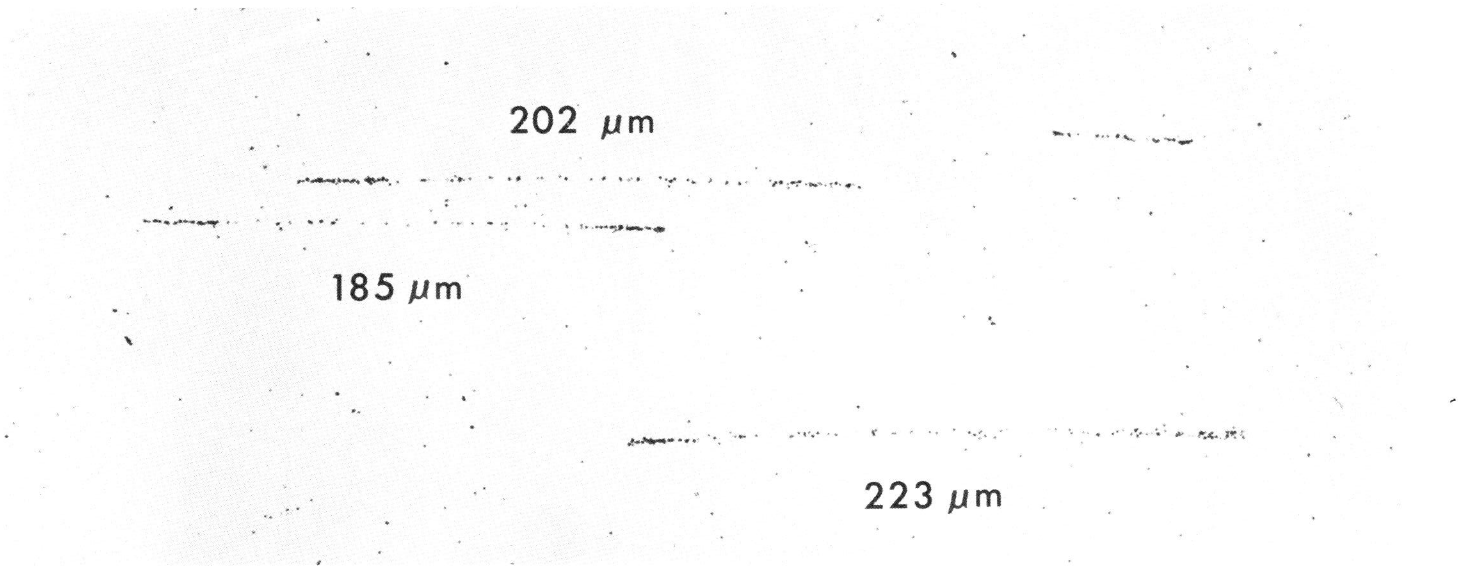

Fig. 1. Bidirectional replication of *Escherichia coli* chromosomes. The autoradiographic patterns were produced by chromosomes labeled with ³H-thymine (10 Ci/mM) for the first 16 min after release from amino acid deprivation, followed by labeling with both ³H-thymine and ³H-thymidine (52 Ci/mM) for 2.5 min. Replication forks are present at the two ends of the patterns. The autoradiographs are produced by the two daughter duplexes, which have collapsed against one another in these preparations.

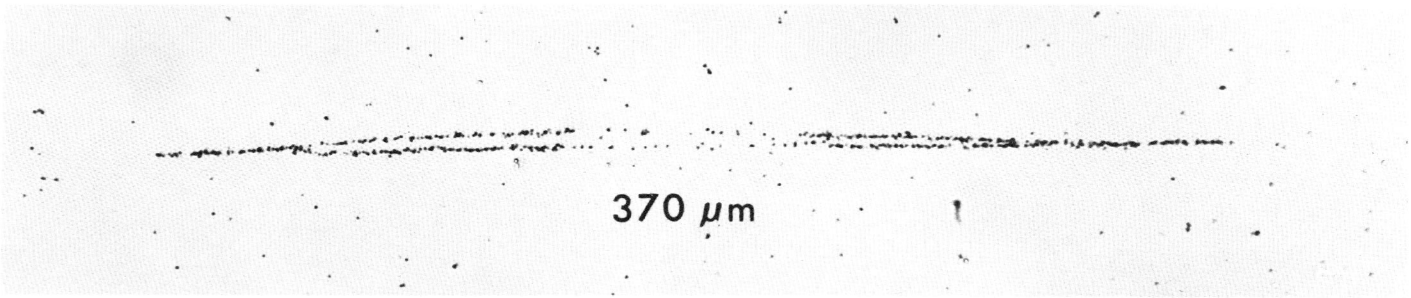

370 μm

Fig. 2. Bidirectional replication of the *Escherichia coli* chromosome. Autoradiograpic pattern produced by a chromosome labeled for 13 min with ³H-thymine (5 Ci/m*M*) and then for 6 min with ³H-thymine and ³H-thymidine (52 Ci/m*M*), after release from amino acid deprivation. In this example the daughter duplexes have separated from one another.

should be 15.5 min old at the end of the 2.5-min ^{3}H-thymidine pulse. These would produce grain tracks approximately 460 μm long if a replication cycle requires 40 min (1) and the chromosome is 1200 μm long (3).

We have conducted a variation of the above experiment to test whether both replication forks travel at the same rate. The cells were labeled with ^{3}H-thymine (5 Ci/mM) for 13 min following readdition of the amino acids, and then labeled with ^{3}H-thymine and ^{3}H-thymidine (52 Ci/mM) for 6 min. A typical grain track is shown in Fig. 2. We have measured 16 pairs of well extended grain tracks produced by the ^{3}H-thymidine, and the lengths of the two densely grained regions in any given chromosome never differ by more than 8% (Table 1). In view of this small difference we conclude that the two forks in a given chromosome travel at approximately the same rate, at least in the early stages of the replication cycle.

Although the lengths of the densely grained tracks in a given chromosome were almost identical, the lengths of the densely grained tracks in different chromosomes varied from 79 to 134 μm. Assuming little or no lag in the incorporation of the ^{3}H-thymidine, these track lengths indicate a minimum rate of travel per replication fork of 13 μm/min, a maximum rate of 22 μm/min, and a mean rate of 16.5 μm/min (Table

Table 1*

Length (μm) for pairs of dense tracks in bidirectional patterns		Mean length (μm)	Rate of fork travel (μm/min)
a	b		
79	81	80	13
79	84	82	14
81	87	84	14
87	94	91	15
87	94	91	15
89	92	91	15
89	92	91	15
89	94	92	15
89	94	92	15
92	92	92	15
100	102	101	17
102	102	102	17
113	121	117	19
114	122	118	19
123	131	127	21
132	134	133	22
Mean 97	101	99	16.5

*Cells were labeled with ^{3}H-thymine (5 Ci/mM) for 13 min after amino acid readdition, followed by labeling with ^{3}H-thymidine (52 Ci/mM) for 6 min.

1). It is not known at present whether this variability is the result of varying degrees of chromosome spreading, a varying lag in the incorporation of ^{3}H-thymidine, or reflects true differences in the rates of fork travel.

We have also observed bidirectional replication in exponentially growing cultures. Cells were labeled with ^{3}H-thymine (5 Ci/mM) for 5 min followed by labeling with ^{3}H-thymine and ^{3}H-thymidine (52 Ci/mM) for 2 or 4 min. Chromosomes that initiate replication during the incubation with ^{3}H-thymine should show bidirectional patterns, and such patterns were readily found (Fig. 3). This demonstrates that bidirectional replication is a general phenomenon, and is not a unique consequence of initiation following amino acid starvation.

The most common autoradiographic pattern obtained from exponentially growing cells was produced by individual replication forks (Fig. 4). Presumably these individual forks have a counterpart within the same chromosome. Such pairs could not be identified, however, since most chromosomes did not initiate replication cycles during the labeling period and the forks were consequently separated from one another by hundreds of microns of unlabeled DNA.

We have observed several grain tracks that may represent replication forks that are approaching each other at the terminus (Fig. 5). Such patterns would be expected to occur at one-half the frequency with which the initiation patterns were observed, but they were much less frequent than that. The rarity of these configurations could be due to several factors.

1. The terminus of the chromosome may be particularly sensitive to the sodium dodecyl sulfate used to lyse the cells or to the shear forces exerted when the DNA is spread on the slide.

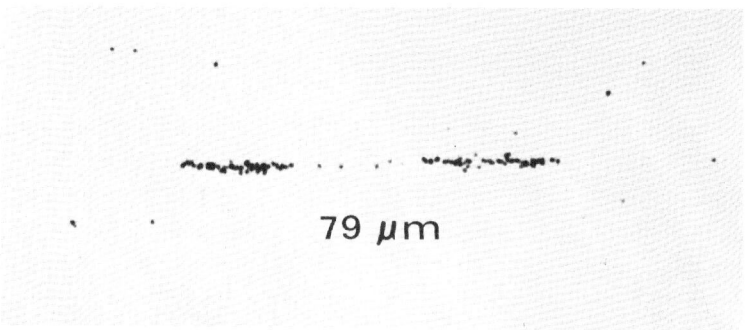

Fig. 3. Bidirectional replication of the *Escherichia coli* chromosome. Autoradiograph of chromosome from log phase cell labeled with ^{3}H-thymine (5 Ci/mM) for 5 min and with ^{3}H-thymine and ^{3}H-thymidine (52 Ci/mM) for 2 min.

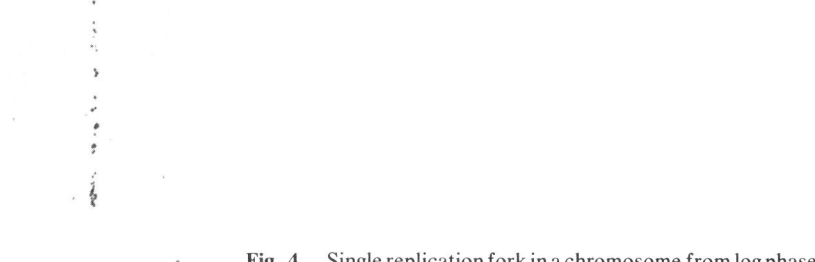

Fig. 4. Single replication fork in a chromosome from log phase cell labeled with ^{3}H-thymine (5 Ci/mM) for 5 min and with ^{3}H-thymine and ^{3}H-thymidine (52 Ci/mM) for 2 min.

2. The presence of two nearly completed daughter chromosomes may complicate chromosome spreading and obscure the relationship between approaching forks.

3. One fork might stop before the other at a genetically determined terminus. The forks would be expected to stop at different times if the terminus were not directly opposite the origin, as has been proposed by Masters and Broda (9). Differences in the rates of travel of the two forks could also result in termination at different times, even if the replication terminus were directly opposite the origin (2). Experiments conducted with phased cells will be required to test these possibilities and to study termination more thoroughly.

In summary, chromosome replication in *E. coli* is initiated in a bidirectional manner in cells phased by amino acid deprivation and in cells growing exponentially with a generation time of 60 min. Bidirectional replication continues until at least 40% (480 μ or 9.5×10^8 daltons) of the chromosome has been replicated, with the two forks traveling at approximately the same rate. The average rate of travel of individual forks is 16.5 μm/min. A few autoradiographic configurations of chromosomes from log cultures appear to reflect the approach of two forks at the terminus of replication. Such configurations are unexpectedly

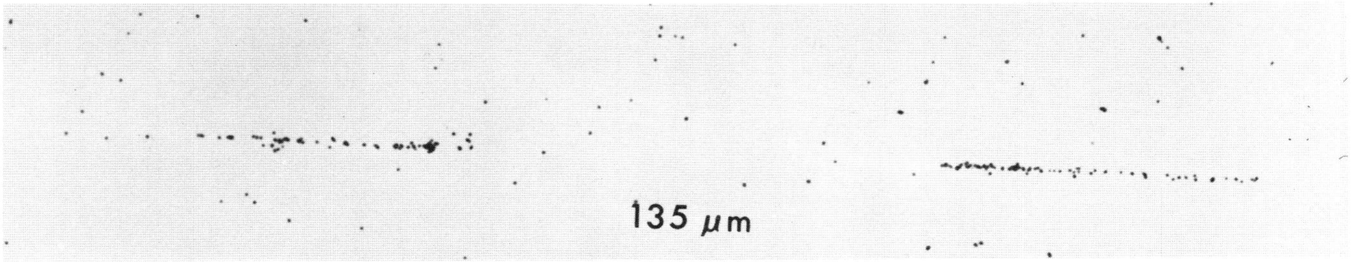

Fig. 5. Autoradiographic pattern of chromosome from log phase cell labeled with ^{3}H-thymine (5 Ci/mM) for 5 min and with ^{3}H-thymine and ^{3}H-thymidine for 4 min. This pattern is presumed to represent replication forks approaching one another at the terminus. The distance between the approaching forks is indicated.

rare, possibly because a genetically determined terminus may not be located directly opposite the origin.

Acknowledgment

This work was supported by National Institutes of Health Grant GM-15905 to P. L. K. and National Science Foundation Grant GB-32232 to D. M. P.

References

1. Bird, R., and Lark, K. G. 1968. Initiation and termination of DNA replication after amino acid starvation of E. coli 15T$^-$. *Cold Spring Harbor Symp. Quant. Biol.* 33: 799.

2. Bird, R. E., Louarn, J., Martuscelli, J., and Caro, L. G. 1972. The origin and sequence of chromosome replication in *Escherichia coli*. *J. Mol. Biol.* 70: 549.

3. Cairns, J. 1963. The chromosome of *Escherichia coli*. *Cold Spring Harbor Symp. Quant. Biol.* 28: 43.

4. Caro, L. G., and Berg, C. M. 1968. Chromosome replication in some strains of *Escherichia coli* K-12. *Cold Spring Harbor Symp. Quant. Biol.* 33: 559.

5. Gyurasits, E. B., and Wake, R. G. 1973. Bidirectional chromosome replication in *Bacillus subtilis*. *J. Mol. Biol.* 73: 55.

6. Huberman, J. A., and Riggs, A. D. 1968. On the mechanism of DNA replication in mammalian chromosomes. *J. Mol. Biol.* 32: 327.

7. Kuempel, P. L. 1972. Molecular weight of deoxyribonucleic acid synthesized during initiation of chromosome replication in *Escherichia coli*. *J. Bacteriol.* 112: 114.

8. Lark, K. G. 1966. Regulation of chromosome replication and segregation in bacteria. *Bacteriol. Rev.* 30: 3.

9. Masters, M., and Broda, P. 1971. Evidence for the bidirectional replication of the *Escherichia coli* chromosome. *Nature New Biol.* 232: 137.

10. Nishioka, Y., and Eisenstark, A. 1970. Sequence of genes replicated in *Salmonella typhimurium* as examined by transduction techniques. *J. Bacteriol.* 102: 320.

11. Prescott, D. M., and Kuempel, P. L. 1972. Bidirectional replication of the chromosome in *Escherichia coli*. *Proc. Nat. Acad. Sci. U. S. A.* 69: 2842.

12. Schnös, M., and Inman, R. B. 1970. Positions of branch points in replicating lambda DNA. *J. Mol. Biol.* 51: 61.

13. Wake, R. G. 1972. Visualization of reinitiated chromosomes in *Bacillus subtilis*. *J. Mol. Biol.* 68: 501.

14. Yahara, I. 1971. On the origin of replication of *Escherichia coli* chromosome. *J. Mol. Biol.* 57: 373.

15. Yahara, I. 1972. The origin and direction of chromosomal replication in *Escherichia coli*. *Jap. J. Genet.* 47: 33.

Discussion

Davern. Bidirectionality is quite the rage these days. I first want to report on an experiment in which I used dna C temperature-sensitive initiation mutants of *Escherichia coli* to terminalize and reinitiate. During this cycle of synchronous replication we periodically mutagenized with nitrosoquanidine and observed a peak of mutagenesis for arginine H at 30 min and a peak of mutagenesis for thymine A at 10 min. These are located at about 11 and 7 o'clock of the *E. coli* map, which would suggest that replication is indeed bidirectional, but the origin would not be where Lucien Caro would have it (74 min), but rather closer to the one suggested by Masters (65 min).

Inman. What proportion of the grain tracks suggests unidirectional replication in your experiments with cells initiating replication after amino acid starvation?

Kuempel. I would say that 70-80% of the grain tracks that we observed are definitely indicative of bidirectionality. This estimate is based on tracks that are clearly displayed. The percentage is lower in the samples taken at the later times, probably because the longer molecules are more susceptible to shear. If we do see something which is unidirectional, we cannot be certain if it is truly unidirectional or if it's a bidirectional molecule that was cut.

Inman. Did Jerry Wake give data about unidirectional objects in his paper?

Sueoka. He did have a minority which could be unidirectional, but statistically the result was consistent with bidirectional replication. There was no correlation between loop size and uni- and bidirectional types.

Cairns. Do you ever see molecules where the hot part is not at the end?

Kuempel. We only saw such patterns at very low frequency.

Index

Actinomycin D, 177, 241-247, 252, 321, 341-356
Antibody, against RNA tumor virus
 DNA polymerase, 243-247
Antiserum against *E. coli* DNA polymerase I, 4
ara-CTP, 260

Bidirectional replication, 386-401, 463-471

Cellophane disc system, 215-228
col E_1, 86, 176
Complementation assays, 137-142, 175-181, 185-192, 195-211, 375-381, 424-430, 451-459
Cross-linked DNA, 126-133
Cycloheximide, 241

Discontinuous replication, 83-101, 215-228
D-loops, in λ DNA, 437-448
dna A, 107, 137-142, 175-181, 375-381
dna B, 107, 137-142, 175-181, 185-192, 215-228, 375-381
dna C, 107, 137-142, 375-381
dna D, 107, 137-142, 375-381
dna E, 65-68, 72, 107, 137-142, 185-192, 375-381
dna F, 107, 137-142, 375-381
dna G, 107, 137-142, 185-192, 215-228, 375-381
DNA
 λ, 77, 131, 292-305
 M. luteus, 169
 M13 (fd), 75, 125-133, 137-142, 175, 287-305
 nucleoids from *E. coli*, 145-160
 φχ-174, 30, 137-142, 164, 175, 215-228
 PM2, 168
 T4, 86, 155
 T7, 35-45, 53, 66, 77, 451-459
DNA polymerase. *See* Polymerase
DNA polymers, 5-9, 15-23, 52, 66, 239-247, 287-305, 309-327, 333-338, 361-366, 415-430

DNA primers, 5, 14-23, 77-79, 175-181, 239-247, 251-282, 309-327, 333-338, 361-366
DNA-RNA hybrids, 21, 237-249, 253-282, 287-305, 309-327, 333-338, 361-366
DNase-related ATPase, 123-133

Escherichia coli, reckless, 32
Excision-repair, 27-32, 405-430
Exonuclease, $3' \to 5'$, 6-10, 16-23
Exonuclease, $5' \to 3'$, 6-10, 15-23, 27-32
Exonuclease III from *E. coli*, 27-32, 66

F factor, 176, 375-381
Folded DNA, 145-160

Hydroxyphenylazouracil and hydroxyphenylazoisocytosine, 435

Lambda phage replication, 185-192, 437-448
Ligase
  involvement in repair, 27-32, 423-430
  involvement in replication, 225
  mutants, 107-119

Mason-Pfizer agent, 253-282
Membranes, involvement in replication, 146-160, 361
Methylmethanesulfonate-sensitive strains, 6, 65-68
Misincorporation, 48, 309-327

Nalidixic acid, 426, 430
Nearest neighbor analysis, 39, 309-327
N-ethylmaleimide, effect on polymerase, 3-10, 259, 298-304, 388, 409
Neurospora endonuclease, 86
Nuclease, UV-specific, 28, 31
Nucleotide sequencing, 35-45

Okazaki fragments, 83-101, 107-119, 215-228, 316
Omega protein (swivelase), 163-172

Pancreatic DNase, 5, 12-17, 127, 199
Phage, P2, 83-101, 437-448
Phage induction, 118
Photoreactivation, 28
Poly A, 251-282
Polymerase, DNA
  avian myeloblastosis virus, 242, 253-282, 287-305, 309-327, 333-338
  Bacillus subtilis, 385-401, 405-430, 435-436
    mutants of, 385-401, 405-430
  chicken cell, 239-247
  E. coli pol I, 3-10, 264-282, 309-327

effect of salt on reaction, 19-23
mutants of, 3-10, 65-68, 83-101, 107, 137-142, 172, 175-181, 185-192, 215-228
nuclease levels in mutant polymerases, 3-10
proteolytic cleavage, 13-23
purification of mutants, 6
E. coli pol II, 4, 31, 65-68, 71-79
mutants of, 65-68
E. coli pol III, 4, 31, 65-68, 71-79, 185-192
purification of, 72
human leukemic blood lymphocytes, 252-282
human lymphocytes, 251-282
intracisternal A-type particles, 361-366
Micrococcus luteus, 14, 27-32, 309-327
associated exonuclease, 27-32
Moloney leukemia virus, 369-371
phytohemagglutinin-stimulated normal blood lymphocytes, 252-282
Rauscher leukemia virus, 254-282
Rous sarcoma virus, 239-247, 253-282, 341-356
T4 infected cells, 4-10, 31, 35-45, 47-61
nuclease associated with, 35-45, 47-61
T7 infected cells, 451-459
Polymerase, RNA, influence on replication, 71-79, 139, 145-160, 388, 451-459
Precursors for DNA replication, 83-101
Provirus, 239, 247

recB and *recC* genes, 123-133
Recombination, 116-119, 124
Repair synthesis, 27-32, 116-119, 393-401, 405-430
Restriction endonuclease
from *E. coli* B, 123-133
from *Hemophilus influenzae*, 36-45, 129
Reverse transcriptase. See DNA polymerase from RNA tumor virus sources
Rifampicin, 139, 176-181, 241, 253-280, 321
RNA
viral 4, 5, and 7 S, 341-356
viral 60-70 S, 239-247, 253-282, 297-305, 309-327, 333-338, 341-356
RNA-DNA covalent bonds, 83-101, 175-181, 309-327, 333-338, 341-356
RNA polymerase. See Polymerase
RNA primer for DNA synthesis, 77-79, 83-101, 175-181, 252-282, 287-305, 309-327, 333-338, 341-356, 451-459
RNase H
avian myeloblastosis virus, 287-305, 309-327, 333-338
E. coli, 298-305

Sodium fluoride, 321-327
Spermidine, 175-181
Stimulatory protein, avian myeloblastosis virus, 287-305
Strand separation, 38-45

Streptolydigin, 177, 253
Supercoiled DNA, 145-160, 163-172

T4 genetic map, 196
T4 phage replication, 195-211, 233
Thymine dimers, 30
Toluenized cells, 388-401, 408
Transforming activity, *Bacillus subtilis*, 385-401
Trimethylpsoralen, 126

Unwinding protein
  E. coli, 74-79, 175-181
  T4, 196-211, 233, 303
UV damage, 27, 422

Vinyl analogs of polynucleotides, 369-371